Lecture Notes in Computer Science   14899

Founding Editors

Gerhard Goos
Juris Hartmanis

Editorial Board Members

Elisa Bertino, *Purdue University, West Lafayette, IN, USA*
Wen Gao, *Peking University, Beijing, China*
Bernhard Steffen, *TU Dortmund University, Dortmund, Germany*
Moti Yung, *Columbia University, New York, NY, USA*

The series Lecture Notes in Computer Science (LNCS), including its subseries Lecture Notes in Artificial Intelligence (LNAI) and Lecture Notes in Bioinformatics (LNBI), has established itself as a medium for the publication of new developments in computer science and information technology research, teaching, and education.

LNCS enjoys close cooperation with the computer science R & D community, the series counts many renowned academics among its volume editors and paper authors, and collaborates with prestigious societies. Its mission is to serve this international community by providing an invaluable service, mainly focused on the publication of conference and workshop proceedings and postproceedings. LNCS commenced publication in 1973.

Zsuzsanna Lipták · Edleno Moura ·
Karina Figueroa · Ricardo Baeza-Yates
Editors

String Processing and Information Retrieval

31st International Symposium, SPIRE 2024
Puerto Vallarta, Mexico, September 23–25, 2024
Proceedings

Editors
Zsuzsanna Lipták ⓘ
University of Verona
Verona, Italy

Edleno Moura ⓘ
Federal University of Amazonas
Manaus, Brazil

Karina Figueroa ⓘ
Michoacan University of Saint Nicholas
of Hidalgo
Morelia, Mexico

Ricardo Baeza-Yates ⓘ
Northeastern University
Boston, MA, USA

ISSN 0302-9743   ISSN 1611-3349 (electronic)
Lecture Notes in Computer Science
ISBN 978-3-031-72199-1   ISBN 978-3-031-72200-4 (eBook)
https://doi.org/10.1007/978-3-031-72200-4

© The Editor(s) (if applicable) and The Author(s), under exclusive license
to Springer Nature Switzerland AG 2025
Chapter "Simultaneously Building and Reconciling a Synteny Tree" is licensed under the terms of the Creative Commons Attribution 4.0 International License (http://creativecommons.org/licenses/by/4.0/). For further details see license information in the chapter.

This work is subject to copyright. All rights are solely and exclusively licensed by the Publisher, whether the whole or part of the material is concerned, specifically the rights of translation, reprinting, reuse of illustrations, recitation, broadcasting, reproduction on microfilms or in any other physical way, and transmission or information storage and retrieval, electronic adaptation, computer software, or by similar or dissimilar methodology now known or hereafter developed.
The use of general descriptive names, registered names, trademarks, service marks, etc. in this publication does not imply, even in the absence of a specific statement, that such names are exempt from the relevant protective laws and regulations and therefore free for general use.
The publisher, the authors and the editors are safe to assume that the advice and information in this book are believed to be true and accurate at the date of publication. Neither the publisher nor the authors or the editors give a warranty, expressed or implied, with respect to the material contained herein or for any errors or omissions that may have been made. The publisher remains neutral with regard to jurisdictional claims in published maps and institutional affiliations.

This Springer imprint is published by the registered company Springer Nature Switzerland AG
The registered company address is: Gewerbestrasse 11, 6330 Cham, Switzerland

If disposing of this product, please recycle the paper.

Preface

The 31st International Symposium on String Processing and Information Retrieval (SPIRE) was held on September 23–25, 2024, in Puerto Vallarta (Mexico), followed by the 18th Workshop on Compression, Text, and Algorithms (WCTA) held on September 26, 2024.

SPIRE started in 1993 as the South American Workshop on String Processing. It was held in Latin America until 2000. Then, SPIRE moved to Europe, and from then on, it has been held in Australia, Japan, the UK, Spain, Italy, Finland, Portugal, Israel, Brazil, Chile, Colombia, Mexico, Argentina, Bolivia, Peru, the USA, and France. SPIRE continues the long and well-established tradition of encouraging high-quality research at the broad nexus of string processing, information retrieval, and computational biology.

This volume contains the accepted papers presented at SPIRE 2024. SPIRE 2024 received a total of 41 submissions, 34 full papers and 7 short papers. Each submission received at least three single-blind reviews. After the discussion phase, the Scientific Program Committee accepted 22 full papers and 4 short papers. We thank all the authors for their valuable contributions and presentations at the conference and thank the Program Committee members and additional reviewers for their valuable work during the review and discussion phases. We also thank the members of the Local Organizing Committee for their support in organizing SPIRE.

We appreciate the high-quality talks included in the scientific program from three renowned researchers: Juliana Freire (New York University, USA), Marinella Sciortino (University of Palermo, Italy), and Gerardo Sierra (National Autonomous University of Mexico, Mexico). This edition also had a Best Paper Award, sponsored by Springer. The award was announced during the conference.

We thank our sponsors: ACM SIGIR, Web4Good, Springer, Dipartimento di Informatica of Università di Verona, and Universidad Michoacana de San Nicolás de Hidalgo. Their generous support has been instrumental in making this conference a reality, fostering academic excellence and enabling us to bring together a diverse group of researchers and students. Finally, we thank Springer for publishing the proceedings of SPIRE 2024 in the LNCS series.

August 2024

Zsuzsanna Lipták
Edleno Moura
Karina Figueroa
Ricardo Baeza-Yates

Organization

General Chairs

Karina Figueroa-Mora   Universidad Michoacana, Mexico
Ricardo Baeza-Yates   Northeastern University, USA, & University of Chile, Chile

Program Committee Chairs

Zsuzsanna Lipták   University of Verona, Italy
Edleno S. de Moura   Federal University of Amazonas and Jusbrasil, Brazil

Steering Committee

Diego Arroyuelo   Pontificia Universidad Católica de Chile and Millennium Institute for Foundational Research on Data, Chile
Ricardo Baeza-Yates   Northeastern University, USA & University of Chile, Chile
Thierry Lecroq   University of Rouen Normandy, France
Franco Maria Nardini   ISTI-CNR, Pisa, Italy
Nadia Pisanti   University of Pisa, Italy
Barbara Poblete   University of Chile and Amazon, Chile
Berthier Ribeiro-Neto Google Inc. and Federal University of Minas Gerais, Brazil
Hélène Touzet   CNRS Lille, France
Rossano Venturini   University of Pisa, Italy
Nivio Ziviani   Universidade Federal Minas Gerais, Brazil

Program Committee

Omar Alonso   Amazon, USA
Diego Arroyuelo   Pontificia Universidad Católica de Chile and Millennium Institute for Foundational Research on Data, Chile

Golnaz Badkobeh — University of London, UK
Djamal Belazzougui — CERIST (Research Centre for Scientific and Technical Information), Algeria
Giulia Bernardini — Università di Trieste, Italy
Marilia Braga — Bielefeld University, Germany
Laurent Bulteau — CNRS - Université Gustave Eiffel, France
Edgar Chavez — CICESE, Mexico
Manuel Cáceres — Aalto University, Finland
Edleno S. De Moura (Co-chair) — Federal University of Amazonas and Jusbrasil, Brazil
Nadia El-Mabrouk — University of Montreal, Canada
Jonas Ellert — ENS - PSL, France
Antonio Fariña — University of A Coruña, Spain
Paweł Gawrychowski — University of Wrocław, Poland
Daniel Gibney — University of Texas at Dallas, USA
Inge Li Gørtz — Technical University of Denmark, Denmark
Meng He — Dalhousie University, Canada
Katharina T. Huber — University of East Anglia, UK
Tomohiro I — Kyushu Institute of Technology, Japan
Tomász Kociumaka — Max Planck Institute for Informatics, Germany
M. Oguzhan Kulekci — Indiana University Bloomington, USA
Dominik Köppl — University of Yamanashi, Japan
Susana Ladra — University of A Coruña, Spain
Moshe Lewenstein — Bar Ilan University, Israel
Zsuzsanna Lipták (Co-chair) — University of Verona, Italy
Felipe A. Louza — Universidade Federal de Uberlândia, Brazil
Camille Marchet — CNRS - Université de Lille, France
Viviane P. Moreira — Instituto de Informatica - UFRGS, Brazil
Nadia Pisanti — University of Pisa, Italy
Cinzia Pizzi — University of Padua, Italy
Svetlana Puzynina — Saint Petersburg State University, Russia
Narad Rampersad — University of Winnipeg, Canada
Kunihiko Sadakane — University of Tokyo, Japan
Leena Salmela — University of Helsinki, Finland
Blerina Sinaimeri — LUISS University of Rome, Italy
Jouni Sirén — University of California, Santa Cruz, USA
Tatiana Starikovskaya — École Normale Supérieure, France
Rossano Venturini — Università di Pisa, Italy
Aaron Williams — Williams College, USA
Michal Ziv-Ukelson — Ben-Gurion University of the Negev, Israel

Additional Reviewers

Aksenov, Vitaly
Amadini, Roberto
Ascone, Rocco
Benslimane, Seddik
Carmel, Amir
Cenzato, Davide
Chakraborty, Sankardeep
Cordeiro, Lucas
Cotumaccio, Nicola
Deharbe, David
Delabre, Mattéo
Fici, Gabriele
Ganardi, Moses
Ganguly, Arnab
Gascon, Mathieu
Ghazawi, Samah
Gómez-Brandón, Adrián
Hossen, Md Helal
Maity, Anuran
Martayan, Igor
Mhaskar, Neerja
Mäkinen, Veli
Nishimoto, Takaaki
Olbrich, Jannik
Parmigiani, Luca
Pirola, Yuri
Prezza, Nicola
Radoszewski, Jakub
Saad, Daniel
Steiner, Teresa Anna
Stoye, Jens
Tiskin, Alexander
Waleń, Tomasz
Wu, Kaiyu
Zuba, Wiktor

Abstracts of Invited Talks

v

Dataset Search for Data Discovery, Augmentation, and Explanation

Juliana Freire [iD]

New York University
`juliana.freire@nyu.edu`

Abstract. In recent years, we have witnessed an explosion in our capacity to collect and catalog vast amounts of data about our environment, society, and populace. Moreover, with the push towards transparency and open data, scientists, governments, and organizations are increasingly making structured data available on the Web and in various repositories and data lakes. Combined with advances in analytics and machine learning, the availability of such data should, in theory, allow us to make progress on many of our most important scientific and societal questions.

However, this opportunity is often unrealized due to a central technical barrier: it is remains nearly impossible for domain experts to sift through the overwhelming amount of available information to discover datasets they need for their specific applications. While search engines have addressed the discovery problem for Web documents, supporting the discovery of structured data presents new challenges. These include crawling the Web in search of datasets, indexing datasets and supporting dataset-oriented queries, and creating new techniques to rank and display results.

In this talk, I will discuss these challenges and present our recent work in this area. Specifically, I will describe strategies for finding relevant datasets on the web and deriving metadata to be indexed. Additionally, I will introduce a new class of data-relationship queries and outline a collection of methods that efficiently support various types of relationships, demonstrating how they can be used for data explanation and augmentation. Finally, I will showcase Auctus, an open-source dataset search engine that we have developed at the NYU Visualization, Imaging, and Data Analysis (VIDA) Center. I will conclude by highlighting open problems and suggesting directions for future research.

Exploring Repetitiveness in Texts: From BWT to Morphisms

Marinella Sciortino

Dipartimento di Matematica e Informatica, University of Palermo, Italy
marinella.sciortino@unipa.it

Abstract. The notion of repetitiveness plays a fundamental role in processing very large collections of texts. In many applications, massive and highly repetitive data need to be stored, analyzed, and queried. Therefore, having good measures capable of capturing repetitiveness implies having effective parameters to evaluate the performance of compressed indexing data structures for such types of data.

Many repetitiveness measures are defined using compression schemes. One of these measures, denoted r, is the number of maximal equal-letter runs in the output produced by the Burrows-Wheeler Transform (BWT), a transformation which permutes the characters of a text to boost the effects of run-length encoding. Besides having a crucial role in the definition of recent compressed indexing data structures, such as the r-index, the measure r has attracted attention in Combinatorics on Words because it has allowed for defining and recognizing properties of repetitive strings. A pioneering result is the characterization of finite Sturmian words as the binary strings for which r assumes its minimum value.

From a complementary perspective, morphisms are classic tools in Combinatorics on Words for generating collections of repetitive texts. Injective morphisms, known as codes, are widely used in Information Theory. Recently, morphisms, combined with copy-paste mechanisms, have been used to define new repetitiveness measures and compressors, called NU-systems.

In this talk, I will explore our recent results on the properties of the measure r that allow analysis of the combinatorial characteristics of input texts. I will then show very recent interesting findings on the identification of collections of generic highly repetitive strings using the measure r. Next, we will see recent results on the evaluation of some compression-based repetitiveness measures for collections of strings generated by morphisms. I will close with our latest research on the close correlations between morphisms and the measure r, with exciting implications in the theory of codes.

Preservation and Accessibility of Documentary Heritage

Gerardo Sierra

National Autonomous University of Mexico (UNAM), Mexico

Abstract. Preservation and accessibility of documentary heritage are essential for maintaining and disseminating the cultural and historical wealth of a society. These concepts encompass a set of actions and strategies aimed at conserving historical documents and ensuring their availability for future generations, fostering research and knowledge across various disciplines.

National libraries play a crucial role as the primary reservoir of a country's documentary heritage. They store and protect a vast collection of documents, both printed and digital, that reflect a nation's cultural diversity and legacy.

Printed documents include codices, manuscripts, documents in indigenous languages, and multimodal texts. Each type presents unique preservation challenges due to its fragility, rarity, and linguistic and material diversity. The preservation of printed documents faces several challenges, such as the need for specialized techniques for physical conservation, the digitization of multimodal texts, and the translation and cataloging of documents in indigenous languages. These tasks require an interdisciplinary approach and advanced technologies to ensure the integrity and accessibility of these materials.

Natural language processing (NLP) and artificial intelligence (AI) offer powerful tools to address these challenges. These technologies can support, among others: Metadata extraction, cataloging and classification, and summary generation.

The use of NLP and AI not only enhances preservation but also increases the accessibility of documentary heritage. These technologies enable the creation of digital access platforms, vectorized databases, and advanced search tools, which are essential for research in digital humanities, stylometry, literary studies, and more.

Contents

Linear Time Reconstruction of Parameterized Strings from Parameterized
Suffix and LCP Arrays for Constant-Sized Alphabets 1
    Amihood Amir, Eitan Kondratovsky, Shoshana Marcus, and Dina Sokol

Bijective BWT Based Compression Schemes 16
    Golnaz Badkobeh, Hideo Bannai, and Dominik Köppl

Indexing Finite-State Automata Using Forward-Stable Partitions 26
    Ruben Becker, Sung-Hwan Kim, Nicola Prezza, and Carlo Tosoni

Burst Edit Distance 41
    Itai Boneh, Shay Golan, Avivit Levy, Ely Porat, and B. Riva Shalom

Generalization of Repetitiveness Measures for Two-Dimensional Strings 57
    *Lorenzo Carfagna, Giovanni Manzini, Giuseppe Romana,
    Marinella Sciortino, and Cristian Urbina*

On Computing the Smallest Suffixient Set 73
    Davide Cenzato, Francisco Olivares, and Nicola Prezza

Revisiting the Folklore Algorithm for Random Access
to Grammar-Compressed Strings 88
    Alan M. Cleary, Joseph Winjum, Jordan Dood, and Shunsuke Inenaga

Logarithmic-Time Internal Pattern Matching Queries in Compressed
and Dynamic Texts 102
    Anouk Duyster and Tomasz Kociumaka

Bounded-Ratio Gapped String Indexing 118
    *Arnab Ganguly, Daniel Gibney, Paul MacNichol,
    and Sharma V. Thankachan*

Simultaneously Building and Reconciling a Synteny Tree 127
    Mathieu Gascon, Mattéo Delabre, and Nadia El-Mabrouk

Quantum Algorithms for Longest Common Substring with a Gap 143
    Daniel Gibney and Md Helal Hossen

Online Computation of String Net Frequency 159
    Peaker Guo, Seeun William Umboh, Anthony Wirth, and Justin Zobel

On the Number of Non-equivalent Parameterized Squares in a String 174
  Rikuya Hamai, Kazushi Taketsugu, Yuto Nakashima,
  Shunsuke Inenaga, and Hideo Bannai

Another Virtue of Wavelet Forests 184
  Aaron Hong, Christina Boucher, Travis Gagie, Yansong Li,
  and Norbert Zeh

All-Pairs Suffix-Prefix on Dynamic Set of Strings 192
  Masaru Kikuchi and Shunsuke Inenaga

Adaptive Dynamic Bitvectors 204
  Gonzalo Navarro

Compressed Graph Representations for Evaluating Regular Path Queries 218
  Gonzalo Navarro and Josefa Robert

Greedy Conjecture for the Shortest Common Superstring Problem and Its
Strengthenings 233
  Maksim S. Nikolaev

Faster Computation of Chinese Frequent Strings and Their Net Frequencies 249
  Enno Ohlebusch, Thomas Büchler, and Jannik Olbrich

Faster Algorithms for Ranking/Unranking Bordered and Unbordered Words ... 257
  Jakub Radoszewski, Wojciech Rytter, and Tomasz Waleń

Computing String Covers in Sublinear Time 272
  Jakub Radoszewski and Wiktor Zuba

LZ78 Substring Compression with CDAWGs 289
  Hiroki Shibata and Dominik Köppl

2d Side-Sharing Tandems with Mismatches 306
  Shoshana Marcus, Dina Sokol, and Sarah Zelikovitz

Faster and Simpler Online/Sliding Rightmost Lempel-Ziv Factorizations 321
  Wataru Sumiyoshi, Takuya Mieno, and Shunsuke Inenaga

Space-Efficient SLP Encoding for $O(\log N)$-Time Random Access 336
  Akito Takasaka and Tomohiro I

Simple Linear-Time Repetition Factorization 348
  Yuki Yonemoto and Shunsuke Inenaga

Author Index 363

Linear Time Reconstruction of Parameterized Strings from Parameterized Suffix and LCP Arrays for Constant-Sized Alphabets

Amihood Amir[1,2], Eitan Kondratovsky[3], Shoshana Marcus[4(✉)], and Dina Sokol[5,6]

[1] Department of Computer Science, Bar-Ilan University, Ramat Gan, Israel
amir@esc.biu.ac.il
[2] College of Computing, Georgia Institute of Technology, 801 Atlantic Drive, 30318 Atlanta, GA, USA
[3] Department of Mathematics and Computer Science, Open University, Raanana, Israel
eitan.k@openu.ac.il
[4] Department of Mathematics and Computer Science, Kingsborough Community College of the City University of New York, 2001 Oriental Blvd, Brooklyn, NY, USA
shoshana.marcus@kbcc.cuny.edu
[5] Department of Computer and Information Science, Broklyn College, Brooklyn, NY, USA
sokol@sci.brooklyn.cuny.edu
[6] Department of Computer Science, The Graduate Center of the City University of New York, New York, NY, USA
https://u.cs.biu.ac.il/~amir/, http://www.sci.brooklyn.cuny.edu/~sokol

Abstract. A parameterized string (p-string) is a string that can contain two kinds of characters, static symbols and parameter characters. Parameterized pattern matching is a form of pattern matching that allows parameters to be renamed by applying a one-to-one function. The parameterized suffix array is a data structure that is useful in efficient parameterized pattern matching when accompanied by the parameterized longest common prefix (LCP) array. Reconstructing input from a given instance of a data structure is the task of determining whether the instance is valid or not, and if valid, producing a plausible set of data that it can represent. In this paper we consider parameterized suffix and LCP arrays and reconstruct a corresponding p-string that they can represent. In previous work, an algorithm can determine in $O(n^2)$ time whether a p-string can be constructed to correspond to the input parameterized suffix and LCP arrays of size n. In this work, we develop an algorithm that accomplishes this in $O(n)$ time for constant-sized alphabets, and $O(n \log n)$ time for general alphabets. Furthermore, when reconstruction is possible, we demonstrate that a p-string can be reconstructed over the *minimal* alphabet in $O(n^2)$ time.

Keywords: Strings · Parameterized Strings · Suffix Array · Longest Common Prefix Array · Reverse Engineering

1 Introduction

Parameterized pattern matching was introduced by Baker [4,5]. A parameterized string (p-string) consists of characters from both a static alphabet Σ and a parameterized alphabet Π. Two p-strings of the same length are said to parameterize match (p-match) if one p-string can be transformed into the other by applying a one-to-one function that renames the parameter symbols and maintains the static symbols. For example, let $\Sigma = \{X, Y, Z\}$, $\Pi = \{a, b, c\}$, $r = $ bcbXbcZ, $s = $ abaXabY and $t = $ bcbXbcY. We can say that p-strings s and t p-match each other, while r and s do not p-match.

Recent work in parameterized matching includes parameterized DAWGs [18], variations of the parameterized longest previous factor [7], longest common parameterized subsequence [16], parameterized matching in the streaming model [14], and improved parameterized longest common prefix array construction [1].

The *prev encoding* of a p-string maintains each static character $\in \Sigma$ and replaces each parameterized character $\in \Pi$ with a number, the distance to its previous occurrence in the p-string (or 0 if it is the first occurrence) [5]. Two p-strings p-match if and only if their prev encodings are equivalent, For example, the prev encodings of both $s = $ abaXabY and $t = $ bcbXbcY are 002X24Y. Thus, the parameterized matching problem amounts to efficiently comparing the prev encodings of p-strings.

The parameterized suffix array (pSA) of a p-string is an index that sorts the suffixes of a p-string by their prev encodings, with numbers considered lexicographically smaller than the static letters [8]. The parameterized LCP (pLCP) array of a p-string is an array containing the lengths of the longest common prefixes of prev encoded suffixes that are adjacent in the pSA. The parameterized suffix and LCP arrays can be directly computed in linear time for constant-sized alphabets [10].

There has been much work recently on reverse engineering data structures [2,3,6,9,12,15,19]. Reverse engineering consists of determining whether a given input is a valid instance of a particular data structure. In this paper, we focus on the problem of reverse engineering a parameterized suffix array along with its corresponding parameterized longest common prefix (pLCP) array. We find a p-string whose p-suffix array and pLCP array are equal to the given arrays of integers, if there is a suitable p-string.

Amir et al. [2] reconstruct a p-string from the p-suffix and pLCP arrays, when it is possible to reconstruct one. Their algorithm[1] requires $O(n^2)$ time and space for input arrays of size n, as it potentially reconstructs all of the prev encoded suffixes. In this paper, we develop a *linear time* algorithm that inserts values directly into the prev encoding of the entire p-string without first reconstructing

---
[1] It it not necessary to be familiar with the authors' earlier result in order to understand the current paper.

the individual prev encoded suffixes. Our new algorithm runs in linear time for constant-sized alphabets and $O(n \log n)$ time for general alphabets. In this paper, we also contribute the first algorithm that reconstructs a p-string from p-suffix and pLCP arrays which minimizes the alphabet size, in $O(n^2)$ time.

This paper is organized as follows. In Sect. 2, we begin with an overview and introduce the *dist query*, which lies at the core of the new techniques presented in this paper. In Sect. 3, we present a linear time algorithm to reconstruct a p-string for the input, if the input is reconstructible. In Sect. 4, we adapt the algorithm to reconstruct the p-string over minimal alphabet. The proofs of lemmas and theorems were omitted from this paper due to lack of space and will appear in the journal version.

2 Overview

2.1 Problem Definition

Input: A parameterized suffix array (pSA) of size n and its corresponding parameterized LCP (pLCP) array of size $n - 1$. We used 0-based indexing of the arrays.
Output: Does there exist a parameterized string S of size n such that the given pSA and pLCP arrays are the pSA and pLCP arrays, respectively, of S? If yes, construct one such parameterized string S.

We can incorporate the linear time technique developed by Amir et al. [2] to resolve all static characters and place them in the appropriate locations in the p-string. Thus, in this work, we focus on recovering p-strings that consist of only parameter characters and no static characters.

2.2 Main Idea

We use the term p-suffix and notation p_i to refer to the prev encoding of suffix i in a p-string. We refer to two p-suffixes that occur consecutively in the pSA as a *neighboring pair* of p-suffixes. For a neighboring pair of p-suffixes, we call the position of mismatching characters that follow immediately after the pLCP the *point of mismatch*. If the pLCP encompasses the entire p-suffix that is first in the neighboring pair, the pair has no point of mismatch.

Let $prev_S$ denote the prev encoding for the p-string S that we are attempting to reconstruct. We note that $prev_S = p_0$. The goal of our algorithm is to reconstruct the prev encoding $prev_S$ so that we can construct a p-string S from it, if one exists. In theory, $prev_S$ can be reconstructed by "copying" all values greater than 0 from all p-suffixes, and then placing zeros in the remaining locations, since each value $v > 0$ in a p-suffix implies the same value in $prev_S$. Of course, the p-suffixes are not available to us, however, we use the pSA and pLCP arrays to find the important values. More specifically, as shown in [2], only the values in a p-suffix that are included in a pLCP, up to and including the point of mismatch, are necessary. In this paper we further narrow down the values that are necessary to compute. The following seminal lemma proves that we only care about the mismatching value at a location in the second p-suffix in a pair of neighboring p-suffixes, and furthermore, that it is only necessary to compute the points of mismatch of the second p-suffix that are exactly equal to the distance from the start of the p-suffix, denoted by *dist*.

Lemma 1. *Every value v that is a point of mismatch in the pSA has to appear as point of mismatch in location $v = dist$ of the second p-suffix in a neighboring pair of p-suffixes.*

2.3 Dist Query

The p-suffixes that are sorted by the pSA are the prev encoded suffixes of S. We index the locations within each such p-suffix beginning with 0, thus, $p_k[j]$ refers to the value in the prev encoding of suffix k at index j. When the value of $p_k[j] = j$ we call j "*dist*" since it is the distance from the beginning of the p-suffix, in effect pointing to the first character. The following query is the critical component of our new algorithm, as it answers the question: is the value at the mismatch location in the second p-suffix equal to *dist*?

Dist Query: Given two neighboring locations i and $i+1$ in pSA (and the corresponding pLCP array). Let $k = pSA[i]$ and $k' = pSA[i+1]$, that is, p_k and $p_{k'}$ are the p-suffixes that are neighbors in the pSA. (See Fig. 1 for illustration.) Let j be the location of the point of mismatch between p_k and $p_{k'}$, i.e. $j = pLCP[i]$. Is $p_{k'}[j] = j$?

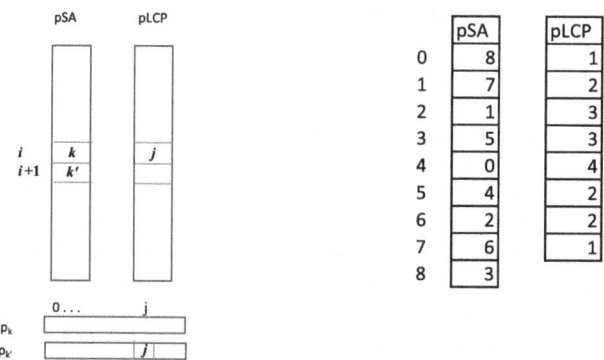

Fig. 1. (left) Illustration of Dist Query description. (right) Example input to our algorithm: pSA and pLCP arrays, with indices listed at the left.

Consider p_{k+1} and $p_{k'+1}$. The answer to the query is yes if and only if either:

1. $p_{k'+1}$ comes before p_{k+1} in the pSA, i.e., the order switches
2. $\text{pLCP}(p_{k+1}, p_{k'+1}) \geq \text{pLCP}(p_k, p_{k'})$, i.e., the pLCP grows

The input provides the pLCP between p-suffixes that are *adjacent* in the pSA. We can use a range-minimum query (RMQ) to obtain the pLCP of *any* two p-suffixes. RMQ returns the position of the smallest value in a subarray. A range-minimum query is answered in constant time after linear time preprocessing of the input array [11,13].

We refer to Fig. 1 for a set of input to our algorithm and explain how the Dist Query works on this data set. When $i = 0$, $i = 1$, $i = 2$, $i = 4$, and $i = 5$, the Dist Query returns *no* since neither condition is met. In greater detail, when $i = 1$, the neighbors in the pSA are p-suffixes 7 and 1, with $p_7 < p_1$. The first condition is not met since $p_8 < p_2$, i.e. p-suffixes 8 and 2 appear in the same order as p-suffixes 7 and 1 in the pSA. The second condition is not satisfied either since pLCP$(p_7, p_1) = 2$ and pLCP$(p_8, p_2) = 1$.

When $i = 3$, the Dist Query returns *yes* since the first condition is met, i.e. the order switches (although the pLCP shrinks). That is, $p_5 < p_0$ and $p_6 > p_1$. Thus, we conclude that $p_0[3] = 3$. When $i = 6$, the Dist Query also returns *yes* since the first condition is met.

When $i = 7$, the Dist Query returns *yes* since the second condition is met, that the pLCP extends further (although the order of the p-suffixes remains the same). The first condition is not met since $p_6 < p_3$ and $p_7 > p_4$. The second condition is met since both pLCP$(p_6, p_3) = 1$ and LCP$(p_8, p_4) = 1$. Thus, we conclude that $p_3[1] = 1$.

We first prove correctness of our answer to the query and then explain how to easily compute the answer to the query.

Lemma 2. *The two conditions specified in the dist query answer are sufficient to always correctly answer the query.*

Steps to Compute the Answer to the Dist Query (Algorithm 1):
First use the inverse p-suffix array (pISA) to find the locations of $k+1$ and $k'+1$ in the pSA and then check if either condition is satisfied:

1. Check if the location of $k' + 1$ is before the location of $k + 1$ in the pSA.
2. Use a range minimum query (RMQ) on the found locations to obtain the pLCP of p_{k+1} and $p_{k'+1}$ and check whether it is greater than or equal to the pLCP of p_k and $p_{k'}$, i.e. pLCP$[i]$.

Lemma 3. *The dist query is answered in $O(1)$ time after $O(n)$ time preprocessing of the input arrays.*

3 Algorithm

Algorithm Outline:

1. Reconstruct the prev encoding at *points of mismatch* between adjacent entries in the pSA.
2. Propagate each value found in Step 1 to all other *matching* locations.

Algorithm 1. Dist Query

Input: pSA, pISA, pLCP preprocessed for RMQ, indexes i and $i+1$
Output: returns true or false
                           ▷ Suppose pSA$[i+1] = k'$ and pLCP$[i] = j$. Query: Is $p_{k'}[j] = j$?
$k \leftarrow$ pSA$[i]$
$k' \leftarrow$ pSA$[i+1]$ ▷ k precedes k' in pSA
if $k+$pLCP$[i] = n$ **then**   ▷ no point of mismatch
    return false
end if
$loc_{k+1} \leftarrow$ pISA$[k+1]$ ▷ $pSA[loc_{k+1}] = k+1$
$loc_{k'+1} \leftarrow$ pISA$[k'+1]$ ▷ $pSA[loc_{k'+1}] = k'+1$
if $loc_{k'+1} < loc_{k+1}$ **then**   ▷ order switches
    return true
end if
                                                ▷ loc_{k+1} precedes $loc_{k'+1}$ in pSA
$i_r \leftarrow$ RMQ$(loc_{k+1}, loc_{k'+1})$ in pLCP array
if pLCP$[i_r] \geq$ pLCP$[i]$ **then**   ▷ pLCP does not shrink
    return true
end if
return false   ▷ has not met either condition to return true

3. Place zeros at all other locations of $prev_S$.
4. Reconstruct p-string S from $prev_S$.
5. Verify that the pSA and pLCP arrays of S match the input.

The following two paragraphs describe Steps 1 - 2 of the outline, and correspond to Algorithm 2.

The algorithm iterates through the pSA and pLCP arrays, calculating values at points of mismatch. Since, by Lemma 1, all values will eventually appear as the *dist* value in the second p-suffix of the neighboring p-suffixes, these are the only values that the algorithm must compute. A call to the query for each location i is sufficient to calculate all *dist* values. Once calculated, as shown in Lemma 4, the value j figured out by the Dist Query can be placed at location $k' + j$ in $prev_S$. Since *dist* is the largest value that can occur at a point of mismatch, the preceding value at that position is strictly smaller than *dist*. Thus, dist values are propagated down the pSA, and propagation is indicated by subsequent non-decreasing pLCP value(s).

Each non-zero entry in $prev_S$ provides an offset to the closest earlier occurrence of the same character in the p-string, and can be thought of as pointing at that earlier location. In other words, if $prev_S[i] = k > 0$, $0 < i < n$, we say that position i *points at* position $i - k$. We ensure that no position in $prev_S$ is pointed at by more than one subsequent location, so that it is a valid prev encoding. We use a Boolean pointedAt array of size n and initialize each of its elements to `false`. Then, we go through each location $prev_S[m]$ set to value v and set pointedAt$[m - v]$ to `true`. If we attempt to reset a value to `true`, or if we attempt to point at a location that is before location 0, we conclude that a valid p-string cannot be constructed.

Then, we set the remaining unknowns in $prev_S$ to 0, to ensure that the choice of character will be in agreement with the input. Then we reconstruct S from $prev_S$ and verify that its pSA and pLCP arrays match the input, as described in [2]. If not, we conclude that the input is not reconstructible.

Preprocessing of the input should ensure that the algorithm fails if a pLCP extends to the end of k' or if a pLCP is longer than either p_k or $p_{k'}$. Furthermore, we can only perform a dist query when there is a point of mismatch.

Lemma 4. *If the answer to the dist query: does $p_{k'}[j] = j$? is 'yes', then the value at location $k' + j$ in $prev_S$ equals to j, i.e. $prev_S[k' + j] = j$.*

We refer back to the example in Fig. 1 and show how our algorithm reconstructs $prev_S$ and S. We know from the dist queries that $p_0[3] = 3$, $p_6[2] = 2$, and $p_3[1] = 1$. Thus, we set $prev_S[3] = 3$, $prev_S[8] = 2$, and $prev_S[4] = 1$. Based on the pLCP array, We propagate $p_0[3] = 3$ to $p_4[3]$ so we set $prev_S[7] = 3$. In this example, the other values provided by dist queries are not propagated. We insert zeros at the unknown positions in $prev_S$ and arrive at $prev_S = 000310032$. This can correspond to bcabbdebe, among other possibilities.

The following two lemmas prove the correctness of our algorithms.

Lemma 5. *Every value implied by the pSA/pLCP arrays will be placed in $prev_S$.*

Lemma 6. *If the input is reconstructible then our algorithm will reconstruct it.*

The following lemma is key to proving that the algorithm's running time is linear.

Lemma 7. *Propagating is done over all values in linear time.*

To clarify Lemma 7 with an example, consider the value $v = 1$. In the first place that the p-suffixes switch from beginning with 00 to 01, the pLCP equals 1. In all p-suffixes beginning with 01, the value 1 is $dist$. It is true that 1 appears in many different places in many p-suffixes, but the only time that the algorithm deals with 1 is when the pLCP is 1, i.e. $dist = 1$. After the first position is set in $prev_S$, we propagate the 1 value. Propagating consists of going to all positions that match with pLCP>1 and placing a 1 there, which in this case involves all subsequent p-suffixes until the end of the pSA. If the algorithm returns to any one of the locations that got a 1 during propagation, and tries to put another value there, then there is a contradiction in the input.

Theorem 1. *The algorithm correctly reconstructs a p-string from the pSA and pLCP arrays when it is possible to reconstruct one. The algorithm runs in linear time for constant-sized alphabet.*

3.1 General Alphabets

The verification step of the algorithm runs in linear time only when the alphabet size is constant, since the verification process relies on the linear time algorithm that directly constructs the pSA and pLCP arrays for a p-string [10].

All other steps of the algorithm run in linear time for general alphabet sizes. The verification step ensures that the reconstructed string indeed produces identical pSA and pLCP arrays.

We can compute the pSA and pLCP arrays for a p-string of a general alphabet indirectly via the parameterized suffix tree. The pSA is the set of ordered leaves in the parameterized suffix tree and the pLCP array can be derived by longest common ancestor queries among the leaves. Kosaraju [17] constructs the parameterized suffix tree in $O(n(\log |\Pi| + \log |\Sigma|))$. Thus, for general alphabets, verification can be performed in $O(n \log n)$ time. With this, the entire reconstruction process completes in $O(n \log n)$ time for general alphabets.

Algorithm 2. Reconstruct $prev_S$

Input: pSA and pLCP arrays
Output: $prev_S[0 \ldots n-1]$
                    ▷ preprocessing for dist query
preprocess pLCP array for RMQ
populate inverse p-suffix array, pISA$[0 \ldots n-1]$
                     ▷ populate $prev_S$ array
$prev_S[0] \leftarrow 0$ ▷ prev encoding begins with 0
for $i \leftarrow 1$ to $n-1$ **do**   ▷ rest of array is initialized to unknown ∞
 $prev_S[i] \leftarrow \infty$
end for
for $i \leftarrow 1$ to $n-2$ **do**
 $j \leftarrow$ pLCP$[i]$ ▷ figure out j and k' based on i
 $k' \leftarrow$ pSA$[i+1]$
 if DistQuery(i, i+1) **then**   ▷ $p_{k'}[j] = j$
  if $prev_S[k'+j] \neq \infty$ **then**
   Exit with Failure   ▷ invalid input - should not set same location twice
  end if
  $prev_S[k'+j] \leftarrow j$ ▷ exit with failure if invalid index
  next $\leftarrow i+1$
            ▷ propagate each value as soon as it is calculated
  while pLCP[next] $> j$ **do**
   $k \leftarrow$ pSA[next+1]
   $prev_S[k+j] \leftarrow j$ ▷ exit with failure if invalid index
   next++   ▷ go down
  end while
 end if
end for
                 ▷ initialize and populate pointedAt array
for $i \leftarrow 0$ to $n-1$ **do**
 pointedAt$[i] \leftarrow$ false
end for
for $m \leftarrow 0$ to $n-1$ **do** $p \leftarrow prev_S[v]$
 if $m - v \geq 0$ and pointedAt$[m-v] =$ false **then**
  pointedAt$[m-v] \leftarrow$ true
 else
  Exit with Failure   ▷ invalid prev encoding
 end if
end for

4 Minimizing the Alphabet Size

In this section, we call an array element $prev_S[k]$, $0 < k < n$, an *unknown element* if $prev_S[k]$ did not receive a value from dist queries or by propagation of values obtained from dist queries.

The algorithm presented in Sect. 3 places zeros at all locations in $prev_S$ whose values do not have a role in satisfying the input pSA and pLCP arrays. Each zero in $prev_S$ introduces a new character to S. Thus, setting all the unknowns in $prev_S$ to zero can cause S to have *maximal* possible alphabet size.

In this section we focus on the problem of reconstructing a p-string with *minimal* alphabet size. The structure of the algorithm remains the same; we merely replace the step that sets unknown elements of $prev_S$ to zero with more complex procedures. We begin with a linear-time algorithm in Sect. 4.1 to decrease the alphabet size. Then, in Sect. 4.2 we present an algorithm that fully minimizes the alphabet size, albeit in $O(n^2)$ time.

4.1 Reducing the Alphabet Size

Observation 1. *An unknown element $prev_S[k]$, $0 < k < n$, can be set to v as long as $pointedAt[k-v] = $ `false` and v is outside of all the pLCP's and points of mismatch that span k, i.e. no p-suffix from location 0 to location $k-v$ includes location k in its pLCP or point of mismatch.*

As stated in Observation 1, for each p-suffix we only care about the prefix pLCP that it shares with a neighbor in the pSA, and the points of mismatch, since the prev encoding of this prefix determines the position of the p-suffix in the pSA. We can reuse characters prior to this range of interest, and they are still encoded as zeros in the particular p-suffix.

Suppose pSA$[i'] = i$. We say that position y *is included* in a pLCP for position i if either $i + pLCP[i'] - 1 \geq y$ or $i + pLCP[i' - 1] - 1 \geq y$. Similarly, position y *is included* in a point of mismatch for position i if there is a point of mismatch for $pLCP[i']$ or for $pLCP[i' - 1]$ that occurs at position y.

We construct an integer array of size n called furthestSuff to represent for each position in $prev_S$ the earliest p-suffix in which that position is included in a pLCP or point of mismatch. The value in the array furthestSuff$[k]$ is the smallest j, i.e. earliest p-suffix p_j, that is involved in a pLCP that extends through location k or has a point of mismatch at location k. All p-suffixes and positions before furthestSuff$[k]$ have no bearing on the determination of $prev_S[k]$. Hence, we can choose to set $prev_S[k]$ to point to any position (and reuse any character) prior to furthestSuff$[k]$, so long as pointedAt contains `false` at that position.

To populate the furthestSuff array, we start with p_0, and initialize to 0 each location from the beginning of the array to the point of mismatch following the longest pLCP that p_0 is involved with. We perform this process for each subsequent p-suffix, and for each p-suffix we begin with the index that the previous p-suffix left off with. In general, for each i (from 0 to $n-1$), let i' be the location of p_i in the pSA (obtained from the inverse suffix array), i.e. $pSA[i'] = i$. We set

m to represent the longest pLCP that p_i is involved with, extending to the point of mismatch if neither pLCP ends at the end of the p-string. For a specific i, we consider pLCP$[i']$ and pLCP$[i'-1]$ to obtain m. Then we set to the value i all locations in furthestSuff that occur after the positions relevant to p_{i-1} through the positions that are relevant to p_i.

Observation 2. *furthestSuff is a non-decreasing sequence of integers 0 to $n-1$.*

Observation 3. *Allocating the earliest available positions for reuse in order of increasing furthestSuff values results in maximum reuse of characters occurring prior to the pLCPs that span the unknown elements.*

We use Observations 2 and 3 to set the maximum number of unknown elements in $prev_S$ to refer to positions of false values in the pointedAt array occurring prior to their furthestSuff values. Those that cannot be resolved in this manner are simply set to 0 and introduce a new character to the p-string. We resolve the unknowns in $prev_S$ sequentially, and for each unknown, we reuse the earliest available position that is before its furthestSuff value. This way, we save later positions at which pointedAt is false for unknowns that may have larger furthestSuff entries.

Referring back to the example in Fig. 1, when we run the algorithm of the previous section, and insert zeros at unknown elements in $prev_S$, we derive $prev_S$= 000310032 with parameterized alphabet of size 5. This can correspond to S = bcabbdebe. If we instead use the furthestSuff array to reduce the alphabet size, we derive $prev_S$ = 000314432 with parameterized alphabet of size 3. ($prev_S[5]$ makes use of the false value in pointedAt[1] and $prev_S[6]$ makes use of the false value in pointedAt[2].) This can correspond to S = bcabbcaba. The corresponding pointedAt and furthestSuff arrays are depicted in Table 1.

Another example is presented in Table 2. When we run the algorithm of the previous section on this set of input, and insert zeros at unknown positions in $prev_S$, we derive $prev_S$= 0011003110100110 with parameterized alphabet of size 8. This can correspond to S = abbbcdbbbeefgggh. If we instead use the furthestSuff array to reduce the alphabet size, we derive $prev_S$ = 0011003119177117 with parameterized alphabet of size 4, which can correspond to S=abbbcdbbbaacdddb. The alphabet size shrinks from 8 to 4 by reusing the characters at the positions of the first 4 false values in the pointedAt array.

Table 1. Auxiliary arrays that facilitate the reduction of Π for input in Fig. 1

index	0	1	2	3	4	5	6	7	8
$prev_S$ $|\Pi|=5$	0	0	0	3	1	0	0	3	2
$prev_S$ with ∞	0	∞	∞	3	1	∞	∞	3	2
pointedAt	T	F	F	T	T	F	T	F	F
furthestSuff	0	0	0	0	0	4	4	4	4
$prev_S$ $|\Pi|=3$	0	0	0	3	1	4	4	3	2

Table 2. Sample input and solution using furthestSuff array to reduce Π

index	0	1	2	3	4	5	6	7	8	9	10	11	12	13	14	15
pSA	15	14	10	4	3	8	11	0	5	13	9	2	7	12	1	6
pLCP		1	2	6	3	2	3	5	5	1	3	4	3	2	4	4
$prev_S$ $\|\Pi\|=8$	0	0	1	1	0	0	3	1	1	0	1	0	0	1	1	0
$prev_S$ with ∞	0	∞	1	1	∞	∞	3	1	1	∞	1	∞	∞	1	1	∞
pointedAt	F	T	T	T	F	F	T	T	F	T	F	F	T	T	F	F
furthestSuff	0	0	0	0	0	0	2	4	4	4	5	8	9	9	10	10
$prev_S$ $\|\Pi\|=4$	0	0	1	1	0	0	3	1	1	9	1	7	7	1	1	7

Lemma 8. *The pointedAt and furthestSuff arrays are set up in linear time.*

Lemma 9. *If the input is reconstructible then we will reconstruct it in linear time with our algorithm that uses the pointedAt and furthestSuff arrays.*

4.2 Minimizing the Alphabet Size

The method presented in Sect. 4.1 can reduce the alphabet size but does not necessarily arrive at the *minimal* alphabet. Sometimes, an unknown element $prev_S[k]$, $1 \leq k < n$, can point to a position closer than furthestSuff$[k]$, by reusing characters *within* a pLCP, and obtain a p-string over smaller alphabet. In this section we present an algorithm that fully minimizes Π in $O(n^2)$ time.

Observation 4. *An unknown element $prev_S[k]$, $0 < k < n$, can be set to v as long as pointedAt$[k-v]$ = false and no p-suffix from location 0 to location $k-v$ needs to have a zero in location k to satisfy the pSA / pLCP arrays.*

It follows from Observation 4 that for an unknown element, we can reuse characters prior to the furthest p-suffix that is required to be zero by the given input. We introduce the furthestSuff_Zero array to store for each position the furthest suffix that requires a zero at the position.

Definition 1. *For $0 \leq k < n$, furthestSuff_Zero$[k] = \min \ell$ such that $p_\ell[k-\ell]$ must be 0 to satisfy the input pSA / pLCP arrays.*

We initialize each entry in the furthestSuff_Zero array to the latest p-suffix that spans that position, i.e., furthestSuff_Zero$[k] = k$, $0 \leq k < n$. As we identify and propagate zeros, we only update furthestSuff_Zero$[k]$ if the new value is smaller than the existing value of furthestSuff_Zero$[k]$, $0 < k < n$. When we change a value furthestSuff_Zero$[k]$ from α to $\beta < \alpha$, $0 < k < n$, we propagate the zero to corresponding positions based on pLCPs of $p_\beta, \ldots, p_{\alpha-1}$.

We identify the points of mismatch that evaluate to zero as they correspond to positions of unknown elements. For each element $0 < pSA[k] < n-1$ with zero at its point of mismatch, we propagate the zero towards the beginning of the pSA as long as the pLCP entry remains larger than pLCP[k]. We note that a value of zero does not propagate towards the end of the pSA since the mismatch is greater than 0 and follows immediately in the pSA.

For each position $0 \le k < n$ such that $prev_S[k] = dist$, we consider the shorter p-suffixes that span that position and are closer than dist, and make sure that we are aware that their shared pLCPs have zeros at those points.

Lemma 10. *The furthestSuff_Zero array is set up in $O(n^2)$ time.*

We split the unknown elements into two groups. In Group 1, furthestSuff[k] < furthestSuff_Zero[k], and in Group 2, furthestSuff[k] = furthestSuff_Zero[k]. First we resolve the unknowns in Group 1 that can reuse a character *within* a pLCP. Then we resolve the remaining unknowns and reuse a character that occurred prior to the pLCP, if possible, and set the values to 0 otherwise.

Group 1: furthestSuff[k] < furthestSuff_Zero[k]

In this case, we are attempting to reuse a character within the pLCP, and must ensure that the value is valid in all instances of the shared pLCP. In the remainder of this section, we say that pLCP[i] *is shared* by r p-suffixes if pLCP[i] occurs within a consecutive set of $r-1$ entries in the pLCP array in which the values, excluding pLCP[i], are all larger than pLCP[i].

Lemma 11. *Suppose p-suffixes p_i and $p_{i'}$ have pLCP $\gamma > \ell$ in common, $prev_S[i+\ell]$ is unknown and in Group 1 (i.e., furthestSuff[i + ℓ] < furthestSuff_Zero[i + ℓ]). Then, $prev_S[i' + \ell]$ is unknown, in Group 1, and $i + \ell$ − furthestSuff_Zero[i + ℓ]= $i' + \ell$ − furthestSuff_Zero[i' + ℓ].*

Corollary 1. *Suppose $pLCP[i] > \ell$ is shared by r p-suffixes and $prev_S[i+\ell]$ is unknown and in Group 1. Then, all r occurrences contain unknown elements at position ℓ and are in Group 1. Furthermore, the furthestSuff_Zero entries within the shared pLCPs are the same offset from each of these unknown elements.*

Observation 5. *Allocating the earliest available positions for reuse in order of increasing furthestSuff_Zero values results in maximum reuse of characters occurring within pLCPs that span the unknown elements.*

We follow increasing order of the furthestSuff_Zero array to maximize the number of unknowns in Group 1 that can be resolved by reusing characters within the pLCP. For an unknown at position k, we can reuse a character between furthestSuff[k] and furthestSuff_Zero[k] if it is not pointed at subsequently. By Lemma 11 and Corollary 1, corresponding positions in pLCPs will have the same offset to their furthestSuff_Zero values. We resolve simultaneously all occurrences of an unknown that is shared by r p-suffixes. We consult r subarrays of the pointedAt array corresponding to the shared pLCPs, and identify the furthest offset at which pointedAt is `false` for all occurrences.

When we reuse a character at distance d within the pLCP, we set $prev_S[k]$ to d in all r occurrences of the pLCP and set the corresponding r `false` entries of the pointedAt array to `true`. If we do not find a position to reuse between furthestSuff[k] and furthestSuff_Zero[k], we resolve this unknown together with those in Group 2.

Lemma 12. *In $O(n^2)$ time, the algorithm resolves all the unknown elements $prev_S[k]$ $0 < k < n$ that can reuse characters within the pLCPs or points of mismatch that span their positions k.*

Group 2: furthestSuff[k] = furthestSuff_Zero[k]

In this case, we are not reusing a character *within* the pLCP so the unknown is set to zero within the pLCP. In this case, each of the unknowns with common pLCP can be resolved independently of each other. Thus, we apply the technique of Sect. 4.1 and use the updated pointedAt with furthestSuff.

Table 3 depicts the continuation of the example in Table 2. In the previous section, we derived a p-string with an alphabet of size 4. Using the furthestSuff_Zero array, we arrive at the minimal alphabet of size 3. The prev encoding 0011403114174114 can correspond to the p-string abbbacbbbccabbba.

Table 3. Sample input and solution using furthestSuff_Zero array to minimize Π

index	0	1	2	3	4	5	6	7	8	9	10	11	12	13	14	15
pSA	15	14	10	4	3	8	11	0	5	13	9	2	7	12	1	6
pLCP	1	2	6	3	2	3	5	5	1	3	4	3	2	4	4	
$prev_S$ with ∞	0	∞	1	1	∞	∞	3	1	1	∞	1	∞	∞	1	1	∞
pointedAt	F	T	T	T	F	F	T	T	F	T	F	F	T	T	F	F
furthestSuff	0	0	0	0	0	0	2	4	4	4	5	8	9	9	10	10
furthestSuff_Zero	0	0	2	3	2	0	4	7	8	7	10	8	9	13	14	13
$prev_S$ $\|\Pi\| = 3$	0	0	1	1	4	0	3	1	1	4	1	7	4	1	1	4

Theorem 2. *If the input is reconstructible then our algorithm reconstructs a p-string S over the minimal alphabet in $O(n^2)$ time.*

Acknowledgments. Amihood Amir was partially supported by ISF grant 168-23 and BSF grant 2018141. Shoshana Marcus was partially supported by PSC-CUNY Awards 66369-00 54 and 67488-00 55, jointly funded by The Professional Staff Congress and The City University of New York. Dina Sokol was partially supported by BSF grant 2018141.

Disclosure of Interests. The authors have no competing interests to declare that are relevant to the content of this article.

References

1. Amir, A., Kondratovsky, E.: Towards a real time algorithm for parameterized longest common prefix computation. Theoret. Comput. Sci. **852**, 132–137 (2021). https://doi.org/10.1016/j.tcs.2020.11.023

2. Amir, A., Kondratovsky, E., Landau, G.M., Marcus, S., Sokol, D.: Reconstructing parameterized strings from parameterized suffix and LCP arrays. Theoret. Comput. Sci. **981**, 114230 (2024). https://doi.org/10.1016/j.tcs.2023.114230
3. Amir, A., Kondratovsky, E., Levy, A.: On suffix tree detection. In: Nardini, F.M., Pisanti, N., Venturini, R. (eds.) String Processing and Information Retrieval - 30th International Symposium, SPIRE 2023, Pisa, Italy, September 26-28, 2023, Proceedings. Lecture Notes in Computer Science, vol. 14240, pp. 14–27. Springer (2023). https://doi.org/10.1007/978-3-031-43980-3_2
4. Baker, B.S.: Parameterized pattern matching: Algorithms and applications. J. Comput. Syst. Sci. **52**(1), 28–42 (1996). https://doi.org/10.1006/jcss.1996.0003
5. Baker, B.S.: Parameterized duplication in strings: Algorithms and an application to software maintenance. SIAM J. Comput. **26**(5), 1343–1362 (1997). https://doi.org/10.1137/S0097539793246707
6. Bannai, H., Inenaga, S., Shinohara, A., Takeda, M.: Inferring strings from graphs and arrays. In: Rovan, B., Vojtás, P. (eds.) Mathematical Foundations of Computer Science 2003, 28th International Symposium, MFCS 2003, Bratislava, Slovakia, August 25–29, 2003, Proceedings. Lecture Notes in Computer Science, vol. 2747, pp. 208–217. Springer (2003). https://doi.org/10.1007/978-3-540-45138-9_15
7. Beal, R., Adjeroh, D.: Variations of the parameterized longest previous factor. J. Discret. Algorithm. **16**, 129–150 (2012)
8. Deguchi, S., Higashijima, F., Bannai, H., Inenaga, S., Takeda, M.: Parameterized suffix arrays for binary strings. In: Holub, J., Zdárek, J. (eds.) Proceedings of the Prague Stringology Conference 2008, Prague, Czech Republic, September 1-3, 2008. pp. 84–94. Prague Stringology Club, Department of Computer Science and Engineering, Faculty of Electrical Engineering, Czech Technical University in Prague (2008). http://www.stringology.org/event/2008/p08.html
9. Duval, J., Lefebvre, A.: Words over an ordered alphabet and suffix permutations. RAIRO Theor. Inf. Appl. **36**(3), 249–259 (2002). https://doi.org/10.1051/ita:2002012
10. Fujisato, N., Nakashima, Y., Inenaga, S., Bannai, H., Takeda, M.: Direct linear time construction of parameterized suffix and LCP arrays for constant alphabets. In: Brisaboa, N.R., Puglisi, S.J. (eds.) String Processing and Information Retrieval - 26th International Symposium, SPIRE 2019, Segovia, Spain, October 7-9, 2019, Proceedings. Lecture Notes in Computer Science, vol. 11811, pp. 382–391. Springer (2019). https://doi.org/10.1007/978-3-030-32686-9_27
11. Gabow, H.N., Bentley, J.L., Tarjan, R.E.: Scaling and related techniques for geometry problems. In: DeMillo, R.A. (ed.) Proceedings of the 16th Annual ACM Symposium on Theory of Computing, April 30 - May 2, 1984, Washington, DC, USA. pp. 135–143. ACM (1984). https://doi.org/10.1145/800057.808675
12. Gawrychowski, P., Jez, A., Jez, L.: Validating the Knuth-Morris-Pratt failure function, fast and online. Theory Comput. Syst. **54**(2), 337–372 (2014). https://doi.org/10.1007/s00224-013-9522-8
13. Harel, D., Tarjan, R.E.: Fast algorithms for finding nearest common ancestors. SIAM J. Comput. **13**(2), 338–355 (1984). https://doi.org/10.1137/0213024
14. Jalsenius, M., Porat, B., Sach, B.: Parameterized matching in the streaming model. In: Portier, N., Wilke, T. (eds.) 30th International Symposium on Theoretical Aspects of Computer Science, STACS 2013, February 27 - March 2, 2013, Kiel, Germany. LIPIcs, vol. 20, pp. 400–411. Schloss Dagstuhl - Leibniz-Zentrum für Informatik (2013). https://doi.org/10.4230/LIPIcs.STACS.2013.400

15. Kärkkäinen, J., Piatkowski, M., Puglisi, S.J.: String inference from longest-common-prefix array. Theoret. Comput. Sci. **942**, 180–199 (2023). https://doi.org/10.1016/j.tcs.2022.11.032
16. Keller, O., Kopelowitz, T., Lewenstein, M.: On the longest common parameterized subsequence. Theoret. Comput. Sci. **410**(51), 5347–5353 (2009)
17. Kosaraju, S.R.: Faster algorithms for the construction of parameterized suffix trees (preliminary version). In: 36th Annual Symposium on Foundations of Computer Science, Milwaukee, Wisconsin, USA, 23-25 October 1995. pp. 631–637. IEEE Computer Society (1995). https://doi.org/10.1109/SFCS.1995.492664
18. Nakashima, K., Fujisato, N., Hendrian, D., Nakashima, Y., Yoshinaka, R., Inenaga, S., Bannai, H., Shinohara, A., Takeda, M.: Parameterized DAWGs: Efficient constructions and bidirectional pattern searches. Theoret. Comput. Sci. **933**, 21–42 (2022)
19. Nakashima, Y., Okabe, T., I, T., Inenaga, S., Bannai, H., Takeda, M.: Inferring strings from lyndon factorization. Theoret. Comput. Sci. **689**, 147–156 (2017). https://doi.org/10.1016/j.tcs.2017.05.038

Bijective BWT Based Compression Schemes

Golnaz Badkobeh[1], Hideo Bannai[2(✉)], and Dominik Köppl[3]

[1] City, University of London, London, UK
Golnaz.Badkobeh@city.ac.uk
[2] Tokyo Medical and Dental University, Tokyo, Japan
hdbn.dsc@tmd.ac.jp
[3] University of Yamanashi, Kofu, Japan
dkppl@yamanashi.ac.jp

Abstract. We investigate properties of the bijective Burrows-Wheeler transform (BBWT). We show that for any string w, a bidirectional macro scheme of size $O(r_B)$ can be induced from the BBWT of w, where r_B is the number of maximal character runs in the BBWT. We also show that $r_B = O(z \log^2 n)$, where n is the length of w and z is the number of Lempel-Ziv 77 factors of w. Then, we show a separation between BBWT and BWT by a family of strings with $r_B = \Omega(\log n)$ but having only $r = 2$ maximal character runs in the standard Burrows–Wheeler transform (BWT). However, we observe that the smallest r_B among all cyclic rotations of w is always at most r. While an $o(n^2)$ algorithm for computing an optimal rotation giving the smallest r_B is still open, we show how to compute the Lyndon factorizations – a component for computing BBWT – of all cyclic rotations in $O(n)$ time. Furthermore, we conjecture that we can transform two strings having the same Parikh vector to each other by BBWT and rotation operations, and prove this conjecture for the case of binary alphabets and permutations.

Keywords: Repetitiveness measure · Burrows–Wheeler Transform

1 Introduction

The Burrows–Wheeler transform (BWT) [11] has seen numerous applications in data compression and text indexing, and is used heavily by various tools in the field of bioinformatics. For any string w, BWT(w) is defined as the string obtained by concatenating the last characters of all cyclic rotations of w, in the lexicographic order of the cyclic rotations. BWT is not injective, as two strings are transformed to the same string if they are cyclic rotations of each other. Also, BWT is not surjective since BWT preserves the string length and by the pigeonhole principle, there exist strings that are not in the image of BWT.

For a string x, the inverse BWT transform is induced from the LF-mapping function $\psi_x(i)$ which maps position i in x to its rank among all positions ordered by $(x[i], i)$, i.e. $\psi_x(i) = |\{j \in [1, |x|] \mid x[j] < x[i]\}| + |\{j \in [1, i] \mid x[j] = x[i]\}|$. For a primitive string w, the standard BWT always constructs a string $x = \mathsf{BWT}(w)$ such that ψ_x forms a single cycle (i.e., $\forall i, \exists j$ s.t. $\psi_x^j(i) = 1$), and $x[\psi_x^{|w|-1}(i)] \cdots x[\psi_x^0(i)]$ is a cyclic rotation of w. As an example, for $w[1..6] =$ banana and $x = \mathsf{BWT}(w) = $ nnbaaa, $\psi_x(1) = 5, \psi_x(2) = 6, \psi_x(3) = 4, \psi_x(4) = 1, \psi_x(5) = 2, \psi_x(6) = 3$, and thus $\psi_x^6(i) = i$ for all $i \in [1, 6]$.

In general, ψ_x can form several cycles (e.g., when x is not in the image of BWT), and it is more natural to view the inverse BWT transform as a mapping from a string to a multiset of primitive cyclic strings. The bijective BWT (BBWT) [13,15,17] exploits this to define a bijection on strings. By selecting the lexicographically smallest rotation of each cyclic string and concatenating them in non-increasing lexicographic order, this mapping becomes a bijection that maps a string to another string. The forward transform $\mathsf{BBWT}(w)$ can then be defined as a transform that first computes the Lyndon factorization [12] of w, and then taking the last symbol of all cyclic rotations of all the Lyndon factors, sorted in ω-order $\prec_\omega$, which is an order defined, when x, y are primitive, as $x \prec_\omega y \iff x^\infty \prec y^\infty$, and $\prec$ denotes the standard lexicographic order.

It is known that the BBWT can be computed in linear time [6,8]. It can also be used as an index similar to the BWT [5,6], or as an index for a set of circular strings [10]. While the size r of the run-length compressed BWT (RLBWT) has been a focus of study in various contexts and is known to be small for highly-repetitive texts [20], the size r_B of the run-length compressed BBWT (RLBBWT) has not yet been studied rigorously. Biagi et al. [7] study the sensitivity [1] of r_B with respect to the *reverse* operation, and present an infinite family of strings such that r_B of a string and its reverse can differ by a factor of $\Omega(\log n)$.

In this paper, we investigate properties of BBWT and r_B. In detail, we show that we can induce a bidirectional macro scheme (BMS) [23] of size $O(r_B)$ for w, from the RLBBWT of w (Lemma 1). We further show that $r_B = O(z \log^2 n)$ where z is the number of Lempel-Ziv 77 (LZ77) factors of w (Theorem 1). Then, we show a separation between r_B and r, by a family of strings with $r_B = \Omega(\log n)$ but $r = 2$ (Theorem 2).

Noticing that $r_B = r$ for Lyndon words, the smallest r_B among all cyclic rotations of w is always at most r. While we do not yet know how to compute, in subquadratic time, such an optimal rotation that gives the smallest r_B, we show that we can compute the Lyndon factorizations of all rotations of w in linear time (Theorem 3).

Finally, we conjecture that two strings having the same Parikh vector can be transformed to each other by BBWT and rotation operations (Conjecture 1); we prove this conjecture for special cases (Theorem 4).

2 Preliminaries

Let Σ be a set of symbols referred to a the *alphabet*, and Σ^* the set of strings over Σ. For a string $x \in \Sigma^*$, $|x|$ denotes x's length. The empty string (the string of length 0) is denoted by ε. For integer $i \in [1, |x|]$, $x[i]$ is the ith symbol of x, and for integer $j \in [i, |x|]$, $x[i..j] = x[i] \cdots x[j]$. For convenience, let $x[i..j] = \varepsilon$ if $i > j$. Let $x = x^1$, and for integer $k \geq 2$, $x^k = xx^{k-1}$ A string is *primitive*, if it cannot be represented as x^k for some string x and integer $k \geq 2$.

Let $rot(x) = x[|x|]x[1..|x|-1]$. A string y is a *cyclic rotation* (or simply a *rotation*) of x if there exists i such that $y = rot^i(x)$, where $rot^1(x) = rot(x)$, and for integer $k \geq 2$, $rot^k(x) = rot^{k-1}(rot(x))$.

Given a total order $\prec$ on Σ, the lexicographic order (also denoted by $\prec$) induced by $\prec$ is a total order on Σ^* such that $x \prec y$ if and only if x is a prefix of y, or, $x[i] \prec y[i]$ where $i = \min\{k \geq 1 \mid x[k] \neq y[k]\}$. A string w is a *Lyndon word*, if it is lexicographically smaller than all of its proper suffixes [16]. Lyndon words must therefore be primitive. Also, any string w can be partitioned into a unique sequence of lexicographically non-increasing Lyndon words, called the *Lyndon factorization* [12] of w, i.e., $w = f_1^{k_1} \cdots f_{\ell(w)}^{k_{\ell(w)}}$ where each f_i ($1 \leq i \leq \ell(w)$) is a Lyndon word, and $f_i \succ f_{i+1}$ for all $1 \leq i < \ell(w)$. We call $f_i^{k_i}$ the i-th *Lyndon necklace* of w. The ω-order $\prec_\omega$ is a total order over primitive strings, defined as: $x \prec_\omega y$ if and only if $x^\infty \prec y^\infty$.[1]

Given a string w, the Burrows–Wheeler transform $\mathsf{BWT}(w)$ is a string obtained by concatenating the last symbol of all cyclic rotations of w, in lexicographic order. The bijective BWT $\mathsf{BBWT}(w)$ is a string obtained by concatenating the last symbol of all cyclic rotations of all Lyndon factors in the Lyndon factorization of w, in ω-order. The number of maximal same-character runs in $\mathsf{BWT}(w)$ and $\mathsf{BBWT}(w)$ will be denoted by $r(w)$, and $r_B(w)$ respectively. Although $r(w)$, $r_B(w)$ and $\ell(w)$ are functions on strings to non-negative integers, we will omit writing the considered string and just write r, r_B, ℓ, if the context is clear.

3 Properties of r_B

We here analyze r_B as a repetitiveness measure for a string w. We first confirm that r_B corresponds to the size of a *bidirectional macro scheme (BMS)*, and is a repetitiveness measure for a form of dictionary compression, as is RLBWT. A BMS [23], the most expressive form of dictionary compression, partitions w into phrases, such that each phrase of length at least 2 can be represented as a reference to another substring of w. The referencing of the phrases induces a referencing forest over the positions: any position in a phrase of length at least 2 references another position in w, such that all positions in the same phrase have the same offset and thus adjacent positions point to adjacent positions, and there are no cycles.

Lemma 1. *There exists a BMS of size $O(r_B(w))$ that represents the string w.*

---

[1] Mantaci et al. [17] define the ω-order as a total order over arbitrary strings (including non-primitive strings), but as it is not relevant in our presentation, we omit this for simplicity.

Proof. We follow the existence proof for a BMS of size $O(r(w))$ by Navarro et al. [21]. We consider a BMS such that each text position that does not correspond to a beginning of a same-character run in BBWT(w) will reference the text position corresponding to the preceding character in the BBWT(w). It is clear that there are no cycles in such a referencing, and we claim that this allows the string to be partitioned into $O(r_B)$ phrases, such that references of adjacent positions in a given phrase of length at least two point to adjacent positions.

Focus on the i-th Lyndon necklace $f_i^{k_i}$ of w. For any cyclic rotation of f_i, its k_i copies occur adjacently when writing all cyclic conjugates of all Lyndon factors in the ω-order, and thus correspond to adjacent characters in a run in BBWT(w). It follows that for any position in the last $k-1$ copies of f_i, the reference points to the corresponding position in the preceding copy. Thus, adjacent positions reference adjacent positions, and can be contained in the same phrase.

Next, consider a position in the first copy of f_i that does not correspond to a beginning of a same-character run in BBWT(w). Then, since the character at this position and the preceding (in ω-order) cyclic string is the same, their preceding positions must also correspond to adjacent positions in BBWT(w). This implies that, as long as the corresponding position is again not a beginning of a same-character run, adjacent positions will refer to adjacent positions, albeit, in the cyclic sense. Being the beginning of a same-character run in BBWT(w) can happen at most r_B times. Being adjacent in the cyclic sense, but not being adjacent in text-order can happen at most once per referenced Lyndon necklace. Thus, the number of times adjacent text positions can be in a different phrase is bounded by $O(r_B + \ell(w))$. Since $\ell(w) \leq r_B$ can be shown from Corollary 2 of [9], this concludes the proof. □

Example 1. For the string $w = abbbabbababab$, we give below an example for the BMS computed from its BBWT. The Lyndon factorization of w is $abbb$, abb, ab, ab, ab.

i	1	2	3	4	5	6	7	8	9	10	11	12	13
$w[i]$	a	b	b	b	a	b	b	a	b	a	b	a	b
BBWT$[i]$	b	b	b	b	b	a	a	a	b	b	a	b	a
CSA$[i]$	8	10	12	5	1	9	11	13	7	4	6	3	2
CSA$[i]-1$	9	11	13	7	4	8	10	12	6	3	5	2	1
ref?		✓	✓	✓	✓		✓	✓		✓			

The number of runs r_B is 6. BBWT positions belonging to a referencing phrase are marked with ✓ in the last row. We therefore have 6 non-referencing phrases.

The referencing phrases are computed as follows: In the table, CSA is the circular suffix array whose entry CSA$[i]$ denotes the text position after the one from which we took BBWT$[i]$ (or the first text position if BBWT$[i]$ belongs to the last character of a Lyndon factor). The row CSA$[i]-1$ denotes the text position corresponding to BBWT[i]. By construction of our BMS, the CSA entry positions 2-5, 7-8, and 10 correspond to referencing phrases, i.e., the text positions after 10,12,5,1,11,13,4 (applying CSA on these entry positions), i.e., the text positions 11, 13, 7, 4, 10, 12, 3. These positions refer to 9, 11, 13, 7, 8, 10, 6, respectively. If we group together neighboring positions that have the same offset to its reference, we obtain the following 9 phrases. Figure 1 gives a visualization of this BMS.

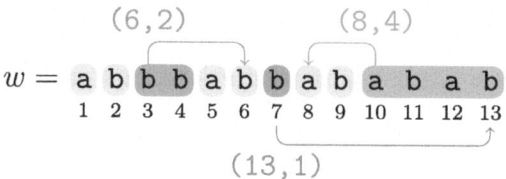

Fig. 1. The BMS factorization of Example 1

Theorem 1. *For any string, $r_B = O(z \log^2 n)$.*

Proof. (Sketch) We follow the proof of $r = O(z \log^2 n)$ by Kempa and Kociumaka [14]. The LCP array of a string w is an array of integers such that its i-th entry is the length of the longest common prefix (lcp) between the lexicographically ($i-1$)-th and i-th cyclic rotation of w. An *irreducible LCP position* is a position i such that $i = 1$ or BWT(w)[$i-1$] $\neq$ BWT(w)[i], and thus the number r of BWT runs is the number of irreducible LCP positions. For a multiset of primitive cyclic strings, we analogously define the ω-LCP array such that its i-th entry is the lcp between x^∞ and y^∞, where x and y are respectively the ($i-1$)-th and i-th string, in ω-order, among all cyclic rotations of all Lyndon factors of the Lyndon factorization of w. By construction, the number r_B of BBWT runs is the number of irreducible ω-LCP positions, i.e., $i = 1$ or BBWT(w)[$i-1$] $\neq$ BBWT(w)[i]. Note that ω-LCP values can be infinite when there are Lyndon necklaces in the Lyndon factorization with exponent at least 2, but they can be safely disregarded since they are not irreducible. The theorem follows if we can show that for any value k, the number of irreducible ω-LCP values in $[k, 2k)$ is $O(z \log n)$ and considering $k = 2^i$ for $i = 0, \ldots, \lfloor \log n \rfloor$.

The arguments in the proof in [14] proceed by first asserting that for any integer k, a string contains at most $3kz$ distinct strings of length $3k$. Then, each irreducible LCP value in $[k, 2k)$ is associated with a cost of k, which are charged to positions in the at most $3kz$ strings that have an occurrence crossing the corresponding suffix array position, and it is shown that each substring can be charged at most $2 \log n$ times. The total cost is thus at most $6kz \log n$ and thus the number of irreducible LCP values is $O(z \log n)$.

For ω-LCP, the corresponding length $3k$ substring associated with the suffix array position may not occur in the original string but instead will correspond to a substring of some Lyndon necklace of the Lyndon factorization. Note that there are at most $3k$ distinct substrings of length $3k$ that are not substrings of the original string but a substring of a given Lyndon necklace. Since $\ell(w) < 4z$ [24], we have that the total number of such distinct substrings of length $3k$ that occur in this context is still bounded by $O(kz)$, and that the arguments still hold. □

Despite sharing common traits, r and r_B can be asymptotically different:

Theorem 2. *There exists a family of strings with $r_B = \Omega(\log n)$ and $r = 2$.*

Proof. Define the Fibonacci words as follows: $F_0 = \mathtt{b}, F_1 = \mathtt{a}, F_i = F_{i-1}F_{i-2}$. The infinite Fibonacci word is $\lim_{k\to\infty} F_k$. Melançon [19] showed that the k-th factor (the first factor being the 0-th) of the Lyndon factorization of the infinite Fibonacci word, has length f_{2k+2}, where $f_i = |F_i|$. Now, $\sum_{k=0}^{i} f_{2k+2} = -f_1 + (\cdots(((f_1 + f_2) + f_4) + f_6) + \cdots) + f_{2i+2} = f_{2i+3} - 1$. Therefore, the word obtained by deleting the last symbol of F_{2i+3} has $i + 1$ distinct Lyndon factors. Noticing that the last symbol of F_{2i+3} must be 'a' and will form a distinct Lyndon factor, we have that the size of the Lyndon factorization of F_{2i+3} is $i + 2$. Since $\ell(w) \leq r_B$ [9], the BBWT of the k-th Fibonacci word F_k for odd k has $\Omega(k)$ runs, while the BWT of any Fibonacci word has $r(F_k) = 2$ runs [18]. □

We have not yet been able to find a family of strings where $r_B = o(r)$.

4 RLBBWT and Rotation

Theorem 2 may give the impression that r may be a smaller measure compared to r_B. However, if we are to incorporate a rotation operation, which can be encoded as a single $\log n$-bit integer, we could possibly obtain a representation smaller than r using BBWT. This is because we have $r(x) = r_B(x)$ for the Lyndon rotation x of any primitive word. The Lyndon rotation of a primitive word can be computed in linear time [22].

For any string w, let $\hat{w} = \arg\min_{uv=w}\{r_B(vu)\}$ be the *optimal* rotation with respect to r_B. We observe that $\hat{w}$ is not always the Lyndon rotation of w. For example, for the Lyndon word $w = \mathtt{aaabaabaaabaabb}$, we have that $\mathsf{BWT}(w) = \mathsf{BBWT}(w) = \mathtt{bbbaabaaaabaaaa}$, thus $r_B(w) = 6$. However, we have that $\hat{w} = rot(w) = \mathtt{baaabaabaaabaab}$ and $\mathsf{BBWT}(\hat{w}) = \mathtt{bbbbaaaaaaaaaab}$, thus $r_B(\hat{w}) = 3$.

Since $\mathsf{BBWT}(w)$ (and hence $r_B(w)$) can be computed in $O(n)$ time, it is straightforward to compute $\hat{w}$ (and hence $r_B(\hat{w})$) in $O(n^2)$ time. A subquadratic time algorithm for this problem would be very interesting. (For LZ77, it was recently shown that indeed subquadratic time computation is possible [3].) While we have not yet been able to achieve this, we give a partial result: a linear time algorithm for computing the Lyndon factorizations, a precursor to computing BBWT, of all cyclic rotations.

Theorem 3. *We can compute the sizes of the Lyndon factorizations of all cyclic rotations of w in time linear in the length of w.*

Proof. (Sketch) Assume that w is Lyndon, consider the string $W = ww$, and view the cyclic rotations of w as substrings of length $|w|$ of W. Any Lyndon factorization of such a substring consists of the Lyndon factorization of a suffix of w and a prefix of w, since any $x = uv$ such that u is a suffix of w and v is a prefix of w cannot be Lyndon: it would imply $uv \prec v \prec w \prec u$, a contradiction.

We observe that the factors of the Lyndon factorization for any suffix of w, are the sequence of maximal right sub-trees of the standard (right) Lyndon tree [4] of w that are contained in the suffix. Similarly, for any prefix of w, they are the maximal left sub-trees of the left Lyndon tree [2] of w that are contained in the prefix.

The right Lyndon tree of a Lyndon word w is a binary tree defined recursively as follows: if w is a single letter, it is a leaf, otherwise, the left and right child are respectively the right Lyndon trees of u, v where $w = uv$ and v is the longest proper suffix of w that is a Lyndon word. Note that it can be shown that this choice of v implies that u is a Lyndon word. The left Lyndon tree is defined analogously, but u is the longest proper prefix of w that is a Lyndon word. Similarly, it can be shown that this choice of u implies that v is a Lyndon word.

By definition of the Lyndon trees, and the property of the Lyndon factorization which states that the first (resp. last) factor is the longest prefix (resp. suffix) that is a Lyndon word, it is a simple observation that the Lyndon factorization of a suffix of w is exactly the sequence of maximal right nodes of the right Lyndon tree that are contained in the suffix, and the Lyndon factorization of a prefix of w is exactly the sequence of maximal left nodes of the left Lyndon tree that are contained in the prefix.

Both trees can be computed in linear time [2,4]. It is not difficult to see that the changes in the sequences, and thus the sizes of the Lyndon factorizations can be computed in total linear time for each of the suffixes and prefixes, by a left-to-right traversal on the trees. □

5 BBWT Reachability

Further developing the idea of combining rotation and BBWT in order to obtain a smaller representation of a string, we give the following conjecture.

Conjecture 1. Given two words with the same Parikh vector, we can transform one to the other by using only rotation and BBWT operations.

If true, this would suggest that it is possible, for example, to represent a string w based on w's Parikh vector, and a sequence of integers where each integer alternately represents the offset of the rotation or the number of times BBWT is applied, to reach w from the lexicographically smallest string with the same Parikh vector.

We have computationally confirmed the conjecture for ternary strings of up to length 17, with code available at https://github.com/koeppl/bbwtreachability, and have proved it for the specific cases where the alphabet is binary, or, when all symbols are distinct. We first give the following lemma which shows that, under the condition of the lemma, we can obtain a lexicographically smaller string using rotation operations and an inverse BBWT operation.

Lemma 2. *Let x be a word of length n whose smallest rotation is itself (i.e., a necklace), over the alphabet $\{c_1, \ldots, c_\sigma\}$ where $c_1 \prec \cdots \prec c_\sigma$, and let $(e_1, \ldots, e_\sigma)$ be the Parikh vector of x. Let $y = c_1^{e_1} \cdots c_\sigma^{e_\sigma}$, i.e., the lexicographically smallest string with the same Parikh vector as x, and $i = lcp(x, y)$. If $x[i] \neq x[n]$, then, there exists k such that $rot^k(\text{BBWT}^{-1}(rot(x))) \prec x$.*

Proof. Note that $x[i] \neq x[n]$ implies $y \prec x$ since $y = x$ implies $i = n$. Let $x[1..i] = c_1^{e_1} \cdots c_k^{e'}$, where $e' \leq e_k$. This implies that symbols smaller than c_k are all used up in $x[1..i] = y[1..i]$, and cannot occur in $x[i+1..n]$ nor $y[i+1..n]$. Thus, for all $j \in [i+1, n]$, it holds that $x[j] \succeq x[i] = c_k$, in particular, for all $j \in [1, i]$, it holds that $x[n] \succ x[i] \succeq x[j]$ since $x[i] \neq x[n]$ is assumed.

Next, consider traversing the symbols of $\hat{x} = rot(x) = x[n]x[1..n-1]$ using the LF mapping $\psi_{\hat{x}}$ starting from position i' such that $\psi_{\hat{x}}(i') = i+1$ to recover a cyclic substring of $\text{BBWT}^{-1}(\hat{x})$, whose smallest rotation (Lyndon rotation) will be a substring of $\text{BBWT}^{-1}(\hat{x})$. Thus, we start from the symbol $\hat{x}[i'] = y[i+1]$. Since $\hat{x}[1] = x[n] \neq x[j] = y[j]$ for any $j \in [1..i]$ and $\hat{x}[2..i+1] = x[1..i] = y[1..i]$, it follows that $\psi_{\hat{x}}(j) = j - 1$ for any $j \in [2..i+1]$. Therefore, we have that $y[i+1]$ is prefixed by $x[1..i]$, i.e., $x[1..i]y[i+1] \prec x[1..i+1]$ is a cyclic substring of $\text{BBWT}^{-1}(\hat{x})$, where the inequality follows from the definition of y and i. Since the length $i+1$ prefix of the Lyndon rotation of the whole cyclic substring that is retrieved by $\psi_{\hat{x}}$ starting from $i+1$ cannot be larger than $x[1..i]y[i+1]$, it follows that $\text{BBWT}^{-1}(\hat{x})$ contains a length $i+1$ substring that is smaller than $x[1..i+1]$, and the lemma holds. □

Theorem 4. *Given two words of the same length with the same Parikh vector, it is possible to transform one to the other by using only rotations and BBWT transformations if all symbols are distinct, or if the alphabet is binary.*

Proof. Given any word, consider its smallest rotation x, and let y be the smallest word with the same Parikh vector. Since BBWT and rotations are bijections, it is easy to see that $\text{BBWT}^{-1}(x)$ (resp. $rot^{-1}(x)$) can be represented by a sequence of $\text{BBWT}(x)$ (resp. $rot(x)$) operations. Therefore, it suffices to show that we can reach y from x using any of these operations. If $y \prec x$, using Lemma 2, we can always obtain a strictly lexicographically smaller string using rotations and BBWT^{-1} and thus eventually reach y: when all symbols are distinct, it is easy to see that the condition of Lemma 2 holds. If the alphabet is binary, i.e., $\{a, b\}$, we have that $x[n] = b$ since x is a smallest rotation. Furthermore, if $i = lcp(x, y)$, then, since $x[i] = b$ would imply $x = y$, we have $x[i] = a \neq b = x[n]$. □

Acknowledgments. This work was supported by JSPS KAKENHI Grant Numbers JP24K02899 (HB) and JP23H04378 (DK).

References

1. Akagi, T., Funakoshi, M., Inenaga, S.: Sensitivity of string compressors and repetitiveness measures. Inf. Comput. **291**, 104999 (2023). https://doi.org/10.1016/j.ic.2022.104999
2. Badkobeh, G., Crochemore, M.: Linear construction of a left Lyndon tree. Inf. Comput. **285**(Part), 104884 (2022). https://doi.org/10.1016/j.ic.2022.104884
3. Bannai, H., Charalampopoulos, P., Radoszewski, J.: Maintaining the size of LZ77 on semi-dynamic strings. In: Inenaga, S., Puglisi, S.J. (eds.) 35th Annual Symposium on Combinatorial Pattern Matching, CPM 2024, June 25-27, 2024, Fukuoka, Japan. LIPIcs, vol. 296, pp. 3:1–3:20. Schloss Dagstuhl - Leibniz-Zentrum für Informatik (2024). https://doi.org/10.4230/LIPIcs.CPM.2024.3
4. Bannai, H., I, T., Inenaga, S., Nakashima, Y., Takeda, M., Tsuruta, K.: The "runs" theorem. SIAM J. Comput. **46**(5), 1501–1514 (2017)
5. Bannai, H., Kärkkäinen, J., Köppl, D., Piatkowski, M.: Indexing the bijective BWT. In: Pisanti, N., Pissis, S.P. (eds.) 30th Annual Symposium on Combinatorial Pattern Matching, CPM 2019, June 18-20, 2019, Pisa, Italy. LIPIcs, vol. 128, pp. 17:1–17:14. Schloss Dagstuhl - Leibniz-Zentrum für Informatik (2019). https://doi.org/10.4230/LIPIcs.CPM.2019.17
6. Bannai, H., Kärkkäinen, J., Köppl, D., Piatkowski, M.: Constructing and indexing the bijective and extended Burrows-Wheeler transform. Information and Computation **297**, 105153 (2024). https://doi.org/10.1016/j.ic.2024.105153, https://www.sciencedirect.com/science/article/pii/S089054012400018X
7. Biagi, E., Cenzato, D., Lipták, Zs., Romana, G.: On the number of equal-letter runs of the bijective Burrows-Wheeler transform. In: Castiglione, G., Sciortino, M. (eds.) Proceedings of the 24th Italian Conference on Theoretical Computer Science, Palermo, Italy, September 13-15, 2023. CEUR Workshop Proceedings, vol. 3587, pp. 129–142. CEUR-WS.org (2023), https://ceur-ws.org/Vol-3587/4564.pdf
8. Boucher, C., Cenzato, D., Lipták, Zs., Rossi, M., Sciortino, M.: Computing the original eBWT faster, simpler, and with less memory. In: Lecroq, T., Touzet, H. (eds.) String Processing and Information Retrieval - 28th International Symposium, SPIRE 2021, Lille, France, October 4-6, 2021, Proceedings. Lecture Notes in Computer Science, vol. 12944, pp. 129–142. Springer (2021). https://doi.org/10.1007/978-3-030-86692-1_11
9. Boucher, C., Cenzato, D., Lipták, Zs., Rossi, M., Sciortino, M.: r-indexing the eBWT. In: Proc. SPIRE. LNCS, vol. 12944, pp. 3–12 (2021)
10. Boucher, C., Cenzato, D., Lipták, Zs., Rossi, M., Sciortino, M.: r-indexing the eBWT. Information and Computation **298**, 105155 (2024). https://doi.org/10.1016/j.ic.2024.105155, https://www.sciencedirect.com/science/article/pii/S0890540124000208
11. Burrows, M., Wheeler, D.J.: A block sorting lossless data compression algorithm. Tech. Rep. 124, Digital Equipment Corporation, Palo Alto, California (1994)
12. Chen, K.T., Fox, R.H., Lyndon, R.C.: Free differential calculus, IV. The quotient groups of the lower central series. Annals of Mathematics **68**(1), 81–95 (1958)
13. Gil, J.Y., Scott, D.A.: A bijective string sorting transform. CoRR **abs/1201.3077** (2012), http://arxiv.org/abs/1201.3077
14. Kempa, D., Kociumaka, T.: Resolution of the Burrows-Wheeler transform conjecture. Commun. ACM **65**(6), 91–98 (2022). https://doi.org/10.1145/3531445
15. Kufleitner, M.: On bijective variants of the Burrows–Wheeler transform. In: Proc. PSC. pp. 65–79 (2009)

16. Lyndon, R.C.: On Burnside's problem. Trans. Am. Math. Soc. **77**(2), 202–215 (1954)
17. Mantaci, S., Restivo, A., Rosone, G., Sciortino, M.: An extension of the Burrows-Wheeler transform. Theor. Comput. Sci. **387**(3), 298–312 (2007)
18. Mantaci, S., Restivo, A., Sciortino, M.: Burrows-Wheeler transform and Sturmian words. Inf. Process. Lett. **86**(5), 241–246 (2003)
19. Melançon, G.: Lyndon words and singular factors of Sturmian words. Theor. Comput. Sci. **218**(1), 41–59 (1999)
20. Navarro, G.: Indexing highly repetitive string collections, part I: repetitiveness measures. ACM Comput. Surv. **54**(2), 29:1–29:31 (2021)
21. Navarro, G., Ochoa, C., Prezza, N.: On the approximation ratio of ordered parsings. IEEE Trans. Inf. Theory **67**(2), 1008–1026 (2021)
22. Shiloach, Y.: Fast canonization of circular strings. J. Algorithms **2**(2), 107–121 (1981)
23. Storer, J.A., Szymanski, T.G.: Data compression via textual substitution. J. ACM **29**(4), 928–951 (1982). https://doi.org/10.1145/322344.322346
24. Urabe, Y., Nakashima, Y., Inenaga, S., Bannai, H., Takeda, M.: On the size of overlapping Lempel-Ziv and Lyndon factorizations. In: Pisanti, N., Pissis, S.P. (eds.) 30th Annual Symposium on Combinatorial Pattern Matching, CPM 2019, June 18-20, 2019, Pisa, Italy. LIPIcs, vol. 128, pp. 29:1–29:11. Schloss Dagstuhl - Leibniz-Zentrum für Informatik (2019). https://doi.org/10.4230/LIPIcs.CPM.2019.29

Indexing Finite-State Automata Using Forward-Stable Partitions

Ruben Becker, Sung-Hwan Kim, Nicola Prezza, and Carlo Tosoni(✉)

DAIS, Ca' Foscari University of Venice, Venice, Italy
{rubensimon.becker,sunghwan.kim,nicola.prezza,carlo.tosoni}@unive.it

Abstract. An index on a finite-state automaton is a data structure able to locate specific patterns on the automaton's paths and consequently on the regular language accepted by the automaton itself. Cotumaccio and Prezza [SODA '21], introduced a data structure able to solve pattern matching queries on automata, generalizing the famous FM-index for strings of Ferragina and Manzini [FOCS '00]. The efficiency of their index depends on the width of a particular partial order of the automaton's states, the smaller the width of the partial order, the faster is the index. However, computing the partial order of minimal width is NP-hard. This problem was mitigated by Cotumaccio [DCC '22], who relaxed the conditions on the partial order, allowing it to be a partial preorder. This relaxation yields the existence of a unique partial preorder of minimal width that can be computed in polynomial time. In the paper at hand, we present a new class of partial preorders and show that they have the following useful properties: (i) they can be computed in polynomial time, (ii) their width is never larger than the width of Cotumaccio's preorders, and (iii) there exist infinite classes of automata on which the width of Cotumaccio's preorder is linearly larger than the width of our preorder.

Keywords: Nondeterministic Finite Automata · Graph Indexing · Forward-Stable Partitions · FM-index

1 Introduction

The Burrows-Wheeler transform (BWT) is a famous reversible string transformation [5]. While the BWT was initially conceived as a compression tool, it has subsequently been used to implement the FM index [8] that is able to locate patterns in almost optimal time, while efficiently compressing the strings at the same time. This indexing strategy has been extended to some finite automata (or directed edge-labeled graphs), the so-called *Wheeler graphs*, by Gagie et al. [9]. Wheeler graphs are a particular class of automata that can be succinctly encoded by a representation that allows pattern matching queries on the strings labeling directed paths in the automaton in almost optimal time. This indexing property relies on the fact that the states of the automaton can be totally ordered in a way that is consistent with the co-lexicographic order of the strings accepted by the automaton's states. Such an order of the states is called a *Wheeler order*. As not all automata admit a Wheeler order, this strategy may fail to apply and hence not all finite automata are Wheeler. Cotumaccio and Prezza [7] extended the notion to arbitrary finite automata by not restricting the state order to be total, instead allowing arbitrary partial orders, called *co-lex orders*. As any finite automaton admits such a co-lex order, this enables us to build an efficient index for the language accepted by any automaton in a very similar fashion as one does with the original FM index for a given string. The co-lex order however is not in general unique and, furthermore, the choice of the co-lex order is of crucial importance for the efficiency of the index. More precisely, the efficiency is directly characterized by the *width* of the particular co-lex order: the lower the width of the co-lex order is, the faster pattern matching queries are and the smaller the index is. Unfortunately, computing the co-lex order of minimal width is NP-hard [7]. Cotumaccio [6] introduced *co-lex relations* by relaxing the requirements on co-lex orders, allowing them to be reflexive relations, i.e., unrequiring antisymmetry and transitivity. This relaxation yields that a unique maximum co-lex relation always exists. Moreover, this maximum is a preorder (i.e., transitivity is recovered through maximality). Maybe most importantly though, the maximum co-lex relation can be computed in polynomial time (more precisely in quadratic time in the number of transitions). The width of the maximum co-lex relation is at most the width of every co-lex order and may be asymptotically smaller. Hence, the index built based on the maximum co-lex relation is never slower and can be asymptotically faster than those built based on co-lex orders [7].

Our Contribution. In this paper, we present a new category of preorders that we call *coarsest forward-stable co-lex (CFS) orders*. CFS orders are equally useful as maximum co-lex relations for constructing indices on automata. Similar to co-lex orders being the generalization of Wheeler orders to width larger than one, CFS orders are the generalization of Wheeler preorders (introduced by Becker et al. [3]) to width larger than one. As shown by Cotumaccio for maximum co-lex relations, we can show that our CFS orders always exist and are unique. Furthermore, for a given automaton, the unique CFS order can be computed in quadratic time in the number of transitions as well. Most importantly however, we prove that the width of the CFS order is never larger than the width of the maximum co-lex relation and, in some cases, asymptotically smaller.

In Fig. 1, we show the relationships among the previously mentioned concepts for building indices on automata. All previously known orders shown in Fig. 1 are introduced in Sect. 2, where we describe the preliminaries for this work. In Sect. 3 we introduce coarsest forward-stable co-lex (CFS) orders and in Sect. 4 we prove the claimed properties of these preorders. The full version of this article is available at [4].

2 Preliminaries

Relations and Partitions. Given a set U, a *relation* $R \subseteq U \times U$ over U is a set of ordered pairs of elements from U. For two elements u, v from U, we write

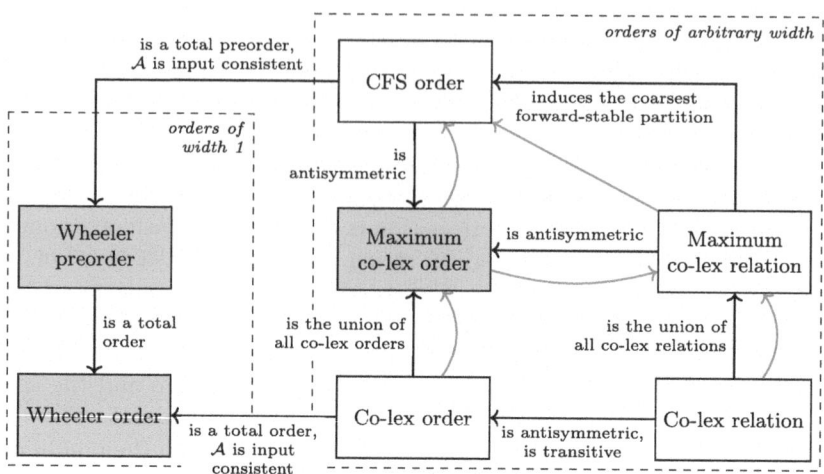

Fig. 1. An NFA $\mathcal{A}$ is *input consistent* if for each state u in $\mathcal{A}$, all incoming edges of u are labeled with the same character. We show the connections between the different relations described. A relation is white, if every automaton always admits an instance of that relation, and it is gray otherwise. Orders on the left, i.e., *Wheeler preorders* and *Wheeler orders*, are of width 1, while the others may be of arbitrary width. A gray edge $A \to B$ means that any relation of type A has always a width larger than or equal to a relation of type B. A black edge $A \xrightarrow{c} B$ means a relation of type A is also a relation of type B if it satisfies the requirements c. In this case, a relation of type B is always also a relation of type A with the following exceptions: (i) The *coarsest forward-stable co-lex order* may not be equal to the *maximum co-lex relation*. (ii) If the *maximum co-lex order* exists, then it may not be equal to the *coarsest forward-stable co-lex order*. (iii) A *Wheeler order* may not be equal to the *Wheeler preorder*. All implications either directly follow from their definitions or are proved in Appendix A of the extended version of this article [4]

uRv to denote that $(u,v) \in R$. A *partial order* over a set U is a relation that satisfies reflexivity, antisymmetry, and transitivity. If a partial order satisfies also connectedness then it is a *total order*. A *partial preorder* over a set U is a relation that satisfies reflexivity and transitivity. A *total preorder* is a partial preorder over a set U that satisfies also connectedness. An *equivalence relation* over a set U is a relation that satisfies reflexivity, symmetry and transitivity. In this paper we use the symbol $\leq$ to denote both partial orders and partial preorders and the symbol $\sim$ for equivalence relations. Given a set U, a partition $\mathcal{U} = \{U_i\}_{i=1}^{k}$ of U is a set of pairwise disjoint non-empty sets $\{U_1, ..., U_k\}$ whose union is U. We call the sets $U_1, ..., U_k$ the parts of $\mathcal{U}$. Given two partitions $\mathcal{U}$ and $\mathcal{U}'$, we say that $\mathcal{U}'$ is a refinement of $\mathcal{U}$ if every part of $\mathcal{U}'$ is contained in a part of $\mathcal{U}$. Note that every partition is a refinement of itself. Given an equivalence relation $\sim$ over a set U, we denote with $[u]_\sim$ the equivalence class of the element $u \in U$ with respect to $\sim$, i.e., $[u]_\sim := \{v \in U : u \sim v\}$. We denote with $U/\sim$ the partition of U consisting of all equivalence classes $[u]_\sim$, for $u \in U$. We may not specify the set U over which the relation is defined, if it is clear from the context. Given a partial order or a partial preorder $\leq$ over a set U, a set $L \subseteq U$ is an *antichain* according to $\leq$ if for every $u, v \in L$, with $u \neq v$, pairs (u, v) and (v, u) do not belong to $\leq$. The *width* of $\leq$, denoted by width($\leq$), is the size of the largest antichain for $\leq$. A partial preorder $\leq$ over a set U induces an equivalence relation $\sim$ over U in the following way; For $u, v \in U$, we define $u \sim v$ if and only if $u \leq v$ and $v \leq u$. Moreover, a partial preorder $\leq$ over a set U induces a partial order $\leq'$ over the set $U/\sim$ (where $\sim$ is the equivalence induced by $\leq$) defined as $[u]_\sim \leq' [v]_\sim$ if and only if $u \leq v$. Given a partial order or a partial preorder $\leq$ over a set U, we define the symbol $<$ in the following way: for each $u, v \in U$, $u < v$ holds if and only if $u \leq v$ and $\neg(v \leq u)$.

Nondeterministic Finite Automata and Forward-Stable Partitions. A nondeterministic finite automaton (NFA) is a 4-tuple (Q, δ, Σ, s), where Q represents the set of the states, $\delta : Q \times \Sigma \to 2^Q$ is the automaton's transition function, Σ is the alphabet and $s \in Q$ is the initial state. The standard definition of NFAs includes also a set of final states that we omit since we are not concerned in distinguishing between final and non-final states. We assume the alphabet Σ to be effective, i.e., each character of Σ labels at least one edge of the transition function. A deterministic finite automaton (DFA) is an NFA such that each state has at most one outgoing edge labeled with a given character. Given an NFA $\mathcal{A} = (Q, \delta, \Sigma, s)$, for a state $u \in Q$, and a character $a \in \Sigma$, we use the shortcut $\delta_a(u)$ for $\delta(u, a)$. For a set $S \subseteq Q$, we define $\delta_a(S) := \bigcup_{u \in S} \delta_a(u)$. The set of finite strings over Σ, denoted by Σ^*, is the set of finite sequences of letters from Σ. We extend the transition function δ to the elements of Σ^* in the following way: for $\alpha \in \Sigma^*$ and $u \in Q$ we define $\delta(u, \alpha)$ recursively as follows. If $\alpha = \varepsilon$ (i.e. α is the empty string) then $\delta(u, \alpha) = \{u\}$. Otherwise, if $\alpha = \alpha'a$, with $\alpha' \in \Sigma^*$, $a \in \Sigma$, then $\delta(u, \alpha) = \bigcup_{v \in \delta(u, \alpha')} \delta(v, a)$.

Definition 1 (Strings reaching a state). *Given an NFA $\mathcal{A} = (Q, \delta, \Sigma, s)$, the set of strings reaching a state $u \in Q$ is defined as $I_u = \{\alpha \in \Sigma^* : u \in \delta(s, \alpha)\}$.*

We say that I_u is the regular language recognized by state u, while the regular language recognized by $\mathcal{A}$ is the union of the regular languages recognized by all $\mathcal{A}$'s states. Regarding the NFAs we treat, we make the following assumptions: (i) We assume that every state is reachable from the initial state. (ii) We assume that the initial state has no incoming edges. (iii) We do not require each state to have an outgoing edge for all possible labels. None of these assumptions are restrictive, since any NFA can be modified to satisfy these assumptions without changing its accepted language. Given an NFA $\mathcal{A} = (Q, \delta, \Sigma, s)$ and an equivalence relation $\sim$ over Q, we denote with $\mathcal{A}/\sim$ the quotient automaton of $\mathcal{A}$ obtained by collapsing the equivalence classes of $\sim$ into single states (see [2, Definition 8] for a formal definition). Figure 2 shows an example of a quotient automaton. Given an NFA $\mathcal{A} = (Q, \delta, \Sigma, s)$, and a state $u \in Q$, $u \neq s$, we denote with $\lambda(u)$ the set of the characters of Σ that label the incoming edges of u. If $u = s$, we define $\lambda(u) = \{\#\}$, where $\# \notin \Sigma$. We assume that there exists a total order $\leq$ among the elements of the set $\Sigma \cup \{\#\}$, where $\# < a$ for each $a \in \Sigma$. In addition, given two states u and v of an NFA, we say that $\lambda(u) \leq \lambda(v)$ holds if and only if for each $a \in \lambda(u)$ and for each $a' \in \lambda(v)$, it holds that $a \leq a'$. In this paper we deliberately overload the notation regarding the symbol $\leq$, i.e., we will use the same symbol for orders on integers, the alphabet $\Sigma \cup \{\#\}$ and states Q of automata; It will always be clear from the context which order we are referring to. We use the concept of forward-stable partitions, see also the work of Alanko et al. [2, Section 4.2].

Definition 2 (Forward-Stability). *Given an NFA $\mathcal{A} = (Q, \delta, \Sigma, s)$ and two sets of states $S, T \subseteq Q$, we say that S is forward-stable with respect to T, if, for all $a \in \Sigma$, $S \subseteq \delta_a(T)$ or $S \cap \delta_a(T) = \emptyset$ holds. A partition $\mathcal{Q}$ of $\mathcal{A}$'s states is forward-stable for $\mathcal{A}$, if, for any two parts $S, T \in \mathcal{Q}$, it holds that S is forward-stable with respect to T.*

Given an NFA $\mathcal{A} = (Q, \delta, \Sigma, s)$ we say that $\mathcal{Q}$ is the *coarsest* forward-stable partition for $\mathcal{A}$, if for every forward-stable partition $\mathcal{Q}'$ for $\mathcal{A}$, it holds that $\mathcal{Q}'$ is a refinement of $\mathcal{Q}$. It is easy to demonstrate that for each automaton $\mathcal{A}$ there exists a unique coarsest forward-stable partition. An interesting property about forward-stable partitions is the following: states in the same part of a forward-stable partition are reached by the same set of strings. A proof of this property can be found in [2, Lemma 4.7]. Given an NFA $\mathcal{A} = (Q, \delta, \Sigma, s)$, we define as $\sim_{FS}$ the equivalence relation over Q, such that for each $u, v \in Q$, $u \sim_{FS} v$ holds if and only if u and v belong to the same part of the coarsest forward-stable partition for $\mathcal{A}$. Consequently, partition $Q/\sim_{FS}$ is the coarsest forward-stable partition for $\mathcal{A}$. Figure 2 shows an example of a coarsest forward-stable partition for an automaton $\mathcal{A}$.

Wheeler and Quasi-Wheeler NFAs. The notion of Wheeler NFAs was first introduced by Gagie et al. [9, Definition 1]. Wheeler NFAs are a class of NFAs that can be endowed with a specific total order of their states, known as *Wheeler order*, which enables efficient compression and indexing. In particular, A Wheeler order is a total order over Q that sorts the states of Q according to the set of strings that reach them (see Definition 1). The possibly largest drawback of Wheeler NFAs is that, recognizing whether a given NFA is Wheeler or not is an NP-complete problem [10]. For this reason, Alanko et al. [2] proposed a relaxed version of the problem, introducing the notion of quasi-Wheeler NFAs. An NFA is said to be quasi-Wheeler if it admits a partial preorder $\leq$ over its set of states, such that; (i) $\sim_{FS}$ is the equivalence relation induced by $\leq$. (ii) The total order $\leq'$ induced by $\leq$ is a Wheeler order for the quotient automaton $\mathcal{A}/\sim_{FS}$. A partial preorder that satisfies these two properties is called a *Wheeler preorder*. Due to the fact that states in the same part of a forward-stable partition are reached by the same set of strings, it is easy to observe that the automata $\mathcal{A}$ and $\mathcal{A}/\sim_{FS}$ recognize the same language. Thus, from an indexing perspective the NFA $\mathcal{A}/\sim_{FS}$ is equally useful as $\mathcal{A}$. Moreover, every Wheeler NFA is also a quasi-Wheeler NFA, and there exist NFAs that are quasi-Wheeler but not

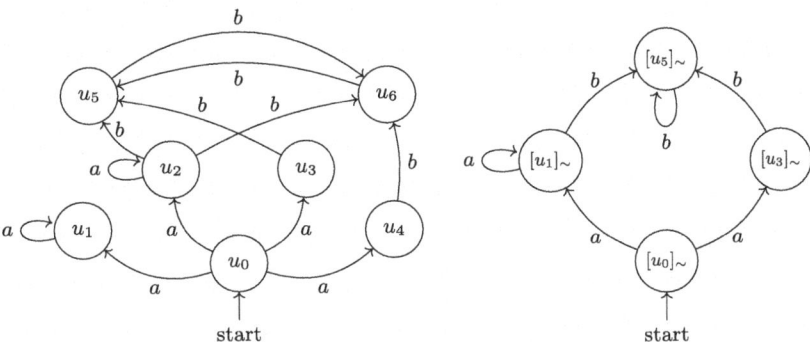

Fig. 2. An NFA $\mathcal{A} = (Q, \delta, \Sigma, s)$ on the left. On the right, the corresponding quotient automaton $\mathcal{A}/{\sim_{FS}} = (Q/{\sim_{FS}}, \delta/{\sim_{FS}}, \Sigma, s_{\sim_{FS}})$ of $\mathcal{A}$ for the coarsest forward-stable partition $Q/{\sim_{FS}} = \{\{u_0\}, \{u_1, u_2\}, \{u_3, u_4\}, \{u_5, u_6\}\}$

Wheeler. This implies that the class of quasi-Wheeler NFAs is strictly larger than the class of Wheeler NFAs (for a formal proof see Lemma 9 and Fig. 1 of the article of Becker et al. [3]). Probably, the most important difference between Wheeler NFAs and quasi-Wheeler NFAs is that the latter can be recognized in polynomial time. In fact, Becker et al. [3] proposed an algorithm that, given in input an NFA $\mathcal{A}$, is able to recognize if $\mathcal{A}$ is quasi-Wheeler in $O(|\delta|\log|Q|)$ time and, if this is the case, compute a Wheeler preorder for $\mathcal{A}$ in the same running time. This result was achieved by extending the partition refinement framework of Paige and Tarjan [11].

p-Sortable NFAs. Despite the fact that Wheeler NFAs can be indexed and compressed almost optimally, a major limitation regarding Wheeler NFAs is that the class of Wheeler NFAs is very limited. In fact, Wheeler languages, i.e., languages that are recognized by a Wheeler NFA, are star-free and closed only under intersection [2]. Moreover, also the amount of "nondeterminism" within a Wheeler NFA is bounded, since every Wheeler NFA admits an equivalent Wheeler DFA of linear size [1] (while in the general case there is an exponential blow-up of the number of states). For these reasons, Cotumaccio and Prezza [7, Definition 3.1] have extended the notion of Wheeler orders by introducing the concept of *co-lex orders*. In order to introduce the notion of co-lex order, we first introduce Cotumaccio's *co-lex relations* [6, Definition 1], which will be use later in this paper.

Definition 3 (Co-lex relations). *Let $\mathcal{A} = (Q, \delta, \Sigma, s)$ be an NFA. A co-lex relation of $\mathcal{A}$ is a reflexive relation R over Q that satisfies:*

1. *For every $u, v \in Q$, with $u \neq v$, if $(u, v) \in R$, then $\lambda(u) \leq \lambda(v)$.*
2. *For every pair $u \in \delta_a(u')$ and $v \in \delta_a(v')$, with $u \neq v$, if $(u, v) \in R$, then $(u', v') \in R$.*

A co-lex order is a co-lex relation that is antisymmetric and transitive. A main difference between Wheeler orders and co-lex orders is that every NFA $\mathcal{A} = (Q, \delta, \Sigma, s)$ admits a co-lex order; in fact, the relation $\leq := \{(v, v) : v \in Q\}$ is a co-lex order of $\mathcal{A}$. Given an NFA $\mathcal{A}$ and a co-lex order $\leq$ of $\mathcal{A}$, we say that $\leq$ is the *maximum* co-lex order of $\mathcal{A}$ if $\leq$ is equal to the union of all co-lex orders of $\mathcal{A}$. In general, an NFA does not always admit a maximum co-lex order. Cotumaccio and Prezza also define the notion of co-lexicographic width of an NFA [7, Definition 3.3]:

Definition 4 (Co-lexicographic width). *Let $\mathcal{A}$ be an NFA.*

1. *The NFA $\mathcal{A}$ is p-sortable if there exists a co-lex order $\leq$ of $\mathcal{A}$ of width p.*
2. *The co-lexicographic width $\bar{p}$ of $\mathcal{A}$ is the smallest p for which $\mathcal{A}$ is p-sortable.*

From Definition 4, it is possible to observe that if an NFA $\mathcal{A}$ is a Wheeler NFA, then the co-lexicographic width of $\mathcal{A}$ is equal to 1, since a Wheeler order is a particular type of co-lex order that is also a total order. As Cotumaccio and Prezza shows, the co-lexicographic width $\bar{p}$ of an NFA $\mathcal{A}$ measures the space and time complexity with which $\mathcal{A}$ can be encoded and indexed. However, the problem of determining the co-lexicographic width of a given NFA is known to be NP-hard. This follows from NP-hardness of Wheelerness.

Indexable Partial Preorders. At this point, we have seen that some particular NFAs $\mathcal{A} = (Q, \delta, \Sigma, s)$ (i.e. Wheeler NFAs) can be endowed with a particular total order over Q (i.e. a Wheeler order) in order to build efficient indexes on top of them. Despite the fact that the problem of determining whether or not a general NFA $\mathcal{A}$ admits a Wheeler order is NP-complete, we have seen that it is possible to compute in polynomial time a particular total preorder $\leq$ over Q such that the quotient automaton $\mathcal{A}/_{\sim_{FS}} = (Q/_{\sim_{FS}}, \delta/_{\sim_{FS}}, \Sigma, s/_{\sim_{FS}})$ (where $\sim_{FS}$ is the equivalence relation induced by $\leq$) has the following properties: (*i*) $\mathcal{A}$ and $\mathcal{A}/_{\sim_{FS}}$ accept the same language, (*ii*) If $\mathcal{A}$ is Wheeler, then also $\mathcal{A}/_{\sim_{FS}}$ is Wheeler, (*iii*) if $\mathcal{A}/_{\sim_{FS}}$ is Wheeler, then the total order $\leq'$ over $Q/_{\sim_{FS}}$ induced by $\leq$ is a Wheeler order for $\mathcal{A}/_{\sim_{FS}}$. Now it is natural to wonder whether or not these reasonings can be generalized also to arbitrary NFAs; to this end, we introduce the notion of indexable partial preorders.

Definition 5 (Indexable partial preorders). *Let $\mathcal{A} = (Q, \delta, \Sigma, s)$ be an NFA of co-lexicographic width $\bar{p}$, and let $\leq$ be a partial preorder over Q. Consider $\mathcal{A}/_\sim = (Q/_\sim, \delta/_\sim, \Sigma, s/_\sim)$ the quotient automaton defined by the equivalence relation $\sim$ induced by $\leq$. We say that $\leq$ is an indexable partial preorder for $\mathcal{A}$ if the following requirements are satisfied:*

1. *$\mathcal{A}$ and $\mathcal{A}/_\sim$ accept the same language.*
2. *$\mathcal{A}/_\sim$ has co-lexicographic width $\bar{q}$, with $\bar{q} \leq \bar{p}$.*
3. *The partial order $\leq'$ over $Q/_\sim$ induced by $\leq$ is a co-lex order of $\mathcal{A}/_\sim$ of width $\bar{q}$.*

It is possible to observe that a Wheeler preorder of an automaton $\mathcal{A}$ is also an indexable partial preorder for $\mathcal{A}$. However, in the literature there exists another example of indexable partial preorders, namely the *maximum* co-lex relations of Cotumaccio [6]. Consider Definition 3, Given an NFA $\mathcal{A}$ and a co-lex relation R of $\mathcal{A}$, we say that R is the *maximum* co-lex relation of $\mathcal{A}$ if R is equal to the union of all co-lex relations of $\mathcal{A}$. Although, an NFA $\mathcal{A}$ does not always admit a maximum co-lex order, it has been proved that every NFA admits a maximum co-lex relation, denoted hereafter as $\leq_R$, and that $\leq_R$ satisfies always transitivity, i.e., $\leq_R$ is always a partial preorder [6, Lemma 5]. Moreover, Cotumaccio also demonstrated that the maximum co-lex relation of an NFA can be computed in $O(|\delta|^2)$ time. In this work, we propose a new class of indexable partial preorders, which we term *coarsest forward-stable co-lex orders*. Then we demonstrate that the width of our class of indexable partial preorders is always smaller than or equal to the width of the maximum co-lex relation of Cotumaccio, and in some cases, even linearly smaller with respect to the number of the automaton's states.

3 Coarsest Forward-Stable Co-Lex Orders

Co-Lex Relations. In this section we define a new category of indexable partial preorders (Definition 5). From here on, given an NFA $\mathcal{A} = (Q, \delta, \Sigma, s)$ we will denote by $\mathcal{A}/_{\sim_R} = (Q/_{\sim_R}, \delta/_{\sim_R}, \Sigma, s/_{\sim_R})$ the quotient automaton of $\mathcal{A}$ defined by the maximum co-lex relation $\leq_R$, where $\sim_R$ is the equivalence relation over Q induced by $\leq_R$. The next lemma shows an important property that characterizes the maximum co-lex relation of an NFA.

Lemma 1 [6, Corollary 16]. *Let $\leq_R$ be the maximum co-lex relation of an automaton $\mathcal{A} = (Q, \delta, \Sigma, s)$, and let $\mathcal{A}/_{\sim_R} = (Q/_{\sim_R}, \delta/_{\sim_R}, \Sigma, s/_{\sim_R})$ be the quotient automaton of $\mathcal{A}$ defined by $\leq_R$. Then the partial order $\leq$ over $Q/_{\sim_R}$ induced by $\leq_R$ is the maximum co-lex order of $\mathcal{A}/_{\sim_R}$.*

Lemma 1 is important for our purposes, because it demonstrates that the quotient automaton $\mathcal{A}/_{\sim_R}$ always admits a maximum co-lex order. Moreover, in Lemma 7 (1) we will observe that if an NFA $\mathcal{A}$ admits a maximum co-lex order $\leq$, then width($\leq$) is smaller than or equal to the width of any co-lex order of $\mathcal{A}$, since $\leq$ must be a superset of every co-lex order of $\mathcal{A}$. This in turn demonstrates that the co-lexicographic width of $\mathcal{A}$ must be equal to the width of $\leq$. In the next lemma, we demonstrate that the partition $Q/_{\sim_R}$ is always a forward-stable partition for $\mathcal{A}$, though we will see later that $Q/_{\sim_R}$ may not necessarily be the coarsest forward-stable partition for $\mathcal{A}$.

Lemma 2. *Let $\mathcal{A} = (Q, \delta, \Sigma, s)$ be an NFA and $\leq_R$ its maximum co-lex relation. Consider the partition $Q/_{\sim_R}$, where $\sim_R$ is the equivalence relation over Q induced by $\leq_R$. Then $Q/_{\sim_R}$ is a forward-stable partition for $\mathcal{A}$.*

Proof. Let S, T be two arbitrary parts of partition $Q/_{\sim_R}$. We want to demonstrate that for each $a \in \Sigma$ either $S \subseteq \delta_a(T)$ or $S \cap \delta_a(T) = \emptyset$ holds. Let $a \in \Sigma$

and suppose for contradiction that there exists $u \in S \cap \delta_a(T)$ and $v \in S \setminus \delta_a(T)$. Then there exists $u' \in T$ such that $u \in \delta_a(u')$ and $S = [u]_{\sim_R}$, $T = [u']_{\sim_R}$. Since v and u belong to the same part S of $Q/_{\sim_R}$, it must hold that $v \leq_R u$ and $u \leq_R v$. It implies that, due to Axiom 1 of Definition 3, $\lambda(v) \leq \lambda(u)$ and $\lambda(u) \leq \lambda(v)$, which can simultaneously hold if and only if $\lambda(v) = \lambda(u) = \{a\}$. It follows that there exists $v' \in Q$ such that $v \in \delta_a(v')$. Due to Axiom 2 of Definition 3, since $v \leq_R u$ we have that $v' \leq_R u'$, moreover, since $u \leq_R v$ we have that $u' \leq_R v'$ and so $u' \sim_R v'$. Therefore $v' \in [u']_{\sim_R}$ and so $v' \in T$. Therefore, it follows that $v \in \delta_a(T)$, a contradiction. □

Quotient Automaton $\mathcal{A}/_{\sim_{FS}}$. Given an NFA $\mathcal{A}$, in the following part of this article, we will show some properties of the quotient automaton $\mathcal{A}/_{\sim_{FS}} = (Q/_{\sim_{FS}}, \delta/_{\sim_{FS}}, \Sigma, s/_{\sim_{FS}})$, where $Q/_{\sim_{FS}}$ is the coarsest forward-stable partition for $\mathcal{A}$. In particular, in the next lemma we show that $\mathcal{A}/_{\sim_{FS}}$ always admits a maximum co-lex order.

Lemma 3. *For an NFA* $\mathcal{A} = (Q, \delta, \Sigma, s)$ *let* $\mathcal{A}/_{\sim_{FS}} = (Q/_{\sim_{FS}}, \delta/_{\sim_{FS}}, \Sigma, s/_{\sim_{FS}})$ *be the quotient automaton of* $\mathcal{A}$ *such that* $Q/_{\sim_{FS}}$ *is the coarsest forward-stable partition of* $\mathcal{A}$ *and let* $\leq'_R$ *be the maximum co-lex relation of* $\mathcal{A}/_{\sim_{FS}}$. *Then* $\leq'_R$ *is the maximum co-lex order of* $\mathcal{A}/_{\sim_{FS}}$.

Proof. As the set of co-lex relations is a superset of the set of co-lex orders, it suffices to show that $\leq'_R$ is a co-lex order. By Definition $\leq'_R$ is reflexive and by maximality $\leq'_R$ is transitive [6, Lemma 5]. Hence, it remains to show antisymmetry. Therefore, let us suppose for the purpose of contradiction that $\leq'_R$ is not antisymmetric, and let us consider the equivalence relation $\sim'_R$ over $Q/_{\sim_{FS}}$ induced by $\leq'_R$. Moreover, let us consider the partition $\mathcal{P}$ of Q such that states $u, v \in Q$ belong to the same part of $\mathcal{P}$ if and only if $[u]_{\sim_{FS}} \sim'_R [v]_{\sim_{FS}}$ holds. Clearly, partition $Q/_{\sim_{FS}}$ is a refinement of $\mathcal{P}$ since every part of $Q/_{\sim_{FS}}$ is contained in a part of $\mathcal{P}$. However, as $\leq'_R$ is not antisymmetric, it is easy to observe that the reverse does not hold, and therefore $\mathcal{P}$ is not a refinement of $Q/_{\sim_{FS}}$.

At this point, we want to prove that $\mathcal{P}$ is a forward-stable partition for $\mathcal{A}$. Let S, T be two arbitrary parts of $\mathcal{P}$. We have to prove that $\forall a \in \Sigma$, $S \subseteq \delta_a(T)$ or $S \cap \delta_a(T) = \emptyset$ hold. Let $a \in \Sigma$ and suppose for contradiction that there exists $u \in \delta_a(T) \cap S$, and $v \in S \setminus \delta_a(T)$, so we know that there exists $u' \in T$, such that $u \in \delta_a(u')$. Clearly, $v \notin [u]_{\sim_{FS}}$, since if $v \in [u]_{\sim_{FS}}$, then, due to the fact that $Q/_{\sim_{FS}}$ is forward-stable, it must exist a state $v' \in [u']_{\sim_{FS}}$, such that $v \in \delta_a(v')$. Therefore we have $[v]_{\sim_{FS}} \neq [u]_{\sim_{FS}}$. Since $[v]_{\sim_{FS}} \sim'_R [u]_{\sim_{FS}}$, and due to Axiom 1 of Definition 3, we have that $\lambda([v]_{\sim_{FS}}) = \lambda([u]_{\sim_{FS}}) = \{a\}$. It follows that there exists a state $[v']_{\sim_{FS}} \in Q/_{\sim_{FS}}$ such that $[v]_{\sim_{FS}} \in \delta/_{\sim_{FS}}([v']_{\sim_{FS}}, a)$. Due to Axiom 2 of Definition 3, since $[u]_{\sim_{FS}} \leq'_R [v]_{\sim_{FS}}$, it follows that $[u']_{\sim_{FS}} \leq'_R [v']_{\sim_{FS}}$, and since $[v]_{\sim_{FS}} \leq'_R [u]_{\sim_{FS}}$, it follows that $[v']_{\sim_{FS}} \leq'_R [u']_{\sim_{FS}}$; therefore $[v']_{\sim_{FS}} \sim'_R [u']_{\sim_{FS}}$ holds. Due to the fact that $Q/_{\sim_{FS}}$ is forward-stable, there exists $v'' \in [v']_{\sim_{FS}}$ such that $v \in \delta_a(v'')$, moreover, since $[u']_{\sim_{FS}} \sim'_R [v'']_{\sim_{FS}}$, it follows that $v'' \in T$, which in turn proves that $v \in \delta_a(T)$. Therefore $\mathcal{P}$ is a forward-stable partition for $\mathcal{A}$; however, $\mathcal{P}$ is not a refinement of $Q/_{\sim_{FS}}$, this contradicts the hypothesis that $Q/{\sim_{FS}}$ is the coarsest forward-stable partition of $\mathcal{A}$. It follows that it was absurd to assume that $\leq'_R$ was not antisymmetric. □

CFS Orders. Lemma 3 not only proves that $\mathcal{A}/{\sim_{FS}}$ always admits a maximum co-lex order, but also that the maximum co-lex order of $\mathcal{A}/{\sim_{FS}}$ is equal to the maximum co-lex relation of $\mathcal{A}/{\sim_{FS}}$. At this point we have all the ingredients to introduce *coarsest forward-stable co-lex (CFS) orders*.

Definition 6 (Coarsest forward-stable co-lex (CFS) order). *Let $\mathcal{A} = (Q, \delta, \Sigma, s)$ be an NFA, and let $\mathcal{A}/{\sim_{FS}} = (Q/{\sim_{FS}}, \delta/{\sim_{FS}}, \Sigma, s/{\sim_{FS}})$ be the quotient automaton of $\mathcal{A}$ such that $Q/{\sim_{FS}}$ is the coarsest forward-stable partition for $\mathcal{A}$. Let $\leq$ be the maximum co-lex order of $\mathcal{A}/{\sim_{FS}}$, then we say that the coarsest forward-stable co-lex (CFS) order of $\mathcal{A}$, denoted as $\leq_{FS}$, is the (unique) relation over Q such that, for each $u, v \in Q$, $u \leq_{FS} v$ holds if and only if $[u]_{\sim_{FS}} \leq [v]_{\sim_{FS}}$.*

We will now demonstrate some properties of the CFS order of an NFA.

Lemma 4. *The CFS order $\leq_{FS}$ of any NFA is a partial preorder over its states.*

Proof. We have to prove that $\leq_{FS}$ satisfies reflexivity and transitivity. Let us consider the maximum co-lex order $\leq$ of the quotient automaton $\mathcal{A}/{\sim_{FS}} = (Q/{\sim_{FS}}, \delta/{\sim_{FS}}, \Sigma, s/{\sim_{FS}})$. By definition of co-lex order, $\leq$ is a partial order, therefore it satisfies reflexivity and transitivity. Let us consider an arbitrary state $u \in Q$. By Definition 6, $u \leq_{FS} u$ if and only if $[u]_{\sim_{FS}} \leq [u]_{\sim_{FS}}$; however $[u]_{\sim_{FS}} \leq [u]_{\sim_{FS}}$ holds since $\leq$ is reflexive, therefore also $u \leq_{FS} u$ must hold. It follows that $\leq_{FS}$ satisfies reflexivity. Let us consider three arbitrary states $u, v, z \in Q$ such that $u \leq_{FS} v$ and $v \leq_{FS} z$. We have to prove that $u \leq_{FS} z$. Since $u \leq_{FS} v$ we have that $[u]_{\sim_{FS}} \leq [v]_{\sim_{FS}}$, and since $v \leq_{FS} z$ we have that $[v]_{\sim_{FS}} \leq [z]_{\sim_{FS}}$. Moreover, due to the fact that $\leq$ is transitive, we have also that $[u]_{\sim_{FS}} \leq [z]_{\sim_{FS}}$, and consequently $u \leq_{FS} z$. It follows that $\leq_{FS}$ satisfies transitivity. □

The following remarks follow from $\leq_{FS}$ being a partial preorder.

Remark 1. For an NFA $\mathcal{A} = (Q, \delta, \Sigma, s)$ let $\mathcal{A}/{\sim_{FS}} = (Q/{\sim_{FS}}, \delta/{\sim_{FS}}, \Sigma, s/{\sim_{FS}})$ be the quotient automaton with $Q/{\sim_{FS}}$ being the coarsest forward-stable partition for $\mathcal{A}$. Moreover, let $\leq_{FS}$ and $\leq$ be the CFS order of $\mathcal{A}$ and the maximum co-lex order of $\mathcal{A}/{\sim_{FS}}$, respectively. Then (1) $\sim_{FS}$ is equal to the equivalence relation induced by $\leq_{FS}$, and (2) $\leq$ is the partial order induced by $\leq_{FS}$.

Proof. For (1), observe that $\leq$ is a partial order and thus satisfies antisymmetry. Therefore, given two states $u, v \in Q$, $u \leq_{FS} v$ and $v \leq_{FS} u$ can both hold if and only if u and v belong to the same part of the coarsest forward-stable partition $Q/{\sim_{FS}}$ for $\mathcal{A}$, i.e., if and only if $u \sim_{FS} v$. For (2), note that $\leq_{FS}$ is a partial preorder over Q, while $\leq$ is a partial order over $Q/{\sim_{FS}}$. Moreover, $u \leq_{FS} v$ holds if and only if $[u]_{\sim_{FS}} \leq [v]_{\sim_{FS}}$ holds. □

4 Relation Between CFS Order and Max Co-Lex Relation

At this point, we have introduced the maximum co-lex relation $\leq_R$ and the CFS order $\leq_{FS}$ of an NFA. In this section we study the relation between $\leq_R$ and $\leq_{FS}$. In particular, we can prove that $\leq_{FS}$ is always a superset of $\leq_R$. Before proving this result, we have to repeat some concepts from the work of Cotumaccio, starting from the notion of preceding pairs [6, Definition 6].

Definition 7 (Preceding pairs). *Let $\mathcal{A} = (Q, \delta, \Sigma, s)$ be an NFA and let $(u', v'), (u, v) \in Q \times Q$ be pairs of distinct states. The pair (u', v') precedes (u, v) in $\mathcal{A}$ if there are $u_1, ..., u_r, v_1, ..., v_r \in Q$ with $r \geq 1$ and $a_1, ..., a_{r-1} \in \Sigma$ s.t. (1) $u_1 = u'$ and $v_1 = v'$, (2) $u_r = u$ and $v_r = v$, (3) for all $i = 1, ..., r-1$, $u_i \neq v_i$, and (4) for all $i = 1, ..., r-1$, $u_{i+1} \in \delta_{a_i}(u_i)$ and $v_{i+1} \in \delta_{a_i}(v_i)$.*

Note that if $u, v \in Q$ are distinct states, then pair (u, v) trivially precedes itself. Cotumaccio characterized the maximum co-lex relation in terms of preceding pairs by showing that $u \leq_R v$ holds if and only if for all pairs (u', v') preceding (u, v) it holds that $\lambda(u') \leq \lambda(v')$ [6, Lemma 7]. We have the following lemma regarding preceding pairs in quotient automata.

Lemma 5. *Let $\mathcal{A} = (Q, \delta, \Sigma, s)$ be an NFA, and let $\mathcal{A}/\sim = (Q/\sim, \delta/\sim, \Sigma, s/\sim)$ be a quotient automaton of $\mathcal{A}$ such that $Q/\sim$ is a forward-stable partition for $\mathcal{A}$. Consider two pairs $([u]_\sim, [v]_\sim), ([u']_\sim, [v']_\sim)$, with $[u]_\sim, [v]_\sim, [u']_\sim, [v']_\sim \in Q/\sim$, such that $([u']_\sim, [v']_\sim)$ precedes $([u]_\sim, [v]_\sim)$ in $\mathcal{A}/\sim$. Then, there exist $u'', v'' \in Q$, with $u'' \in [u']_\sim, v'' \in [v']_\sim$, such that (u'', v'') precedes (u, v) in $\mathcal{A}$.*

Proof. Since $([u']_\sim, [v']_\sim)$ precedes $([u]_\sim, [v]_\sim)$ in $\mathcal{A}/\sim$, we know that there exists a sequence $[u_1]_\sim, ..., [u_r]_\sim, [v_1]_\sim, ..., [v_r]_\sim$ that satisfies the requirements of Definition 7. The proof proceeds by induction over the parameter r. Firstly, let us prove the statement for $r = 1$. We know that u, v are distinct states, because $[u]_\sim, [v]_\sim$ are distinct, and therefore pair (u, v) trivially precedes itself in $\mathcal{A}$. Now let us prove the statement for a general r. Let us consider the sequence $[u_2]_\sim, ..., [u_r]_\sim, [v_2]_\sim, ..., [v_r]_\sim$, due to the inductive hypothesis, we know that there exist $u^* \in [u_2]_\sim$ and $v^* \in [v_2]_\sim$, such that (u^*, v^*) precedes (u, v) in $\mathcal{A}$. Since $[u_2]_\sim \in \delta/\sim([u_1]_\sim, a_1)$ and $[v_2]_\sim \in \delta/\sim([v_1]_\sim, a_1)$, and due to the definition of quotient automaton, we know that there exist $\bar{u}_1, \bar{u}_2, \bar{v}_1, \bar{v}_2 \in Q$ such that: $\bar{u}_1 \in [u_1]_\sim, \bar{u}_2 \in [u_2]_\sim, \bar{v}_1 \in [v_1]_\sim, \bar{v}_2 \in [v_2]_\sim, \bar{u}_2 \in \delta(\bar{u}_1, a_1), \bar{v}_2 \in \delta(\bar{v}_1, a_1)$. Moreover, since $Q/\sim$ is forward-stable, there must exist also $u'' \in [u_1]_\sim$ and $v'' \in [v_1]_\sim$, such that $u^* \in \delta(u'', a_1)$ and $v^* \in \delta(v'', a_1)$ (note that $u'' \neq v''$ because $[u_1]_\sim \neq [v_1]_\sim$). Moreover, since (u^*, v^*) precedes (u, v) in $\mathcal{A}$, then also (u'', v'') must precede (u, v) in $\mathcal{A}$. It follows that the lemma holds. □

After Lemma 5, we are ready to prove that $\leq_{FS}$ is always a superset of $\leq_R$.

Lemma 6. *For an NFA $\mathcal{A} = (Q, \delta, \Sigma, s)$ let $\leq_R$ be the maximum co-lex relation of $\mathcal{A}$ and let $\leq_{FS}$ be the CFS order of $\mathcal{A}$. Then $\leq_{FS}$ is a superset of $\leq_R$.*

Proof. We suppose for contradiction that $\leq_{FS}$ is not a superset of $\leq_R$. Then, there is $(u,v) \in \leq_R$ with $(u,v) \notin \leq_{FS}$. Let $\mathcal{A}/{\sim_{FS}} = (Q/{\sim_{FS}}, \delta/{\sim_{FS}}, \Sigma, s/{\sim_{FS}})$ be the quotient automaton of $\mathcal{A}$ such that $Q/{\sim_{FS}}$ is the coarsest forward-stable partition for $\mathcal{A}$, and let $\leq$ be the maximum co-lex order of $\mathcal{A}/{\sim_{FS}}$ (which exists due to Lemma 3). By Definition 6, $(u,v) \notin \leq_{FS}$ if and only if $([u]_{\sim_{FS}}, [v]_{\sim_{FS}}) \notin \leq$. In addition, $[u]_{\sim_{FS}} \neq [v]_{\sim_{FS}}$, since $\leq$ satisfies reflexivity. Due to Lemma 7 in the article by Cotumaccio [6] and because $\leq$ is also the maximum co-lex relation of $\mathcal{A}/{\sim_{FS}}$ (see Lemma 3), there exist $[u']_{\sim_{FS}}, [v']_{\sim_{FS}} \in Q/{\sim_{FS}}$ such that the pair $([u']_{\sim_{FS}}, [v']_{\sim_{FS}})$ precedes $([u]_{\sim_{FS}}, [v]_{\sim_{FS}})$ in $\mathcal{A}/{\sim_{FS}}$, and $\lambda([u']_{\sim_{FS}}) \leq \lambda([v']_{\sim_{FS}})$ does not hold. Due to Lemma 5, there exist $u'' \in [u']_{\sim_{FS}}$, $v'' \in [v']_{\sim_{FS}}$ such that (u'',v'') precedes (u,v) in $\mathcal{A}$. Moreover, since $u'' \in [u']_{\sim_{FS}}$, we have that $\lambda(u'') = \lambda([u']_{\sim_{FS}})$, and analogously $\lambda(v'') = \lambda([v']_{\sim_{FS}})$. Hence, if $\lambda([u']_{\sim_{FS}}) \leq \lambda([v']_{\sim_{FS}})$ does not hold, then also $\lambda(u'') \leq \lambda(v'')$ does not hold. However, again using Lemma 7 in the article by Cotumaccio [6], this contradicts the assumption $(u,v) \in \leq_R$. Hence, $\leq_{FS}$ is a superset of $\leq_R$. □

The following lemma then allows us to conclude that the width of the CFS order is always at most the width of the maximum co-lex relation.

Lemma 7. *The following two statements hold: (1) Let $\leq$ and $\leq'$ be two partial orders or partial preorders over the same set U. If $\leq$ is a superset of $\leq'$, then* width$(\leq) \leq$ width$(\leq')$. *(2) Let $\leq$ be a partial preorder over a set U and let $\leq'$ be the partial order induced by $\leq$ over $U/{\sim}$. Then,* width$(\leq) =$ width$(\leq')$.

Proof. (1) Since $\leq$ is a superset of $\leq'$, we know that for each $u,v \in U$, if $(u,v) \notin \leq$, then $(u,v) \notin \leq'$. It follows that every antichain according to $\leq$ is also an antichain according to $\leq'$. (2) Let us denote with p and p' the integers width$(\leq)$ and width$(\leq')$, respectively. Let us consider L, an arbitrary antichain according to $\leq$, and let us define the set $L' := \{[u]_\sim : u \in L\}$; clearly $|L| = |L'|$. It is easy to observe that L' is an antichain according to $\leq'$, this proves that $p \leq p'$. Now, let us consider L', an arbitrary antichain according to $\leq'$, and let us consider a possible set L, in such a way that L contains a representative of each equivalence class $[u]_\sim$, such that $[u]_\sim \in L'$. Clearly, L is an antichain according to $\leq$, and $|L'| = |L|$. This proves that $p' \leq p$. It follows that $p = p'$. □

We are now ready to prove that CFS orders are a class of indexable partial preorders, see Definition 5. Moreover, we will show that the quotient automaton $\mathcal{A}/{\sim_{FS}}$ has never more states than $\mathcal{A}/{\sim_R}$, and that the co-lexicographic width of the former is never larger then the co-lexicographic width of the latter.

Theorem 1. *Let $\mathcal{A} = (Q, \delta, \Sigma, s)$ be an NFA, and let $\leq_R$ and $\leq_{FS}$ be the maximum co-lex relation and the CFS order of $\mathcal{A}$, respectively. Finally, let $\mathcal{A}/{\sim_R} = (Q/{\sim_R}, \delta/{\sim_R}, \Sigma, s/{\sim_R})$ and $\mathcal{A}/{\sim_{FS}} = (Q/{\sim_{FS}}, \delta/{\sim_{FS}}, \Sigma, s/{\sim_{FS}})$ be the quotient automata of $\mathcal{A}$ defined by $\leq_R$ and $\leq_{FS}$, respectively. Then:*

1. *$\leq_{FS}$ is an indexable partial preorder for $\mathcal{A}$.*
2. *The co-lexicographic width of $\mathcal{A}/{\sim_{FS}}$ is smaller than or equal to the co-lexicographic width of $\mathcal{A}/{\sim_R}$.*

3. The number of parts in partition $Q/\!\sim_{FS}$ is smaller than or equal to the number of parts in partition $Q/\!\sim_R$.

Proof. 1. By Lemma 4, $\leq_{FS}$ is a partial preorder over Q, thus it remains to prove the three properties in Definition 5. (a) Clearly, $\mathcal{A}$ and $\mathcal{A}/\!\sim_{FS}$ accept the same language, since states in the same part of a forward-stable partition are reached by the same set of strings [2, Lemma 4.7]. (b) Let $\bar{p}$ be the co-lexicographic width of $\mathcal{A}$. We have to prove that the co-lexicographic width of $\mathcal{A}/\!\sim_{FS}$ is smaller than or equal to $\bar{p}$. Let $\leq$ be the maximum co-lex order of $\mathcal{A}/\!\sim_{FS}$. Remark 1 yields that $\leq$ is the partial order induced by $\leq_{FS}$ and hence Lemma 7 (2) implies that width($\leq$) = width($\leq_{FS}$). Then, Lemma 6 yields that $\leq_{FS}$ is a superset of $\leq_R$, the maximum co-lex relation of $\mathcal{A}$, and thus Lemma 7 (1) implies that width($\leq_{FS}$) $\leq$ width($\leq_R$). Now let $\leq'$ be a co-lex order of $\mathcal{A}$ of width $\bar{p}$. Then, $\leq'$ is also a co-lex relation and thus $\leq_R$ is a superset of $\leq'$. Applying Lemma 7 (1) again yields width($\leq_R$) $\leq$ width($\leq'$) = $\bar{p}$. Thus altogether width($\leq$) $\leq \bar{p}$. (c) Let $\bar{q}$ be the co-lexicographic width of $\mathcal{A}/\!\sim_{FS}$, and let $\leq$ be the maximum co-lex order of $\mathcal{A}/\!\sim_{FS}$. Remark 1 yields that $\leq$ is the partial order induced by $\leq_{FS}$. Hence, we have to prove that width($\leq$) = $\bar{q}$. It is clear that width($\leq$) $\geq \bar{q}$. For the other direction, let $\leq'$ be a co-lex order of $\mathcal{A}/\!\sim_{FS}$ with width($\leq'$) = $\bar{q}$. Since $\leq$ is a superset of $\leq'$, Lemma 7 (2) yields width($\leq$) $\leq$ width($\leq'$) = $\bar{q}$.
2. Remark 1 yields that the maximum co-lex order of $\mathcal{A}/\!\sim_{FS}$ is the partial order induced by $\leq_{FS}$. Hence the co-lexicographic width of $\mathcal{A}/\!\sim_{FS}$ is equal to width($\leq_{FS}$) by Lemma 7 (2). Lemma 6 states that $\leq_{FS}$ is a superset of $\leq_R$ and thus width($\leq_{FS}$) $\leq$ width($\leq_R$) according to Lemma 7 (1). From Lemma 1, we conclude that the partial order $\leq$ induced by $\leq_R$ is the maximum co-lex order of $\mathcal{A}/\!\sim_R$. Again using Lemma 7 (2) implies width($\leq_R$) = width($\leq$).
3. By Lemma 2, $Q/\!\sim_R$ is forward-stable for $\mathcal{A}$, while $Q/\!\sim_{FS}$ is the coarsest forward-stable partition for $\mathcal{A}$. Thus $Q/\!\sim_R$ is a refinement of $Q/\!\sim_{FS}$ and the number of parts of $Q/\!\sim_{FS}$ is at most the number of parts of $Q/\!\sim_R$. □

We now prove another important result regarding the automata $\mathcal{A}/\!\sim_{FS}$ and $\mathcal{A}/\!\sim_R$. Namely, that the number of states and the co-lexicographic width of $\mathcal{A}/\!\sim_{FS}$ can be in some cases arbitrary smaller with respect to those of $\mathcal{A}/\!\sim_R$.

Theorem 2. *There exists a class of NFAs such that, for each NFA $\mathcal{A} = (Q, \delta, \Sigma, s)$ in this class; the quotient automaton $\mathcal{A}/\!\sim_R$, defined by the maximum co-lex relation of $\mathcal{A}$, has a number of states and a co-lexicographic width equal to $O(|Q|)$, while the quotient automaton $\mathcal{A}/\!\sim_{FS}$, defined by the CFS order of $\mathcal{A}$, has a number of states and a co-lexicographic width equal to $O(1)$.*

Proof. We describe the structure of this class of NFAs in Fig. 3. In particular, in this figure we can observe that the automaton $\mathcal{A}/\!\sim_{FS}$ has always co-lexicographic width equal to 1 and it is a Wheeler NFA. □

Finally, we show that the CFS order of an automaton $\mathcal{A} = (Q, \delta, \Sigma, s)$ can be computed in $O(|\delta|^2)$ time.

Corollary 1. *Let $\mathcal{A} = (Q, \delta, \Sigma, s)$ be an NFA. We can compute the CFS order $\leq_{FS}$ of $\mathcal{A}$ in $O(|\delta|^2)$ time.*

Proof. It is possible to use the partition refinement framework of Paige and Tarjian [11] to compute the coarsest forward-stable partition of $\mathcal{A}$ and consequently the quotient automaton $\mathcal{A}/\!\sim_{FS} = (Q/\!\sim_{FS}, \delta/\!\sim_{FS}, \Sigma, s/\!\sim_{FS})$ in $O(|\delta|\log|Q|)$ time. After that, we can compute the maximum co-lex relation $\leq'_R$ of $\mathcal{A}/\!\sim_{FS}$ using Cotumaccio's algorithm [6, Theorem 8] in $O(|\delta|^2)$ time, where by Lemma 3 we know that $\leq'_R$ is also the maximum co-lex order of $\mathcal{A}/\!\sim_{FS}$. Once we have computed $\leq'_R$, we can also reconstruct $\leq_{FS}$. □

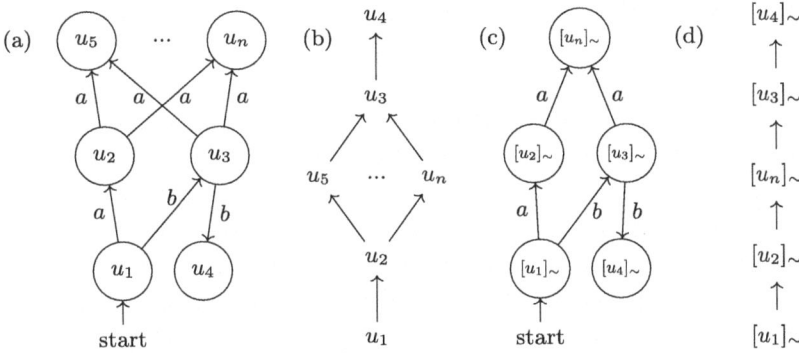

Fig. 3. (a) An NFA $\mathcal{A} = (Q, \delta, \Sigma, s)$ with $\Sigma = \{a, b\}$, $Q = \{u_1, ..., u_n\}$, with $n > 4$, where $u_1 = s$ is the initial state, and δ is s.t. $u_2 \in \delta_a(u_1)$, $u_3 \in \delta_b(u_1)$, $u_4 \in \delta_b(u_3)$ and for each $4 < i \leq n$, $u_i \in \delta_a(u_2)$, $u_i \in \delta_a(u_3)$. We observe that $\mathcal{A}/\!\sim_R$ is equal to $\mathcal{A}$ itself. Note that the example is not trivial, as states u_2 and u_3 may not be merged without changing the language of $\mathcal{A}$ (due to state u_4). (b) The Hasse diagram of the maximum co-lex order of $\mathcal{A}/\!\sim_R$, where $\{u_5, ..., u_n\}$ forms a largest antichain, consequently the co-lexicographic width of $\mathcal{A}/\!\sim_R$ is equal to $n - 4$. (c) The automaton $\mathcal{A}/\!\sim_{FS}$ consisting of five states, where for each $4 < i \leq n$, $u_i \in [u_n]_{\sim_{FS}}$. (d) The Hasse diagram of the maximum co-lex order of $\mathcal{A}/\!\sim_{FS}$. Note that this total order is also a Wheeler order of $\mathcal{A}/\!\sim_{FS}$, i.e., the co-lexicographic width of $\mathcal{A}/\!\sim_{FS}$ is equal to 1

Acknowledgments. We would like to thank Nicola Cotumaccio for insightful discussions on the topic of this paper. Ruben Becker, Sung-Hwan Kim, Nicola Prezza, and Carlo Tosoni are funded by the European Union (ERC, REGINDEX, 101039208). Views and opinions expressed are however those of the author(s) only and do not necessarily reflect those of the European Union or the European Research Council. Neither the European Union nor the granting authority can be held responsible for them.

References

1. Alanko, J., D'Agostino, G., Policriti, A., Prezza, N.: Regular languages meet prefix sorting. In: Chawla, S. (ed.) Proceedings of the 2020 ACM-SIAM Symposium on Discrete Algorithms, SODA 2020, Salt Lake City, UT, USA, 5–8 January 2020, pp. 911–930. SIAM (2020). https://doi.org/10.1137/1.9781611975994.55
2. Alanko, J., D'Agostino, G., Policriti, A., Prezza, N.: Wheeler languages. Inf. Comput. **281**, 104820 (2021). https://doi.org/10.1016/J.IC.2021.104820
3. Becker, R., et al.: Sorting finite automata via partition refinement. In: Gørtz, I.L., Farach-Colton, M., Puglisi, S.J., Herman, G. (eds.) 31st Annual European Symposium on Algorithms, ESA 2023, Amsterdam, The Netherlands, 4–6 September 2023. LIPIcs, vol. 274, pp. 15:1–15:15. Schloss Dagstuhl - Leibniz-Zentrum für Informatik (2023). https://doi.org/10.4230/LIPICS.ESA.2023.15
4. Becker, R., Kim, S.H., Prezza, N., Tosoni, C.: Indexing finite-state automata using forward-stable partitions (2024). https://arxiv.org/abs/2406.02763
5. Burrows, M., Wheeler, D.: A block-sorting lossless data compression algorithm. SRS Res. Rep. **124** (1994)
6. Cotumaccio, N.: Graphs can be succinctly indexed for pattern matching in $O(|E|^2 + |V|^{5/2})$ time. In: Bilgin, A., Marcellin, M.W., Serra-Sagristà, J., Storer, J.A. (eds.) Data Compression Conference, DCC 2022, Snowbird, UT, USA 22–25, March 2022, pp. 272–281. IEEE (2022). https://doi.org/10.1109/DCC52660.2022.00035
7. Cotumaccio, N., Prezza, N.: On indexing and compressing finite automata. In: Marx, D. (ed.) Proceedings of the 2021 ACM-SIAM Symposium on Discrete Algorithms, SODA 2021, Virtual Conference, 10–13 January 2021, pp. 2585–2599. SIAM (2021). https://doi.org/10.1137/1.9781611976465.153
8. Ferragina, P., Manzini, G.: Opportunistic data structures with applications. In: 41st Annual Symposium on Foundations of Computer Science, FOCS 2000, Redondo Beach, California, USA, 12–14 November 2000, pp. 390–398. IEEE Computer Society (2000). https://doi.org/10.1109/SFCS.2000.892127
9. Gagie, T., Manzini, G., Sirén, J.: Wheeler graphs: a framework for BWT-based data structures. Theor. Comput. Sci. **698**, 67–78 (2017). https://doi.org/10.1016/J.TCS.2017.06.016
10. Gibney, D., Thankachan, S.V.: On the hardness and inapproximability of recognizing wheeler graphs. In: Bender, M.A., Svensson, O., Herman, G. (eds.) 27th Annual European Symposium on Algorithms, ESA 2019, Munich/Garching, Germany, 9–11 September 2019. LIPIcs, vol. 144, pp. 51:1–51:16. Schloss Dagstuhl - Leibniz-Zentrum für Informatik (2019). https://doi.org/10.4230/LIPICS.ESA.2019.51
11. Paige, R., Tarjan, R.E.: Three partition refinement algorithms. SIAM J. Comput. **16**(6), 973–989 (1987). https://doi.org/10.1137/0216062

Burst Edit Distance

Itai Boneh[1,2], Shay Golan[1,2], Avivit Levy[3(✉)], Ely Porat[4], and B. Riva Shalom[3]

[1] Reichman University, Herzliya, Israel
{itai.bone,golansh1}@biu.ac.il
[2] University of Haifa, Haifa, Israel
[3] Department of Software Engineering, Shenkar College of Engineering, Design, Art, Ramat Gan, Israel
{avivitlevy,rivash}@shenkar.ac.il
[4] Department of Computer Science, Bar-Ilan University, Ramat Gan, Israel
porately@biu.ac.il

Abstract. In this paper we define two types of burst edit errors that occur in Text Editing scenarios when the communication speed is unstable within a wireless keyboard usage: (1) A *Burst of Errors (BE)* involves a sequence of erroneous *identical* symbols and allows a single edit operation applied to a sequence of identical symbols; (2) A *Burst of Operations (BO)* involves a sequence of erroneous symbols that are *not necessarily identical* and allows a single edit operation applied to a sequence of symbols. In both burst types, every burst operation has a *penalty*, which is a cost function $F(k)$, where k is the burst length. The *burst edit distance* of two strings S and T is: (1) The minimum cost of a sequence of BE operations that transforms S into T in the bursts of errors variant (EDBE); (2) The minimum cost of a sequence of BO operations that transforms S into T in the bursts of operations variant (EDBO). We describe solutions to both problems for general natural penalty functions families. A conditional lower bound for the EDBE problem is also given. The $\mathcal{K}$-bounded versions of the problems are considered as well.

Keywords: String Similarity · Edit Distance · Burst Errors

1 Introduction

String similarity analysis plays a central role in various data science applications for a wide assortment of tasks (e.g., [9,10,12,17,18,20–22,24,27]). The choice of a similarity measure for a specific task within an application domain is of great importance. Therefore, many string similarity measures were suggested over the years to adjust to various application needs, some of which initiated the use of different string operators (e.g., swap [4], interchange [2,19]) or cost models (e.g., element-cost [16], string rearrangements [3,5]) thereby introducing new string distance/similarity measures (e.g., LCSk [7]).

Many common distance measures involve operators on single or pairs of symbols in a string. In this paper errors involving a *sequence* of symbols in the input

© The Author(s), under exclusive license to Springer Nature Switzerland AG 2025
Z. Lipták et al. (Eds.): SPIRE 2024, LNCS 14899, pp. 41–56, 2025.
https://doi.org/10.1007/978-3-031-72200-4_4

strings rather than only single symbols or pairs are considered in settings that are rooted in *the communication process*, where a disruption from the sender to the receiver is regarded as a communication noise. Errors encountered in digital wireless channels are not independent but occur in bursts or clusters. By [26], a burst of noise does not necessarily mean that every bit in these consecutive bits is erroneous. Our Text Editing scenarios lead to a slightly different errors view.

In Text Editing scenario it is not unusual that while typing, one finger lingers a while or that when a wireless keyboard is used and the communication speed is unstable yielding a repetitious typing of the same symbol. Thus, a *Burst of Errors (BE)* involves a sequence of erroneous identical symbols. When considering the Edit Distance between two strings, a burst of errors represents a single edit operation applied to a sequence of identical symbols.

We also define another type of burst errors, which we call a *Burst of Operations (BO)*. In the same Text Editing scenario, if the used keyboard buttons are Delete or Insert, we may also get an erroneous sequence of deletion or insertion operations in which the deleted or inserted symbols are not necessarily identical. When considering the Edit Distance between two strings, a burst of operations represents two mismatching sub-strings, not necessarily of the same length, where in at least one of the sub-strings a sequence of the *same edit operation* can be applied to form a match of the input sub-strings.

In both burst types, every burst edit operation has a *penalty*, which is a cost function $F(k)$, where k is the burst length. This leads to the following definitions of the resulting *Burst Edit Distance* problems. The Burst Edit Distance of two strings S and T is: (1) the minimum cost of a sequence of BE operations that transforms S into T in the bursts of errors variant (EDBE); (2) the minimum cost of a sequence of BO operations that transforms S into T in the bursts of operations variant (EDBO). Note that EDBE/EDBO are generalizations of standard Edit Distance produced by taking $F(k) = k$ as a penalty function.

Paper Organization. Section 2 gives some formal definitions and basic cubic time solutions based on dynamic programming. We then describe in Sect. 3 an improved $O(nm)$-time algorithm to compute $EDBE(S,T)$ given an m-length string S and an n-length string T both over alphabet Σ, for *a general family of sub-additive penalty functions*, which capture the batch characteristic of burst errors. This improved solution cannot be applied for the EDBO problem. Therefore, in Sect. 4 we describe an improved $\tilde{O}(nm)$-time algorithm to compute $EDBO(S,T)$ given an m-length string S and an n-length string T both over alphabet Σ, for *the sub-family of concave penalty functions*, which are of special interest as they enable to better differentiate between adding an error to a long or a short burst. A conditional lower bound for the EDBE problem is given in Sect. 5, showing that there is no $O(n^{2-\varepsilon})$ algorithm to solve the EDBE problem on two strings S, T with $|S|, |T| \leq n$, unless SETH is false. The $\mathcal{K}$-bounded versions are considered in Sect. 6.

2 Preliminaries

In this section we give formal definitions and basic properties.

Definition 1. [A Repetition and a Run] A sub-string $S[i,j]$, $i \leq j$, over an alphabet Σ is called a **repetition** if the sub-string symbols are identical, that is, $S[i] = S[i+1] = \ldots = S[j]$. A *repetition* that cannot be extended, where $S[i-1] \neq S[i]$ and $S[j] \neq S[j+1]$, is called a **run**.

In the classical *Edit Distance* three types of edit operations are considered. (1) An *insertion* operation (σ, i) inserts the character σ before location i. (2) A *deletion* operation (i) deletes the ith character. (3) A *substitution* operation (σ, i) substitutes the ith character with another character σ. The edit distance of two strings S and T is the minimal number of edit operations one has to apply on S to get T. We next formally define the Edit Distance with the BE/BO edit operations applied to repetitions/sub-strings with non-uniform costs.

Definition 2. [A Burst of Errors (BE)] Let S and T be two strings over an alphabet Σ. In the *Burst of Errors (BE)* variant of Edit Distance the three following operations are allowed: *burst insertion*, *burst deletion* and *burst substitution*, each is given with a length k.

(1) A BE *burst insertion* (σ, k, i) inserts σ^k before location i.

(2) A BE *burst deletion* (k, i) on a string S with $S[i..i+k-1] = \sigma^k$ for some σ deletes $S[i..i+k-1]$.

(3) A BE *burst substitution* (σ_1, k, i) on a string S with $S[i..i+k-1] = \sigma_2^k$ for some $\sigma_2 \neq \sigma_1$ substitutes $S[i..i+k-1]$ with σ_1^k.

The repetition $S[i..i+k-1] = \sigma^k$ on which the BE burst operations are applied is called a BE *bursting area*. After applying a BE (k,i) or (σ, k, i) operation, the next operation can only be applied to an index j for $j \geq i+k$.

The Burst of Operations set of edit operations is also defined over a sequence of symbols, yet does not require the consecutive symbols to be repetitions.

Definition 3. [A Burst of Operations (BO)] Let S and T be two strings over an alphabet Σ. In the *Burst of Operations (BO)* variant of Edit Distance the three following operations are allowed: *burst insertion*, *burst deletion* and *burst substitution*, each is given with a length k.

(1) A BO *burst insertion* (i', k, i) inserts $T[i'..i'+k-1]$ before location i of S.

(2) A BO *burst deletion* (k, i) on a string S deletes $S[i..i+k-1]$.

(3) A BO *burst substitution* (i', k, i) on a string S substitutes $S[i..i+k-1]$ with $T[i'..i'+k-1]$.

The sub-sequence $S[i..i+k-1]$ on which the BO burst operations are applied is called a BO *bursting area*. After applying a BO bursting operation (k, i) or (i', k, i), the next operation can only be applied to an index j for $j \geq i+k$.

Note that our motivating application for a burst of operations in text editing only explains a burst of deletions or insertions. Nevertheless, we allow also a burst of mismatches since our algorithms can handle also this generalized problem, which may be of interest to other applications.

On the one hand, a burst of errors/operations cannot be regarded as an edit operation applied to a single symbol and penalized as one error. On the other hand, assigning a penalty for every edited symbol in the bursting area misses the batch characteristic of the operation. We, therefore, assign an edit operation applied to a bursting area a penalty that is smaller than the length of the edited area yet length-relative. To this end we define a *valid penalty function*.

Definition 4. [A Valid Penalty Function] A function $F : \mathbb{R} \to \mathbb{R}$ is called a *valid penalty function* if the following hold: (1) The penalty F is an increasing function; (2) $F(1) = 1$; (3) $F(0) = 0$; (4) F is computable in $O(1)$ time.

Note that the second condition requires that the penalty of an operation on a single character is 1, hence, the traditional edit distance is a special case of EDBE/EDBO. Let S and T be two strings over alphabet Σ, and let F be a valid penalty function. Then, the penalty of a *single* k-length BE/BO used in the sequence of operations transforming S into T is $F(k)$. For example, let $p > 0$, then $F(k) = k^p$ is a valid penalty function. For $p = 0.5$, we have that the penalty of BE *deletion* ($k = 2, i = 1$) on $S = aaab$ is $\sqrt{2}$. This penalty function is *sub-additive*, and, in particular, *concave*. These terms are hereafter defined.

Definition 5. [A Sub-Additive Function] Let F be a function, where $F : \mathbb{R} \to \mathbb{R}$. F is a *sub-additive* function if and only if the following holds for every $k_1, k_2 \in \mathbb{R}$: $F(k_1) + F(k_2) \geq F(k_1 + k_2)$. If $F(k_1) + F(k_2) > F(k_1 + k_2)$ then F is called *strictly sub-additive*.

Definition 6. [A Concave Function] Let F be a differentiable function, where $F : \mathbb{R} \to \mathbb{R}$. F is (strictly) *concave* if and only if its derivative function F' is (strictly) monotonically decreasing.

Note that, every concave function is also sub-additive. Sub-additive penalty functions well capture a burst of errors characterization thus have less penalty than separate errors. Concave penalty functions are of special interest since they enable to better differentiate between adding an error to a long or a short burst.

Edit Distance with Bursts of Errors (EDBE). Let S and T be two strings over alphabet Σ. We denote by $EDBE_{i,j}$ the edit distance with bursts of errors of the prefixes $S[1..i]$ and $T[1..j]$. Lemma 1 below, formally describes its computation for valid penalty functions $F(k)$. We denote by $H(S[i], T[j])$ the Hamming distance of $S[i]$ and $T[j]$, that returns 1 when $S[i] \neq T[j]$ and 0 otherwise.

Lemma 1. *Let S and T be two strings over alphabet Σ. The* Edit Distance with Bursts of Errors *using a valid penalty function F can be computed as follows.*

$$EDBE_{i,j} = \min \begin{cases} EDBE_{i,j-k} + F(k), & \forall 0 < k \leq j \text{ s.t.} \\ & T[j-k+1,j] \text{ is a repetition} \\ EDBE_{i-k,j} + F(k), & \forall 0 < k \leq i \text{ s.t.} \\ & S[i-k+1,i] \text{ is a repetition} \\ EDBE_{i-k,j-k} + H(S[i],T[j]) \cdot F(k), & \forall 0 < k \leq \min\{i,j\} \text{ s.t.} \\ & T[j-k+1,j], S[i-k+1,i] \\ & \text{are repetitions} \end{cases}$$

An example of the EDBE DP-table appears in Fig. 1. It is easy to see that the EDBE of two strings of length n can be computed via dynamic programming (DP) in $O(n^3)$ using Lemma 1.

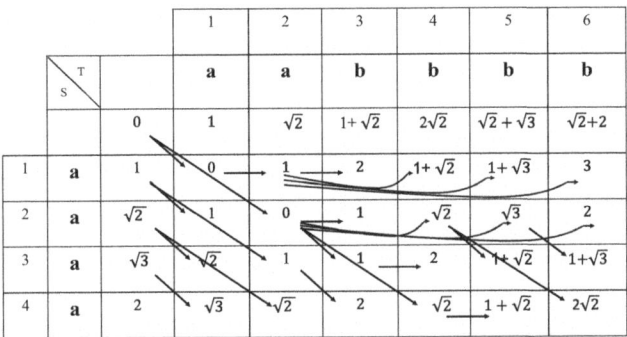

Fig. 1. EDBE DP-table computation example for penalty function $F(k) = \sqrt{k}$. Arrows represent the optimal burst length per entry computation

Edit Distance with Bursts of Operations (EDBO). Let S and T be two strings over alphabet Σ. We denote by $EDBO_{i,j}$ the edit distance with Bursts of Operations of the prefixes $S[1..i]$ and $T[1..j]$. Lemma 2 formally describes its computation for valid penalty functions F.

Lemma 2. *Let S and T be two strings over alphabet Σ. The* Edit Distance with Bursts of Operations *using a valid penalty function $F(k)$ can be computed as follows.*

$$EDBO_{i,j} = \min \begin{cases} EDBO_{i,j-k} + F(k), & \forall 0 < k \leq j \\ EDBO_{i-k,j} + F(k), & \forall 0 < k \leq i \\ EDBO_{i-k,j-k}, & \forall k \leq \min\{i,j\} \text{ s.t.} \\ & S[i-k+1..i] = T[j-k+1..j] \\ EDBO_{i-k,j-k} + F(k), & \forall 0 < k \leq \min\{i,j\} \text{ s.t. } \forall k' \in [0, k-1] \\ & S[i-k'] \neq T[j-k'] \end{cases}$$

It is easy to see that the EDBO of two strings of length n can be computed via DP in $O(n^3)$ using Lemma 2.

3 $O(nm)$-Algorithm for EDBE with Sub-additive Penalty Functions

In this section we describe an improved EDBE algorithm for valid *sub-additive* penalty functions F. Since bursts of errors are applied to repetitions, we divide the DP-table into *blocks* as done in [8] for matching run-length encoded strings.

The Blocks Division. A *block* $[i, i'] \times [j, j']$ is a rectangle in the DP-table defined by its vertical borders $[i, i']$ and horizontal borders $[j, j']$. A block $[i, i'] \times [j, j']$ is called a *Match-block* if $S[i..i']$ and $T[j..j']$ are runs of *the same* symbol. Otherwise, it is a *Mismatch-block*. An example of the DP-table blocks representation appears in Fig. 2.

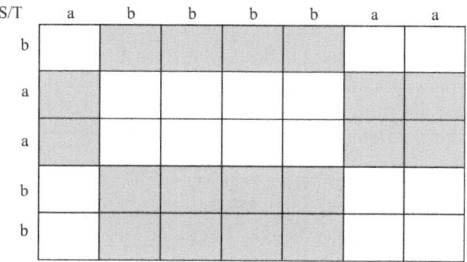

Fig. 2. Division of the DP-table into blocks. Match blocks are in grey. (Color figure online)

Definition 7. [Paths in the DP-table] A sequence of entries $a_1, a_2, \ldots, a_k$ in the DP-table where for every $1 \leq i < k$, if $a_i = [i, j]$ then $a_{i+1} \in \{[i + 1, j], [i, j + 1], [i + 1, j + 1]\}$ is called a *path* in the table. A solution of an EDBE problem is a path in the DP-table starting from entry $[1, 1]$ and ending at entry $[m, n]$. A possible optimal solution is called a *shortest path* in the DP-table.

A path going through block $[i, i'] \times [j, j']$ enters the block either from the bottom row of the block above ($i - 1$th row) or from the rightmost column of the block to its left ($j - 1$th) or from the right-bottom corner of the diagonally preceding block (entry $[i-1, j-1]$) (see light gray entries of Fig. 3). It leaves the block either from its bottom row (i'th row) or from its rightmost column (j'th column) (see gray entries of Fig. 3). Therefore, for each block we compute the EDBE value of the entries on its bottom row and rightmost column using the information of EDBE values of the entries on the bottom row of the block above or on the rightmost column of the block to its left and the right-bottom corner of the diagonally preceding block.

Note that even for sub-additive penalty function that prefer long bursting areas, a greedy computation of the EDBE by taking the longest possible diagonal

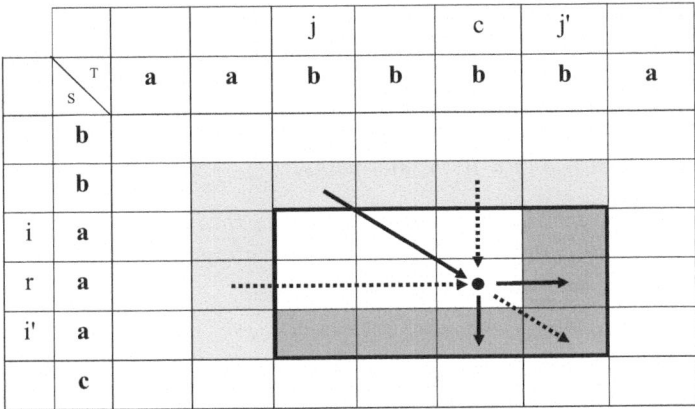

Fig. 3. An example for a block $[i, i'] \times [j, j']$. The EDBE of the gray entries is computed according to the EDBE value of the light gray entries. For a turn point entry $[r, c]$ all four path patterns are depicted. The black arrows represent the *Diagonal-Down* and the *Diagonal-Right* paths. The dotted arrows represent the *Down-Diagonal* and the *Right-Diagonal* paths. (Color figure online)

in a block as a match or mismatch may fail to find the optimal solution, as the optimal solution may choose **not** to start a match/mismatch from the beginning of a run. For example, see the case depicted in Fig. 1, where $S = a^4$, $t - a^2b^4$. If we consider $T[1, 2]$ as a burst deletion and a a burst substitution of $S[1, 4]$ and $T[3, 6]$, we get $EDBE(S, T) = \sqrt{2} + \sqrt{4}$. Yet, considering matching $S[1, 2]$ with $T[1, 2]$, then a burst deletion of $T[3, 4]$ and a burst substitution of $S[3, 4]$ and $T[5, 6]$ we get $EDBE(S, T) = 2\sqrt{2}$. We, thus, exploit the blocks framework differently by exploiting the following characterization of shortest paths going through a block in the DP-table, which mostly derive from Definition 5.

Observation 1. *If the shortest path goes through a block, its sub path within the block does not contain two consecutive bursts of identical edit operations.*

Definition 8. [Sub-Paths Types and Turn Points] A sub-path $a_i, a_{i+1}, \ldots, a_k$ of a path $a_1, \ldots, a_m$ going through a block can be of the following types:

(1) *Down* sub-path, a sub-path from $a_i = [i, j]$ to $a_{i+k} = [i + k, j]$ for $k > 0$, representing a *burst insertion* $(\sigma, k, i + k + 1)$ over s.

(2) *Right* sub-path, a sub-path from $a_i - [i, j]$ to $a_{i+k} - [i, j + k]$ for $k > 0$, representing a *burst deletion* (k, j) over T.

(3) *Diagonal* sub-path, a sub-path from $a_i[i, j]$ to $a_{i+k}[i + k, j + k]$ for $k > 0$, representing a *burst substitution* (σ, k, i) or a match between $S[i..i + k]$ and $T[j..j+k]$. An entry a at which a sub-path of type $X \in \{Down, Right, Diagonal\}$ ends and from which starts a sub-path of type $Y \in \{Down, Right, Diagonal\}$ where $X \neq Y$ is called a *turn point* in the path.

Observation 2. *Let $p = a_1, \ldots, a_k$ and $p' = a'_1, \ldots, a'_k$ be two sub-paths within a block where $a_1 = a'_1$ and $a_k = a'_k$. Let $X, Y \in \{Down, Right, Diagonal\}$,*

$X \neq Y$, where p is a concatenation of two sub-paths the first of type X and the second of type Y and p' is a concatenation of two sub-paths the first of type Y and the second of type X. Then, the penalty of p is equal to the penalty of p'.

Observation 3. *A sub-path of a shortest path within a block cannot be a concatenation of a Down and a Right sub-paths.*

Lemma 3. *For computing shortest paths going through a block it is sufficient to consider only sub-paths of type Diagonal concatenated with a sub-path of type Down or Right, or vice-verse.*

Theorem 1 then follows.

Theorem 1. *Given strings S and T of length m and n, respectively, both over alphabet Σ, $EDBE(S,T)$ for sub-additive penalty functions F can be computed in $O(nm)$ time.*

Proof. For a mismatch block $[i, i'] \times [j, j']$ we fill its bottom row (i.e. entries $[i', j+x]$ for $0 \le x \le j' - j$) and its rightmost column (i.e., entries $[i+y, j']$ for $0 \le y \le i' - i$). In the pre-process phase we initialize these entries with ∞.

By Lemma 3, only four paths patterns going through a block are relevant, each has at most a single turn point. Each block entry can serve as a turn point of these paths. Let $[r, c]$ be the current entry considered as a turn point.

In the case of a Diagonal-Y sub-path pattern for $Y \in \{Down, Right\}$, the length of the diagonal *ending* at $[r, c]$ (representing a burst substitution) is $dlen1_{[r,c]} = \min\{r - i, c - j\} + 1$, implying the diagonal emanates from entry $[x_1, y_1] = [r - dlen1_{[r,c]} + 1, c - dlen1_{[r,c]} + 1]$.

In the case of a Y-Diagonal sub-path pattern for $Y \in \{Down, Right\}$, the length of the diagonal *starting* at $[r, c]$ and ending at the bottom row or at the rightmost column of the block is $dlen2_{[r,c]} = \min\{i' - r, j' - c\} + 1$, implying the diagonal ends at entry $[x_2, y_2] = [r + dlen2_{[r,c]} - 1, c + dlen2_{[r,c]} - 1]$.

A burst crossing blocks is impossible, since bursts are defined for runs and adjacent blocks represent distinct runs. Therefore, knowing the entry indices $[a, b]$, from which the current path enters block $[i, i'] \times [j, j']$, we can compute the $EDBE$ of relevant bottom row or rightmost column entries by increasing $EDBE[a, b]$ by the penalty of the current path through the block. Due to the above arguments, for a turn point entry $[r, c]$ the algorithm assigns the following:

1. For the *Diagonal-Right* path,

$$EDBE[r, j'] \leftarrow \min\{EDBE[r, j'], (EDBE[x_1, y_1] + F(dlen1_{[r,c]}) + F(j' - c))\}$$

2. For the *Diagonal-Down* path,

$$EDBE[i', c] \leftarrow \min\{EDBE[i', c], (EDBE[x_1, y_1] + F(dlen1_{[r,c]}) + F(i' - r))\}$$

3. For the *Right-Diagonal* path,

$$EDBE[x_2, y_2] \leftarrow \min\{EDBE[x_2, y_2], (EDBE[r, j] + F(c - j) + F(dlen2_{[r,c]}))\}$$

4. For the *Down-Diagonal* path,

$$EDBE[x_2, y_2] \leftarrow \min\{EDBE[x_2, y_2], (EDBE[i, c] + F(r - i) + F(dlen2_{[r,c]}))\}$$

Match-blocks computation is similar, except for ignoring of the *Diagonal* sub-path penalty, as it represents a match. Note, that even though sub-paths of patterns *Diagonal-Down* and *Down-Diagonal* have the same penalty due to Observation 2, these cases are required for the EDBE computation as additional possibilities to reach the bottom row or the block rightmost column. The cases of *Diagonal-Right* and *Right-Diagonal* sub-paths are similar.

After considering every entry of the block (including the bottom row and rightmost column) as a turn point, the values in the block's bottom row and rightmost column are set to the best computed values. The process continues for every block in increasing order of rows and columns until the block containing entry $[m, n]$ is processed. Every entry requires a constant time computation, we therefore get an $O(nm)$ time complexity for computing $EDBE[m, n]$.

4 $\tilde{O}(nm)$-Algorithm for EDBO with Concave Penalty Functions

In this section we describe an improved EDBO algorithm for valid *concave* penalty functions F.[1] We use the next property, which follows from Definition 6:

Lemma 4. *Let $F : \mathbb{R} \to \mathbb{R}$ be a concave function on an interval of $\mathbb{R}$, then for every numbers x, y, z, $x < y$, where x, y, $y + z$ belong to the interval, we have that: $F(y) - F(x) \geq F(y + z) - F(x + z)$.*

Our algorithm is similar to the Alignment with Gap Penalties algorithm [11,13]. It uses a dynamic programming (DP) table computation based on Lemma 2, where every possible burst length is represented by the entry from which the burst starts, and is called a candidate for the current EDBO DP-table entry computation. However, instead of trying every value k as a possible burst length, only the set of current candidates is checked and kept for each direction: row (R), column (C) and diagonal (D). We next show that for each EDBO DP-table entry computation the current valuable candidates set size is $O(1)$, and that valuable candidates retrieval can be done efficiently.

Bounding the candidate set size is based on an elimination procedure. For its description the following definitions are needed.

Definition 9. [A *v*-Relevant Entry] For any $0 \leq i \leq m$, $0 \leq j \leq n$, we say that the DP-table entry $[i, j]$ is *v-relevant* with respect to direction $d \in \{R, C, D\}$, if one of the following holds: (1) $v = i$ and $d = R$. In this case, the entry *position with respect to the direction* is j. (2) $v = j$ and $d = C$. In this case, the entry *position with respect to the direction* is i. (3) $v = i - j$ and $d = D$. In this case, the entry *position with respect to the direction* is $\min\{i, j\}$.

---

[1] Note that the blocks division method of Sect. 3 doesn't work in the EDBO scenario, since consecutive errors in a burst of operations are not necessarily repetitions.

Definition 10. [Order Relation and Distance for v-Relevant Entries] Let $c_1 = [i_1, j_1]$ and $c_2 = [i_2, j_2]$ be two DP-table entries that are v-relevant with respect to direction d. We say that c_1 *precedes* c_2 and denote $c_1 < c_2$ if one of the following holds: (1) $d = R$ and $j_1 < j_2$. In this case, the *distance between* c_1 *and* c_2 is $dist(c_1, c_2) = j_2 - j_1$. (2) $d = C$ and $i_1 < i_2$. In this case, the *distance between* c_1 *and* c_2 is $dist(c_1, c_2) = i_2 - i_1$. (3) $d = D$ and $i_1 < i_2$ (thus $j_1 < j_2$). In this case, the *distance between* c_1 *and* c_2 is $dist(c_1, c_2) = i_2 - i_1$ ($= j_2 - j_1$).

Lemma 5. [Candidates Elimination] Let F be a valid concave penalty function and let c_1, c_2, c_3 be EDBO DP-table entries that are v-relevant with respect to direction d, where $c_1 < c_2 < c_3$. Then, if c_1 is preferred over c_2 for the computation of c_3, that is:

$$EDBO[c_1] + F(dist(c_1, c_3)) < EDBO[c_2] + F(dist(c_2, c_3)) \qquad (1)$$

then, c_1 is preferred over c_2 for the computation of *every* EDBO DP-table entry c_4 that is v-relevant with respect to direction d, where $c_3 < c_4$, and we say that c_1 *eliminates* c_2 from c_3.

Our improved EDBO DP-based algorithm considers each of the three values v, for which the last computed entry is v-relevant with respect to the three possible directions d. However, it reduces the possible burst lengths considered for the computation of the current EDBO DP-table entry to *a single* candidate (in each direction) based on Lemma 5 using the *Duels algorithm* described next.

The Duels Algorithm. Let $d \in \{R, C, D\}$ be any direction and suppose that the EDBO current DP-table entry is v-relevant with respect to direction d. The algorithm maintains the set of previously computed candidate entries that are v-relevant with respect to the direction d. For each such candidate we keep an interval $\langle t_1, t_2 \rangle$ specifying the range of future v-relevant entry positions, where this candidate should be considered for the computation, as follows: (1) t_1 is the first v-relevant entry position with respect to the direction d, where this candidate is preferred over any preceding v-relevant candidate with respect to the direction d, and (2) t_2 is the last v-relevant entry position with respect to the direction d, where this candidate is preferred over any preceding v-relevant candidate with respect to the direction d. Determining the values t_1 and t_2 for a new computed candidate entry is done as follows.

The first computed entry that is v-relevant with respect to the direction d is inserted to the set of candidates that are v-relevant with respect to the direction d with the interval $\langle 1, M \rangle$, where M is the maximum position of entries that are v-relevant with respect to the direction d. For $d = R$ we have: $M = n$, for $d = C$, we have: $M = m$, and for $d = D$, we have: $M = \min(m, n)$.

Let k be the position of the last computed entry in the list of DP-table entries that are v-relevant with respect to the direction d. The duels algorithm determines whether the entry at position k should enter the candidates list as well as its corresponding interval $\langle t_1(k), t_2(k) \rangle$ as follows. Let k' be a position of the

DP-table entry in the set of candidates entries that are v-relevant with respect to the direction d with interval $\langle t_1(k'), t_2(k') \rangle$ such that $k+1 \in \langle t_1(k'), t_2(k') \rangle$. We will later show that there is *a single* such candidate k'. The duels algorithm searches in the direction d the *largest* value t, if it exists, for which it holds that:

$$EDBO[c_k] + F(t-k) < EDBO[c_{k'}] + F(t-k') \tag{2}$$

Since both $EDBO[c_k]$ and $EDBO[c_{k'}]$ are already computed and the k, k' are known, the verification in Eq. 2 takes $O(1)$-time for any fixed value t.

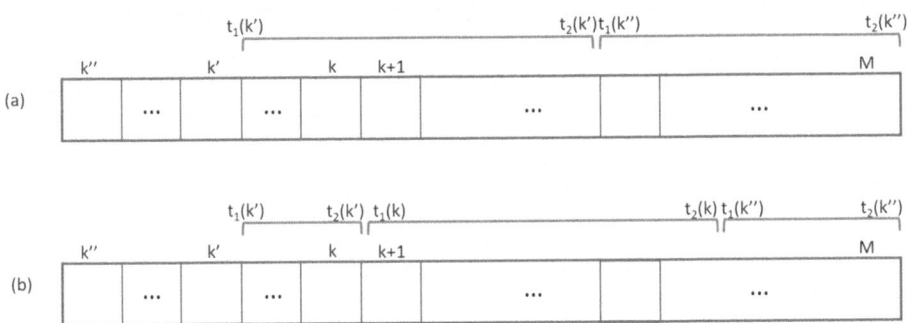

Fig. 4. Feasible intervals of candidates k' and k'': (a) before, and (b) after the duels with k. First, k' is found in the candidate set as the single candidate that $k+1$ belongs to its interval. Since k wins the duel with k' on the position $k+1$ and also the next duel on the position $t_2(k')$ then k' is deleted from the candidates set. Next, k'' is found in the candidate set as the single candidate that $t_2(k') + 1$ belongs to its interval. Since k wins the duel with k'' on the position $t_2(k') + 1$ but k'' wins the next duel on the position $t_2(k'')$ then the binary search to find the value of $t_2(k)$ is performed on the scope $[t_2(k') + 1..(t_2(k''))]$

Duels Set-Up Phase: The search for the largest value t initializes the scope borders $[b_1..b_2]$ to be $b_1 = k+1$, $b_2 = M$ (meaning that $t_1(k)$ is at least $k+1$, since it can only be a candidate for the computation of entries computed after its computation). Next, we set $t = b_1$ in Eq. 2 and check if it holds. If Eq. 2 does not hold for $t = b_1$, we say that k' *wins the duel*, the search ends and k is not inserted to the candidates set of the v-relevant entries with respect to direction d since by Lemma 5 $c_{k'}$ eliminates c_k.

Otherwise, the algorithm performs another duel between k and k' by setting $t = t_2(k')$ in Eq. 2 and checking if it holds. If Eq. 2 holds for $t = t_2(k')$, we say that k *wins the duel*. In this case, k' is deleted from the candidates set of the v-relevant entries with respect to direction d, as c_k is preferred over $c_{k'}$ for all next DP-table entry positions computations. Still, there might be another candidate $c_{k''}$ (see Fig. 4) in the candidates set of the v-relevant entries with respect to direction d with $t_2(k') + 1 \in \langle t_1(k''), t_2(k'') \rangle$. There exists a single such k''. Thus, the duels set-up phase proceeds with k'' instead of k' by repeating the above mentioned two duels with k: the first with $t = b_1 = t_2(k') + 1$ (if k' wins, the duels algorithm ends with case (2) in the next paragraph) and, if k wins the first duel, a second duel is performed with $t = b_2 = t_2(k'')$.

The duels set-up phase is repeated until one of the following conditions hold: (1) k wins the duel for $t = M$. In this case, the duels algorithm ends and k is inserted to the candidates set of the v-relevant entries with respect to direction d with the interval $\langle k+1, M \rangle$; (2) We have candidates k', k'' such that k'' wins the first of the two set-up phase duels with k on the position $b_1 = t_2(k') + 1 > k + 1$ (thus, a second duel of k and k'' is not performed). In this case, the duels algorithm ends and k is inserted to the candidates set of the v-relevant entries with respect to direction d with the interval $\langle k+1, t_2(k') \rangle$; (3) We have a candidate k' such that k wins the first of the two set-up phase duels with k' on the position b_1, but k' wins the second duel with k on the position $b_2 = t_2(k')$. In this case, we know that the *largest* value t for which Eq. 2 holds is in the current scope $[b_1..b_2]$ (see Fig. 4). Thus, the algorithm proceeds to the next phase.

Duels Binary Search Phase: In this phase, the algorithm finds the value of t using a binary search on the current scope $[b_1..b_2]$. While $b_1 \leq b_2$, we set $t = b_1 + \lfloor \frac{b_2 - b_1}{2} \rfloor$ in Eq. 2 and check if it holds. If Eq. 2 holds for t, we say that k *wins the duel* and proceed the search with $b_1 = t + 1$. If Eq. 2 does not hold for t, we say that k' *wins the duel* and proceed the search with $b_2 = t - 1$. The binary search ends when $b_1 > b_2$ and k is inserted to the candidates set of the v-relevant entries with respect to direction d with the interval $\langle k+1, b_2 \rangle$. In addition, the interval of k', $\langle t_1(k'), t_2(k') \rangle$, is updated to be $\langle b_2 + 1, t_2(k') \rangle$.

Observation 4. *For every direction $d \in \{R, C, D\}$ the intervals of candidates in the maintained set of previously computed DP-table v-relevant entries that are v-relevant with respect to the direction d are disjoint.*

Based on Observation 4, we can assure an $O(1)$-bound on the candidates set size for each EDBO DP-table entry computation.

Maintaining the Candidate Set. In order to maintain the candidate set properly for the duels algorithm we need a dynamic data structure that enables to maintain a set of intervals $I_1, \ldots, I_\ell$ over the natural numbers (i.e., for each $k \in \{1, \ldots, \ell\}$, $I_k = \langle t_1(k), t_2(k) \rangle$, where $t_1(k), t_2(k) \in \mathbb{N}$) supporting the following operations: (1) Insertion/Deletion of an interval. (2) Retrieving all intervals that include a specific natural number, that is, *stabbing*.[2] Note that updating a retrieved interval borders can be supported via a deletion and insertion. Thus, a data structure supporting the above operations is enough for implementing the duels algorithm. Describing such a data structure is called *the dynamic interval stabbing problem*, which is well-studied in the literature for many variants (see

---

[2] Actually, since the intervals are disjoint at any point by Observation 4, it suffices to use a simpler predecessor data structure.

e.g. [1,23,25]). It is known that this problem on n intervals can be solved in $O(n)$ space and $O(\log n)$ time for insertion, deletion and stabbing operations. Theorem 2 follows.

Theorem 2. *Given strings S and T of length m and n, respectively, both over alphabet Σ, $EDBO(S,T)$ for concave penalty functions F can be computed in time $O(nm \log \max\{n,m\})$.*

5 EDBE Conditional Lower Bound

In this subsection, we show that the EDBE problem on two n-length input strings is not likely to be solved in $O(n^{2-\epsilon})$-time, for some $\epsilon > 0$. Specifically, we use a conditional lower bound on the problem of computing the Edit Distance between two strings n-length input strings S and T (denoted by $ED(S,T)$), which is the minimum number of insertions, deletions or mismatches of single symbols required so that S matches T. Theorem 3 below assures that any algorithm for this problem with strongly sub-quadratic running time also yields a more efficient algorithm for SAT, breaking the Strong Exponential Time Hypothesis (SETH) introduced in [14,15], which postulates that such algorithms do not exist.

Theorem 3. *[CLB on the Edit Distance (ED) Problem] [6] There is no $O(n^{2-\epsilon})$ algorithm solving the Edit Distance problem on two strings S,T with $|S|, |T| \leq n$, unless SETH is false.*

Note that though the EDBE (similarly, EDBO) problem is a generalization of the ED problem by taking a specific penalty function $F(k) = k$, the above lower bound only applies when using this specific penalty function, whereas *we describe a lower bound for EDBE under any valid penalty function.*

The Reduction from ED to EDBE. Let S and T be two strings with length at most n over alphabet Σ. We show a reduction from the problem of computing the $ED(S,T)$ (the edit distance of S and T) to the problem of computing the $EDBE(S',T')$ for the strings S', T' constructed from S, T with $|S|, |T| \in O(n)$ as follows: (1) A new alphabet Σ' is constructed such that for every symbol $\sigma \in \Sigma$, a new symbol σ' is added to Σ'. (2) The strings S', T' are of length at most $2n$, where for every $1 \leq i \leq n$, if $S[i] = \sigma$ (respectively, $T[i] = \sigma$) then $S'[2i-1] = \sigma$ and $S'[2i] = \sigma'$ (respectively, $T'[2i-1] = \sigma$ and $T'[2i] = \sigma'$). Note that the construction of S' and T' requires $O(n)$-time.

Lemma 6. $ED(S',T') = 2 \cdot ED(S,T)$

Lemma 7. $ED(S',T') = EDBE(S',T')$

Theorem 4 then follows.

Theorem 4. *Given two n-length input strings S, T over Σ, $EDBE(S,T)$ cannot be computed in $O(n^{2-\epsilon})$-time, for some $\epsilon > 0$, unless SETH is false.*

6 EDBE and EDBO with $\mathcal{K}$-Bounded Distance

In this section we describe improved solutions for the $\mathcal{K}$-bounded versions of the EDBE and EDBO problems, formally defined as follows.

Definition 11. [$\mathcal{K}$-Bounded EDBE and EDBO] Let S, T be two strings over Σ of length m and n, respectively, $\mathcal{K} \in \mathbb{R}$ a parameter and F a valid penalty function. The $\mathcal{K}$-Bounded EDBE (respectively, EDBO) problem is to report $EDBE(S, T)$ (respectively, $EDBO(S, T)$), if it is at most $\mathcal{K}$, or ∞ otherwise.

Assume without loss of generality that $n > m$. In what follows we describe an $O(n\mathcal{K}^{1/p})$-time algorithm for EDBE with $\mathcal{K}$-bounded distance and $\tilde{O}(n\mathcal{K}^{1/p})$-time algorithm for EDBO with $\mathcal{K}$-bounded distance for the well-known family of valid penalty functions of the form $F(k) = k^p$, $0 < p < 1$, for a burst of length k. Note that all functions in this family are concave (thus, also sub-additive), therefore, the improved algorithms from Sects. 3 and 4 apply also for the $\mathcal{K}$-bounded EDBE and EDBO problems with this family of valid penalty functions.

We will next show the above mentioned further improved algorithms based on the observation that only a part of the dynamic tables for EDBE (based on the algorithm of Sect. 3) and EDBO (based on the algorithm of Sect. 4) is required for the computation when the distance is bounded by $\mathcal{K}$. We use the term *b-diagonal* as follows. Given an integer b, $-n \leq b \leq m$, the b-diagonal of a DP-table Tab is the set of all entries $Tab[i, j]$ that are $b = (i - j)$-relevant with respect to the direction D. The 0-diagonal is also called the *main* diagonal.

Observation 5. *For any two strings S and T over alphabet Σ of length m and n, respectively, let Tab be the DP-table computed for the EDBE or EDBO. Then, any entry on a b-diagonal of Tab for $|b| > \mathcal{K}^{1/p}$ has value greater than $\mathcal{K}$.*

By Observation 5, it is enough to compute only entries on b-diagonals of Tab for $|b| \leq \mathcal{K}^{1/p}$ while using the algorithm of Sect. 3 for EDBE-computation or the algorithm of Sect. 4 for EDBO-computation. Theorem 5 then follows.

Theorem 5. *Given S, T strings over Σ of length m and n, respectively, w.l.o.g. $n > m$, and a parameter $\mathcal{K} \in \mathbb{R}$, then there exists an $O(n\mathcal{K}^{1/p})$-time algorithm (respectively, $\tilde{O}(n\mathcal{K}^{1/p})$-time algorithm) for the $\mathcal{K}$-bounded EDBE (respectively, $\mathcal{K}$-bounded EDBO) problem with penalty function $F(k) = k^p$ for a length k burst.*

References

1. Agarwal, P.K., Arge, L., Kaplan, H., Molad, E., Tarjan, R.E., Yi, K.: An optimal dynamic data structure for stabbing-semigroup queries. SIAM J. Comput. **41**(1), 104–127 (2012). https://doi.org/10.1137/10078791X
2. Amir, A., et al.: Pattern matching with address errors: rearrangement distances. In: Proceedings of SODA, pp. 1221–1229 (2006)
3. Amir, A., Aumann, Y., Indyk, P., Levy, A., Porat, E.: Efficient computations of l_1 and l_∞ rearrangement distances. Theor. Comput. Sci. **410**(43), 4382–4390 (2009). https://doi.org/10.1016/j.tcs.2009.07.019

4. Amir, A., Aumann, Y., Landau, G., Lewenstein, M., Lewenstein, N.: Pattern matching with swaps. J. Algor. **37**, 247–266 (2000)
5. Amir, A., Levy, A.: String rearrangement metrics: a survey. In: Elomaa, T., Mannila, H., Orponen, P. (eds.) Algorithms and Applications. LNCS, vol. 6060, pp. 1–33. Springer, Heidelberg (2010). https://doi.org/10.1007/978-3-642-12476-1_1
6. Backurs, A., Indyk, P.: Edit distance cannot be computed in strongly subquadratic time (unless SETH is false). In: Proceedings of STOC, pp. 51–58. ACM (2015)
7. Benson, G., Levy, A., Maimoni, S., Noifeld, D., Shalom, B.R.: LCSk: a refined similarity measure. Theor. Comput. Sci. **638**, 11–26 (2016)
8. Bunke, H., Csirik, J.: An algorithm for matching run-length coded strings. Computing **50**, 297–314 (1993)
9. Clifford, R., Gawrychowski, P., Kociumaka, T., Martin, D.P., Uznański, P.: The dynamic k-mismatch problem. In: Proceedings of CPM, vol. 223, pp. 18:1–18:15 (2022)
10. Damerau, F.J.: A technique for computer detection and correction of spelling errors. Commun. ACM **7**(3), 171–176 (1964)
11. Galil, Z., Giancarlo, R.: Speeding up dynamic programming with applications to molecular biology. Theor. Comput. Sci. **64**(1), 107–118 (1989). https://doi.org/10.1016/0304-3975(89)90101-1
12. Gawrychowski, P., Uznański, P.: Towards unified approximate pattern matching for Hamming and L_1 distance. In: Proceedings of ICALP, vol. 107, pp. 62:1–62:13 (2018)
13. Gusfield, D.: Algorithms on Strings, Trees, and Sequences - Computer Science and Computational Biology. Cambridge University Press, Cambridge (1997). https://doi.org/10.1017/cbo9780511574931
14. Impagliazzo, R., Paturi, R.: On the complexity of k-sat. J. Comput. Syst. Sci. **62**(2), 367–375 (2001)
15. Impagliazzo, R., Paturi, R., Zane, F.: Which problems have strongly exponential complexity? J. Comput. Syst. Sci. **63**(4), 512–530 (2001)
16. Kapah, O., Landau, G.M., Levy, A., Oz, N.: Interchange rearrangement: the element-cost model. In: Proceedings of SPIRE, pp. 224–235 (2008)
17. Kociumaka, T., Porat, E., Starikovskaya, T.: Small-space and streaming pattern matching with k edits. In: Proceedings of FOCS, pp. 885–896 (2022)
18. Levenshtein, V.I.: Binary codes capable of correcting deletions, insertions and reversals. Soviet Phys. Doklady **10**, 707–710 (1966)
19. Levy, A.: Exploiting pseudo-locality of interchange distance. In: Lecroq, T., Touzet, H. (eds.) SPIRE 2021. LNCS, vol. 12944, pp. 227–240. Springer, Cham (2021). https://doi.org/10.1007/978-3-030-86692-1_19
20. Levy, A., Porat, E., Shalom, B.R.: Partial permutations comparison, maintenance and applications. In: Bannai, H., Holub, J. (eds.) 33rd Annual Symposium on Combinatorial Pattern Matching, CPM 2022, Prague, Czech Republic, 27–29 June 2022. LIPIcs, vol. 223, pp. 10:1–10:17. Schloss Dagstuhl - Leibniz-Zentrum für Informatik (2022). https://doi.org/10.4230/LIPIcs.CPM.2022.10
21. Levy, A., Shalom, B.R.: A comparative study of dictionary matching with gaps: limitations, techniques and challenges. Algorithmica **84**(3), 590–638 (2022). https://doi.org/10.1007/s00453-021-00851-6
22. Lin, D.: An information-theoretic definition of similarity. In: Proceedings of ICML, pp. 296–304. Morgan Kaufmann Publishers Inc., San Francisco (1998)

23. Nekrich, Y.: A dynamic stabbing-max data structure with sub-logarithmic query time. In: Asano, T., Nakano, S., Okamoto, Y., Watanabe, O. (eds.) ISAAC 2011. LNCS, vol. 7074, pp. 170–179. Springer, Heidelberg (2011). https://doi.org/10.1007/978-3-642-25591-5_19
24. Porat, E., Efremenko, K.: Approximating general metric distances between a pattern and a text. In: Proceedings of SODA, p. 419–427 (2008)
25. Thorup, M.: Space efficient dynamic stabbing with fast queries. In: Proceedings of the Thirty-Fifth Annual ACM Symposium on Theory of Computing. STOC 2003, pp. 649–658. Association for Computing Machinery, New York (2003). https://doi.org/10.1145/780542.780636
26. Wang, C.X., Xu, W.: A new class of generative models for burst-error characterization in digital wireless channels. IEEE Trans. Commun. **55**(3), 453–462 (2007)
27. Winkler, W.E.: String comparator metrics and enhanced decision rules in the Fellegi-Sunter model of record linkage. In: Proceedings of the Section on Survey Research, pp. 354–359 (1990)

Generalization of Repetitiveness Measures for Two-Dimensional Strings

Lorenzo Carfagna[1], Giovanni Manzini[1], Giuseppe Romana[2(✉)], Marinella Sciortino[2], and Cristian Urbina[3,4]

[1] Dipartimento di Informatica, University of Pisa, Pisa, Italy
lorenzo.carfagna@phd.unipi.it, giovanni.manzini@unipi.it
[2] Dipartimento di Matematica e Informatica, University of Palermo, Palermo, Italy
{giuseppe.romana01,marinella.sciortino}@unipa.it
[3] Department of Computer Science, University of Chile, Santiago, Chile
crurbina@dcc.uchile.cl
[4] Centre for Biotechnology and Bioengineering (CeBiB), Santiago, Chile

Abstract. Detecting and measuring repetitiveness of strings is a problem that has been extensively studied in data compression and text indexing. When the data are structured in a non-linear way, as in two-dimensional strings, inherent redundancy offers a rich source for compression, yet systematic studies on repetitiveness measures are still lacking. In this paper, we extend to two dimensions the measures δ and γ, defined in terms of the submatrices of the input, as well as the measures g, g_{rl}, and b, which are based on copy-paste mechanisms. We study their properties and mutual relationships, and we show that the two classes of measures become incomparable when two-dimensional inputs are considered. We also compare our measures with the 2D Block Tree data structure [Brisaboa et al., Computer J., 2024], and provide some insights for the design of effective 2D compressors.

Keywords: Two-dimensional strings · Repetitiveness measures · Text compression

1 Introduction

In the latest decades, the amount of data generated in the world has become massive but it has been observed that, in many fields, most of this data is highly repetitive. For the study of highly repetitive one-dimensional data, an important role is played by the notions of *substring complexity*, *string attractor*, *bidirectional macro scheme*, and *grammar compression*, which lead to the definition of the repetitiveness measures δ, γ, b, g, and g_{rl} (see [18] for the definitions and their remarkable properties). Such measures provide the theoretical basis for the design and analysis of data compressors and compressed indexing data structures for highly repetitive data [17].

Two-dimensional data, ranging from images to matrices, often contains inherent redundancy, wherein identical or similar substructures recur throughout the dataset. This great source of redundancy can be exploited for compression. Very recently, Brisaboa et al. introduced the 2D Block Trees to compress images, graphs, and maps [3]. On the theoretical side in [4], the authors propose two generalizations of the measures δ and γ for square 2D input strings. Such generalized measures are based on properties of the *square* submatrices of the input string. The choice of considering only square submatrices was dictated by purely practical considerations: 2D submatrices can be efficiently handled using the 2D Suffix Tree data structure [12], and the 2D Block Tree is based on repeated occurrences of square submatrices in the input string. Indeed, [4] provides the first theoretical analysis of the space usage of the 2D Block Tree and shows that the generalized δ measure can be computed in linear time. More recently, in [5] the same authors have introduced a 2D generalization of the measure b considering the parsing of an input square matrix into (possibly overlapping) square phrases.

In this paper, we set aside any consideration about algorithmic efficiency and we generalize the 1D measures mentioned above (δ, γ, b, g and g_{rl}) to 2D strings, considering submatrices of any rectangular shape and not only square submatrices as in [4,5]. Our main results can be summarized as follows:

- we show that using rectangular submatrices the above 1D measures can be naturally generalized to 2D strings, and we compare their properties with those of the square-based 2D measures introduced in [4,5];
- we establish some relationships between the new 2D measures and we prove that some properties which are valid in 1D are no longer valid in 2D; other properties are still valid but the gap between some measures can be asymptotically much larger than in 1D;
- we show that the measures δ and γ, which have a simple definition in terms of the submatrices of the input, are not as expressive as in 1D, while the measures g, g_{rl} and b appear to retain their role of capturing the repetitiveness of the input even in 2D;
- we show that the 2D Block Tree data structure, being based on square submatrices, fails to capture the regularities of some inputs which are instead captured by the measures g, g_{rl} and b;
- we use our generalized measures to analyze a frequently used heuristics for 2D compression, namely *linearization* i.e. the transformation of the input matrix into a 1D string which is then compressed. We measure the effectiveness of this technique for the simple row-by-row linearization and the more complex linearization based on the Peano-Hilbert space-filling curve.

Overall our results shed some light on the difficulties of detecting and exploiting repetitiveness in the 2D setting, and show that some concepts/tools introduced in 1D are less effective in 2D. Our results on the 2D Block Tree suggest that it could be worthwhile to make this data structure more flexible by possibly using also non-square submatrices and that 2D compression algorithms based on grammar compression (approaching the measures g and g_{rl}) and "copy and paste" macro schemes (approaching the measure b) should be considered as alternatives.

For space reasons, in this paper, we focus only on 2D strings. The repetitiveness measures defined and studied in this paper can be naturally extended to d-dimensional strings with $d > 2$. In this context, all the results provided in the paper are still valid. More extensive and detailed results on higher dimensional strings are deferred to the full paper.

2 2D Strings and Measures δ and γ

Let $\Sigma = \{a_1, a_2, \ldots, a_\sigma\}$ be a finite ordered set of *symbols*, which we call an *alphabet*. A *2D string* $M_{m \times n}$ is a $(m \times n)$-matrix with m *rows* and n *columns* such that each element $M[i][j]$ belongs to Σ. The size of $M_{m \times n}$ is $N = mn$. Note that a position in $M_{m \times n}$ consists of a pair (i,j), with $1 \leq i \leq m$ and $1 \leq j \leq n$. Throughout the paper, we assume that for each 2D string $M_{m \times n}$ it holds that $m, n \geq 1$. Note that traditional one-dimensional strings are a special case of 2D strings with $m = 1$. We denote by $\Sigma^{m \times n}$ the set of all matrices with m rows and n columns over Σ. A 2D string in $\Sigma^{m \times n}$ is called *square* if $m = n$.

The concatenation between two matrices is a partial operation that can be performed horizontally ($\oplus$) or vertically ($\ominus$), with the constraint that the number of rows or columns coincide respectively. Such operations have been described in [11] where concepts and techniques of formal languages have been generalized to two dimensions.

We denote by $M_{m \times n}[i_1 \ldots i_2][j_1 \ldots j_2]$ the submatrix starting at position (i_1, j_1) and ending at position (i_2, j_2). We say that a matrix F is a *factor* or *substring* of $M_{m \times n}$ if there exist two positions (i_1, j_1) and (i_2, j_2) such that $F = M_{m \times n}[i_1 \ldots i_2][j_1 \ldots j_2]$. Given a 2D string $M_{m \times n}$, the *2D substring complexity* function P_M counts for each pair of positive integers (k_1, k_2) the number of distinct $(k_1 \times k_2)$-factors in $M_{m \times n}$. The following definition extends the notion of δ measure [7] to 2D strings.

Definition 1. *Let $M_{m \times n}$ be a 2D string and P_M be the 2D substring complexity of $M_{m \times n}$. Then, $\delta(M_{m \times n}) = \max\{P_M(k_1, k_2)/k_1 k_2, 1 \leq k_1 \leq m, 1 \leq k_2 \leq n\}$.*

Note that for 1D strings (i.e. $m = 1$) the above definition coincides with the one used in the literature [14, 18]. Recently, in [4] Carfagna and Manzini introduced an alternative extension of δ, here denoted by $\delta_\square$, limited to square 2D input strings and using only *square* factors for computing the substring complexity. Below we report the definition of such a measure, applied to a generic two-dimensional string.

Definition 2. *Let $M_{m \times n}$ be a 2D string and P_M be the 2D substring complexity of $M_{m \times n}$. Then, $\delta_\square(M_{m \times n}) = \max\{P_M(k, k)/k^2, 1 \leq k \leq \min\{m, n\}\}$.*

From the definitions of $\delta_\square$ and δ, the following lemma easily follows.

Lemma 1. *For every 2D string $M_{m \times n}$ it holds that $\delta(M_{m \times n}) \geq \delta_\square(M_{m \times n})$.*

Although the two measures δ and $\delta_\square$ may seem similar, considering square factors instead of rectangular ones may result in very different values. Example 1 shows how different the two measures can be when applied to one-dimensional strings, while Example 2 shows that there exist families of square 2D strings for which $\delta_\square = o(\delta)$.

Example 1. Given a 1D string $S \in \Sigma^n$, let $M_{1 \times n} \in \Sigma^{1 \times n}$ be the matrix such that $M_{1 \times n}[1][1 \ldots n] = S[1 \ldots n]$. Since the only squares that occur in $M_{1 \times n}$ are the factors of size 1×1, it is $\delta_\square(M_{1 \times n}) = P_M(1,1)/1^2 \leqslant |\Sigma|$. On the other hand, $\delta(M_{1 \times n}) = \delta(S)$.

Example 2. Let $M_{n \times n}$ be the square 2D string in [4, Lemma 4]. Assuming n is a perfect square, the first row of $M_{n \times n}$ is the string $S = B_1 B_2 \ldots B_{\sqrt{n}/2}$ composed by $\sqrt{n}/2$ blocks, each one of size $2\sqrt{n}$, with $B_i = 1^i 0^{(2\sqrt{n}-i)}$. The remaining rows of $M_{n \times n}$ are all #^{n}. In [4, Lemma 4] it is shown that $\delta_\square(M_{n \times n}) = O(1)$. On the other hand, notice that for $i \in [2 \ldots \sqrt{n}/2]$ and $j \in [0 \ldots \sqrt{n} - i]$, the strings $0^j 1^i 0^{\sqrt{n}-j-i}$ are all different substrings of length $\sqrt{n}$ of S. Since these substrings are in total $\Omega(n)$, it is $\delta(M_{n \times n}) = \Omega(\sqrt{n})$.

The following definition generalizes to 2D strings the notion of attractor [13].

Definition 3. *An attractor for a 2D string $M_{m \times n}$ is a set $\Gamma \subseteq [1 \ldots m] \times [1 \ldots n]$ with the property that any substring $M[i \ldots j][k \ldots l]$ of $M_{m \times n}$ has an occurrence $M[i' \ldots j'][k' \ldots l']$ such that $\exists (x,y) \in \Gamma$ with $i' \leq x \leq j'$ and $k' \leq y \leq l'$. The size of the smallest attractor for $M_{m \times n}$ is denoted by $\gamma(M_{m \times n})$.*

When $m = 1$ the above definition coincides with the one for 1D strings, hence the measure γ inherits the properties for the one-dimensional case [13,16]. In particular: γ is not monotone and computing $\gamma(M_{m \times n})$ is NP-hard. In addition, the following property holds.

Proposition 1. *For every 2D string $M_{m \times n}$, it is $\delta(M_{m \times n}) \leqslant \gamma(M_{m \times n})$.*

Proof. Reasoning as in the 1D case, we get that for any 2D string M it is $P_M(k_1, k_2) \leqslant k_1 k_2 \gamma(M)$. □

The next proposition shows that in the 2D context, the gap between δ and γ can be larger than the one-dimensional case, where it is logarithmic [14].

Proposition 2. *For all $m, n \geqslant 1$ there exists a 2D string $M_{m \times n}$ such that $\delta(M_{m \times n}) = O(1)$ and $\gamma(M_{m \times n}) = \Omega(\min(m,n))$.*

Proof. Let I_k be the $k \times k$ identity matrix. For all $m, n \geqslant 1$, let us consider the 2D string $M_{m \times n}$ such that $M_{m \times n}[1 \ldots \min(m,n)][1 \ldots \min(m,n)] = I_{\min(m,n)}$, and all the remaining symbols are 0's. When either $m = 1$ or $n = 1$ the proof is trivial from known results on 1D strings, so let us assume $m, n \geqslant 2$. Let us further assume $m < n$, next we show that $\Gamma = \{(2,2) \ldots (m-1, m-1)\} \cup \{(1,m), (m,1), (m, m+1)\}$ is an attractor for $M_{m \times n}$. All the substrings of $M_{m \times n}$

that contain at least two occurrences of 1's have an occurrence crossing the position (i,i), for some $1 < i < m$, while all the substrings that consist of only 0's have an occurrence aligned with one of the 0's at position $(1,m)$, $(m,1)$, or $(m, m+1)$. The factors left contain only one occurrence of 1's that do not cross any position in Γ. These factors of size $k_1 \times k_2$ have to cross either the 1 in position $(1,1)$ or in position (m,m), and therefore it is either $k_1 = 1$ and $k_2 < m$, or vice versa. Observe that all these factors have another occurrence either starting in $(2,2)$ or ending in $(m-1, m-1)$, and therefore Γ is an attractor of $M_{m \times n}$. Since $M_{m \times n}$ has $m+1$ distinct columns the above attractor has minimum size, i.e. $\gamma(M_{m \times n}) = m + 1$. On the other hand, there exist at most $k_1 + k_2$ distinct substrings of size $k_1 \times k_2$ in $M_{m \times n}$: $k_1 + k_2 - 1$ correspond to substrings where the diagonal of $M_{m \times n}$ touches ones of the positions in the left or upper borders of the factor; the last one is the string of only 0's. Hence $\delta(M_{m \times n}) \leqslant 2$. The case $m > n$ is treated symmetrically by considering the attractor $\Gamma' = \{(2,2)..(n-1, n-1)\} \cup \{(1,n), (n,1), (n+1,n)\}$. For the case $n = m$ it is $M_{m \times n} = I_n$ and reasoning as above it is easy to see that $\Gamma'' = \{(2,2)..(n-1, n-1)\} \cup \{(1,n), (n,1)\}$ of size n is a minimal attractor for I_n and that $\delta(I_n) \leqslant 2$. □

In [4] the authors introduced an alternative definition of string attractors for square 2D input strings in which they consider only square factors. We can define such a measure, denoted by $\gamma_\square$, also for generic 2D strings, by simply considering only square substrings of $M_{m \times n}$ in Definition 3. From the definitions of γ and $\gamma_\square$ we immediately get the following relationship:

Lemma 2. *For every 2D string $M_{m \times n}$ it holds that $\gamma(M_{m \times n}) \geqslant \gamma_\square(M_{m \times n})$.*

The following example shows that γ and $\gamma_\square$ can be asymptotically different.

Example 3. Consider again the $m \times m$ identity matrix I_m. For each $k \leqslant m$, a $k \times k$ square factor of I_m either consists of i) all 0's, or ii) all 0's except only one diagonal composed by 1's. Hence, all square factors of type i) have an occurrence that includes position $(m,1)$ (i.e. the bottom left corner), while all those of type ii) have an occurrence that includes the position $(\lfloor m/2 \rfloor, \lfloor m/2 \rfloor)$ (i.e. the 1 at the center). It follows that $\gamma_\square(I_m) = 2 \in O(1)$, while from the proof of Proposition 2 it can be deduced that $\gamma(I_m) = \Theta(m)$.

The measures δ and γ inherit from the 1D case the property that δ is unreachable and γ is unknown to be reachable [18]. In the next sections, we show how to generalize to the 2D case the measures g, g_{rl}, and b which are reachable both in 1D and 2D (details in the full paper).

3 (Run-Length) Straight-Line Programs for 2D Strings

In this section we consider a generalization of SLPs for the two-dimensional space introduced in [2] and use it to generalize the measures g and g_{rl} to 2D strings.

Definition 4. Let $M_{m \times n}$ be a 2D string. A 2-dimensional straight-line program (2D SLP) for $M_{m \times n}$ is a context-free grammar (V, Σ, R, S) that uniquely generates $M_{m \times n}$ and where the definition of the right-hand side of a variable can have the form

$$A \to a,\ A \to B \oplus C,\ or\ A \to B \ominus C,$$

where $a \in \Sigma$, $B, C \in V$. We call these definitions terminal rules, horizontal rules, and vertical rules, respectively. The expansion of a variable is defined as

$$\exp(A) = a,\ \exp(A) = \exp(B) \oplus \exp(C),\ or\ \exp(A) = \exp(B) \ominus \exp(C),$$

respectively. The size of a 2D-SLP is the sum of the sizes of all right-hand sides of the rules of the grammar, where we assume the terminal rules have size 1, and the horizontal and vertical rules have size 2. The measure $g(M_{m \times n})$ is defined as the size of the smallest 2D SLP generating $M_{m \times n}$.

It is easy to see that the above definition coincides with that of the measure g for one-dimensional strings if only horizontal concatenations are considered.

Proposition 3. *It always holds that $g(M_{m \times n}) = \Omega(\log(mn))$.*

Proof. From the starting variable S, each substitution step can double the size of the current 2D string of variables. Hence, a 2D SLP of size g can produce a 2D string of size at most $2^{\lfloor g/2 \rfloor}$. Therefore, a string of size $N = mn$ needs a grammar of size $\Omega(\log_2 N)$. □

Proposition 4. *The problem of determining if there exists a 2D SLP of size at most k generating a text $M_{m \times n}$ is NP-complete.*

Proof. Clearly, the problem belongs to NP. Observe that the 1D version of the problem, known to be NP-complete [6], reduces to the 2D version by considering 1D strings as matrices of size $1 \times n$. □

Given any 2D SLP, we can support direct access to an arbitrary cell of the input in $O(h)$ time, where h is the height of the parse tree[1] of the 2D SLP. Note that the balancing procedure of Ganardi et al. [10] can be easily adapted in the 2D setting, so $h = O(\log N)$, where N is the size of the 2D string.

As in the 1D case, we can extend 2D SLPs with *run-length rules* obtaining more powerful grammars.

---

[1] Analogously to the 1D setting, the *parse tree* of 2D SLP is an ordered labeled tree where S is the root, and the children of a variable A are the variables in its right-hand side (possibly repeated).

Definition 5. *A 2-dimensional run-length straight-line program (2D RLSLP) is a 2D SLP that in addition allows special rules, which are assumed to be of size 2, of the form*

$$A \to \oplus^k B \text{ and } A \to \ominus^k B$$

for $k > 1$, with their expansions defined as

$$\exp(A) = \exp(\underbrace{B \oplus B \oplus \cdots \oplus B}_{k \text{ times}}) \quad \text{and} \quad \exp(A) = \exp(\underbrace{B \ominus B \ominus \cdots \ominus B}_{k \text{ times}})$$

respectively. The measure $g_{rl}(M_{m \times n})$ is defined as the size of the smallest 2D RLSLP generating $M_{m \times n}$.

Proposition 5. *For every 2D string $M_{m \times n}$ it holds that $g_{rl}(M_{n \times n}) \leqslant g(M_{m \times n})$. Moreover, there are infinite string families where $g_{rl} = o(g)$.*

Proof. The first claim is trivial by definition. The second claim is proven by considering the family of $1 \times n$ matrices $M_{1 \times n} = 1^n$ for which $g = \Omega(\log n)$ and $g_{rl} = O(1)$. □

4 Macro Schemes for 2D Strings

The notion of macro scheme and the corresponding measure b can be naturally generalized to 2D strings with the following definition.

Definition 6. *A 2D macro scheme for a string $M_{m \times n}$ is any factorization of $M_{m \times n}$ into a set of disjoint phrases such that any phrase is either a square of dimension 1×1 called an* explicit symbol/phrase, *or is a copied phrase with source in $M_{m \times n}$ starting at a different position. For a 2D macro scheme to be* valid *or* decodable, *there must exist a function* $\text{map} : ([1..m] \times [1..n]) \cup \{\bot\} \to ([1..m] \times [1..n]) \cup \{\bot\}$ *such that: i)* $\text{map}(\bot) = \bot$, *and if $M[i][j]$ is an explicit symbol, then $\text{map}(i,j) = \bot$; ii) for each copied phrase $M[i_1..j_1][i_2..j_2]$, it must hold that $\text{map}(i_1+t_1, i_2+t_2) = \text{map}(i_1,i_2) + (t_1,t_2)$ for $(t_1,t_2) \in [0..j_1-i_1] \times [0..j_2-i_2]$, where $\text{map}(i_1,i_2)$ is the upper left corner of the source for $M[i_1..j_1][i_2..j_2]$; iii) for each $(i,j) \in [1..m] \times [1..n]$ there exists $k > 0$ such that $\text{map}^k(i,j) = \bot$. We define $b(M_{m \times n})$ as the size of the smallest valid 2D macro scheme for $M_{m \times n}$.*

Example 4. Let I_n be the $n \times n$ identity matrix. A macro scheme for I_n consists of the phrases $\{X_1, X_2, X_3, X_4, X_5, X_6\}$ where: i) $X_1 = I_n[1][1]$ is an explicit symbol (the 1 in the top-left corner); ii) $X_2 = I_n[1][2]$ is an explicit symbol; $X_3 = I_n[2][1]$ is an explicit symbol; $X_4 = I_n[1][3..n]$ is a phrase with source $(1,2)$; $X_5 = I_n[3..n][1]$ is a phrase with source $(2,1)$; and $X_6 = I_n[2..n][2..n]$ is a phrase with source $(1,1)$. The underlying function map is defined as $\text{map}(1,1) = \text{map}(1,2) = \text{map}(2,1) = \bot$, $\text{map}(1,j) = (1, j-1)$ for $j \in [3..n]$, $\text{map}(i,1) =$

$(i-1, 1)$ for $i \in [3..n]$, and $\mathtt{map}(i, j) = (i-1, j-1)$ for $i, j \in [2..n] \times [2..n]$. One can see that $\mathtt{map}^n(i, j) = \bot$ for each i and j. Hence, the macro scheme is valid and $b(I_n) \leq 6$. Figure 1 shows this macro scheme for I_7.

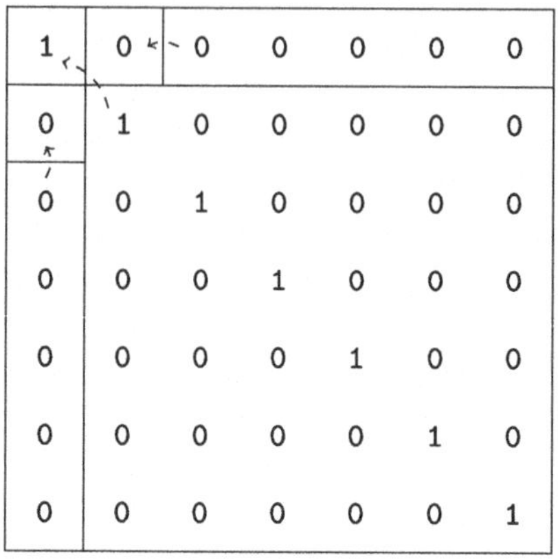

1	0	0	0	0	0	0
0	1	0	0	0	0	0
0	0	1	0	0	0	0
0	0	0	1	0	0	0
0	0	0	0	1	0	0
0	0	0	0	0	1	0
0	0	0	0	0	0	1

Fig. 1. Macro scheme with 6 phrases for I_7. The entries $(1,1)$, $(1,2)$, and $(2,1)$ are explicit symbols. The remaining phrases point to the source from where they are copied

The following two propositions show that the computability properties of b and its relationship with the measures g_{rl} and g are preserved in the 2D context.

Proposition 6. *The problem of determining if there exists a valid 2D macro scheme of size at most k for a text $M_{m \times n}$ is NP-complete.*

Proof. The 1D version of the problem, which is known to be NP-complete [9], reduces to the 2D version of the problem in constant time. □

Proposition 7. *For every 2D string $M_{m \times n}$ it holds that $b(M_{m \times n}) \leq g_{rl}(M_{m \times n})$. Moreover, there are string families where $b = o(g_{rl})$.*

Proof. We show how to construct a macro scheme from a 2D RLSLP, representing the same 2D string and having the same asymptotic size. Let G be a 2D RLSLP generating $M_{m \times n}$ and consider its parse tree. The first occurrence of each variable B that expands to a single symbol at position $M[i][j]$, becomes an explicit phrase of the parsing at that position. For each occurrence of a variable B that is not the first one in this tree, we create a phrase—exactly where this occurrence of $\exp(B)$ should be in M—that maps to the expansion of the first occurrence of B in M. For a rule $A \to \mathbb{O}^k B$, we construct at most two phrases: one phrase for the leftmost B of $\mathbb{O}^k B$ pointing to the first occurrence of B, or if the expansion of this B is a single symbol (or no phrase if this B is the first occurrence but its expansion is not a single symbol), and another phrase for

$\mathbb{O}^{k-1}B$ pointing to the expansion of the occurrence of the B before in M. We do analogously, for 2D vertical run-length rules. It is easy to see that this parsing is decodable and that its size is bounded by the size g_{rl} of the grammar. For a family where $b = o(g_{rl})$, this holds for the 1D version [19]. □

In [5] the authors introduce a notion of 2D macro scheme for square strings that differs from the one in Definition 6 in that 1) phrases must all be square and 2) phrases are allowed to overlap (see [5, Section 4] for details). Given a 2D string M, in the following we call $b_\square(M)$ the minimum number of phrases in a macro scheme with square, possibly overlapping, phrases. We now show that the use of square phrases can limit significantly the power of a macro scheme since there are string families for which $b = o(b_\square)$. This result will have important consequences also for other measures.

For any $k > 0$, let D_k denote a binary de Bruijn sequence of length $n = 2^k + k - 1$ containing all the possible binary substrings of length k exactly once. We define the $n \times n$ matrix B_k over the alphabet $\Delta = \{\langle 0,0 \rangle, \langle 0,1 \rangle, \langle 1,0 \rangle, \langle 1,1 \rangle\}$ by the relationship

$$B_k[i][j] = \langle D_k[i], D_k[j] \rangle.$$

Notice that each row and each column of B_k is a de Bruijn sequence over a binary alphabet which is a subset of Δ. For example, if $D_k[i] = 1$, then row i of B_k is a de Bruijn sequence over the alphabet $\{\langle 1,0 \rangle, \langle 1,1 \rangle\}$. Similarly, if $D_k[j] = 0$, then column j of B_k is a de Bruijn sequence over the alphabet $\{\langle 0,0 \rangle, \langle 1,0 \rangle\}$. Notice that B_k contains only two distinct rows/columns since rows/columns i and j are different if and only if $D_k[i] \neq D_k[j]$.

Lemma 3. *For any $k > 0$ it is $g(B_k) = O(n \log \log n / \log n)$.*

Proof. In [8] the authors show that $g(D_k) = O(n \log \log n / \log n)$, that is, there exists a (one dimensional) SLP G of size $O(n \log \log n / \log n)$ generating the de Bruijn sequence D_k. If in G we replace the terminal symbols $\{0, 1\}$ with $\{\langle 0,0 \rangle, \langle 0,1 \rangle\}$ we obtain a SLP G_0 that generates all rows i in B_k such that $D_k[i] = 0$. Similarly, if in G we replace the terminals $\{0, 1\}$ with $\{\langle 1,0 \rangle, \langle 1,1 \rangle\}$ we obtain a SLP G_1 that generates all rows i in B_k such that $D_k[i] = 1$. Finally, if in G we make all rules vertical and we replace the terminal 0 with the starting symbol of G_0 and the terminal 1 with the starting symbol of G_1, we obtain a 2D SLP G_2 that combined with G_0 and G_1 generates the matrix B_k. The thesis follows since the size of $G_2 \cup G_1 \cup G_0$ is $O(n \log \log n / \log n)$. □

Lemma 4. *Every square submatrix of size k or more appears in B_k at most once.*

Proof. It suffices to prove the result for the square submatrices of size k. Assume that the $k \times k$ submatrix with upper left corner in (i, j) is identical to the submatrix with upper left corner in (u, v). The crucial observation is that $B_k[i][j] = B_k[u][v]$ implies $D_k[i] = D_k[u]$ and $D_k[j] = D_k[v]$. Considering also the other entries, we get that if the two submatrices are equal then we must have $D_k[i..i+k-1] = D_k[u..u+k-1]$ and $D_k[j..j+k-1] = D_k[v..v+k-1]$. Since D_k is a de Bruijn sequence, this implies $i = u$ and $j = v$ as claimed. □

Proposition 8. *For the 2D string B_k of size $N = n \times n$, with $n = 2^k + k - 1$, it is $b_\square = \Omega(g\sqrt{N}/(\log N \log \log N)) = \Omega(b\sqrt{N}/(\log N \log \log N))$.*

Proof. Since $b \leqslant g$, by Lemma 3 $b = O(\sqrt{N} \log \log N / \log N)$. Lemma 4 implies that there cannot be square phrases of size $k \times k$ or larger. Hence, the number of phrases is at least n^2/k^2, so $b_\square = \Omega(N/\log^2 N)$ and the lemma follows. □

5 On the Relative Power of 2D Measures

A remarkable property of the measures considered in this paper is that, for 1D strings, they can be totally ordered in terms of their relative power at capturing regularities in the input; indeed, for any 1D string S it is $\delta(S) \leqslant \gamma(S) \leqslant b(S) \leqslant g_{rl}(S) \leqslant g(S)$ (see [13,18]). In the previous sections, we have shown that when we consider also 2D strings the relationships $\delta \leqslant \gamma$ and $b \leqslant g_{rl} \leqslant g$ still hold. In this section, however, we prove that the two classes of measures, i.e. δ and γ from one side, based on the counting and distribution of distinct factors, and b, g_{rl}, and g from the other side, based on copy-paste mechanisms, become incomparable when also 2D strings are considered. In particular g can be asymptotically smaller than δ:

Proposition 9. *There exists an infinite family of 2D strings of size N with $\delta = \Omega(gN/\log^3 N)$.*

Proof. For $k \geqslant 1$, consider the 2D binary string E_k of size $N = k \times 2^k$ such that, for all $1 \leqslant i \leqslant 2^k$, the ith column of E_k is the binary representation of $i - 1$ in k digits (with the top row containing the least significant bits). Since all columns of E_k are distinct, F_k contains 2^k distinct factors of size $k \times 1$ and therefore it is $\delta(E_k) \geqslant 2^k/k = kN/k^3$. We prove that $g(E_k) = O(k)$ by exhibiting a 2D SLP for E_k of size $O(k)$ and the result immediately follows because $k < \log N$.

Consider the 2D SLP $G_k = (V_k, \{0,1\}, R'_k, S_k)$ having the following rules R'_k:

- $X_0 \to 0$ and $X_h \to X_{h-1} \oplus X_{h-1}$ for all $1 \leqslant h \leqslant k - 1$;
- $Y_0 \to 1$ and $Y_h \to Y_{h-1} \oplus Y_{h-1}$ for all $1 \leqslant h \leqslant k - 1$;
- $C_h \to X_{h-1} \oplus Y_{h-1}$ for all $2 \leqslant h \leqslant k$;
- $S_1 = X_0 \oplus Y_0$ and $S_h \to R_h \ominus C_h$ for all $2 \leqslant h \leqslant k$.
- $R_h \to S_{h-1} \oplus S_{h-1}$ for all $2 \leqslant h \leqslant k$;

G_k has size $\Theta(k)$ and it is easy to see that $\exp(X_h) = 0^{(2^h)}$, $\exp(Y_h) = 1^{(2^h)}$ and therefore $\exp(C_h) = 0^{(2^{h-1})}1^{(2^{h-1})}$. In the following, we show by induction on k that G_k is a 2D SLP for E_k, that is $\exp(S_k) = E_k$. For the base case $k = 1$ it is $\exp(S_1) = \exp(X_0) \oplus \exp(Y_0) = 0 \oplus 1 = E_1$. For the inductive step we assume that $\exp(S_k) = E_k$ and we note that for $k \geqslant 1$ the bottom row $E_{k+1}[k+1][1..2^{k+1}]$ of E_{k+1} is the string $0^{(2^k)}1^{(2^k)} = \exp(C_{k+1})$ and the remaining rectangular submatrix $E_{k+1}[1..k][1..2^{k+1}]$ is $E_k \oplus E_k$.

By taking two expansion steps starting from S_{k+1}, we obtain $\exp(S_{k+1}) = (\exp(S_k) \oplus \exp(S_k)) \ominus \exp(C_{k+1})$ which by inductive hypotheses and the definition of C_{k+1} expands to $(E_k \oplus E_k) \ominus E_{k+1}[k+1][1..2^{k+1}] = E_{k+1}$. □

From the above proposition, we get that the measure g (and therefore also g_{rl} and b) can be much smaller than both δ and γ. Intuitively, the reason is that the matrix E_k is hard to compress by columns (they are all distinct) but easily compressible by rows. The measure g is defined in terms of the best grammar compressor, so it fully exploits row compressibility. In contrast, the measure δ is based on occurrences of factors of any shape and is therefore "hindered" by the difficulty of compressing columns. For 1D strings it is always $\delta \leqslant g$, but in the 2D setting, because of the greater freedom in choosing the shape of the copied patterns, the measures b, g_{rl} and g become mutually incomparable with both γ and δ.

Given the above observation, it is worthwhile to compare g with the measures $\delta_\square$ and $\gamma_\square$ whose definitions are based on square factors. In some sense, using square factors can be seen as a method to capture both horizontal and vertical compressibility. Unfortunately, the following proposition shows that g can be asymptotically smaller than $\delta_\square$ even for input square matrices. Note that the same result also holds for $\gamma_\square$, δ, and γ, since $\delta_\square \leqslant \gamma_\square$ and $\delta_\square \leqslant \delta \leqslant \gamma$.

Proposition 10. *There exists an infinite family of square 2D strings with $\delta_\square = \Omega(g\sqrt{N}/(\log N \log \log N))$, where N is the size of the input string.*

Proof. Consider the matrix B_k defined in Sect. 4. By Lemma 3 it is $g(B_k) = O(n \log \log n / \log n) = O(\sqrt{N} \log \log N / \log N)$. By Lemma 4 B_k contains $\Theta(n^2) = \Theta(N)$ distinct $k \times k$ factors, hence $\delta_\square = \Omega(N/\log^2 N)$ and the bound follows. □

Note that Propositions 8, 9 and 10 show that in the 2D setting there can be a significant gap between b, g_{rl} and g from one side, and $b_\square$, $\delta_\square$, $\gamma_\square$, δ, γ from the other. Furthermore, since the square matrix $0_{n \times n}$ consisting of only zeros has constant measures $b_\square$, $\delta_\square$, $\gamma_\square$ but it is $g(0_{n \times n}) = \Omega(\log n)$ (see Proposition 3), Propositions 8 and 10 imply that g is also mutually incomparable with each of the above square measures, even if we consider only square input matrices.

We now show that even the recently introduced 2D Block Tree data structure [3], which is also based on square factors, can fail to capture the regularity of certain two-dimensional strings. The 2D Block Tree is a tree-like compressed representation of a square matrix supporting random access to individual entries in logarithmic time. Given an $n \times n$ input matrix M, an integer parameter $c > 1$, and assuming that n is a power of c, the root of the 2D Block tree at level $\ell = 0$ represents the whole matrix M. To build the level $\ell \geqslant 1$ of the tree we recursively partition (some of) the submatrices represented at level $\ell - 1$ into c^2 smaller non-overlapping submatrices of size $n/c^\ell \times n/c^\ell$ called *blocks*; for each of these blocks, the tree stores a corresponding descending node at level ℓ. The 2D Block Tree attempts to compress the input matrix by avoiding the storage of redundant submatrices: if a block has a prior occurrence in row-major order (RMO), its corresponding subtree is candidate to be pruned and replaced with $O(1)$ pointers to the nodes overlapping its first occurrence in RMO. The pruned blocks are not partitioned into smaller matrices, and their corresponding nodes are leaves in the 2D block tree. See [3,5] for details.

Unfortunately, the following example exhibits a family of 2D strings that are significantly more compressible when represented as an SLP compared to their 2D Block Tree representation: for these matrices, the 2D Block Tree fails to achieve a compression close to the measure g (and therefore to g_{rl} and b).

Proposition 11. *There exists a family of square 2D strings such that the number of nodes of the corresponding 2D Block Tree is $\Omega(g\sqrt{N}/(\log N \log \log N))$, where N is the size of the input string.*

Proof. Consider again the matrix B_k defined in Sect. 4. By Lemma 4 until we reach the tree level in which the blocks are smaller than $k \times k$, all blocks are first occurrences, and therefore the corresponding tree nodes cannot be pruned. Hence, the 2D Block Tree nodes are $\Omega(n^2/k^2) = \Omega(N/\log^2 N)$. □

6 Effectiveness of Linearization Techniques

A classical heuristic for compressing 2D strings is to transform a matrix $M_{m \times n}$ into a 1D string S and use a one-dimensional compressor on S. Having generalized 1D measures to 2D strings, it is natural to measure the effectiveness of linearization techniques by comparing, for a given measure μ, the values $\mu(M_{m \times n})$ and $\mu(S)$. Clearly, for each matrix, there exists a linearization that makes the 2D string highly compressible: we can visit in order from left to right and from top to bottom all the occurrences of $a_1 \in \Sigma$, followed by all the occurrences of $a_2 \in \Sigma$, and so on, obtaining a string consisting in $|\Sigma|$ equal-letter runs. However, this method requires an ad-hoc linearization for each matrix which may require substantial additional information to retrieve the original input. It is therefore customary in the literature to consider linearization techniques that can be inverted efficiently in terms of both time and space.

The simplest linearization technique consists in mapping $M_{m \times n}$ to the string $\texttt{rlin}(M_{m \times n}) = \bigodot_{i=1}^{i=m} M[i][1..n] = M[1][1..n] \cdots M[m][1..n]$, obtained by concatenating its rows. The (lack of) effectiveness of this simple technique with respect to grammar compression has been already shown in [2, Theorem 2] with an example of a matrix T_n of size $(2^{n+1}-1) \times (2^n+1)2^n$ such that $g(T_n) = O(n)$, while $g(\texttt{rlin}(T_n)) = \Omega(2^n)$. The following example shows a similar result for the measure δ.

Example 5. Let $M_{n \times n}$ be obtained by appending to the identity matrix I_{n-1} a row of 0's at the bottom, and then a column of 1's at the right. For each k_1, k_2, $P_M(k_1, k_2)$ is at most $3(k_1 + k_2)$. We can see this by considering three cases: the submatrices that do not intersect the last row or column, the submatrices intersecting the last row, and the submatrices intersecting the last column. In each case, the distinct submatrices are associated to where the diagonal of 1's intersects a submatrix (if it does so). This can happen in at most $k_1 + k_2$ different ways. As $3(k_1 + k_2)/k_1 k_2 \leq 6$, we obtain $\delta(M_{n \times n}) = O(1)$. On the other hand, for each $k \in [0..n-2]$ and $i \in [0..n-k-2]$, each factor $0^i 10^k 10^{n-k-i-2}$ appears in $\texttt{rlin}(M_{n \times n})$. There are $n - k - 1$ of these factors for each k. Summing over all k, we obtain $P_M(n)/n = (n-1)/2 = \Omega(n)$. Thus, $\delta(\texttt{rlin}(M_{n \times n})) = \Omega(n)$.

Somewhat surprisingly, in some settings, the linearized matrix has a smaller measure. The following example shows a family of 2D strings E_k for which $\gamma(\mathtt{rlin}(E_k))$ is asymptotically smaller than $\gamma(E_k)$.

Example 6. Consider the matrix E_k having size $N = k \times 2^k$ of Proposition 9. We note that the i-th row of E_k is the periodic string $(0^{(2^{i-1})}1^{(2^{i-1})})^{(2^{k-i})}$ and therefore $\mathtt{rlin}(E_k) = \bigodot_{i=1}^{i=k}(0^{(2^{i-1})}1^{(2^{i-1})})^{(2^{k-i})}$. We define the set $A = \bigcup_{i=1}^{i=k}\{(i-1)2^k + 1, (i-1)2^k + 1 + 2^{i-1}, (i-1)2^k + 2^i\}$, that is the set of positions of $\mathtt{rlin}(E_k)$ where respectively the first 0 and the first/last 1 of the leftmost occurrence of $0^{(2^{i-1})}1^{(2^{i-1})}$ in row i are mapped during the linearization. We claim that A is an attractor for $\mathtt{rlin}(E_k)$. If a substring S of $\mathtt{rlin}(E_k)$ spans more than one row of E_k, it includes a 0 from the first column of E_k, and therefore it crosses a position of A. Otherwise S is a substring of the ith row and since the rows of E_k are periodic, the leftmost occurrence S' of S starts inside the first group of $0^{(2^{i-1})}1^{(2^{i-1})}$ i.e. in $E_k[i][1..2^i]$. Suppose that S' does not include any attractor position. Then, S' has to be shorter than the maximum distance between two adjacent attractor positions in the same row i.e. it must be $l = |S'| \leq 2^{i-1} - 1$, and therefore $S' = 0^a 1^{l-a}$ or $S' = 1^a 0^{l-a}$ for some $0 \leq a < l$ because S' cannot overlap two distinct groups of 1's or 0's. If $a = 0$ then S' must include respectively the first 1 or the first 0 in the run, otherwise it must include respectively the first or the last 1, therefore we conclude that A is an attractor for $\mathtt{rlin}(E_k)$.

Since A has size $3k - 1$, it is $\gamma(\mathtt{rlin}(E_k)) = O(k)$, on the other hand since each column of E_k is a distinct non-overlapping $k \times 1$ factor it is $\gamma(E_k) \geq 2^k$.

Another well-known linearization technique is the use of a plane-filling curve, such as the Peano-Hilbert curve. Lempel and Ziv [15, Lemma 1] showed that this technique, here denoted by $\mathtt{phlin}$, is effective for compressing 2D strings using a finite-state encoder. We show that this is not necessarily true for the measure b, as shown in the following proposition. We postpone the formal definition of the Peano-Hilbert linearization $\mathtt{phlin}$ to the full paper, as well as the proof the following characterization of the string $\mathtt{phlin}(I_{2^k})$ obtained by applying $\mathtt{phlin}$ to the identity matrix I_{2^k}: it is $\mathtt{phlin}(I_2) = 1010$, and, for $k \geq 1$,

$$\mathtt{phlin}(I_{2^{k+1}}) = \mathtt{phlin}(I_{2^k})0^{4^k}\mathtt{phlin}(I_{2^k})0^{4^k}.$$

Proposition 12. *It is $b(I_{2^k}) = O(1)$ and $b(\mathtt{phlin}(I_{2^k})) = \Omega(k)$.*

Proof. We have already observed in Example 4 that $b(I_n) = O(1)$ for any $n > 0$. To prove the second bound, for $\ell \geq 1$ define $t_\ell = \sum_{i=0}^{\ell-1} 4^i$. We preliminary prove by induction on k that: a) $\mathtt{phlin}(I_{2^k})$ starts with 1, b) $\mathtt{phlin}(I_{2^k})$ ends with 10^{t_k}, c) $\mathtt{phlin}(I_{2^k})$ contains the substrings $10^{t_\ell}1$ for $\ell = 1, .., k$.

For $k = 1$ $\mathtt{phlin}(I_2) = 1010$ satisfies all conditions. For the inductive step a) is trivial. Since $\mathtt{phlin}(I_{2^{k+1}})$ ends with $\mathtt{phlin}(I_{2^k})0^{4^k}$ also b) is immediate. To prove c) observe that since $\mathtt{phlin}(I_{2^{k+1}})$ contains $\mathtt{phlin}(I_{2^k})$, by induction it contains all substrings $10^{t_\ell}1$ for $\ell = 1, .., k$. To prove that it also contains $10^{t_{k+1}}1$

observe that by *a)* $\texttt{phlin}(I_{2^{k+1}})$ starts with $\texttt{phlin}(I_{2^k})0^{4^k}1$ and by *b)* this string ends with $10^{t_k}0^{4^k}1 = 10^{t_{k+1}}1$, and therefore is a substring of $\texttt{phlin}(I_{2^{k+1}})$.

Having established that $\texttt{phlin}(I_{2^k})$ contains the *distinct* substrings $10^{t_\ell}1$ for $\ell = 1, \ldots, k$, since a single string position can be contained in at most two such substrings, we conclude that $b(\texttt{phlin}(I_{2^k})) \geq \gamma(\texttt{phlin}(I_{2^k})) = \Omega(k)$. □

7 Conclusions and Future Works

In this paper, we have proposed the first complete extension of repetitiveness measures previously used in the 1D context to generic two-dimensional strings. In particular, we have introduced extensions of the measures δ and γ to the two-dimensional case based on distinct factors of arbitrary rectangular shape, as well as the extensions of the measures g, g_{rl}, and b, which are based on copy-paste mechanisms.

We have studied the mutual relationships between these measures and we have shown that $\delta \leq \gamma$ and $b \leq g_{rl} \leq g$. We point out that, unlike in the 1D context where $\delta \leq \gamma \leq b \leq g_{rl} \leq g$, the two classes of measures become incomparable when 2D strings are considered. Indeed, we have shown that, depending on the 2D input, the measures g, g_{rl}, and b can be asymptotically smaller than δ and γ.

The results presented in the paper highlight that in the 2D case, the measures δ and γ (as well as their square-based versions introduced in [4,5]) are not satisfactory for capturing the regularities of a generic two-dimensional string, which are instead effectively detected by g, g_{rl}, and b measures. More importantly, the problem of finding a time-efficient 2D compression scheme approaching the measures g, g_{rl}, or b is still open. Indeed, in this paper we have proven that the 2D-Block Tree data structure, introduced in [3] to compress square matrices while supporting efficient random access to its elements, fails to achieve a compression close to g (and g_{rl} and b) for some 2D strings. At the same time, we have shown that the use of linearization strategies as preprocessing to compress two-dimensional input does not seem to be effective, even when considering approaches based on the Peano-Hilbert space-filling curve. For the above reasons, we believe it would be worthwhile to explore possible approximation strategies for b and g, as well as 2D versions of greedy grammar construction algorithms like the ones described in [1,19].

Acknowledgments. LC and GM are partially funded by the PNRR ECS00000017 Tuscany Health Ecosystem, Spoke 6, CUP I53C22000780001, funded by the NextGeneration EU programme, by the spoke "FutureHPC & BigData" of the ICSC—Centro Nazionale di Ricerca in High-Performance Computing, Big Data and Quantum Computing, funded by the NextGeneration EU programme.

GR and MS are partially funded by the MUR PRIN Project "PINC, Pangenome INformatiCs: from Theory to Applications" (Grant No. 2022YRB97K).

LC, GM, MS, and GR are partially funded by the INdAM-GNCS Project CUP E53C23001670001.

CU is partially funded by ANID-Subdirección de Capital Humano/Doctorado Nacional/2021-21210580, ANID, Chile, partially funded by Basal Funds FB0001, ANID, Chile, and partially funded by Fondecyt Grant 1-230755.

Disclosure of Interests. The authors have no competing interests to declare that are relevant to the content of this article.

References

1. Bannai, H., et al.: The smallest grammar problem revisited. IEEE Trans. Inf. Theory **67**(1), 317–328 (2021)
2. Berman, P., Karpinski, M., Larmore, L.L., Plandowski, W., Rytter, W.: On the complexity of pattern matching for highly compressed two-dimensional texts. J. Comput. Syst. Sci. **65**(2), 332–350 (2002)
3. Brisaboa, N.R., Gagie, T., Gómez-Brandón, A., Navarro, G.: Two-dimensional block trees. Comput. J. **67**(1), 391–406 (2024)
4. Carfagna, L., Manzini, G.: Compressibility measures for two-dimensional data. In: Proceedings of the 30th International Symposium on String Processing and Information Retrieval, SPIRE 2023. LNCS, vol. 14240, pp. 102–113. Springer, Heidelberg (2023). https://doi.org/10.1007/978-3-031-43980-3_9
5. Carfagna, L., Manzini, G.: The landscape of compressibility measures for two-dimensional data. IEEE Access **12**, 87268–87283 (2024)
6. Charikar, M., et al.: The smallest grammar problem. IEEE Trans. Inf. Theory **51**(7), 2554–2576 (2005)
7. Christiansen, A.R., Ettienne, M.B., Kociumaka, T., Navarro, G., Prezza, N.: Optimal-time dictionary-compressed indexes. ACM Trans. Algor. **17**(1), 8:1–8:39 (2021)
8. Gagie, T., Navarro, G., Prezza, N.: On the approximation ratio of lempel-ziv parsing. In: Bender, M.A., Farach-Colton, M., Mosteiro, M.A. (eds.) LATIN 2018. LNCS, vol. 10807, pp. 490–503. Springer, Cham (2018). https://doi.org/10.1007/978-3-319-77404-6_36
9. Gallant, J.K.: String Compression Algorithms. Ph.D. thesis, Princeton University (1982)
10. Ganardi, M., Jez, A., Lohrey, M.: Balancing straight-line programs. J. ACM **68**(4), 27:1–27:40 (2021)
11. Giammarresi, D., Restivo, A.: Two-dimensional languages. In: Rozenberg, G., Salomaa, A. (eds.) Handbook of Formal Languages, pp. 215–267. Springer, Heidelberg (1997). https://doi.org/10.1007/978-3-642-59126-6_4
12. Giancarlo, R.: A generalization of the suffix tree to square matrices, with applications. SIAM J. Comput. **24**(3), 520–562 (1995)
13. Kempa, D., Prezza, N.: At the roots of dictionary compression: string attractors. In: STOC, pp. 827–840. ACM (2018)
14. Kociumaka, T., Navarro, G., Prezza, N.: Toward a definitive compressibility measure for repetitive sequences. IEEE Trans. Inf. Theory **69**(4), 2074–2092 (2023)
15. Lempel, A., Ziv, J.: Compression of two-dimensional data. IEEE Trans. Inf. Theory **32**(1), 2–8 (1986)
16. Mantaci, S., Restivo, A., Romana, G., Rosone, G., Sciortino, M.: A combinatorial view on string attractors. Theor. Comput. Sci. **850**, 236–248 (2021)

17. Navarro, G.: Indexing highly repetitive string collections, part II: compressed indexes. ACM Comput. Surv. **54**(2), 26 (2021)
18. Navarro, G.: Indexing highly repetitive string collections, part I: repetitiveness measures. ACM Comput. Surv. **54**(2), 29 (2021)
19. Navarro, G., Ochoa, C., Prezza, N.: On the approximation ratio of ordered parsings. IEEE Trans. Inf. Theory **67**(2), 1008–1026 (2021)

On Computing the Smallest Suffixient Set

Davide Cenzato[1](✉) , Francisco Olivares[2,3](✉) , and Nicola Prezza[1](✉)

[1] DAIS, Ca' Foscari University of Venice, Venice, Italy
{davide.cenzato,nicola.prezza}@unive.it
[2] Centre for Biotechnology and Bioengineering, Santiago, Chile
folivares@uchile.cl
[3] Department of Computer Science, University of Chile, Santiago, Chile

Abstract. Let $T \in \Sigma^n$ be a text over alphabet Σ. A *suffixient set* $S \subseteq [n]$ for T is a set of positions such that, for every one-character right-extension $T[i,j]$ of every right-maximal substring $T[i, j-1]$ of T, there exists $x \in S$ such that $T[i,j]$ is a suffix of $T[1,x]$. It was recently shown that, given a suffixient set of cardinality q and an oracle offering fast random access on T (for example, a straight-line program), there is a data structure of $O(q)$ words (on top of the oracle) that can quickly find all Maximal Exact Matches (MEMs) of any query pattern P in T with high probability. The paper introducing suffixient sets left open the problem of computing the smallest such set; in this paper, we solve this problem by describing a simple quadratic-time algorithm, a $O(n + \bar{r}|\Sigma|)$-time algorithm running in compressed working space ($\bar{r}$ is the number of runs in the Burrows-Wheeler transform of T reversed), and an optimal $O(n)$-time algorithm computing the smallest suffixient set. We present an implementation of our compressed-space algorithm and show experimentally that it uses a small memory footprint on repetitive text collections.

Keywords: Suffixient sets · Right-maximal substrings · Text indexing · Compressed-space algorithms

1 Introduction

In [5], Depuydt et al. described a new compressed index based on the following principle. Let $T \in \Sigma^n$ be a text of length n. Choose a set $S \subseteq [n]$ such that, for every one-character right-extension $T[i,j]$ of every right-maximal substring $T[i, j-1]$ of T, there exists $x \in S$ such that $T[i,j]$ is a suffix of $T[1,x]$ (α is said to be *right-maximal* if both $\alpha \cdot a$ and $\alpha \cdot b$ appear in T for some characters $a \neq b$ or if α is a suffix of T). The idea behind the index is that the set of prefixes

---

D. Cenzato and F. Prezza—Funded by the European Union (ERC, REGINDEX, 101039208). Views and opinions expressed are however those of the author(s) only and do not necessarily reflect those of the European Union or the European Research Council. Neither the European Union nor the granting authority can be held responsible for them.

F. Olivares—Funded by Ph.D Scholarship 21210579, ANID, Chile.

$\{T[1,x] : x \in S\}$, together with random access on T, can be used to locate one occurrence of any pattern P in T via the following strategy. Assuming that P occurs in T and that the prefix $P[1,j]$ is right-maximal in T: (i) we find one of the selected prefixes $T[1,x]$, for $x \in S$, suffixed by $P[1,j+1]$, and (ii) we right-extend the match by comparing $T[x+1,\ldots]$ and $P[j+2,\ldots]$. This way, we either match until the end of P or we find a mismatch $T[x+1+t] \neq P[j+2+t]$, for some $t \geq 0$. In the former case, we are done. In the latter case, we know that $P[j+1+t]$ is right-maximal in T so we can repeat steps (i) and (ii). Starting with the empty prefix of P, this procedure yields one occurrence of P in T and can be easily extended to return all Maximal Exact Matches (MEMs) of P in T. Depuydt et al. [5] called *suffixient* a set S with the property above outlined. They moreover exhibited a suffixient set of cardinality $2\bar{r}$, where $\bar{r}$ is the number of runs in the Burrows-Wheeler transform of T reversed. To conclude, they observed that the set $\{T[1,x] : x \in S\}$ can be stored in $O(|S|)$ words of space using z-fast tries while supporting fast suffix searches (thereby supporting step (i)), and that the right-extension at step (ii) can be performed in logarithmic time and compressed space using a straight-line program of size g on T. As a result, they obtained an index of $O(\bar{r}+g)$ words able to efficiently locate all MEMs of P in T.

This new indexing strategy raises a natural question: how to compute the smallest suffixient set, of size χ? Observe that χ is independent on the alphabet ordering, while $\bar{r}$ is. Since re-ordering the alphabet may reduce asymptotically $\bar{r}$ [1], from the bound $\chi \leq 2\bar{r}$ we immediately obtain that there exist string families on which $\chi = o(\bar{r})$. In other words, χ is a better repetitiveness measure than $\bar{r}$, even though the problem of determining whether χ is *reachable* (i.e. if $O(\chi)$ words are always sufficient to store the text) is still open. In this respect, Depuydt et al. [5] prove that any suffixient set is a string attractor [6], entailing in particular that T can be compressed in $O(\chi \log(n/\chi))$ words of space.

In this paper, we start by characterizing the smallest suffixient set. Our main contribution is to show that the smallest suffixient set can be computed efficiently. We describe three algorithms solving the problem, all based on the Burrows-Wheeler transform (BWT), the Longest Common Prefix array (LCP), and the Suffix Array (SA) of T reversed. The first algorithm runs in quadratic time and is conceptually very simple. The other two algorithms are optimizations of the first one. The second algorithm runs in $O(n+\bar{r}\sigma)$ time (which can be easily reduced to $O(n+\bar{r}\log\sigma)$) and requires only one pass on those three arrays. The third algorithm requires random access on those arrays and runs in optimal $O(n)$ time. While being slower, the one-pass property of the second algorithm makes it ideal to be combined with Prefix Free Parsing (PFP) [2], a technique able to stream those three arrays in compressed space. We show experimentally that this combination of PFP with our one-pass algorithm can quickly compute the smallest suffixient set in compressed space on repetitive text collections.

2 Preliminaries

Let Σ be a finite ordered alphabet of size σ. We use the following notation: $[i,j]$ for an interval indicating all values $\{i, i+1, \ldots, j\}$ (if $j < i$, then $[i,j] = \emptyset$),

$[n]$ for $[1,n]$, $T = T[1,n]$ for a string T of length n over Σ, $T[i]$ for the ith character of T, $T[i,j]$ for the *substring* $T[i] \cdots T[j]$ of T, where $1 \leq i,j \leq n$, $|T|$ for the length of T, and ϵ for the empty string. We assume that T begins with an end-of-string character $T[1] = \$$ being smaller than all the other characters in Σ and appearing only in $T[1]$. We moreover assume $n = |T| \geq 2$, so T contains at least two distinct characters, and that $\sigma \leq n$, since the alphabet can always be re-mapped to $[n]$ without changing the solutions to the problems considered in this paper. A *run* for a string T is a maximal substring $T[i,j]$ containing only one distinct character; we denote by $\text{runs}(T)$ the number of runs of T. We refer to $T^{\text{rev}} = T[n]T[n-1]\cdots T[1]$ as the reverse of the string T.

A *right-maximal substring* (or *repeat*) $T[i,j]$ ($j \geq i-1$) of a string $T[1,n]$ is a substring of T such that either $j = n$ or for which there exist at least two distinct characters $a, b \in \Sigma$ such that both $T[i,j] \cdot a$ and $T[i,j] \cdot b$ are substrings of T. Note that the empty string ϵ is right-maximal.

For space reasons, we assume the reader to be familiar with the notions of lexicographic order, suffix array (SA), Longest Common Prefix array (LCP), and Burrows-Wheeler Transform (BWT). See the full version [4] for the formal definitions. Given a set $B \subseteq [n]$, $\text{LCP}[B]$ indicates the set $\{\text{LCP}[i] : i \in B\}$.

Given a $-terminated string $T = T[1..n]$, the Burrows-Wheeler Transform of T [3], $\text{BWT}(T)$, is a permutation of the characters of T defined by concatenating the characters following the lexicographically-sorted suffixes of T. We indicate with $\bar{r}$ the number of runs in $\text{BWT}(T^{\text{rev}})$. For example, if $T = \$\text{BANANA}$, $\text{BWT}(T^{\text{rev}}) = \text{BNN}\$\text{AAA}$, and $\bar{r} = 4$.

We say that position $1 < i \leq n$ is a *c-run break*, for $c \in \Sigma$, if $\text{BWT}[i-1,i] = ac$ or $\text{BWT}[i-1,i] = ca$, with $a \in \Sigma$ and $a \neq c$. Note that, if i is a c-run break, then it is also an a-run break for some $a \neq c$.

3 The Smallest Suffixient Set

We start from the definition of *suffixient set* [5]:

Definition 1 (Suffixient set [5]). *A set $S \subseteq [n]$ is* suffixient *for a string T if, for every one-character right-extension $T[i,j]$ ($j \geq i$) of every right-maximal string $T[i, j-1]$, there exists $x \in S$ such that $T[i,j]$ is a suffix of $T[1,x]$.*

Next, we characterize the smallest suffixient set. We first define the concept of *supermaximal extensions* (Definition 2). Then, in Definition 3 we define a set (actually, a family of sets) $\mathcal{S}$ containing all positions in T "capturing" all (and only the) supermaximal extensions of T. Finally, in Lemma 1 we prove that $\mathcal{S}$ is a suffixient set of minimum cardinality.

Definition 2. *We say that $T[i,j]$ (with $j \geq i$) is a* supermaximal extension *if $T[i, j-1]$ is right-maximal and, for each right-maximal $T[i', j'-1] \neq T[i, j-1]$ (with $i' \leq j' < n$), $T[i,j]$ is not a suffix of $T[i', j']$.*

The following definition depends on a particular total order $<_t$ on $[n]$. We use $<_t$ to break ties between equivalent candidate positions of the suffixient set, choosing the position of maximum rank according to $<_t$ in case of a tie. In order to make notation lighter, we do not parameterize the set on the particular tie-breaking strategy ($<_t$ will always be clear from the context).

Definition 3. *Let $<_t$ be any total order on $[n]$. We define a set $\mathcal{S} \subseteq [n]$ as follows: $x \in \mathcal{S}$ if and only if there exists a supermaximal extension $T[i, j]$ such that (i) $T[i, j]$ is a suffix of $T[1, x]$, and (ii) for all prefixes $T[1, y]$ suffixed by $T[i, j]$, if $y \neq x$ then $y <_t x$.*

We prove that $\mathcal{S}$ is a suffixient set of minimum cardinality:

Lemma 1. *For any tie-breaking strategy (total order) $<_t$, $\mathcal{S}$ is a suffixient set of minimum cardinality for T.*

Proof. To see that $\mathcal{S}$ is suffixient, consider any right-maximal substring $T[i, j-1]$ ($i \leq j \leq n$). We want to prove that there exists $x \in \mathcal{S}$ such that $T[i, j]$ is a suffix of $T[1, x]$. Let $T[1, x]$ be a prefix of T being suffixed by $T[i, j]$, breaking ties (on its endpoint x) by $<_t$. If $T[i, j]$ is a supermaximal extension, then by Definition 3 it holds $x \in \mathcal{S}$ and we are done. Otherwise, let $T[i', j'] \neq T[i, j]$ be a one-character extension of a right-maximal string $T[i', j' - 1]$ such that $T[i, j]$ is a suffix of $T[i', j']$. Without loss of generality, let $T[i', j']$ be the string of maximum length $|T[i', j']| = j' - i' + 1$ with this property. Observe that there cannot be a right-extension $T[i'', j''] \neq T[i', j']$ of a right-maximal string $T[i'', j'' - 1]$ such that $T[i', j']$ is a suffix of $T[i'', j'']$, otherwise $T[i, j]$ would be a suffix of $T[i'', j'']$ and $|T[i'', j'']| > |T[i', j']|$, contradicting the fact that $T[i', j']$ is the longest such string. It follows that $T[i', j']$ is a supermaximal extension. Let $T[1, y]$ be a prefix of T being suffixed by $T[i', j']$, breaking ties (on y) by $<_t$. Then, by Definition 3, $y \in \mathcal{S}$. Also in this case we are done, since $T[i, j]$ suffixes $T[i', j']$ and $T[i', j']$ suffixes $T[1, y]$, so by transitivity of the suffix relation, $T[i, j]$ suffixes $T[1, y]$.

To see that $\mathcal{S}$ is of minimum cardinality, let S be a suffixient set. We prove $|\mathcal{S}| \leq |S|$ by exhibiting an injective function from $\mathcal{S}$ to S.

Let $SE \subseteq \Sigma^+$ be the set of all supermaximal extensions. By definition of supermaximal extension, note that any prefix $T[1, j]$ can be suffixed by at most *one* supermaximal extension $T[i, j]$: assume, for a contradiction, that $T[1, j]$ is suffixed by two distinct supermaximal extensions $T[i, j] \neq T[i', j]$. Without loss of generality, $i' < i$. But then, $T[i, j]$ is a suffix of $T[i', j]$, hence $T[i, j]$ cannot be a supermaximal extension, a contradiction.

We first define a relation $h : \mathcal{S} \to SE$ mapping every $x \in \mathcal{S}$ to a supermaximal extension $h(x) = T[i, j]$ such that $T[i, j]$ is a suffix of $T[1, x]$. From the observation above, there is at most one such string $T[i, j]$. From Definition 3, there exists one such $T[i, j]$. We conclude that h is a function. We now prove that h is injective.

To see that h is injective, assume for a contradiction $h(x) = h(y)$ for some $x \neq y$ with $x, y \in \mathcal{S}$. Let $h(x) = h(y) = T[i, j]$. By Definition 3, there exists a supermaximal extension $T[i_x, x]$ such that $T[1, x]$ is the prefix of T being suffixed by $T[i_x, x]$ with largest endpoint x, according to the total order $<_t$. Following the same reasoning on position y, we can associate an analogous supermaximal extension $T[i_y, y]$ to y. On the other hand, by the definition of h, also string $h(x) = h(y) = T[i, j]$ is a supermaximal extension that suffixes both $T[1, x]$ and $T[1, y]$. Since $T[i, j]$ and $T[i_x, x]$ are supermaximal extensions that suffix $T[1, x]$ and (as proved above) there can exist at most one supermaximal extensions suffixing $T[1, x]$, we conclude that $T[i, j] = T[i_x, x]$. Similarly, we conclude $T[i, j] = T[i_y, y]$, hence $T[i_x, x] = T[i_y, y]$. We obtained a contradiction, since Definition 3 requires both $x <_t y$ and $y <_t x$ and $<_t$ is a total order. We conclude that h is an injective function.

Next, we define a relation $g : SE \to S$ from the set SE of supermaximal extensions to positions in S. For any supermaximal extension $T[i, j] \in SE$, we define $g(T[i, j]) = x$ to be the largest element of S (according to the standard order between integers) such that $T[1, x]$ is suffixed by $T[i, j]$. Such a position must exist since S is a suffixient set, so g is a function. To see that g is also injective, assume for a contradiction $g(T[i, j]) = g(T[i', j']) = x$, for $T[i, j] \neq T[i', j']$ (both supermaximal extensions). By the definition of g, both $T[i, j]$ and $T[i', j']$ are suffixes of $T[1, x]$. But then, since $T[i, j] \neq T[i', j']$, either $T[i, j]$ is a suffix of $T[i', j']$ (so $T[i, j]$ is not a supermaximal extension) or the other way round (so $T[i', j']$ is not a supermaximal extension). In both cases we get a contradiction, so we conclude that g is injective.

We conclude the proof by observing that $g \circ h : \mathcal{S} \to S$ is the composition of two injective functions and is therefore injective. □

From now on, we denote with $\chi = |\mathcal{S}|$ the cardinality of a suffixient set of minimum size. Note that, by Lemma 1, Definition 3 yields a suffixient set of cardinality χ for *any* tie-breaking strategy (total order) $<_t$.

4 Computing the Smallest Suffixient Set

In this section, we present three algorithms for computing a suffixient set of smallest cardinality. We start in Subsect. 4.1 by giving an "operative" version of Definition 3. This "operative" definition will automatically yield a simple quadratic-time algorithm for computing a smallest suffixient set. In the next subsections, we will optimize this algorithm and reach compressed working space (Subsect. 4.2) and optimal linear time (Subsect. 4.3).

4.1 A Simple Quadratic-Time Algorithm

For brevity, in the rest of the paper LCP, BWT, SA denote LCP(T^{rev}), BWT(T^{rev}), and SA(T^{rev}), respectively.

Definition 4. *Given a position $i \in [n]$, we define $box(i) = [\ell, r]$ to be the maximal interval such that $i \in [\ell, r]$ and $\text{LCP}[j] \geq \text{LCP}[i]$ for all $j \in [\ell, r]$.*

Example 1. See Fig. 1: $box(10) = [3, 13]$ (shown with a blue box), because LCP$[3, 13] \geq$ LCP$[10] = 1$, and $box(19) = [17, 20]$ (shown with an orange box).

Next, we define the set of c-run breaks in $box(i)$. Recall that when we say that i is a c-run break, we mean that either $\text{BWT}[i-1,i] = ac$ or $\text{BWT}[i-1,i] = ca$, for some $a \neq c$.

Definition 5. *Given $i \in [n]$ and $c \in \Sigma$, we define*

$$B_{i,c} = \{j \ : \ j \text{ is a } c\text{-run break in } \text{BWT}(T^{rev}) \text{ and } j \in box(i)\}.$$

Example 2. In Fig. 1, we have $B_{10,A} = \{3, 6, 7, 10, 11, 12\}$.

We indicate with $text(i) = n - SA[i] + 2$ the position in T corresponding to character $\text{BWT}[i]$, and with $bwt(j) = SA^{-1}[n+2-j]$ the inverse of function $text()$, i.e. the position in BWT corresponding to character $T[j]$.

The following property stands at the core of all our algorithms for computing a smallest suffixient set:

Lemma 2. *The following hold:*

1. *Let i be a c-run break and $i' \in \{i, i-1\}$ be such that $\text{BWT}[i'] = c$. Let $j' = text(i')$, and let $\ell = \text{LCP}[i]$. If $\ell = \max \text{LCP}[B_{i,c}]$, then the string $T[j' - \ell, j']$ is a supermaximal extension, and*
2. *conversely, for any supermaximal extension $\alpha \cdot c$ ($\alpha \in \Sigma^*$, $c \in \Sigma$) there exists an occurrence $\alpha \cdot c = T[j' - \ell, j']$ and a c-run break $i \in [n]$ such that $\text{BWT}[i'] = c$, with $i' = bwt(j') \in \{i, i-1\}$ and $\ell = \text{LCP}[i] = \max \text{LCP}[B_{i,c}]$.*

Proof. (1) Let i be a c-run break. Let $i', i'' \in \{i, i-1\}$ be such that $\text{BWT}[i'] = c$ and $\text{BWT}[i''] = a \neq c$. Let $j' = text(i')$ and $j'' = text(i'')$. Let moreover $\ell = \text{LCP}[i]$, and assume $\ell = \max \text{LCP}[B_{i,c}]$.

Note that $T[j' - \ell, j' - 1]$ is right-maximal: $\ell = \text{LCP}[i]$ implies that $T[j' - \ell, j' - 1] = T[j'' - \ell, j'' - 1]$ and, by definition of i', i'', j', and j'' we have that $T[j'] = \text{BWT}[i'] \neq \text{BWT}[i''] = T[j'']$.

Assume, for a contradiction, that $T[j' - \ell, j']$ is not a supermaximal extension. Then, since $T[j' - \ell, j' - 1]$ is right-maximal, it must be the case that $T[j' - \ell, j']$ is a suffix of some string $T[\hat{j} - \hat{\ell}, \hat{j}]$ with $\hat{\ell} > \ell$ and such that $T[\hat{j} - \hat{\ell}, \hat{j} - 1]$ is right-maximal. Let $\hat{i} = bwt(\hat{j})$. Since $T[j' - \ell, j']$ is a suffix of $T[\hat{j} - \hat{\ell}, \hat{j}]$ we have that $T[j' - \ell, j' - 1]$ is a suffix of $T[\hat{j} - \hat{\ell}, \hat{j} - 1]$, hence it must be the case that $\hat{i} \in box(i)$. Since $T[\hat{j} - \hat{\ell}, \hat{j} - 1]$ is right-maximal and $T[\hat{j}] = c$, without loss of generality we can assume that $\hat{i} \in \{k-1, k\}$, where k is a c-run break (such a c-run break must exist in $box(i)$) and $\text{LCP}[k] \geq \hat{\ell}$. In particular, $k \in box(i)$. Then, $\text{LCP}[k] \geq \hat{\ell} > \ell$ and $k \in box(i)$ yield $\max \text{LCP}[B_{i,c}] \geq \hat{\ell} > \ell$, a contradiction.

(2) Let $T[j' - \ell, j']$ be a supermaximal extension and let $c = T[j']$. Let $i' = bwt(j')$. Since $T[j' - \ell, j' - 1]$ is right-maximal, we can assume without loss of generality that either $\text{BWT}[i'] \neq \text{BWT}[i' - 1]$ or $\text{BWT}[i'] \neq \text{BWT}[i' + 1]$. Let therefore $i' \in \{i, i-1\}$, where i is a c-run break. By definition of $T[j' - \ell, j']$, it holds $\ell = \text{LCP}[i]$ (because $T[j' - \ell, j' - 1]$ is the longest right-maximal string suffixing $T[1, j' - 1]$). We need to prove that $\ell = \max \text{LCP}[B_{i,c}]$. Assume, for a contradiction, that $\ell < \hat{\ell} = \max \text{LCP}[B_{i,c}]$. This means that there exists a

supermaximal extension $T[\hat{j} - \hat{\ell}, \hat{j}]$ with (i) $T[\hat{j}] = c$, (ii) $\hat{i} \in \{k-1, k\}$, where $\hat{i} = bwt(\hat{j})$ and $k \in B_{i,c}$ is a c-run break in $box(i)$, and (iii) $\text{LCP}[k] = \hat{\ell}$. But then, since $k \in box(i)$ and $T[\hat{j}-\hat{\ell}, \hat{j}-1]$ is a right-maximal string of length $\hat{\ell} > \ell$, we have that $T[j'-\ell, j'-1]$ is a (proper) suffix of $T[\hat{j}-\hat{\ell}, \hat{j}-1]$. This fact and $T[\hat{j}] = c$ contradict the fact that $T[j'-\ell, j']$ is a supermaximal extension. □

Intuitively, Lemma 2 states that supermaximal extensions can be identified by looking at local maxima among c-run breaks contained in LCP ranges of the form $\text{LCP}[box(i)]$. Since, by Definition 3, supermaximal extensions are those characterizing sufficient sets of smallest cardinality, Lemma 2 gives us a tool for computing such a set using the arrays BWT, LCP, and SA. See Fig. 1.

We immediately obtain a simple quadratic algorithm computing a sufficient set of smallest cardinality: see Algorithm 1. Note that, in Lemma 2, if $i'' \in B_{i,c}$ and $\text{LCP}[i''] = \text{LCP}[i]$, then $box(i) = box(i'')$. Since there could be multiple c-run breaks $i'' \in B_{i,c}$ such that $\text{LCP}[i''] = \text{LCP}[i]$, if $\text{LCP}[i] = \text{LCP}[i''] = \max \text{LCP}[B_{i,c}]$ then we have to break ties and insert only one of the corresponding text positions in the sufficient set. In Line 7 of Algorithm 1 we choose the largest such c-run break: as a matter of fact, this defines a particular tie-breaking strategy $<_t$ between positions $[n]$ (as required by Definition 3).

Algorithm 1: Quadratic algorithm to compute a smallest suffixient set

    input : A string $T[1, n] \in \Sigma^n$
    output: A suffixient set $\mathcal{S}$ of smallest cardinality for T.
1 BWT $\leftarrow$ BWT(T^{rev}); LCP $\leftarrow$ LCP(T^{rev}); SA $\leftarrow$ SA(T^{rev});
2 $\mathcal{S} \leftarrow \emptyset$;
3 **for** $i = 2, \ldots, n$ **do**
4   **if** BWT$[i-1] \neq$ BWT$[i]$ **then**
5   **for** $i' \in \{i-1, i\}$ **do**
6             $c \leftarrow$ BWT$[i']$;   /* i is a c-run break */
7   **if** $i = \max\{i'' \,:\, i'' \in B_{i,c} \wedge \text{LCP}[i''] = \max \text{LCP}[B_{i,c}]\}$ **then**
8                 $\mathcal{S} \leftarrow \mathcal{S} \cup \{n - \text{SA}[i'] + 2\}$
9   **end**
10   **end**
11   **end**
12 **end**
13 **return** $\mathcal{S}$;

Computing BWT, LCP, and SA in Line 1 takes $O(n)$ time. Additionally, for each $i = 2, \ldots, n$, Algorithm 1 computes the set $B_{i,c}$ in line 7 by scanning $\text{LCP}[\ldots, i-1, i, i+1, \ldots]$ in order to identify $\text{LCP}[box(i)]$ ($O(n)$ time for each such scan). It follows that the total running time is bounded by $O(n^2)$. Correctness follows immediately from Lemma 2, Definition 3, and Lemma 1.

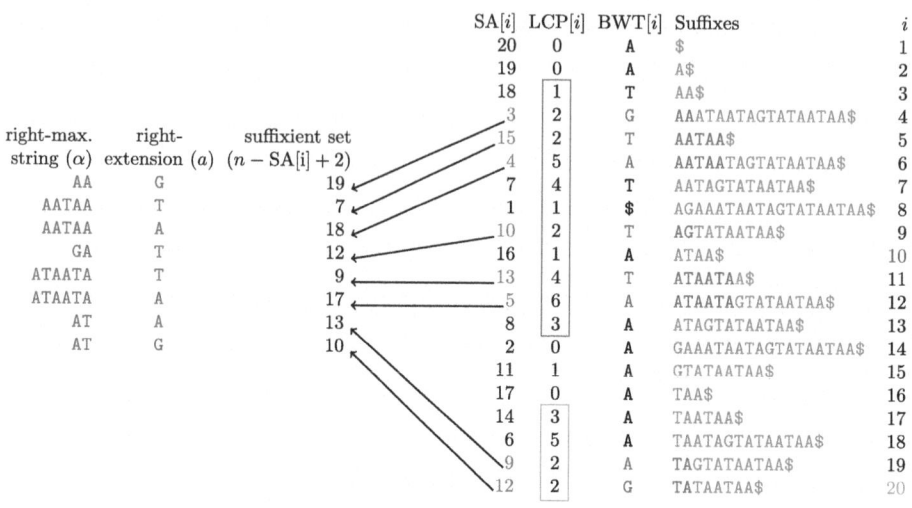

Fig. 1. The figure shows how to construct a smallest suffixient set $\mathcal{S}$ for a string $T[1, n]$ with $n = 20$ following Lemma 2. *To the left of the black arrows*: all supermaximal extensions $\alpha \cdot a$ of T and their selected ending positions in T, forming a suffixient set. *To the right of the black arrows*: SA, LCP, BWT and the sorted suffixes of T^{rev}. In column SA[i], we highlight in red all positions that are selected to be included in $\mathcal{S}$. In columns *Suffixes* and BWT[i], we highlight in green the (reverses of) the supermaximal extensions of T. Black arrows show how the selected SA positions are converted to positions in T using the formula $n - \text{SA}[i] + 2$. *How to identify positions of $\mathcal{S}$*: for each c-run break i, we decide if the two ranks $i' \in \{i-1, i\}$ should contribute to $\mathcal{S}$ (i.e. if they correspond to a supermaximal extension) as described in Lemma 2. For brevity, we show this decisional procedure only on two run breaks. Consider the A-run break at position $i = 10$ (highlighted in blue in column i). The blue box depicts the corresponding LCP interval, LCP[$box(10)$] = LCP[3, 13]. We observe that in [3, 13] there are other A-run breaks i'', such that LCP[i''] > LCP[10]: those are $i'' = 6, 7, 11, 12$. We conclude that text position $n - \text{SA}[10] + 2 = 6$ should not be included in $\mathcal{S}$. Consider now the A-run break $i = 20$, highlighted in orange in column i. Position $i' = 20 - 1 = 19$ is such that BWT[i'] = A. The orange box depicts the corresponding LCP interval, LCP[$box(20)$] = LCP[17, 20]. In this case, there is no other A-run break in [17, 20] with an LCP value larger than LCP[20] = 2. We therefore insert text position $n - SA[i'] + 2 = n - SA[19] + 2 = 13$ in $\mathcal{S}$. Notice that $i = 20$ is also a G-run break; repeating the above reasoning, one can verify that position $i' = 20$ is indeed associated with the supermaximal extension AT · G ending in text position 10.

4.2 A One-Pass Algorithm Working in Compressed Space

In this section, we speed up the simple quadratic algorithm provided in the previous section. We will refer to this version as "one-pass" since it only requires one scan of the BWT, SA, and LCP arrays. The algorithm is summarized in Algorithms 2 and 3. See below for a description.

Algorithm 2: One-pass algorithm

input : A string $T[1..n]$ over a finite alphabet Σ.
output: A suffixient set $\mathcal{S}$ of smallest cardinality for T.

1 $\mathcal{S} \leftarrow \emptyset$; /* initialize the empty suffixient set $\mathcal{S}$ */
2 BWT $\leftarrow$ BWT(T^{rev}); LCP $\leftarrow$ LCP(T^{rev}); SA $\leftarrow$ SA(T^{rev});
3 $R[1, \sigma] \leftarrow ((-1, 0, false), \ldots, (-1, 0, false))$; /* LCP local maxima */
4 $m \leftarrow \infty$; /* LCP minimum inside current BWT run */
5 **for** $i = 2, \ldots, n$ **do**
6      $m \leftarrow \min\{m, \text{LCP}[i]\}$;
7   **if** $\text{BWT}[i] \neq \text{BWT}[i-1]$ **then**
8          $eval(\Sigma, m, R, \mathcal{S})$;
9   **for** $i' \in \{i-1, i\}$ **do**
10   **if** $\text{LCP}[i] > R[\text{BWT}[i']].len$ **then**
11                 $R[\text{BWT}[i']] \leftarrow (\text{LCP}[i], n - \text{SA}[i'] + 2, true)$;
12   **end**
13   **end**
14         $m \leftarrow \infty$;
15   **end**
16 **end**
17 $eval(\Sigma, -1, R, \mathcal{S})$; /* evaluate last active candidates */
18 **return** $\mathcal{S}$;

Algorithm 3: Procedure $eval(C, l, R, \mathcal{S})$

input : A set of characters C, a LCP value l, the candidate's list R and the suffixient set $\mathcal{S}$.
output: The updated suffixient set $\mathcal{S}$.

1 **foreach** $c \in C$ **do**
2   **if** $l < R[c].len$ **then**
3   **if** $R[c].active$ **then**
4              $\mathcal{S} \leftarrow \mathcal{S} \cup \{R[c].pos\}$;
5   **end**
6          $R[c] \leftarrow \{l, 0, false\}$;
7   **end**
8 **end**

Observe that Algorithm 1 runs in quadratic time due to the fact that it needs to scan $\text{LCP}[box(i)]$ for each run break i, and those boxes overlap. Intuitively, Algorithm 2 avoids this inefficiency by storing information about local LCP maxima on c-run breaks (for each $c \in \Sigma$) in a data structure R (Line 3). Intuitively, this allows us to detect if a previous box $box(j)$, with $j < i$, ends in position i by simply checking if $\text{LCP}[j]$ drops below the current LCP maxima associated with each character c.

Assume we scanned the arrays BWT, SA, and LCP up to position i. R is a map associating each $c \in \Sigma$ with information related with some c-run break $j \leq i$ corresponding to a candidate supermaximal extension $\alpha \cdot c$. Each entry

$R[c] = (len, pos, active)$ is composed of three values: $R[c].len = \text{LCP}[j]$ (i.e., the length of the right-maximal string α), $R[c].pos = text(j') = n - \text{SA}[j'] - 2$, where $j' \in \{j-1, j\}$ is such that $\text{BWT}[j'] = c$ (i.e. the position of the last character of the supermaximal extension $\alpha \cdot c$ in T), and a boolean flag $R[c].active$ which is set to $true$ if and only if $R[c].len$ is the maximum value in $\text{LCP}[B_{j,c} \cap [i]]$ (breaking ties by smaller j: we keep the first encountered such maximum). Depending on the value of flag $active$, we will distinguish between *active* and *inactive* candidates.

Example 3. See Fig. 1 and assume we scanned BWT up to position $j = 4$. The candidate associated with letter T is $R[\text{T}] = (2, 3, true)$, where $R[\text{T}].len = \text{LCP}[4] = 2$. This candidate is active since $\text{LCP}[4]$ is the largest value in $\text{LCP}[B_{4,\text{T}} \cap [4]] = \{2\}$. When processing position $j = 6$, we find another T-run break. Here we update $R[\text{T}] = (5, 7, true)$ since $6 \in B_{4,\text{T}} \cap [6]$, and $5 = \text{LCP}[6] > \text{LCP}[4]$.

Algorithm 2 works as follows. We scan the BWT left-to-right: $\text{BWT}[2], \ldots, \text{BWT}[i]$ (for $i = 2, \ldots, n$). For $c \in \Sigma$, let $j < i$ be the c-run break whose information is stored in $R[c]$, i.e. such that $\text{LCP}[j] = R[c].len$. If $\text{LCP}[i] < R[c].len$, then $i \notin box(j)$. But then, if $R[c].active = true$, position $R[c].pos$ must belong to the output set so we have to insert it in $\mathcal{S}$. We also set $R[c].active = false$ in order to record that the maximum in $\text{LCP}[B_{j,c}]$ has been found (if $R[c].active$ is already equal to $false$, $\mathcal{S}$ does not need to be updated since the maximum in $\text{LCP}[B_{j,c}]$ had already been found). Observe that these operations, performed by the call to $eval(\Sigma, m, R, \mathcal{S})$ (lines 8,17) (see below for the meaning of variable m), have to be repeated for each $c \in \Sigma$; however, since i can possibly replace the candidate in $R[c]$ only if i is a c-run break, we can avoid calling $eval(\cdot)$ inside equal-letter BWT runs as follows: we compute the minimum LCP value m inside LCP intervals corresponding to BWT equal-letter runs (line 6), and perform the check $m < R[c].len$ for each $c \in \Sigma$ by calling procedure $eval(\Sigma, m, R, \mathcal{S})$ (costing time $O(\sigma)$) only when the current position i is a run break. This ultimately reduces the algorithm's running time from $O(n \cdot \sigma)$ to $O(n + \bar{r} \cdot \sigma)$.

On the other hand, if i is a c-run break and $\text{LCP}[i] > R[c].len$, then $i \in B_{j,c} \cap [i]$ and $R[c].len$ is not a local maximum in $\text{LCP}[B_{j,c}]$ (line 10). In this case, we have to replace the candidate stored in $R[c]$ with the information associated with the new candidate i: letting $i' \in \{i-1, i\}$ be such that $\text{BWT}[i'] = c$, we replace $R[c] \leftarrow (\text{LCP}[i], n - \text{SA}[i'] + 2, true)$ (line 11).

Example 4. Continuing Example 3. When processing the T-run break $j = 7$, we see that $\text{LCP}[7] < R[\text{T}].len$. Thus, we reached the end of $box(6) = [6,6]$, where $\max \text{LCP}[B_{6,\text{T}}] = R[\text{T}].len = 5$, and insert $R[\text{T}].pos$ in $\mathcal{S}$. Next, we update $R[\text{T}] = (4, 0, false)$. On $j = 8$, we again update $R[\text{T}] = (1, 0, false)$. Finally, on $j = 9$, we get $\text{LCP}[9] > R[\text{T}].len$ since $box(6)$ and $box(9)$ are now disjoint. We update $R[\text{T}] = (\text{LCP}[9], 12, true)$ to the active state.

We obtain (see the full version [4] for a more detailed proof):

Lemma 3. *Given a text $T[1, n]$ over alphabet of size σ, Algorithm 2 computes the smallest suffixient set $\mathcal{S}$ in $O(n + \bar{r} \cdot \sigma)$ time and $O(n)$ words of space.*

Observe that in Algorithm 3 (procedure $eval(\cdot)$), only entries such that $l < R[c].len$ are possibly modified. This suggests that R could be sorted in order to speed up operations. Indeed, it turns out that R can be replaced with a data structure based on dynamic partial sums, such that operation $eval(\Sigma, m, R, \mathcal{S})$ at Line 8 of Algorithm 2 costs amortized time $O(\log \sigma)$. With this modification, Algorithm 2 runs in $O(n + \bar{r} \log \sigma)$ time. We do not enter into the details of this solution, since (i) in the next section we will present an optimal $O(n)$-time solution, and (ii) our one-pass algorithm is mainly of practical interest (see also next paragraph) and in practice the map R fits in cache (σ is small).

One-Pass Algorithm in Compressed Space. One important feature of our algorithm is that we only need one scan of the BWT, SA, and LCP to compute $\mathcal{S}$. This means that our algorithm also works in the streaming scenario where these arrays are provided one element at a time, from first to last. We now show that this feature can be exploited to run the algorithm in compressed working space.

Prefix-free parsing (PFP) is a technique introduced by Boucher et al. [2] to ease the computational burden of computing the BWT of large and repetitive texts. Briefly, with one linear-time scan, PFP divides the input T into overlapping segments, called *phrases*, of variable length, which are then used to construct what is referred to as the *dictionary* D and *parse* P of the text. Then, with a separate linear-time algorithm, the BWT of T is directly computed from D and P; thus using space proportional to the combined size of the two data structures. In [7,8], it was shown how to modify PFP in order to compute also the SA and the LCP arrays. This version streams the three arrays BWT, LCP, and SA in $O(|D|+|P|)$ compressed space, from their first to last entry; this is sufficient for running Algorithm 2 in compressed space, without affecting its running time. The only major change consists in modifying PFP to compute the dictionary and parse of T^{rev} instead of T. This can actually be achieved very easily with a left-to-right scan of T; for space reasons we do not enter into these details (see the full version [4] for more details). In Sect. 5 we present an implementation of this PFP-based algorithm.

4.3 Linear-Time Algorithm

In this section, we further speed up the one-pass algorithm provided in the previous section and achieve linear time. Our new algorithm is summarized in Algorithm 4.

As previously discussed, in Algorithm 2, for the one-character extension of every BWT equal-letter $\text{BWT}[i^*, i] = cc \ldots ccx$ (with $x \neq c$) we run procedure *eval* (Algorithm 3) to check if $\min \text{LCP}[i^* + 1, i]$ drops below the current LCP maxima associated to y-run breaks, for all $y \in \Sigma$. In the end, this step charges an additional $O(\bar{r}\sigma)$ term (which we mentioned can be reduced to $O(\bar{r} \log \sigma)$ by using opportune data structures). Intuitively, Algorithm 4 avoids this cost by calling the *eval* procedure only on the two candidates $R[c']$, where $c' \in \text{BWT}[i-1, i]$. We achieve this by introducing a new vector $\text{LF}[1, \sigma]$

updated on-the-fly, implementing the *LF-mapping* property of the Burrows-Wheeler transform. More formally: assume we have scanned the BWT up to the c-run break i, and let $i' \in \{i-1, i\}$ be such that $\text{BWT}[i'] = c$. Then, LF is such that $\text{LF}[\text{BWT}[i']] = \text{LF}[c] = j$, with $\text{SA}[j] = \text{SA}[i'] + 1$. In other words, $\text{BWT}[j]$ is the character preceding $\text{BWT}[i']$ in T^{rev}. Our linear-time algorithm is based on the following idea. Assume for simplicity that i is not the first c-run break. Let $i^* < i$ be the largest integer such that i^* is a c-run break as well (i.e. i^* is the c-run break immediately preceding i). As in Algorithm 2, our new Algorithm 4 stores in entry $R[c].len$ the value $R[c].len = \text{LCP}[i^*]$. By Lemma 2, we have to discover if $i \in B_{i^*,c}$; if this is the case, then we compare $\text{LCP}[i]$ and $\text{LCP}[i^*]$ and decide if $\text{LCP}[i]$ is the new maximum in $B_{i^*,c} \cap [i]$ (i.e. $\text{LCP}[i] > \text{LCP}[i^*]$) or if the local maximum among c-run breaks in $B_{i^*,c} \cap [i]$ had already been found (i.e. $\text{LCP}[i] \le \text{LCP}[i^*]$). If, on the other hand, $i \notin B_{i^*,c}$, then we insert in $\mathcal{S}$ the local maximum $R[c].pos$ relative to $B_{i^*,c}$ if and only if $R[c].active = true$ (if $R[c].active = false$, then the suffixient position corresponding to the local maximum of $B_{i^*,c}$ had already been stored in $\mathcal{S}$).

In order to discover if $i \in B_{i^*,c}$, we distinguish two cases.

(i) If $\text{BWT}[i-1] = c$ then, since i^* is the previous c-run break, it must be the case that $\text{BWT}[i^*-1, i^*, \ldots, i-1, i] = ycc\ldots ccx$, for some $x \ne c$ and $y \ne c$. It follows that $i \in B_{i^*,c}$ if and only if $\min \text{LCP}[i^*, i] \ge R[c].len = \text{LCP}[i^*]$. Algorithm 4 computes $\min \text{LCP}[i^*, i]$ analogously as Algorithm 2, i.e. by updating a variable m storing the minimum LCP inside intervals corresponding to BWT equal-letter runs.

(ii) If, on the other hand, $\text{BWT}[i-1] = x \ne c$, then since i^* is the previous c-run break, it must be that $\text{BWT}[i^*-1, i^*, \ldots, i-1, i] = cy\ldots xc$, for some $x \ne c$ and $y \ne c$ (and there are no other occurrences of c between $\text{BWT}[i^*-1]$ and $\text{BWT}[i]$). As in the previous case, the goal is to compute $\min \text{LCP}[i^*, i]$. We achieve this by using the LF array. Due to the way the array LF is defined and constructed, we know that $\min \text{LCP}[i^*, i] = \text{LCP}[\text{LF}[c]] - 1$. As a result, we have that $i \in B_{i^*,c}$ if and only if $\text{LCP}[\text{LF}[c]] - 1 \ge R[c].len = \text{LCP}[i^*]$.

In the full version [4] we provide all the details behind the above intuition and prove:

Lemma 4. *Given a text $T[1, n]$ over Σ, Algorithm 4 computes the smallest suffixient set $\mathcal{S}$ in $O(n)$ time and $O(n)$ words of space.*

5 Experimental Results

We implemented the one-pass and linear-time algorithms in C++ and made them publicly available at https://github.com/regindex/suffixient. The one-pass algorithm uses Prefix Free Parsing (PFP) as discussed in Sect. 4.2 and runs in compressed working space. We divide our experimental evaluation into two parts: first, we empirically study the new repetitiveness measure χ by comparing it with $\bar{r}$ using different alphabet orderings; then, we assess the performance of our PFP-based one-pass algorithm on massive genomics datasets. We do not provide comparisons with our linear-time algorithm since the necessary data structures would not fit in the internal memory for such large datasets.

The repetitiveness measure χ. In this experiment, we used nine biological datasets divided into two corpora. The first corpus contains six datasets containing DNA sequences from the Pizza&Chilli collection[1], while the second contains three datasets made by concatenating genomic sequences of different types. The three datasets of the second corpus are formed by 3,600 *SARS-CoV-2* assembled viral genomes, 220 *Salmonella enterica* assembled bacterial genomes, and 17 copies of the human *Chromosome 19*, respectively. We downloaded data from different sources: the viral genomes from the COVID-19 Data Portal[2], the bacterial genomes from NCBI[3], while the Chromosome 19 sequences belong to the

---

Algorithm 4: Linear-time algorithm

input : A string $T[1..n]$ over a finite alphabet Σ.
output: A smallest suffixient set for T.

1 $\mathcal{S} \leftarrow \emptyset$; /* initialize the empty suffixient set $\mathcal{S}$ */
2 BWT $\leftarrow$ BWT(T^{rev}); LCP $\leftarrow$ LCP(T^{rev}); SA $\leftarrow$ SA(T^{rev});
3 $R[1, \sigma] \leftarrow ((-1, 0, false), \ldots, (-1, 0, false))$; /* LCP local maxima */
4 LF$[1, \sigma] \leftarrow (0, \ldots, 0)$; /* LF mapping */
5 $m \leftarrow \infty$; /* LCP minimum inside current BWT run */
6 **for all** $c = 2, \ldots, \sigma$: LF$[c] \leftarrow$ LF$[c-1] + occ(T, c-1)$;
7 LF[BWT[1]] $\leftarrow$ LF[BWT[1]] + 1;
8 **for** $i = 2, \ldots, n$ **do**
9 $\quad$ LF[BWT[i]] $\leftarrow$ LF[BWT[i]] + 1;
10 $\quad$ $m \leftarrow \min\{m, \text{LCP}[i]\}$;
11 $\quad$ **if** BWT[i] $\neq$ BWT[$i-1$] **then**
12 $\quad\quad$ **for** $i' \in \{i-1, i\}$ **do**
13 $\quad\quad\quad$ **if** $i' = i-1$ **then** $eval(\{\text{BWT}[i']\}, m, R, \mathcal{S})$;
14 $\quad\quad\quad$ **else if** $R[\text{BWT}[i']].len \neq -1$ **then**
15 $\quad\quad\quad\quad$ $eval(\{\text{BWT}[i']\}, \text{LCP}[\text{LF}[\text{BWT}[i']]] - 1, R, \mathcal{S})$;
16 $\quad\quad$ **end**
17 $\quad\quad$ **if** LCP[i] $> R[\text{BWT}[i']].len$ **then**
18 $\quad\quad\quad$ $R[\text{BWT}[i']] \leftarrow (\text{LCP}[i], n - \text{SA}[i'] + 2, true)$;
19 $\quad\quad$ **end**
20 $\quad$ **end**
21 $\quad$ $m \leftarrow \infty$;
22 **end**
23 **end**
24 $eval(\Sigma, -1, R, \mathcal{S})$; /* evaluate last active candidates */
25 **return** $\mathcal{S}$;

---

[1] https://pizzachili.dcc.uchile.cl.
[2] https://www.covid19dataportal.org.
[3] https://www.ncbi.nlm.nih.gov/assembly/?term=Salmonella+enterica.

1,000 Genome project[4]. All datasets were filtered in order to keep only nucleotide characters A, C, G, T, in addition to the string terminator $. As a consequence, the alphabet size in our experiments was $\sigma = 5$.

We compared the behavior of our new measure χ with the well-known measure $\bar{r}$: the number of equal-letter runs in the BWT of the reversed text. Since, unlike χ, measure $\bar{r}$ depends on a specific alphabet ordering, we explicitly computed $\bar{r}$ for all $(\sigma - 1)! = 24$ possible orderings (character $ is always required to be the lexicographically-smallest one). Table 1 shows our results. Among all possible alphabet orderings, we only show the default one A < C < G < T, and the ones yielding the largest and smallest $\bar{r}$.

Table 1. Summary of the results on the nine datasets. From left to right, we report the corpus name, the dataset name, the dataset length (number of characters), the size of the smallest suffixient set (χ), and the number of runs of the BWT of the reversed text ($\bar{r}$) for three different alphabet ordering: the default lexicographic order, and the two orderings leading to the minimum and maximum $\bar{r}$.

corpus	dataset	dataset length	χ	default $\bar{r}$	min. $\bar{r}$	max. $\bar{r}$
Pizza&Chili	Cere	428,111,842	9,945,553	11,556,075	11,512,279	11,578,575
	E. Coli	112,682,447	13,126,726	15,041,926	14,974,687	15,043,686
	Influenza	154,795,431	2,234,085	3,015,173	2,979,820	3,023,774
	Para	412,279,605	13,399,378	15,581,823	15,475,176	15,635,533
	dna	403,919,018	215,201,641	243,480,086	243,179,189	243,574,132
	dna.001.1	104,857,499	1,414,711	1,717,155	1,715,622	1,718,418
biological	Chrom. 19	1,005,131,693	28,224,334	32,442,979	32,355,166	32,457,646
	Salmonella	1,043,921,849	19,071,791	22,408,284	22,255,198	22,417,175
	SarsCov2	949,959,504	1,068,951	1,351,234	1,347,182	1,352,671

We observe that, in practice, χ is very close to (and always smaller than) $\bar{r}$ under any alphabet ordering; notice that theory only predicts $\chi \leq 2\bar{r}$ (see [5] and the end of Sect. 4.1). In this experiment, the minimum of $\bar{r}$ was always between 1.13 (dna) and 1.33 (Influenza) times larger than χ. We also observe that, while the alphabet ordering does not have a big influence on $\bar{r}$, the default alphabet ordering never yields the smallest $\bar{r}$. This suggests experimentally that χ is a better repetitiveness measure than $\bar{r}$, since it is not affected by the alphabet order (while always being close to $\bar{r}$).

Smallest Suffixient Set of Massive Datasets. In this experiment, we tested our PFP-based one-pass algorithm on two large genomic datasets: the first dataset is formed by 1,000 human Chromosome 19 copies (59GB), while the second contains 2,005,773 Sars-CoV-2 viral sequences (60GB). Both input and output were streamed from/to disk, keeping in RAM only the PFP data structures. In

---

[4] https://github.com/koeppl/phoni.

both cases, our algorithm performed very well with a maximum resident set size of 20.6 GB and 29.5 GB and a wall clock time of 55:55 (min:sec) and 2:16:48 (h:min:sec), respectively.

References

1. Bentley, J.W., Gibney, D., Thankachan, S.V.: On the complexity of BWT-runs minimization via alphabet reordering. In: Proceedings of the 28th Annual European Symposium on Algorithms (ESA 2020). LIPIcs, vol. 173, pp. 15:1–15:13 (2020). https://doi.org/10.4230/LIPICS.ESA.2020.15
2. Boucher, C., Gagie, T., Kuhnle, A., Langmead, B., Manzini, G., Mun, T.: Prefix-free parsing for building big BWTs. Algor. Mol. Biol. **14**(1), 13:1–13:15 (2019). https://doi.org/10.1186/S13015-019-0148-5
3. Burrows, M., Wheeler, D.J.: A block sorting lossless data compression algorithm. Technical Report 124, Digital Equipment Corporation (1994)
4. Cenzato, D., Olivares, F., Prezza, N.: On computing the smallest suffixient set (2024). https://arxiv.org/abs/2407.18753
5. Depuydt, L., Gagie, T., Langmead, B., Manzini, G., Prezza, N.: Suffixient sets (2024). https://arxiv.org/abs/2312.01359
6. Kempa, D., Prezza, N.: At the roots of dictionary compression: string attractors. In: Proceedings of the 50th Annual ACM SIGACT Symposium on Theory of Computing (STOC 2018). STOC 2018, pp. 827–840. Association for Computing Machinery, New York (2018). https://doi.org/10.1145/3188745.3188814,
7. Kuhnle, A., Mun, T., Boucher, C., Gagie, T., Langmead, B., Manzini, G.: Efficient construction of a complete index for pan-genomics read alignment. J. Comput. Biol. **27**(4), 500–513 (2020). https://doi.org/10.1089/CMB.2019.0309
8. Rossi, M., Oliva, M., Langmead, B., Gagie, T., Boucher, C.: MONI: a pangenomic index for finding maximal exact matches. J. Comput. Biol. **29**(2), 169–187 (2022). https://doi.org/10.1089/CMB.2021.0290

Revisiting the Folklore Algorithm for Random Access to Grammar-Compressed Strings

Alan M. Cleary[1](✉)[iD], Joseph Winjum[2][iD], Jordan Dood[3][iD], and Shunsuke Inenaga[4][iD]

[1] National Center for Genome Resources, Santa Fe, NM, USA
acleary@ncgr.org
[2] Montana State University, Bozeman, MT, USA
joseph.winjum@ecat1.montana.edu
[3] Hyalite Technologies LLC, Bozeman, MT, USA
hyalitetechnologies@gmail.com
[4] Department of Informatics, Kyushu University, Fukuoka, Japan
inenaga.shunsuke.380@m.kyushu-u.ac.jp

Abstract. Grammar-based compression is a widely-accepted model of string compression that allows for efficient and direct manipulations on the compressed data. Most, if not all, such manipulations rely on the primitive *random access* queries, a task of quickly returning the character at a specified position of the original uncompressed string without explicit decompression. While there are advanced data structures for random access to grammar-compressed strings that guarantee theoretical query time and space bounds, little has been done for the *practical* perspective of this important problem. In this paper, we revisit a well-known folklore random access algorithm for grammars in the Chomsky normal form, modify it to work directly on general grammars, and show that this modified version is fast and memory efficient in practice.

Keywords: grammar-based compression · random access · straight-line programs

1 Introduction

Random access on grammar-compressed strings has been used as a key primitive in a number of efficient algorithms that directly work on compressed strings, including pattern matching [3,21,23,27,35], compressed-string indexing [10], q-gram frequencies [16,17], detection of palindromes and repetitions [19,20,26], convolutions [32], finger searches [2], and Lempel-Ziv factorizations in compressed space [11].

Given a grammar $\mathcal{G}$ in the Chomsky normal form for a text T, a folklore algorithm for this problem first computes and stores the length of the string that each non-terminal derives in $O(\text{size}(\mathcal{G}))$-time and space, where $\text{size}(\mathcal{G})$ denotes the total size of the productions in $\mathcal{G}$. Then, given a position p in T, one can climb down the corresponding path to $T[p]$ in the derivation tree for $\mathcal{G}$ in $O(\mathsf{h}(\mathcal{G}))$-time, where $\mathsf{h}(\mathcal{G})$ denotes the height of the derivation tree of $\mathcal{G}$. While $\mathsf{h}(\mathcal{G})$ can be as small as $\Theta(\log n)$ for some highly repetitive strings of length n with balanced grammars $\mathcal{G}$, $\mathsf{h}(\mathcal{G})$ can be as large as $\Theta(n)$ in the worst case. Bille et al. [4] showed how to preprocess $\mathcal{G}$ in $O(\mathsf{size}(\mathcal{G}))$-time and space so that later random access queries can be answered in $O(\log n)$-time, irrespective of $\mathsf{h}(\mathcal{G})$. Garnardi et al. [14] showed how to convert a given grammar $\mathcal{G}$ into another grammar $\mathcal{G}'$, with $\mathsf{size}(\mathcal{G}') = O(\mathsf{size}(\mathcal{G}))$ and $\mathsf{h}(\mathcal{G}') = O(\log n)$, that derives the same string as the original grammar $\mathcal{G}$, thus achieving $O(\log n)$-time random access using $O(\mathsf{size}(\mathcal{G}))$-space. Garnardi et al. [14] also presented an $O(\log n/\log \log n)$-time random access data structure with $O(\mathsf{size}(\mathcal{G}) \log^\epsilon n)$-space for any $\epsilon > 0$, by generalizing the result of Belazzougui et al. [1]. This matches the cell-probe lower bound shown by Verbin and Yu [33], for strings that are only polynomially compressible in n.

While random access on grammars has been extensively studied in the theoretical perspective, as shown above, the only practical results that we are aware of are the works by Maruyama et al. [25] and Gagie et al. [12]. However, these approaches require specific grammar encodings and only work on RePair style grammars, making them incompatible with recent grammar-based compression algorithms [7,9,29]. In fact, we are not aware of a general random access algorithm that will work on any grammar.

In this work, we revisit the folklore algorithm for random access to grammar compressed strings and show that it can be improved to use significantly less space and generalized to operate directly on any grammar. Our experiments show that this modified folklore algorithm achieves state-of-the-art performance in both its space requirements and run-time.

2 Preliminaries

In this section, we define syntax and review information related this paper. Indexes start at 1.

2.1 Strings

Let Σ be an alphabet of size σ. An element in Σ^* is called a *string*. The length of a string T is denoted by $|T|$ and $n = |T|$. The empty string ε is the string of length 0, i.e. $|\varepsilon| = 0$. The pth character in a string T is denoted by $T[p]$ for $1 \leq p \leq n$, and the substring of T that begins at position p and ends at position q is denoted by $T[p..q]$ for $1 \leq p \leq q \leq n$. For convenience, let $T[p..q] = \varepsilon$ for $p > q$.

2.2 Grammar-Based Compression

A *context-free grammar* is a set of recursive rules that describe how to form strings from a language's alphabet. A context-free grammar is called an *admissible grammar* if the language it generates consists only of a single string.

Grammar-based compression is a compression technique that computes an admissible grammar for a given string such that the computed grammar can be stored in less space than the original string. In what follows, we will call admissible grammars simply *grammars*.

Let $\mathcal{G} = \langle X, \Sigma, R, S \rangle$ be a grammar that generates T, where X is a set of non-terminal characters, Σ is the alphabet of T (i.e. terminal characters) and is disjoint from X, R is a finite relation in $X \times (X \cup \Sigma)^*$, and S is the symbol in X that should be used as the *start rule* when using $\mathcal{G}$ to generate T. R defines the *rules* of $\mathcal{G}$ as a set of m productions $\{X_i \to \text{expr}_i \mid 1 \leq i \leq m\}$ such that each X_i is a non-terminal in X and expr_i is a non-empty sequence from $(\Sigma \cup \{X_1, \ldots, X_{i-1}\})^+$. The *size* of grammar $\mathcal{G}$ is the total length of the right-hand sides of the productions and is denoted $\text{size}(\mathcal{G}) = \sum_{i=1}^{m} |\text{expr}_i|$. We say that a non-terminal X_i *represents* a string w if w is the (unique) string that X_i derives. We only consider grammars with no useless rules and symbols, unless stated otherwise. $\mathsf{h}(\mathcal{G})$ denotes the height of the derivation tree of grammar $\mathcal{G}$.

In this work we will discuss three types of grammars: *straight-line programs (SLPs)*, *Chomsky normal form (CNF) grammars*, and *RePair grammars*. An SLP $\mathcal{G}_{\text{SLP}}$ is simply an admissible grammar. A CNF grammar $\mathcal{G}_{\text{CNF}}$ is an SLP in the Chomsky normal form, i.e. every rule (including start rule S) is of the form $X_i \to c$ ($c \in \Sigma$) or $X_i \to X_\ell X_r$ ($\ell, r < i$)[1]. A RePair grammar $\mathcal{G}_{\text{RePair}}$, proposed in [22], is a CNF grammar in which the start rule can have arbitrarily many symbols in its righthand side, i.e. $S \to \text{expr}$ with $\text{expr} \in ((X \setminus S) \cup \Sigma)^+$. Any SLP can be converted to an equivalent CNF grammar [24] and any CNF grammar can be converted to an equivalent balanced CNF grammar with height $O(\log n)$ [14], where n is the length of the uncompressed string. See Fig. 1 (a) and (b) for example SLP and RePair grammars.

Let $\mathcal{G}$ be a grammar that represents a string T of length n. A *random access query* on grammar $\mathcal{G}$ is, given a query position p ($1 \leq p \leq n$), return the pth character $T[p]$ in the uncompressed string T. The *random access problem on grammar-compression* is to preprocess a given grammar $\mathcal{G}$ and build a space-efficient (i.e. compressed) data structure on $\mathcal{G}$ that can quickly return the desired character $T[p]$ for query positions p. In practice, random access queries can be for entire substrings $T[p..q]$, where $1 \leq p \leq q \leq n$. Our experiments include results for substring queries but our algorithm descriptions only consider the single character version of the problem. This is because in our approach only the first character of the substring needs to be located in $\mathcal{G}$'s derivation tree; the rest of the substring can be generated by simply traversing the derivation tree from that location.

---

[1] While the term SLP is often used for grammar-compression in the Chomsky normal form, in this paper, for clarity, we use SLP to denote general admissible grammars and CNF to denote grammars in the Chomsky normal form.

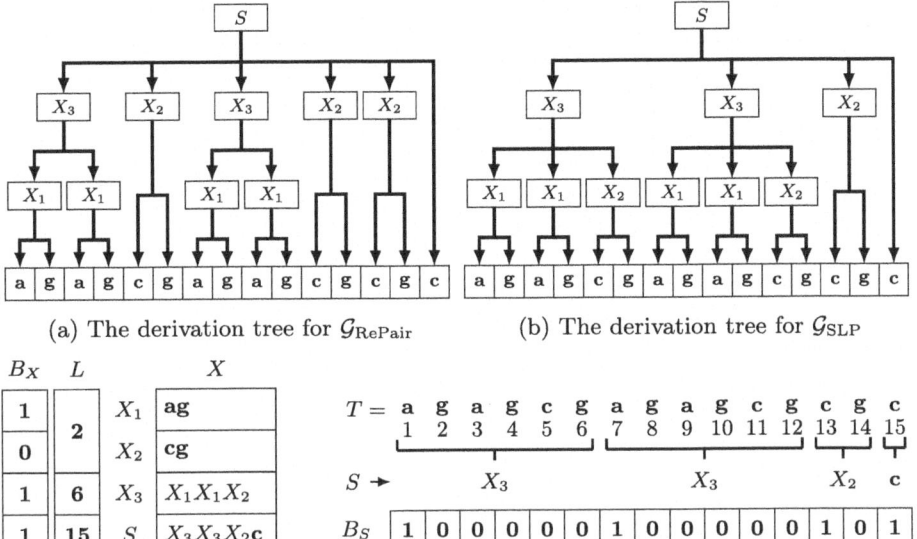

Fig. 1. Example grammars and data structures for the modified folklore algorithm on string $T = \text{agagcgagagcgcgc}$. (a) and (b) depict the derivation trees for a RePair grammar $\mathcal{G}_\text{RePair}$ and an SLP grammar $\mathcal{G}_\text{SLP}$, respectively. (c) depicts the data structures used by our modified version of the folklore random access algorithm. Note that the subscript i for each non-terminal X_i conveys the array index of the non-terminal in the array of arrays representation of the grammars. For simplicity, terminal characters are used directly without a proxy non-terminal character.

3 Algorithms

In this section, we present novel algorithms for random access to grammar-compressed strings. As before, indexes start at 1.

3.1 The Folklore Algorithm

The folklore algorithm for random access to grammar-compressed strings is for CNF grammars. Given a grammar $\mathcal{G}_\text{CNF}$, the algorithm works by first computing an array A that stores the length of the string that each non-terminal represents. Then, given a position p in T, the corresponding path to $T[p]$ in the derivation tree is followed by looking up the string length of each non-terminal's left and right characters in A to determine which character contains the position, requiring $O(\log n)$-time when the grammar is balanced and $O(\mathsf{h}(\mathcal{G}_\text{CNF}))$-time otherwise. See Algorithm 1 for details.

While this algorithm is fast both asymptotically and in practice, it requires much additional space. For instance, if the algorithm is given a non-CNF grammar $\mathcal{G}$, then an equivalent CNF grammar $\mathcal{G}_\text{CNF}$ must be computed, which may

---
Algorithm 1: Folklore Random Access

Data: Grammar $\mathcal{G}_{\text{CNF}}$, i.e. an array of integer arrays
Data: Start rule S, i.e. an index in $\mathcal{G}_{\text{CNF}}$
Data: Length lookup table A
Input: Position p in T
Output: Character $T[p]$

1    $X_i \leftarrow S$
2   **while** X_i *is a non-terminal* $(X_i \rightarrow X_\ell X_r)$ **do**
3        $leftLength \leftarrow A[X_i]$
4   **if** $p \leq leftLength$ **then**
5           $X_i \leftarrow \mathcal{G}_{\text{CNF}}[X_i]_\ell$ `//`X_ℓ `denotes the first character in `X_i`'s array`
6   **else**
7           $p \leftarrow p - leftLength$
8           $X_i \leftarrow \mathcal{G}_{\text{CNF}}[X_i]_r$ `//`X_r `denotes the second character in `X_i`'s array`
9   **end**
10 **end**
11 **return** X_i

---

require introducing useless rules that increase the size of the grammar. Additionally, using the naïve array of arrays grammar encoding [31] where each non-terminal character is an array index, the folklore algorithm requires:

(a) $2m \lg(m + \sigma)$ bits to represent the grammar and
(b) $m \lg(n)$ bits to represent the rule string lengths,

where m denotes the number of non-terminals in $\mathcal{G}_{\text{CNF}}$, in which there are σ rules of form $X_i \rightarrow c$ $(c \in \Sigma)$. Note that the bits to represent the rule string lengths require half as much space as the grammar itself!

3.2 Modified Folklore Algorithm

In our modified folklore algorithm to follow, the non-terminals are required to be sorted (and subsequently numbered) in increasing order of their expansion lengths. Given a grammar $\mathcal{G}$ with m non-terminals, we can simply sort them in $O(m \log m)$-time and renumber them in $O(\text{size}(\mathcal{G}))$-time with $O(m \log(n))$ bits of working space by any suitable comparison-based sorting algorithm. The $O(m \log(n))$ bits of information is discarded after this preprocessing. If the non-terminals in $\mathcal{G}$ are already sorted, this step can be skipped.

To improve the folklore algorithm, we first observe that it can be easily extended to work on any SLP. Specifically, given an SLP $\mathcal{G}_{\text{SLP}}$, the length of the string each non-terminal represents is computed and stored, as before. Then the position in T of each non-terminal character in start rule S is computed and stored as well. Now, given a position p in T, the algorithm will first look up the character in the start rule that contains p and then descend the derivation tree from this character. However, since each rule in an SLP can have arbitrarily many

Algorithm 2: Modified Folklore Random Access

Data: Ordered grammar $\mathcal{G}_{\text{SLP}}$, i.e. an array of integer arrays
Data: Start rule S, i.e. an index in $\mathcal{G}_{\text{SLP}}$
Data: Bitvector B_X
Data: Bitvector B_S
Data: Unique length array L
Input: Position p in T
Output: Character $T[p]$

1   $r \leftarrow \text{rank}(B_S, p)$
2   $X_i \leftarrow \mathcal{G}_{\text{SLP}}[S][r]$
3   $p \leftarrow p - \text{select}(B_S, r)$
4   **while** X_i *is a non-terminal* $(X_i \to \text{expr}_i \text{ with } \text{expr}_i \in (\Sigma \cup \{X_1, \ldots, X_{i-1}\})^+)$ **do**
5   **for** X_j *in* $\mathcal{G}_{\text{SLP}}[X_i]$ **do**
6          $r \leftarrow \text{rank}(B_X, X_j)$
7          $length \leftarrow L[r]$
8   **if** $p \leq length$ **then**
9              $X_i \leftarrow X_j$
10   break
11   **else**
12              $p \leftarrow p - length$
13   **end**
14   **end**
15 **end**
16 **return** X_i

symbols in its righthand side, instead of checking the left and right characters to determine which character contains the query position, a rule's characters are iterated until the character containing the position is found.

Using the naïve array of arrays grammar encoding, this extended version of the folklore algorithm requires:

(c) $\text{size}(\mathcal{G}_{\text{SLP}}) \lg(m + \sigma)$ bits to represent the grammar and
(d) $(|S| + m) \lg(n)$ bits to represent the rule string lengths and start character positions.

Note that when $\mathcal{G}_{\text{SLP}}$ is a CNF grammar, (c) is equivalent to (a).

We observe that the following changes can be made to significantly reduce the space requirements of (d):

1. Using the naïve grammar representation of (c) where each non-terminal character is an array index, order (and subsequently renumber) the rules by string length – shortest to longest – with the start rule S last.
2. Create a sorted array L of the unique rule string lengths.
3. Create a length m bitvector B_X with a bit i set for the first rule X_i of each string length in the grammar array.

4. Create a length n bitvector B_S with a bit p set at the text position of each character in start rule S.

The modified folklore algorithm can then be used by first looking up the start rule character via a paired *rank-select* query on B_S. The algorithm can then descend the derivation tree by using *rank* queries on B_X to look up each rule's string length in L. See Algorithm 2 for details.

Using a standard bitvector the *rank* and *select* operations can be done in $O(1)$-time [6,34]. However, a standard bitvector requires $64\lceil |B|/64 + 1 \rceil$ bits of space, where $|B|$ is the length of the bitvector. This is impractical since $|B_S| = n$.

To minimize the space required by bitvectors B_S and B_X, we propose using the sparse bitvector [30]. This bitvector answers *rank* queries in $O(\log \frac{|B|}{b})$-time and select queries in $O(1)$-time while requiring no more than $b(2+\log \frac{|B|}{b})$ bits of space, where b is the number of set bits. Using the sparse bitvector, the modified folklore algorithm requires no more than $|S|(2+\log \frac{n}{|S|})+|L|(2+\log \frac{m}{|L|})+|L|\lg|L|$ bits to represent the rule string lengths and start character positions.

We observe that the modified folklore algorithm is effectively equivalent to the original folklore algorithm when given a CNF grammar. Since any SLP can be converted to an equivalent balanced CNF grammar [14,24], this implies that when using standard bitvectors the modified folklore algorithm runs in $O(\log n)$-time, and when using sparse bitvectors it runs in $O(\log \frac{n}{|S|} + \log n \log \frac{m}{|L|})$-time. Without converting to the Chomsky normal form or balancing the grammar, the worst-case run-time is $O(n)$. However, algorithms that produce RePair grammars tend to create balanced grammars, and so the expected run-time on RePair grammars is $O(\log n)$. Although the nature of SLPs in general is less predictable, the grammars of the MR-RePair algorithm we use in our experiments have been shown to be isomorphic to grammars produced by the actual RePair algorithm [9], so we expect these grammars to have an $O(\log n)$ run-time as well, despite having to iterate rules' characters when descending the derivation tree.

4 Results

In this section, we describe our implementation and experimental results.

4.1 Implementation

Our implementation of the modified folklore algorithm is called *FRAS* - Folklore Random Access for SLPs. We implemented FRAS in C++. The bitvectors and their respective *rank* and *select* data structures were implemented using the Succinct Data Structure Library (SDSL) [15]. We used the naïve array of arrays encoding to represent grammars [31] and implemented the sparse bitvector variation of the folklore algorithm described in Sect. 3. The source code is available at https://github.com/alancleary/FRAS.

4.2 Experiments

We performed experiments on two corpora of data: the *Pizza&Chili* corpus[2] and a collection of pangenomes. For each corpus, we generated grammars and benchmarked our modified folklore algorithm (FRAS) against the original folklore algorithm (Folklore), the algorithm of [25] (FOLCA), and the algorithm of [12] (ShapedSLP[3]), measuring encoding size and random access run-time. For each grammar generated, we accessed substrings of length 1, 10, 100, and 1,000 at pseudo-random positions.[4] We performed this procedure 10,000 times for each substring length and computed the average run-time. See the following subsections for details.

Experiments were run on a server with two AMD EPYC 7543 32-Core 2.8GHz (3.7GHz max boost) processors and 2TB of 8-channel DDR4 3200MHz memory running CentOS Stream release 9. Note that this server is excessively overpowered for these experiments and was used for the purpose of stability and accuracy of measurement. Similar results can be achieved on a consumer laptop.

Pizza and Chili. For the Pizza&Chili corpus, we generated grammars for all data sets from the *real* and *artificial* collections. For each data set, we generated grammars using Gonzalo Navarro's implementation[5] of RePair [22] and Isamu Furuya's implementation[6] of MR-RePair [9]. MR-RePair generates SLPs that are isomorphic to RePair grammars. We included these grammars to test our hypothesis that the run-times of our FRAS algorithm should be approximately equivalent on an SLP as on an equivalent RePair grammar. The Folklore, FOLCA, and ShapedSLP algorithms were not benchmarked on MR-RePair SLPs as they only work on RePair grammars. See Table 5 in Appendix A for information about the Pizza&Chili data sets and their respective grammars. The space used by each algorithm is listed in Table 1 and the run-times of each algorithm are listed in Table 2.

We found that FRAS consistently used less space than Folklore, especially on the MR-RePair grammars. However, FRAS was also consistently slower than Folklore. This is expected as both FOLCA and ShapedSLP also use less space than Folklore but are slower. FRAS, however, consistently used more space than FOLCA and ShapedSLP but was also much faster. Moreover, FRAS was faster than FOLCA and ShapedSLP on every data set and query size, and it was the only algorithm of the three to achieve sub-microsecond run-times.

Interestingly, FRAS on MR-RePair grammars did not use much more space than FOLCA and ShapedSLP but it was the fastest of all the algorithms, exclud-

---

[2] https://pizzachili.dcc.uchile.cl/.
[3] Note that in [12] "SLP" refers to RePair grammars, thus ShapedSLP is only compatible with RePair grammars.
[4] Pseudo-random numbers were generated on a uniform distribution using the xoroshiro128+ generator [5]. The generator was seeded so that the same numbers were used by every algorithm.
[5] https://users.dcc.uchile.cl/~gnavarro/software/repair.tgz.
[6] https://github.com/izflare/MR-RePair.

Table 1. The space used by the random access algorithms benchmarked on the *Pizze&Chili* corpus. **Data Set** is the names of the data sets and what collection they belong to. The **MR-FRAS, FRAS, Folklore, FOLCA,** and **ShapedSLP** columns are the space used by each algorithm, where **MR-FRAS** is **FRAS** run on MR-RePair grammars; all other results are on RePair grammars. All space is in megabytes.

Data Set		MR-FRAS	FRAS	Folklore	FOLCA	ShapedSLP
real	Escherichia_Coli	17.2	22.9	41.1	13.8	14.0
	cere	16.3	23.7	37.4	13.1	14.0
	coreutils	9.8	16.6	22.7	7.9	9.0
	einstein.de.txt	0.4	0.5	0.7	0.3	0.3
	einstein.en.txt	0.9	1.2	1.9	0.7	0.6
	influenza	8.0	8.8	17.6	6.1	5.5
	kernel	5.6	9.2	13.0	4.5	5.0
	para	21.3	29.5	49.3	17.4	18.2
	world_leaders	1.7	2.1	3.4	1.1	1.1
artificial	fib41	< 0.1	< 0.1	< 0.1	< 0.1	< 0.1
	rs.13	< 0.1	< 0.1	< 0.1	< 0.1	< 0.1
	tm29	< 0.1	< 0.1	< 0.1	< 0.1	< 0.1

Table 2. Random access run-times for grammars built on the *Pizze&Chili* corpus. **Data Set** is the names of the data sets and what collection they belong to. The **MR-FRAS, FRAS, Folklore, FOLCA,** and **ShapedSLP** columns are the algorithms benchmarked and their query run-times, where **MR-FRAS** is **FRAS** run on MR-RePair grammars; all other results are on RePair grammars. For the run-times, **1, 10, 100,** and **1,000** are the lengths of the substrings queried. All run-times are in microseconds.

Data Set		MR-FRAS				FRAS				Folklore				FOLCA				ShapedSLP			
		1	10	100	1,000	1	10	100	1,000	1	10	100	1,000	1	10	100	1,000	1	10	100	1,000
real	Escherichia_Coli	1.8	1.9	3.4	17.7	5.8	6.0	8.4	29.2	1.1	1.2	2.2	12.0	25.0	26.3	42.0	191.2	21.0	22.7	42.5	238.7
	cere	1.3	1.5	3.0	16.2	4.4	4.7	7.2	28.5	1.0	1.1	2.1	10.9	17.7	19.4	34.7	182.3	14.2	16.3	36.5	235.8
	coreutils	17.5	17.8	19.5	34.5	42.2	42.7	46.3	69.1	6.4	7.2	9.5	20.9	156.0	158.3	180.7	355.4	242.9	247.2	276.7	491.9
	einstein.de.txt	0.8	1.0	2.4	15.2	1.4	1.6	3.4	19.0	0.5	0.7	1.6	10.0	10.6	12.3	28.1	176.7	6.6	8.6	27.8	217.6
	einstein.en.txt	1.0	1.2	2.8	16.5	1.3	1.5	3.4	19.2	0.7	0.8	1.8	10.2	11.8	13.6	30.2	184.2	6.2	8.3	27.7	217.9
	influenza	0.3	0.5	2.2	17.1	0.5	0.6	2.3	17.0	0.6	0.7	1.6	10.4	10.6	12.3	27.3	172.1	3.0	5.0	24.1	213.6
	kernel	5.1	5.3	7.0	21.6	15.2	15.9	18.1	36.0	1.9	2.2	3.4	13.3	51.7	54.6	72.3	241.3	60.5	64.3	85.7	289.7
	para	0.9	1.1	2.6	17.4	1.8	2.2	4.8	27.4	0.8	1.0	2.1	11.9	13.2	14.9	30.7	182.3	5.8	7.7	27.0	218.8
	world_leaders	0.5	0.7	1.9	13.0	1.1	1.3	2.7	14.5	0.6	0.7	1.3	7.6	11.8	13.3	28.6	166.7	5.7	7.7	26.7	214.0
artificial	fib41	0.9	1.0	1.2	7.4	1.0	1.0	1.2	7.4	0.2	0.3	0.6	3.6	2.6	3.3	10.4	81.3	4.2	5.7	20.8	172.1
	rs.13	0.9	1.0	1.3	7.9	0.9	1.0	1.2	7.4	0.3	0.3	0.6	3.8	3.1	3.8	11.2	85.8	4.7	6.2	22.0	179.1
	tm29	0.9	0.9	1.2	7.0	0.8	0.9	1.3	7.6	0.2	0.3	0.6	3.7	2.8	3.6	10.6	80.9	4.7	6.3	22.5	183.1

ing Folklore. This suggests that although the size of the rules in MR-RePair grammars is unbounded FRAS is acheiving the expected run-time and is faster than the FRAS on RePair grammars simply because MR-RePair grammars are typically smaller.

Table 3. The space used by the random access algorithms benchmarked on the *Pangenome* corpus. **Data Set** is the names of the data sets and the **FRAS**, **Folklore**, **FOLCA**, and **ShapedSLP** columns are the space used by each algorithm. All results are on BigRePair grammars. All space is in megabytes.

Data Set	FRAS	Folklore	FOLCA	ShapedSLP
YPRP	49.9	75.4	25.9	31.4
Maize	3779.6	6400.4	2529.8	3020.2
c1000	122.3	217.4	86.5	80.6

Pangenomes. A *pangenome* is a collection of genomes from the same species. The ability to efficiently store and access these data at scale is an important problem in bioinformatics [8]. Grammar-based compression is particularly well suited to compressing these data as the size of the grammars scales relative to information content, rather than input size [28]. To demonstrate the practicality of our algorithm, we generated grammars for three pangenomes: the 12 yeast assemblies from the Yeast Population Reference Panel (YPRP) [36]; the 25 Maize assembles from the nested association mapping (NAM) population [18]; and 1000 copies of Human chromosome 19 (c1000) used by Gagie et al. in [12]. Grammars were generated using BigRePair as it is currently the only grammar-based compression algorithm that can generate grammars for the NAM and c1000 data sets [13]. See Table 6 in Appendix A for information about these data sets and their respective grammars. The space used by each algorithm is listed in Table 3 and the run-times of each algorithm are listed in Table 4.

As with the Pizza&Chili corpus, we found that FRAS consistently used less space than Folklore and was consistently slower. And, again, FRAS consistently used more space than FOLCA and ShapedSLP but was also much faster, beating FOLCA and ShapedSLP on every data set and query size.

Table 4. Random access run-times for grammars built on the *Pangenome* corpus. **Data Set** is the names of the data sets and the **FRAS**, **Folklore**, **FOLCA**, and **ShapedSLP** columns are the algorithms benchmarked and their query run-times. For the run-times, **1**, **10**, **100**, and **1,000** are the lengths of the substrings queried. All results are on BigRePair grammars. All run-times are in microseconds.

Data Set	FRAS				Folklore				FOLCA				ShapedSLP			
	1	10	100	1,000	1	10	100	1,000	1	10	100	1,000	1	10	100	1,000
YPRP	1.8	2.2	5.7	37.4	0.9	1.1	2.5	14.7	11.5	13.3	29.6	187.1	2.9	4.9	24.9	220.1
Maize	3.9	4.4	8.5	44.5	3.1	3.6	5.7	24.0	22.3	24.3	44.3	233.6	9.0	11.2	35.1	269.5
c1000	3.2	3.6	7.4	41.0	2.5	2.7	4.8	21.2	16.4	18.3	34.8	192.2	6.9	9.2	31.1	242.2

5 Conclusion

In this work, we showed how the folklore algorithm for random access to grammar-compressed strings can be modified to work on any grammar and achieve good space and run-time performance in practice. We believe this is the first random access algorithm for grammar-compressed strings that works directly on any grammar, thus further enhancing the usefulness of the algorithm. In future work we would like to further improve the modified folklore algorithm by representing the grammar in a manner that uses less space with minimal effect on run-time performance.

Acknowledgments. The work of Alan M. Cleary, Joseph Winjum, and Jordan Dood was support by NSF award number 2105391. The work of Shunsuke Inenaga was supported by JSPS KAKENHI grant numbers JP 20H05964, JP23K24808, and JP23K18466. We would like to thank the authors of [12] for sharing the c1000 data set with us. You saved us much time and computation.

A Appendix

Table 5. Data sets used from the *Pizze&Chili* corpus. **Data Set** is the names of the data sets and what collection they belong to, **Size** is the number of characters in each data set, and **MR-RePair** and **RePair** are information about the grammars generated for these data sets. For the grammars, **Rules** is the number of rules in the grammars, **Depth** is the maximum depth of the grammars, **Start** is the size of the start rules, and **Size** is the total lengths of the right-hand sides of the rules in each grammar, excluding the start rule.

Data Set		Size	MR-RePair				RePair			
			Rules	Depth	Start	Size	Rules	Depth	Start	Size
real	Escherichia_Coli	112,689,515	712,228	23	712,484	1,595,881	2,012,087	3,279	1,601,482	4,024,174
	cere	461,286,644	836,956	29	648,659	3,392,707	2,561,292	1,359	655,298	5,122,584
	coreutils	205,281,778	437,054	30	153,775	2,270,187	1,821,734	43,728	153,346	3,643,468
	einstein.de.txt	92,758,441	21,787	42	12,683	71,709	49,949	269	12,665	99,898
	einstein.en.txt	467,626,544	49,565	48	62,591	150,233	100,611	1,355	62,473	201,222
	influenza	154,808,555	429,027	28	897,657	1,088,872	643,587	366	886,836	1,287,174
	kernel	257,961,616	246,596	34	69,537	1,304,343	1,057,914	5,822	69,427	2,115,828
	para	429,265,758	1,079,287	30	1,134,361	4,157,167	3,093,873	487	1,147,650	6,187,746
	world_leaders	46,968,181	100,293	30	98,397	309,222	206,508	463	94,327	413,016
artificial	fib41	267,914,296	38	40	3	76	38	40	3	76
	rs.13	216,747,218	55	45	121	26	66	47	24	132
	tm29	268,435,456	51	29	6	126	75	45	6	150

Table 6. Data sets used from the *Pangenome* corpus. **Data Set** is the names of the data sets, **Size** is the number of characters in each data set, and **BigRePair** is information about the grammars generated for these data sets. For the grammars, **Rules** is the number of rules in the grammars, **Depth** is the maximum depth of the grammars, **Start** is the size of the start rules, and **Size** is the total lengths of the right-hand sides of the rules in each grammar, excluding the start rule.

Data Set	Size	BigRePair			
		Rules	Depth	Start	Size
YPRP	143,169,450	5,911,887	37	642,828	11,823,774
Maize	55,270,577,570	432,651,138	47	79,378,700	865,302,276
c1000	59,125,115,010	12,898,128	45	4,495,360	21,300,896

References

1. Belazzougui, D., Cording, P.H., Puglisi, S.J., Tabei, Y.: Access, rank, and select in grammar-compressed strings. In: Bansal, N., Finocchi, I. (eds.) ESA 2015. LNCS, vol. 9294, pp. 142–154. Springer, Heidelberg (2015). https://doi.org/10.1007/978-3-662-48350-3_13
2. Bille, P., Cording, P.H., Gørtz, I.L.: Compressed subsequence matching and packed tree coloring. Algorithmica **77**(2), 336–348 (2017)
3. Bille, P., Gørtz, I.L., Cording, P.H., Sach, B., Vildhøj, H.W., Vind, S.: Fingerprints in compressed strings. J. Comput. Syst. Sci. **86**, 171–180 (2017)
4. Bille, P., Landau, G.M., Raman, R., Sadakane, K., Satti, S.R., Weimann, O.: Random access to grammar-compressed strings and trees. SIAM J. Comput. **44**(3), 513–539 (2015)
5. Blackman, D., Vigna, S.: Scrambled linear pseudorandom number generators. ACM Trans. Math. Softw. **47**(4) (2021)
6. Clark, D.: Compact pat trees (1997)
7. Cleary, A., Dood, J.: Constructing the CDAWG CFG using LCP-intervals. In: 2023 Data Compression Conference (DCC), pp. 178–187 (2023)
8. The Computational Pan-Genomics Consortium: Computational pan-genomics: status, promises and challenges. Briefings Bioinform. **19**(1), 118–135 (2016)
9. Furuya, I., Takagi, T., Nakashima, Y., Inenaga, S., Bannai, H., Kida, T.: Practical grammar compression based on maximal repeats. Algorithms **13**(4), 103 (2020)
10. Gagie, T., Gawrychowski, P., Kärkkäinen, J., Nekrich, Y., Puglisi, S.J.: A faster grammar-based self-index. In: Dediu, A.-H., Martín-Vide, C. (eds.) LATA 2012. LNCS, vol. 7183, pp. 240–251. Springer, Heidelberg (2012). https://doi.org/10.1007/978-3-642-28332-1_21
11. Gagie, T., Goga, A., Jez, A., Navarro, G.: Space-efficient conversions from slps. In: LATIN 2024. LNCS, vol. 14578, pp. 146–161 (2024). https://doi.org/10.1007/978-3-031-55598-5_10
12. Gagie, T., et al.: Practical random access to SLP-compressed texts. In: Boucher, C., Thankachan, S.V. (eds.) SPIRE 2020. LNCS, vol. 12303, pp. 221–231. Springer, Cham (2020). https://doi.org/10.1007/978-3-030-59212-7_16
13. Gagie, T., I, T., Manzini, G., Navarro, G., Sakamoto, H., Takabatake, Y.: Rpair: rescaling RePair with Rsync. In: Brisaboa, N.R., Puglisi, S.J. (eds.) SPIRE 2019.

LNCS, vol. 11811, pp. 35–44. Springer, Cham (2019). https://doi.org/10.1007/978-3-030-32686-9_3
14. Ganardi, M., Jez, A., Lohrey, M.: Balancing straight-line programs. J. ACM **68**(4), 27:1–27:40 (2021)
15. Gog, S., Beller, T., Moffat, A., Petri, M.: From theory to practice: Plug and play with succinct data structures. In: Gudmundsson, J., Katajainen, J. (eds.) Experimental Algorithms, pp. 326–337. Springer International Publishing, Cham (2014). https://doi.org/10.1007/978-3-319-07959-2_28
16. Goto, K., Bannai, H., Inenaga, S., Takeda, M.: Computing q-gram non-overlapping frequencies on SLP compressed texts. In: Bieliková, M., Friedrich, G., Gottlob, G., Katzenbeisser, S., Turán, G. (eds.) SOFSEM 2012. LNCS, vol. 7147, pp. 301–312. Springer, Heidelberg (2012). https://doi.org/10.1007/978-3-642-27660-6_25
17. Goto, K., Bannai, H., Inenaga, S., Takeda, M.: Fast q-gram mining on SLP compressed strings. J. Discrete Algorithms **18**, 89–99 (2013)
18. Hufford, M.B., et al.: De novo assembly, annotation, and comparative analysis of 26 diverse maize genomes. Science **373**(6555), 655–662 (2021)
19. Tomohiro, I., et al.: Detecting regularities on grammar-compressed strings. Inf. Comput. **240**, 74–89 (2015)
20. Tomohiro, I., Nishimoto, T., Inenaga, S., Bannai, H., Takeda, M.: Compressed automata for dictionary matching. Theor. Comput. Sci. **578**, 30–41 (2015)
21. Karpinski, M., Rytter, W., Shinohara, A.: An efficient pattern-matching algorithm for strings with short descriptions. Nord. J. Comput. **4**(2), 172–186 (1997)
22. Larsson, N., Moffat, A.: Off-line dictionary-based compression. Proc. IEEE **88**(11), 1722–1732 (2000)
23. Lifshits, Y.: Processing compressed texts: a tractability border. In: Ma, B., Zhang, K. (eds.) CPM 2007. LNCS, vol. 4580, pp. 228–240. Springer, Heidelberg (2007). https://doi.org/10.1007/978-3-540-73437-6_24
24. Lohrey, M.: Algorithmics on slp-compressed strings: a survey. Groups - Complexity - Cryptol. **4**(2), 241–299 (2012)
25. Maruyama, S., Tabei, Y., Sakamoto, H., Sadakane, K.: Fully-online grammar compression. In: Kurland, O., Lewenstein, M., Porat, E. (eds.) SPIRE 2013. LNCS, vol. 8214, pp. 218–229. Springer, Cham (2013). https://doi.org/10.1007/978-3-319-02432-5_25
26. Matsubara, W., Inenaga, S., Ishino, A., Shinohara, A., Nakamura, T., Hashimoto, K.: Efficient algorithms to compute compressed longest common substrings and compressed palindromes. Theor. Comput. Sci. **410**(8–10), 900–913 (2009)
27. Miyazaki, M., Shinohara, A., Takeda, M.: An improved pattern matching algorithm for strings in terms of straight line programs. J. Dis. Algorithms **1**(1), 187–204 (2000)
28. Navarro, G.: Indexing highly repetitive string collections, part ii: compressed indexes. ACM Comput. Surv. **54**(2) (2021)
29. Nunes, D.S.N., Louza, F., Gog, S., Ayala-RincÃn, M., Navarro, G.: A grammar compression algorithm based on induced suffix sorting. In: 2018 Data Compression Conference, pp. 42–51 (2018)
30. Okanohara, D., Sadakane, K.: Practical entropy-compressed rank/select dictionary. In: ALENEX 2007, pp. 60–70
31. Tabei, Y., Takabatake, Y., Sakamoto, H.: A succinct grammar compression. In: Fischer, J., Sanders, P. (eds.) Combinatorial Pattern Matching, pp. 235–246. Springer, Berlin Heidelberg, Berlin, Heidelberg (2013). https://doi.org/10.1007/978-3-642-38905-4_23

32. Tanaka, T., Tomohiro, I., Inenaga, S., Bannai, H., Takeda, M.: Computing convolution on grammar-compressed text. In: DCC 2013, pp. 451–460. IEEE (2013)
33. Verbin, E., Yu, W.: Data structure lower bounds on random access to grammar-compressed strings. In: Fischer, J., Sanders, P. (eds.) CPM 2013. LNCS, vol. 7922, pp. 247–258. Springer, Heidelberg (2013). https://doi.org/10.1007/978-3-642-38905-4_24
34. Vigna, S.: Broadword implementation of rank/select queries. In: McGeoch, C.C. (ed.) WEA 2008, pp. 154–168. Springer, Berlin Heidelberg, Berlin, Heidelberg (2008). https://doi.org/10.1007/978-3-540-68552-4_12
35. Yamamoto, T., Bannai, H., Inenaga, S., Takeda, M.: Faster subsequence and don't-care pattern matching on compressed texts. In: Giancarlo, R., Manzini, G. (eds.) CPM 2011. LNCS, vol. 6661, pp. 309–322. Springer, Heidelberg (2011). https://doi.org/10.1007/978-3-642-21458-5_27
36. Yue, J.X., et al.: Contrasting evolutionary genome dynamics between domesticated and wild yeasts. Nat. Genet. **49**(6), 913–924 (2017)

Logarithmic-Time Internal Pattern Matching Queries in Compressed and Dynamic Texts

Anouk Duyster[1](✉) [iD] and Tomasz Kociumaka[2](✉) [iD]

[1] Max Planck Institute for Software Systems, SIC, Saarbrücken, Germany
aduyster@mpi-sws.org
[2] Max Planck Institute for Informatics, SIC, Saarbrücken, Germany
tomasz.kociumaka@mpi-inf.mpg.de

Abstract. Internal Pattern Matching (IPM) queries on a text T, given two fragments X and Y of T such that $|Y| < 2|X|$, ask to compute all exact occurrences of X within Y. IPM queries have been introduced by Kociumaka, Radoszewski, Rytter, and Waleń [SODA'15], who showed that they can be answered in $\mathcal{O}(1)$ time using a data structure of size $\mathcal{O}(n)$ and used this result to answer various queries about fragments of T. In this work, we study IPM queries on compressed and dynamic strings. Our result is an $\mathcal{O}(\log n)$-time query algorithm applicable to any balanced recompression-based run-length straight-line program (RLSLP). In particular, one can use it on top of the RLSLP of Kociumaka, Navarro, and Prezza [IEEE TIT'23], whose size $\mathcal{O}(\delta \log \frac{n \log \sigma}{\delta \log n})$ is optimal (among all text representations) as a function of the text length n, the alphabet size σ, and the substring complexity δ. Our procedure does not rely on any preprocessing of the underlying RLSLP, which makes it readily applicable on top of the dynamic strings data structure of Gawrychowski, Karczmarz, Kociumaka, Łącki and Sankowski [SODA'18], which supports fully persistent updates in logarithmic time with high probability.

Keywords: Internal Pattern Matching · Recompression · Text Indexing

1 Introduction

Given two fragments $X = T[s_x..t_x)$ and $Y = T[s_y..t_y)$ of a text T such that $|Y| < 2|X|$[1], an Internal Pattern Matching (IPM) query reports all exact occurrences of X within Y, i.e., all fragments matching X and contained within Y. Kociumaka, Radoszewski, Rytter, and Waleń [18] introduced IPM queries as a central building block for answering various further queries about fragments of T. They showed that, for every text of length n, IPM queries can be answered

---
[1] The restriction $|Y| < 2|X|$ ensures that the starting positions of the reported fragments form an arithmetic progression and thus can be represented in constant space.

© The Author(s), under exclusive license to Springer Nature Switzerland AG 2025
Z. Lipták et al. (Eds.): SPIRE 2024, LNCS 14899, pp. 102–117, 2025.
https://doi.org/10.1007/978-3-031-72200-4_8

in $\mathcal{O}(1)$ time using a data structure of size $\mathcal{O}(n)$ that can be constructed in $\mathcal{O}(n)$ expected time. The journal version of their work [19] provides an improved data structure that takes $\mathcal{O}(n/\log_\sigma n)$ space and can be deterministically constructed in $\mathcal{O}(n/\log_\sigma n)$ time if the characters of T are integers in $[0..\sigma)$ for some $\sigma = n^{\mathcal{O}(1)}$. Hence, in the standard setting, IPM queries admit a compact data structure with optimal query and construction time.

Due to their multiple applications (see [19, Sections 1.2–1.3]), IPM queries have also been considered in further settings, including the compressed setting, where the goal is to exploit compressibility of the text, and the dynamic setting, where the text may change over time. Along with the Longest Common Extension (LCE) queries [20], IPM queries constitute an elementary operation of the PILLAR model, introduced by Charalampopoulos, Kociumaka, and Wellnitz [5] with the aim of unifying approximate pattern-matching algorithms across different settings. Whereas LCE queries can be answered $\mathcal{O}(\log n)$ time in the compressed [11,16] and dynamic settings [1,10], the algorithms for IPM queries have been slower so far. In this work, we propose a novel procedure that answers IPM queries in logarithmic time both in the compressed and dynamic settings, and thus we eliminate the bottleneck in the state-of-the-art PILLAR model implementations and multiple approximate pattern matching algorithms [4–7,9].

Our Result. State-of-the-art implementations of LCE and IPM queries represent compressed and dynamic texts using a run-length straight-line program (RLSLP) constructed using a locally consistent parsing scheme. Such a scheme repeatedly partitions the text into blocks and replaces blocks with individual symbols (with identical blocks replaced by matching symbols) until the text consists of a single symbol. The schemes alternate between run-length encoding (with blocks consisting of multiple copies of the same symbol) in odd rounds and another partitioning method (that reduces the text length by a constant factor while ensuring that matching fragments are partitioned consistently, except for blocks at the endpoints) in even rounds. At each even round, the *recompression* technique of Jeż [12,13] classifies the symbols into *left* and *right* symbols and creates a length-two block out of every left symbol followed by a right symbol; the remaining blocks are of length one. A more recent *restricted* variant [17,19] further classifies some symbols as inactive (so that they form length-one blocks).

In this work, we show how to efficiently answer IPM queries using any RLSLP obtained via a (possibly restricted) recompression scheme. Our algorithm does not need any preprocessing; it only assumes that every non-terminal stores its production, the length of its expansion, and the round when it has been created.

Theorem 1. *IPM queries on a text T represented using an r-round (restricted) recompression run-length straight-line program can be answered in $\mathcal{O}(r)$ time.*

IPM Queries in Compressed Texts. Kociumaka, Navarro, and Prezza [17] proved that every non-empty text T admits an $\mathcal{O}(\log n)$-round restricted recompression RLSLP of size $\mathcal{O}(\delta \log \frac{n \log \sigma}{\delta \log n})$, where n is the length of the text, σ is the alphabet size, and δ is the substring complexity of T. More recently, Kempa and Kociumaka [16] showed how to construct such an RLSLP in $\mathcal{O}(\delta \log^7 n)$ time from the LZ77 parsing of T. Combining this with Theorem 1, we obtain the following:

Corollary 2. *For every non-empty text T, there is a data structure of size $\mathcal{O}(\delta \log \frac{n \log \sigma}{\delta \log n})$ that answers IPM queries in $\mathcal{O}(\log n)$ time. Such a data structure can be constructed in $\mathcal{O}(\delta \log^7 n)$ time given the LZ77 parsing of T.*

Analogous results are already known for LCE queries [16, Theorem 5.25], which means that the PILLAR model can be implemented in $\mathcal{O}(\log n)$ time per operation using $\mathcal{O}(\delta \log \frac{n \log \sigma}{\delta \log n})$ space and $\mathcal{O}(\delta \log^7 n)$ preprocessing time.[2] Prior to this work, state-of-the-art implementations of IPM queries in the compressed setting required $\mathcal{O}(\log^3 n)$ time [14] or $\mathcal{O}(\log^2 n \log \log n)$ time [5] per query.

IPM Queries in Dynamic Texts. The dynamic strings data structure of Gawrychowski, Karczmarz, Kociumaka, Łącki, and Sankowski [10] maintains a collection of non-empty strings, allowing one to add new strings to the collection explicitly (`make_string`), by non-destructively concatenating two existing strings (`concat`), or by non-destructively splitting an existing string into two pieces (`split`). These updates take $\mathcal{O}(\log N)$ time with high probability, where N is the total length of the strings of the collection, except for `make_string`, which takes $\mathcal{O}(n + \log N)$ time for creating a string of length n. Internally, the strings are represented using an r-round (non-restricted) recompression RLSLP, where $r = \mathcal{O}(\log N)$ with high probability. Hence, Theorem 1 yields the following

Corollary 3. *The dynamic strings data structure of [10] supports IPM queries in $\mathcal{O}(\log N)$ time w.h.p., where N is the total length of the stored strings.*

The dynamic strings data structure supports LCE queries in $\mathcal{O}(\log N)$ time w.h.p., so these yields a PILLAR model implementation in $\mathcal{O}(\log N)$ time per operation. Previous implementations supported IPM queries in $\mathcal{O}(\log^2 N)$ time [5]. Faster IPM queries in the dynamic setting were only known for an alternative dynamic strings implementation [15] that takes $\mathcal{O}(\log N \log^{2-o(1)} \log N)$ time per update as well as for both LCE and IPM queries; see [6] for a discussion.

2 Preliminaries

In this section, we formally define restricted recompression and the underlying concepts and notations. Our exposition closely follows [16, Section 5.1].

Straight-Line Grammars. For a fixed context-free grammar $\mathcal{G}$, we denote by Σ and $\mathcal{N}$ the sets of terminals and non-terminals, respectively. The set of *symbols* is $\mathcal{S} := \Sigma \cup \mathcal{N}$. We say that $\mathcal{G}$ is a *straight-line grammar* (SLG) if:

---

[2] Using the results of [11] instead of [16], one can achieve a faster $\mathcal{O}(g \log \frac{n}{g})$-time construction from a size-g straight-line grammar generating T at the expense of a larger data structure of size $\mathcal{O}(z \log \frac{n}{z})$, where z is the size of the LZ77 parsing of T.

- each $A \in \mathcal{N}$ has a unique production $A \to \mathsf{rhs}(A)$, whose right-hand side is a non-empty sequence of symbols, i.e., $\mathsf{rhs}(A) \in \mathcal{S}^+$, and
- there is a partial order $\prec$ on $\mathcal{S}$ such that $B \prec A$ if B occurs in $\mathsf{rhs}(A)$.

A straight-line grammar $\mathcal{G}$ is a *run-length straight-line program* (RLSLP) if each non-terminal A satisfies $\mathsf{rhs}(A) = BC$ for some symbols $B, C \in \mathcal{S}$ such that $B \neq C$ or $\mathsf{rhs}(A) = B^m$ for a symbol $B \in \mathcal{S}$ and an integer $m \geq 2$.

Every straight-line grammar $\mathcal{G}$ yields an *expansion* function $\exp : \mathcal{S}^* \to \Sigma^*$:

$$\exp(A) = \begin{cases} A & \text{if } A \in \Sigma, \\ \exp(\mathsf{rhs}(A)) & \text{if } A \in \mathcal{N}, \\ \exp(A[1])\exp(A[2])\cdots\exp(A[a]) & \text{if } A \in \mathcal{S}^a \text{ for } a \neq 1. \end{cases}$$

The expansion of the starting symbol of $\mathcal{G}$ is the string *represented* by $\mathcal{G}$.

Restricted Recompression. Both recompression and restricted recompression, given a string $T \in \Sigma^+$, construct a sequence of strings $(T_k)_{k=0}^{\infty}$ over an infinite alphabet $\mathcal{A}$ defined as the least fixed point of the following equation:

$$\mathcal{A} = \Sigma \cup (\mathcal{A} \times \mathcal{A}) \cup (\mathcal{A} \times \mathbb{Z}_{\geq 2}),$$

i.e., $\mathcal{A} = \bigcup_{k=0}^{\infty} \mathcal{A}_k$, where $\mathcal{A}_0 = \Sigma$ and $\mathcal{A}_k = \mathcal{A}_{k-1} \cup (\mathcal{A}_{k-1} \times \mathcal{A}_{k-1}) \cup (\mathcal{A}_{k-1} \times \mathbb{Z}_{\geq 2})$ for $k \in \mathbb{Z}_{>0}$. Symbols in $\mathcal{A} \setminus \Sigma$ are non-terminals with $\mathsf{rhs}((A_1, A_2)) = A_1 A_2$ for $(A_1, A_2) \in \mathcal{A} \times \mathcal{A}$ and $\mathsf{rhs}((A_1, m)) = A_1^m$ for $(A_1, m) \in \mathcal{A} \times \mathbb{Z}_{\geq 2}$. Intuitively, $\mathcal{A}$ is a universal RLSLP: for every RLSLP with symbols $\mathcal{S}$ and terminals $\Sigma \subseteq \mathcal{S}$, there is a unique homomorphism $f : \mathcal{S} \to \mathcal{A}$ such that $f(A) = A$ if $A \in \Sigma$ and $\mathsf{rhs}(f(A)) = f(A_1)\cdots f(A_a)$ if $\mathsf{rhs}(A) = A_1 \cdots A_a$. As a result, $\mathcal{A}$ provides a convenient formalism for reasoning about procedures generating RLSLPs.

Restricted recompression uses the following transformations of $\mathcal{A}^*$ to $\mathcal{A}^*$.

Definition 4 (Restricted Run-Length Encoding [17,19]). Given $T \in \mathcal{A}^*$ and $\mathcal{B} \subseteq \mathcal{A}$, we define $\mathsf{rle}_{\mathcal{B}}(T) \in \mathcal{A}^*$ to be the string obtained as follows by decomposing T into blocks and collapsing these blocks:

1. For $i \in [1..|T|)$, place a *block boundary* between $T[i]$ and $T[i+1]$ unless $T[i] = T[i+1] \in \mathcal{B}$.
2. Replace each block $T[i..i+m) = A^m$ of length $m \geq 2$ with a symbol (A, m).

Definition 5 (Restricted Pair Compression [17,19]). Given $T \in \mathcal{A}^*$ and disjoint sets $\mathcal{L}, \mathcal{R} \subseteq \mathcal{A}$, we define $\mathsf{pc}_{\mathcal{L}, \mathcal{R}}(T) \in \mathcal{A}^*$ to be the string obtained as follows by decomposing T into blocks and collapsing these blocks:

1. For $i \in [1..|T|)$, place a *block boundary* between $T[i]$ and $T[i+1]$ unless $T[i] \in \mathcal{L}$ and $T[i+1] \in \mathcal{R}$.
2. Replace each block $T[i..i+1]$ with a symbol $(T[i], T[i+1])$.

Restricted recompression repeatedly transforms the input text alternating between restricted run-length encoding and restricted pair compression. Unlike in the classic recompression [12,13], they allow $\mathcal{B} \subsetneq \mathcal{A}$ and $\mathcal{L} \cup \mathcal{R} \subsetneq \mathcal{A}$. By construction, the intermediate strings satisfy $\exp(T_k) = T$ for all $k \in \mathbb{Z}_{\geq 0}$.

Definition 6 (Restricted Recompression [17,19]). For a string $T \in \Sigma^*$, we define $T_0 = T$ and $T_k = \mathsf{shrink}_k(T_{k-1})$ for every $k \in \mathbb{Z}_{>0}$, where

$$\mathsf{shrink}_k = \begin{cases} \mathsf{rle}_{\mathcal{B}_k} & \text{for some fixed set } \mathcal{B}_k \subseteq \mathcal{A} \text{ if } k \text{ is odd}, \\ \mathsf{pc}_{\mathcal{L}_k, \mathcal{R}_k} & \text{for some fixed disjoint sets } \mathcal{L}_k, \mathcal{R}_k \subseteq \mathcal{A} \text{ if } k \text{ is even}. \end{cases}$$

If $r = \min\{k \in \mathbb{Z}_{\geq 0} : |T_k| = 1\}$ exists,[3] we define an *r-round restricted recompression* RLSLP $\mathcal{G}$ to be an RLSLP generating T with symbols $\mathcal{S} = \bigcup_{k=0}^{r} \mathcal{S}_k$, where $\mathcal{S}_k = \{T_k[j] : j \in [1 . . |T_k|]\}$, and starting symbol $T_r[1]$.

Our query algorithm assumes to be given an r-round restricted recompression RLSLP with every $A \in \mathcal{N}$ storing $\mathsf{rhs}(A)$, $|\exp(A)|$, and $\mathsf{lvl}(A) = \min\{k : A \in \mathcal{S}_k\}$.

Parse Trees. The *parse tree* $\mathcal{T}(A)$ of a symbol $A \in \mathcal{S}$ in a straight-line grammar is a rooted ordered tree with each node ν associated to a symbol $\mathsf{s}(\nu) \in \mathcal{S}$. The root of $\mathcal{T}(A)$ is a node ρ with $\mathsf{s}(\rho) = A$. If $A \in \Sigma$, then ρ has no children. If $A \in \mathcal{N}$ and $\mathsf{rhs}(A) = A_1 \cdots A_a$, then ρ has a children, and the subtree rooted at the ith child is (a copy of) $\mathcal{T}(A_i)$. The parse tree $\mathcal{T}$ of $\mathcal{G}$ is the parse tree of the starting symbol of $\mathcal{G}$ (whose expansion is the text T represented by $\mathcal{G}$).

Each node ν of $\mathcal{T}$ is associated with a fragment $\exp(\nu)$ of T matching $\exp(\mathsf{s}(\nu))$. For the root ρ, we define $\exp(\rho) = T[1 . . |T|]$ to be the whole T. Moreover, if $\exp(\nu) = T[\ell . . r]$, $\mathsf{rhs}(\mathsf{s}(\nu)) = A_1 \cdots A_a$, and $\nu_1, \ldots, \nu_a$ are the children of ν, then $\exp(\nu_i) = T[r_{i-1} . . r_i)$, where $r_i = \ell + \sum_{j=1}^{i} |\exp(A_j)|$ for $0 \leq i \leq a$. This way, the fragments $\exp(\nu_i)$ partition $\exp(\nu)$, and $\exp(\nu_i)$ matches $\exp(\mathsf{s}(\nu_i))$.

Our representation of the restricted recompression RLSLP allows for efficiently traversing the underlying parse tree $\mathcal{T}$: we implement a pointer to a node $\nu \in \mathcal{T}$ as a (persistent) stack containing pairs $(\mathsf{s}(\mu), \exp(\mu))$ for every ancestor μ of ν starting from the root ρ at the bottom of the stack to the node ν itself at the top of the stack.[4] Given such a pointer, we can in constant time retrieve a pointer to the parent of ν or to any child of ν, identified by its index among the children of ν or by an arbitrary position contained in its expansion.

Observation 7. *Given any position $j \in [1 . . |T|]$, a pointer to the j-th leaf of $\mathcal{T}$ (whose expansion is $T[j]$) can be retrieved in $\mathcal{O}(r)$ time.*

We also observe that each symbol in each of the strings $(T_k)_{k=0}^{r}$ can be associated with a node in the parse tree $\mathcal{T}$. This mapping is not injective, though: if a symbol of T_k forms a length-one block with respect to shrink_{k+1}, it remains in T_{k+1} represented by the same node. As in [10], this can be interpreted using the notion of an *uncompressed parse tree* $\overline{\mathcal{T}}$, where the edges of $\mathcal{T}$ are subdivided so that, if a node ν of $\mathcal{T}$ spans multiple levels, in $\overline{\mathcal{T}}$ it gets a separate copy for each level. Still, we can refer to a node in $\overline{\mathcal{T}}$ using a pair (ν, k) consisting of the underlying node ν in $\mathcal{T}$ and the considered level k. An immediate consequence

---

[3] Constructions in [10,17,19] provide several strategies of picking $\mathcal{B}_k$, $\mathcal{L}_k$, and $\mathcal{R}_k$ which guarantee that $|T_k| = 1$ holds for some $k = \mathcal{O}(\log n)$ if $T \in \Sigma^n$.
[4] We represent every fragment $\exp(\mu) = T[\ell . . r]$ using the endpoints ℓ and r.

of [19, Lemma 4.4] is that a node ν of $\mathcal{T}$ corresponds to nodes in $\overline{\mathcal{T}}$ precisely at the levels in $[\mathsf{lvl}(s(\nu))..\mathsf{lvl}(s(\mu)))$, where μ is the parent of ν. Hence, a symbol $T_k[i]$ is created by collapsing a block of two or more symbols in T_{k-1} if and only if $\mathsf{lvl}(T_k[i]) = k$. Based on this property, given a pointer to a node in $\overline{\mathcal{T}}$, we can in constant time retrieve a pointer to the parent or any of its children.

As shown in [10], a more sophisticated implementation of pointers (which requires additional auxiliary data structures) also allows moving in $\mathcal{O}(1)$ time between subsequent symbols of the strings T_k (so that, given a pointer to the node of $\overline{\mathcal{T}}$ representing $T_k[i]$, one can in $\mathcal{O}(1)$ time construct pointers to the nodes representing $T_k[i \pm 1]$). In this work, the following simpler result suffices:

Lemma 8. *Given a pointer to a node of $\overline{\mathcal{T}}$, any online sequence of s operations asking to move one level up (to the parent), one level down (to a child), or one position to the right[5] (at the same level) can be implemented in $\mathcal{O}(r + s)$ time.*

Proof. For $i \in [0..s]$, let (ν_i, k_i) be the node of $\overline{\mathcal{T}}$ visited at the ith operation. As noted above, moving a pointer one level up or down can be implemented in constant time. In order to move from a node (ν_i, k_i) to the next node (ν_{i+1}, k_{i+1}) at the same level $k_{i+1} = k_i$, we traverse the path from ν_i to ν_{i+1} in $\mathcal{T}$. For this, we go up the parse tree until we encounter a node that has at least one right sibling. Once we identify such a node, we move to the sibling immediately to the right and keep descending to the leftmost child until we reach a node whose symbol is at level k_i or smaller. The running time of this step is proportional to the length of the path from ν_i to ν_{i+1} in $\mathcal{T}$. Across the whole sequence, these paths form edge-disjoint fragments of the Euler tour of the subtree of $\mathcal{T}$ consisting of all the nodes $\nu_0, \ldots, \nu_s$ and their ancestors. Since each internal node in $\mathcal{T}$ has at least two children, the subtree size is $\mathcal{O}(s + r)$: besides nodes $\nu_0, \ldots, \nu_s$ and their lowest common ancestors, it only contains $\mathcal{O}(r)$ ancestors of ν_0 and ν_s. □

3 Popped Sequences

Efficient implementation of LCE queries on recompression RLSLPs relies on a notion of *popped sequences* introduced by I [11]. We use a slightly simpler construction that is not tailored to any specific version of recompression.

Definition 9 (Popped Sequence). Given a string $X \in \Sigma^*$, we define strings $L_k, R_k, \bar{X}_k, \bar{X}'_k \in \mathcal{A}^*$ for $k \in \mathbb{Z}_{\geq 0}$ as follows. We set $\bar{X}_0 = X$. For $k > 0$, we set $\bar{X}'_k = \mathsf{shrink}_{k+1}(\bar{X}_k)$ and define L_k, R_k, and $\bar{X}_{k+1}$ as follows:

- If $|\bar{X}'_k| \geq 2$, then L_k consists of the first block of $\bar{X}'_k$ with respect to shrink_{k+1}, R_k consists of the last block of $\bar{X}'_k$, and $\bar{X}_{k+1} = \bar{X}'_k[2..|\bar{X}'_k| - 1]$.
- If $|\bar{X}'_k| = 1$, then $L_k = \bar{X}_k$ and $R_k = \bar{X}_{k+1} = \varepsilon$;
- If $|\bar{X}'_k| = 0$, then $L_k = R_k = \bar{X}_{k+1} = \varepsilon$.

The *popped sequence* of the string X is defined as $\mathsf{PSeq}(X) = L_0 \cdots L_q \cdot R_q \cdots R_0$ for any $q \in \mathbb{Z}_{\geq 0}$ such that $\bar{X}_{q+1} = \varepsilon$.

---
[5] One can instead allow moving to the left, but the two directions cannot be combined.

In other words, to construct the sequence $(\bar{X}_k)_{k=0}^{\infty}$, we proceed as in Definition 6, but we remove the leftmost and the rightmost block of $\bar{X}_k$ before we collapse the remaining blocks to obtain $\bar{X}_{k+1}$. The removed blocks are concatenated (in the appropriate order) to form $\mathsf{PSeq}(X)$; see Fig. 1. The following observation captures the immediate consequences of the Definition 9.

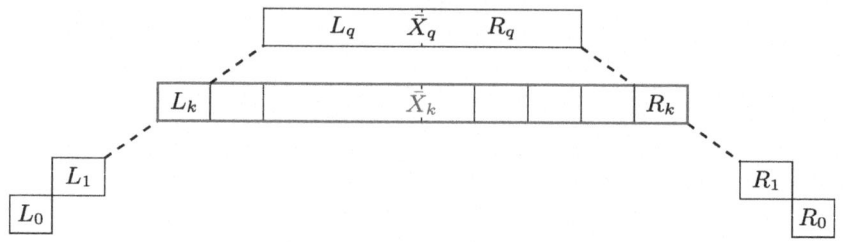

Fig. 1. The popped sequence is build from the blocks L_0 through R_0. In every layer k, the string $\bar{X}_k$ spans from L_k to R_k (inclusive).

Observation 10. *For every $k \in \mathbb{Z}_{\geq 0}$, we have $|\mathsf{rle}(L_k)| \leq 2$ and $|\mathsf{rle}(R_k)| \leq 2$, as well as $X = \exp(L_0 \cdots L_{k-1} \cdot \bar{X}_k \cdot R_{k-1} \cdots R_0)$.*

A crucial property of the popped sequence is that every occurrence of X in a text T induces the occurrences of each non-empty string $\bar{X}_k$ in the string T_k, and thus an "occurrence" of the entire popped sequence in the parse tree of T.

Lemma 11. *If $X \in \Sigma^+$ occurs in a text T at position i and $\bar{X}_k \neq \varepsilon$, then $\bar{X}_k$ occurs in T_k at a position i_k such that $|\exp(T_k[1..i_k))| = i - 1 + |\exp(L_0 \cdots L_{k-1})|$.*

Proof. We proceed by induction on k. The base case is trivial since $T_0 = T$ and $\bar{X}_0 = X$. For $k \in \mathbb{Z}_{>0}$, the inductive hypothesis yields an occurrence $\bar{X}_{k-1}$ starting at position i_{k-1}. If $\bar{X}_k$ is non-empty, then L_{k-1} and R_{k-1} are defined as the leftmost and the rightmost block of $\bar{X}_{k-1}$ with respect to shrink_k, whereas $\bar{X}_k$ is obtained by collapsing all the blocks in between. By Definitions 4 and 5, shrink_k places block boundaries based on the identities of the two surrounding symbols. In particular, the block boundaries strictly inside $\bar{X}_{k-1}$ are placed in the same way regardless of whether $\bar{X}_{k-1}$ is processed as a standalone string or as a fragment of T_{k-1}. Thus, whereas the leftmost and the rightmost block of $\bar{X}_{k-1}$ may (in the latter case) contain symbols outside $\bar{X}_{k-1}$, the blocks collapsed to $\bar{X}_k$ are processed in the same way and induce an occurrence of $\bar{X}_k$ in T_k. The starting position i_k of this occurrence satisfies $|\exp(T_k[1..i_k))| = |\exp(T_{k-1}[1..i_{k-1}) \cdot L_{k-1})| = |\exp(T_{k-1}[1..i_{k-1}))| + |\exp(L_{k-1})|$, which is $i - 1 + |\exp(L_0 \cdots L_{k-1})|$ by the inductive hypothesis. □

Next, we address the problem of efficiently computing a popped sequence.

Lemma 12. *Let $\mathcal{G}$ be an r-round restricted recompression RLSLP representing a text T. For any fragment $X = T[i..j]$, the run-length encoding of the popped sequence $\mathsf{PSeq}(X)$ can be computed in $\mathcal{O}(r)$ time, along with a decomposition $\mathsf{PSeq}(X) = L_0 \cdots L_q \cdot R_q \cdots R_0$ for $q = \max\{k \in \mathbb{Z}_{\geq 0} : \bar{X}_k \neq \varepsilon\}$.*

Proof. For each round k such that $\bar{X}_k \neq \varepsilon$, our algorithm assumes to be given pointers (ν_k, k) and (μ_k, k) to the leftmost and the rightmost symbol within the occurrence of $\bar{X}_k$ in T_k stipulated by Lemma 11. For $k = 0$, the nodes ν_0 and μ_0 corresponding to $T[i]$ and $T[j]$, respectively, can be obtained by Observation 7. At each round, the algorithm first moves the pointer (ν_k, k) one level up to get a pointer $(\nu'_k, k+1)$ corresponding to the symbol of T_{k+1} that represents the block of T_k (with respect to shrink_{k+1}) containing the leftmost character of the occurrence of $\bar{X}_k$. At that stage, we can also determine the symbols constituting the leftmost block of $\bar{X}_k$, that is, $\mathsf{s}(\nu'_k)$ and, if $\nu'_k \neq \nu_k$, the symbols at the right siblings of ν_k; these symbols constitute L_k. Then, to obtain a pointer $(\nu_{k+1}, k+1)$ to the leftmost symbol of $\bar{X}_{k+1}$, we need to move one symbol to the right in T_{k+1}. A symmetric subroutine lets us derive R_k and $(\mu_{k+1}, k+1)$ via $(\mu'_k, k+1)$. In the special case when $\bar{X}_{k+1} = \varepsilon$, we have $\mu'_k = \nu'_k$, in which case $R_k = \varepsilon$ and L_k consists of the symbols between the siblings ν_k and μ_k (inclusive), or $\mu'_k = \nu_{k+1}$, in which case the original construction of L_k and R_k remains valid. By Lemma 8 and its symmetric counterpart, the overall runtime is $\mathcal{O}(q + r) = \mathcal{O}(r)$. □

3.1 LCE Queries in $\mathcal{O}(r)$ Time

Recall that an LCE query, given positions i and i' in $T \in \Sigma^n$, returns

$$\mathsf{LCE}_T(i, i') = \max\{d \in [0..n - \max(i, i') + 1] : T[i..i+d) = T[i'..i'+d)\}.$$

As a sample application of our variant of popped sequences, we adapt [16, Theorem 5.25] to obtain $\mathcal{O}(r)$-time LCE queries in our setting: we re-analyze the original algorithm to prove that its running time is $\mathcal{O}(r)$ in our setting.

Proposition 13 (Adapted from [16, Theorem 5.25]). *LCE queries on a text T represented using an r-round restricted recompression run-length straight-line program can be answered in $\mathcal{O}(r)$ time.*

Proof. We assume $i \neq i'$; otherwise, $\mathsf{LCE}_T(i, i) = n - i + 1$ can be computed trivially. The query algorithm maintains pointers to nodes $\nu, \nu' \in \mathcal{T}$ whose expansions $\exp(\nu) = T[\ell..r]$ and $\exp(\nu') = T[\ell'..r']$ satisfy $T[i..\ell) = T[i'..\ell')$.

We initialize ν as the highest node such that $\exp(\nu)$ starts at position i, and ν' as the highest node such that $\exp(\nu')$ starts at position i'. For this, we follow the paths from the root to the leaves representing $T[i]$ and $T[i']$, respectively; see Observation 7. At each step, we proceed as follows:

1. If $|\exp(\nu)| = |\exp(\nu')| = 1$ and $\mathsf{s}(\nu) \neq \mathsf{s}(\nu')$, return $\ell - i$.
2. If $|\exp(\nu)| > |\exp(\nu')|$, we replace ν by its leftmost child.
3. If $|\exp(\nu')| > |\exp(\nu)|$, we replace ν' by its leftmost child.

4. If $\mathsf{s}(\nu) \neq \mathsf{s}(\nu')$, we replace both ν and ν' by their leftmost children.
5. Suppose that ν is the jth among the d children of its parent and ν' is the j'th among the d' children of its parent. Let $\lambda = \min(d-j, d'-j')$. If $\lambda \leq 1$, we replace ν and ν' by the highest nodes whose expansions start at position r and r', respectively.
6. Otherwise, we replace ν by its $(j+\lambda)$th sibling and ν' by its $(j'+\lambda)$th sibling.

Let us justify the correctness of the algorithm. In case 1, we have $T[i..\ell] = T[i'..\ell']$ yet $T[\ell] = \mathsf{s}(v) \neq \mathsf{s}(\nu') = T[\ell']$. Hence, $\mathsf{LCE}_T(i,i') = \ell - i = \ell' - i'$ is computed correctly. In cases 2–4, the invariant is still satisfied because the leftmost positions in $\exp(\nu)$ and $\exp(\nu')$ do not change. Moreover, whenever we replace ν or ν' with its leftmost child, we have $|\exp(\nu)| > 1$ or $|\exp(\nu')| > 1$, respectively, so the child exists. In the remaining cases, we have $\mathsf{s}(\nu) = \mathsf{s}(\nu')$ and thus $T[i..r] = T[i'..r']$. Thus, the invariant still holds in case 5, where we replace ν and ν' by the highest nodes whose expansions start at positions r and r', respectively. If $\lambda > 1$, then $d, d' \geq 3$, so all siblings of ν and ν' have the same symbol. The invariant holds because we shift ν and ν' by λ siblings to the right.

It remains to analyze the query time. For this, we say that nodes (ν, ν') form a *matching pair* if $\mathsf{s}(\nu) = \mathsf{s}(\nu')$ and the expansions $\exp(\nu) = T[\ell..r]$ and $\exp(\nu') = T[\ell'..r']$ satisfy $T[i..\ell] = T[i'..\ell']$. Denote by N and N' the set of nodes ν and ν' participating in matching pairs; these pairs form a perfect matching between N and N'. We claim that the query algorithm satisfies the following additional invariant: None of the ancestors of ν belong to N and none of the ancestors of ν' belong to N'. The invariant is satisfied at the beginning because all the ancestors of ν and ν' have their expansions starting before position i and i', respectively. In cases 2 and 4, the node ν does not form a matching pair with any ancestor of ν' (because they do not belong to N'), with ν' itself (because $\mathsf{s}(\nu) \neq \mathsf{s}(\nu')$), nor with any descendant of ν' (because their expansions are shorter than $\exp(\nu)$). Thus, ν does not belong to N. By symmetry, the invariant remains satisfied in cases 3 and 4 when we replace ν' by its leftmost child. In cases 5 and 6, all ancestors of the new nodes ν and ν' are ancestors of the old nodes ν and ν'.

Let $\hat{N}$ consist of nodes that do not belong to N yet have descendants in N, and define $\hat{N}'$ analogously based on N' instead of N. Our algorithm visits only nodes in $\hat{N}$ and $\hat{N}'$, their children, and the ancestors of the two leaves whose mismatch causes the algorithm to return. Moreover, if a node in $\hat{N}$ or $\hat{N}'$ contains three or more children, then Case 6 guarantees that we visit at most two of them. Thus, the total runtime of the algorithm is $\mathcal{O}(|\hat{N}| + |\hat{N}'| + r)$.

It remains to prove that $|\hat{N}| = \mathcal{O}(r)$ and, by symmetry, $|\hat{N}'| = \mathcal{O}(r)$. For this, observe that Lemma 11 applied to $T[i..i+d] = T[i'..i'+d]$ implies that the nodes participating in the popped sequence $\mathsf{PSeq}(T[i..i+d])$ belong to N, so $\hat{N}$ may only contained ancestors of these nodes. By Definition 9, the nodes in the popped sequence have $\mathcal{O}(r)$ parents in total (since nodes within each L_k and R_k have a single parent). These nodes also have $\mathcal{O}(r)$ proper ancestors because the parse tree is of height $\mathcal{O}(r)$ and does not contain degree-one vertices. □

We remark that a representation of T as an r-round restricted recompression RLSLP constitutes an analogous representation of the reverse text $\overline{T}$ (it suffices to reverse the right-hand side of every production), so Proposition 13 can also be used to answer the following queries $\overline{\mathsf{LCE}}_T$ queries:

$$\overline{\mathsf{LCE}}_T(i,i') := \max\{d \in [0..\min(i,i') - 1] : T[i-d..i] = T[i'-d..i')\}.$$

4 IPM Query Algorithm: Overview

Recall that the query algorithm is given two fragments X and Y of the text T satisfying $|Y| < 2|X|$, and the goal is to identify exact occurrences of X contained within Y. Our $\mathcal{O}(r)$-time query algorithm performs the following steps:

1. Define $\ell = \max\{k \in \mathbb{Z}_{\geq 0} : |\bar{X}_k| > k\}$, where $(\bar{X}_k)_{k=0}^{\infty}$ is introduced in Definition 9 along with the popped sequence $\mathsf{PSeq}(X)$. As proved in Sect. 5, $\mathsf{rle}(\bar{X}_\ell)$ can be constructed in $\mathcal{O}(r)$ time.
2. Define an appropriate fragment Y'_ℓ of T_ℓ such that every occurrence of X within Y yields an occurrence of $\bar{X}_\ell$ within Y'_ℓ. As proved in Sect. 6, we can ensure that $|Y'_\ell| = \mathcal{O}(|\bar{X}_\ell|)$ and $\mathsf{rle}(Y'_\ell)$ can be constructed in $\mathcal{O}(r)$ time.
3. Use pattern matching in run-length encoded strings to find the occurrences of $\bar{X}_\ell$ in Y'_ℓ, represented as $\mathcal{O}(1)$ arithmetic progressions; see Sect. 7.
4. For each progression, use $\mathcal{O}(1)$ LCE queries (in $\mathcal{O}(r)$ time each) to find the occurrences of $\bar{X}_\ell$ in Y'_ℓ that extend to occurrences of X in Y; see Sect. 8.

5 Constructing the Proxy Pattern $\bar{X}_\ell$

Observe that, for every non-empty fragments X of T, we have $|\bar{X}_0| = |X| > 0$ and $|\bar{X}_k| = 0$ for $k > r$, so $\ell := \max\{k \in \mathbb{Z}_{\geq 0} : |\bar{X}_k| > k\}$ is well-defined.

Lemma 14. *The level ℓ and the run-length encoding $\mathsf{rle}(\bar{X}_\ell)$ of the proxy pattern can be constructed in $\mathcal{O}(r)$ time.*

Proof. Our goal is to construct $\bar{X}_k$ for all $k > \ell$ and then derive $\mathsf{rle}(\bar{X}_\ell)$ from $\bar{X}_{\ell+1}$. For this, we use Lemma 12 to compute $\mathsf{rle}(\mathsf{PSeq}(X))$ decomposed as $\mathsf{rle}(L_0) \cdots \mathsf{rle}(L_q) \cdot \mathsf{rle}(R_q) \cdots \mathsf{rle}(R_0)$, where $q = \max\{k \in \mathbb{Z}_{\geq 0} : \bar{X}_k \neq \varepsilon\}$.

We start with $\bar{X}_{q+1} = \varepsilon$ and then, for subsequent integers k starting with $q+1$, derive $\bar{X}_{k-1}$ from $\bar{X}_k$. To do this, we need to replace A with $\mathsf{rhs}(A)$ for every symbol A in $\bar{X}_k$ with $\mathsf{lvl}(A) = k$, and then prepend L_{k-1} and append R_{k-1}. We cannot afford to scan $\bar{X}_k$ to identify all symbols with level k, so we store $\bar{X}_k$ as a linked list and additionally maintain the symbols of $\bar{X}_k$ in a priority queue, where each symbol is associated with its level (which constitutes the priority) and a pointer to the appropriate element of the linked list. As long as there is a symbol A with $\mathsf{lvl}(A) = k$, we replace A with $\mathsf{rhs}(A)$, both in the linked list and the underlying priority queue. Once there are no symbols with level k, we proceed by prepending the symbols of L_{k-1} (one by one from right to left) and appending the symbols of R_{k-1} (one by one from left to right). If the length of the maintained string reaches k at any point during this process, we conclude that $|\bar{X}_{k-1}| \geq k > k-1$, so we stop and set $\ell = k-1$. Since $|\mathsf{rhs}(A)| \geq 2$ holds for every $A \in \mathcal{A}$ and $|L_k| + |R_k| \geq 1$ for $k \in [0..q]$, each step of the algorithm can be charged to an increment of the length of the maintained string. As we stop the algorithm when the length exceeds ℓ, the total running time, excluding the priority queue implementation, is $\mathcal{O}(\ell)$. The maximum priority in the queue decreases from q to ℓ (and never increases), so we can use a monotone bucket queue [21], whose total runtime is proportional to the number of operations, which is $\mathcal{O}(\ell)$, plus the size of the universe of priorities, which is $\mathcal{O}(q)$.

Once we determined ℓ, we construct $\mathsf{rle}(\bar{X}_\ell)$ by scanning $\bar{X}_{\ell+1}$ and replacing each symbol A with $\mathsf{lvl}(A) = \ell + 1$ with $\mathsf{rle}(\mathsf{rhs}(A))$, and then we prepend $\mathsf{rle}(L_\ell)$ and append $\mathsf{rle}(R_\ell)$. This process takes $\mathcal{O}(\ell + 1)$ time since $\mathsf{rle}(\mathsf{rhs}(A))$, $\mathsf{rle}(L_\ell)$, and $\mathsf{rle}(R_\ell)$ are all of constant length.

Overall, it takes $\mathcal{O}(r)$ time to build $\mathsf{PSeq}(X)$, $\mathcal{O}(\ell + q)$ time to find ℓ and build $\bar{X}_{\ell+1}$, and $\mathcal{O}(\ell + 1)$ time to construct $\mathsf{rle}(\bar{X}_\ell)$. The total runtime is $\mathcal{O}(r)$. □

6 Constructing the Proxy Text Y'_ℓ

The following Lemma 15 defines the appropriate proxy text Y'_ℓ as a fragment of T_ℓ such that $\exp(Y'_\ell)$ covers a sufficiently large portion of Y. A natural choice would be to pick Y'_ℓ so that $\exp(Y'_\ell)$ covers the entire Y; unfortunately, this may result in a proxy text Y'_ℓ that is much longer than the proxy pattern X'_ℓ, even with respect to run-length encoding: the fact that X is covered by a length-$\mathcal{O}(\ell)$ fragment of $T_{\ell+1}$ does not imply that the same is true for Y. Nevertheless, if X occurs in Y, then the portion of Y containing all the occurrences of X is guaranteed to be covered by a length-$\mathcal{O}(\ell)$ fragment of $T_{\ell+1}$. Our definition assures that Y'_ℓ is long enough so that every occurrence of X in Y yields an occurrence of $\bar{X}_\ell$ in Y'_ℓ and, at the same time, short enough to enable efficient implementation of our query algorithm. Specifically, we define Y'_ℓ so that it covers the middle position Y (which is contained in every occurrence of X in Y) and extends in both directions just enough to cover every occurrence X containing that middle position. We quantify this extension using both the length (so that the occurrences of $\bar{X}_\ell$ in Y'_ℓ form $\mathcal{O}(1)$ arithmetic progressions; see Sect. 7) and the number of runs (so that $\mathsf{rle}(Y'_\ell)$ can be constructed efficiently; see Lemma 16).

Lemma 15. *Let $T[m]$ be the middle character in Y and let $T_\ell[m_\ell]$ be its ancestor in T_ℓ. Define the proxy text $Y'_\ell = T_\ell[b..e]$ by maximally extending $T_\ell[m_\ell]$ so that:*

- $\max(|T_\ell[b..m_\ell]|, |T_\ell[m_\ell..e]|) \leq |\bar{X}_\ell| + \ell$, *and*
- $\max(|\mathsf{shrink}_{\ell+1}(T_\ell[b..m_\ell])|, |\mathsf{shrink}_{\ell+1}(T_\ell[m_\ell..e])|) \leq 2\ell + 3$.

For every occurrence of X contained in Y, the corresponding occurrence of $\bar{X}_\ell$ in T_ℓ stipulated by Lemma 11 is contained within Y'_ℓ.

Proof. Let us fix an occurrence $T[i..j]$ of X contained in Y. For every $k \in [0..\ell]$, Lemma 11 yields a corresponding occurrence $T_k[i_k..j_k] = \bar{X}_k$. Let us also define $T_k[i'_k]$ as the ancestor of $T[i]$ in T_k.

The following inductive argument shows that $|T_k[i'_k..i_k]| \leq k$ holds for $k \in [0..\ell]$. This statement is trivial for $k = 0$ due to $i'_0 = i_0 = i$. For $k > 0$, Lemma 11 implies $T_k[1..i_k] = \mathsf{shrink}_k(T_{k-1}[1..i_{k-1}] \cdot L_{k-1})$. In particular, $T_k[i'_k..i_k] = \mathsf{shrink}_k(L'_{k-1} \cdot T_{k-1}[i'_{k-1}..i_{k-1}] \cdot L_{k-1})$, where L'_{k-1} consists of the children of $T_k[i'_k]$ to the left of $T_{k-1}[i'_{k-1}]$. Observe that shrink_k does not place block boundaries within $L'_{k-1} \cdot T_{k-1}[i'_{k-1}]$ (all those symbols have $T_k[i'_k]$ as their parent) and within L_{k-1} (by Definition 9, L_{k-1} is the leftmost block of $\bar{X}_{k-1}$). The number of remaining possible locations of block boundaries within $L'_{k-1} \cdot T_{k-1}[i'_{k-1}..i_{k-1}] \cdot L_{k-1}$ does not exceed $|T_{k-1}[i'_{k-1}..i_{k-1}]|$, which is at most $k-1$ by the inductive hypothesis. Hence, the number of blocks that $L'_{k-1} \cdot T_{k-1}[i'_{k-1}..i_{k-1}] \cdot L_{k-1}$ is partitioned into satisfies $|T_k[i'_k..i_k]| \leq k$. This completes the inductive proof.

The condition $|Y| < 2|X|$ implies that $T[i..j]$ contains the middle position of Y. Thus, $i \leq m$ and $i'_\ell \leq m_\ell$. Consequently, $i_\ell \leq i'_\ell + \ell \leq m_\ell + \ell$. Hence, we have $j_\ell = i_\ell + |\bar{X}_\ell| - 1 \leq m_\ell + |\bar{X}_\ell| + \ell - 1$ and $|T_\ell[m_\ell..j_\ell]| \leq |\bar{X}_\ell| + \ell$. Therefore, $T_\ell[m_\ell..j_\ell]$ is either a suffix of $\bar{X}_\ell$ or consists of the entire $\bar{X}_\ell$ preceded by at most ℓ symbols. Since $|\mathsf{shrink}_{\ell+1}(\bar{X}_\ell)| = |\mathsf{shrink}_{\ell+1}(L_\ell) \cdot \bar{X}_{\ell+1} \cdot \mathsf{shrink}_{\ell+1}(R_\ell)| \leq |\bar{X}_{\ell+1}| + 2 \leq \ell + 3$, we conclude that $|\mathsf{shrink}_{\ell+1}(T_\ell[m_\ell..j_\ell])| \leq \ell + 3 + \ell = 2\ell + 3$.

Recall that e has been chosen as the largest index satisfying $|T_\ell[m_\ell..e]| \leq |\bar{X}_\ell| + \ell$ and $|\mathsf{shrink}_{\ell+1}(T[m_\ell..e])| \leq 2\ell + 3$. As proved above, the index j_ℓ satisfies both inequalities, so we derive $j_\ell \leq e$. A symmetric argument shows that $b \leq i_\ell$, and thus the occurrence $T_\ell[i_\ell..j_\ell]$ induced by $\bar{X}_\ell$ is indeed contained in Y'_ℓ. □

Lemma 16. *The run-length encoding $\mathsf{rle}(Y'_\ell)$ of the proxy text Y'_ℓ can be constructed in $\mathcal{O}(r)$ time.*

Proof. First, we compute a pointer to $T[m]$ using Observation 7, and then we move up until we arrive at a node of $\overline{\mathcal{T}}$ representing a character $T_{\ell+1}[m_{\ell+1}]$ at level $\ell + 1$. Next, we generate the following fragment of $T_{\ell+1}$:

$$\hat{Y}_{\ell+1} := T_{\ell+1}[\max(1, m_{\ell+1} - 2\ell - 2)..\min(|T_{\ell+1}|, m_{\ell+1} + 2\ell + 2)].$$

To do this, move from $T_{\ell+1}[m_{\ell+1}]$ by $2\ell + 2$ positions in both directions. This phase is implemented in $\mathcal{O}(r + \ell + 1) = \mathcal{O}(r)$ time using Lemma 8 (to move to the right) and its symmetric counterpart (to move to the left).

To derive $\mathsf{rle}(Y'_\ell)$ from $\hat{Y}_{\ell+1}$, we replace A with $\mathsf{rhs}(A)$ for every symbol A with $\mathsf{lvl}(A) = \ell + 1$, and then we trim the resulting string to make sure that it contains at most $|\bar{X}_\ell| + \ell - 1$ symbols on either side of $T_\ell[m_\ell]$. This phase costs $\mathcal{O}(\ell + 1)$ time, for a total of $\mathcal{O}(r)$ time for the entire algorithm. □

7 Finding the Occurrences of $\bar{X}_\ell$ in Y'_ℓ

As observed in [2,3,8], exact pattern matching in run-length encoded strings can be implemented in linear time. Below, we adapt this result to support representing the occurrences as arithmetic progressions.

Proposition 17. (see [8, Theorem 1]). *Given the run-length encodings* $\mathsf{rle}(P)$, $\mathsf{rle}(S)$ *of strings* $P, S \in \Sigma^+$, *the set* $\mathsf{Occ}(P,S) = \{i \in [1..|S| - |P| + 1] : P = S[i..i+|P|)\}$, *represented by at most* $\min(|\mathsf{rle}(S)|, \lfloor |S|/|P| \rfloor)$ *arithmetic progressions with difference at most* $|P|$, *can be constructed in* $\mathcal{O}(|\mathsf{rle}(P)| + |\mathsf{rle}(S)|)$ *time.*

Proof. None of the papers [2,3,8] considers the case of $|\mathsf{rle}(P)| = 1$, when reporting $\mathsf{Occ}(P,S)$ explicitly might be too costly. In this case, every occurrence of P is contained within a single run of S. A run $S[j..j+q) = b^q$ contains an occurrence of $P = a^p$ if and only if $a = b$ and $p \le q$; then, the occurrences form an arithmetic progression $(j, j+1, \ldots, j+q-p)$ with difference one. Each such progression corresponds to a run of length at least $|P|$ in S, so their number does not exceed $\min(|\mathsf{rle}(S)|, \lfloor |S|/|P| \rfloor)$. To construct them in $\mathcal{O}(p+s)$ time, we scan $\mathsf{rle}(S)$ maintaining the length of the already processed prefix of S.

For $|\mathsf{rle}(P)| \ge 2$, the algorithm of [8, Theorem 1] reports elements of $\mathsf{Occ}(P,S)$ one be one; the number of occurrences does not exceed $|\mathsf{rle}(S)| - 1$. By [18, Fact 1.1], if we greedily group occurrences into arithmetic progressions with a difference at most $|P|$ each, we obtain at most $\lfloor |S|/|P| \rfloor$ progressions in total. □

Applying Proposition 17 to $\bar{X}_\ell$ and Y'_ℓ, we obtain the following result:

Corollary 18. *Given* $\mathsf{rle}(\bar{X}_\ell)$ *and* $\mathsf{rle}(Y'_\ell)$, *the occurrences of* $\bar{X}_\ell$ *in* Y'_ℓ, *represented as* $\mathcal{O}(1)$ *arithmetic progressions, each with a difference of at most* $|\bar{X}_\ell|$, *can be computed in* $\mathcal{O}(r)$ *time.*

Proof. By Lemmas 14 and 16, both $\mathsf{rle}(\bar{X}_\ell)$ and $\mathsf{rle}(Y'_\ell)$ are of size $\mathcal{O}(r)$. Moreover, the definition of Y'_ℓ (see Lemma 15) guarantees that $|Y'_\ell| < 2|\bar{X}_\ell| + 2\ell < 4|\bar{X}_\ell|$, so the algorithm of Proposition 17 returns at most four progressions. □

8 Verifying Candidate Occurrences

Lemma 15 shows that every occurrence of X in Y corresponds to a position in $\mathsf{Occ}(\bar{X}_\ell, Y'_\ell)$. Moreover, by Corollary 18, $\mathsf{Occ}(\bar{X}_\ell, Y'_\ell)$ is the union of $\mathcal{O}(1)$ arithmetic progressions, each with a difference of at most $|\bar{X}_\ell|$.

Let $V_\ell := \{\bar{a}_\ell + i \cdot g_\ell : i \in [0..s_\ell]\}$ be one such arithmetic progression. Observe that $Y'_\ell[\bar{a}_\ell + i \cdot g_\ell .. \bar{a}_\ell + (i+1) \cdot g_\ell) = \bar{X}_\ell[1..g_\ell]$ holds for each $i \in [0..s_\ell)$. Thus, the arithmetic progression V_ℓ of positions in Y'_ℓ corresponds to an arithmetic progression $V = \{\bar{a} + i \cdot g : i \in [0..s]\}$ of positions in T, where $s = s_\ell$, $g = |\exp(\bar{X}_\ell[1..g_\ell])|$ and $T[\bar{a}]$ is the leftmost leaf in the subtree of $\overline{T}$ rooted at $Y'_\ell[\bar{a}_\ell]$. Since the non-terminals store their expansion lengths, a single pass over $\mathsf{rle}(\bar{X}_\ell)$ and $\mathsf{rle}(Y'_\ell)$ lets us compute the values g and $\bar{a}$, respectively.

Consequently, we henceforth assume that each arithmetic progression $V_\ell \subseteq \mathsf{Occ}(\bar{X}_\ell, Y'_\ell)$ is already represented by the underlying arithmetic progression V of positions in T; note that each such progression V consists of starting positions of occurrences of $\exp(\bar{X}_\ell)$ in T and its difference g satisfies $g \le |\exp(\bar{X}_\ell)|$.

Define $\bar{c} = |\exp(L_0 \cdots L_{\ell-1})| + 1$ and $c = |X| - |\exp(R_{\ell-1} \cdots R_0)|$ so that $\exp(\bar{X}_\ell) = X[\bar{c}..c]$. Let $V = \{\bar{a} + i \cdot g : i \in [0..s]\}$ be any of the aforementioned arithmetic progressions representing occurrences of $\exp(\bar{X}_\ell)$ in T. Our goal is to test, in bulk, for each position $\bar{a} + i \cdot g \in V$ whether X occurs in T at position $\bar{a} + i \cdot g - \bar{c} + 1$. Then, the only remaining step is to filter out occurrences of X that are not contained within Y. If $|V| = 1$, one can use a single LCE query to verify $T[\bar{a} - \bar{c} + 1..\bar{a} - \bar{c} + |X|] = X$. Thus, we henceforth assume $|V| > 1$.

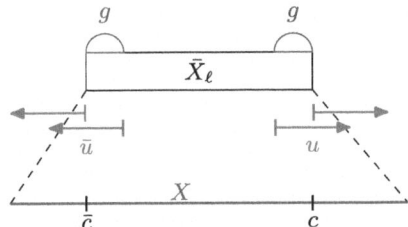 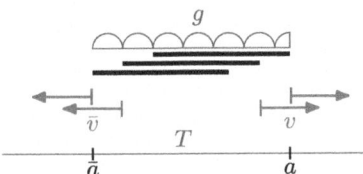

Fig. 2. The queries that compute $\bar{u}$ and u check how far the period g of $\exp(\bar{X}_\ell) = X[\bar{c}..c]$ extends within X.

Fig. 3. The thick black lines are the detected occurrences of $\exp(\bar{X}_\ell)$ in T; these occurrence start g positions apart and their union is $T[\bar{a}..a]$. The queries that compute $\bar{v}$ and v check how far the period g of $T[\bar{a}..a]$ extends within T.

Let $a = \bar{a} + s \cdot g + |\exp(\bar{X}_\ell)| - 1$ be last position of the rightmost occurrence of $\exp(\bar{X}_\ell)$ captured by V. We ask the following LCE queries; see Figs. 2 and 3.

$$\bar{u} := \overline{\mathsf{LCE}}_X(\bar{c}, \bar{c}+g), \qquad u := \mathsf{LCE}_X(c+1, c-g+1),$$
$$\bar{v} := \overline{\mathsf{LCE}}_T(\bar{a}, \bar{a}+g), \qquad v := \mathsf{LCE}_T(a+1, a-g+1)$$

Case 1: $\bar{u} = \bar{c}-1$ *and* $u = |X|-c$. In this case, the period g of $\exp(\bar{X}_\ell) = X[\bar{c}..c]$ extends to the entire X. Therefore, an occurrence of $\exp(\bar{X}_\ell)$ contained in $T[\bar{a}..a]$ extends to an occurrence of X in T if and only if the latter is contained in $T[\bar{a} - \bar{v}..a + v]$, which is the maximal extension of $T[\bar{a}..a]$ with period g. The starting positions of these occurrences of X form an arithmetic progression:
$$\left\{ \bar{a} - \bar{c} + 1 + i \cdot g : i \in \left[\left\lceil \tfrac{\max(0, \bar{u} - \bar{v})}{g} \right\rceil .. s - \left\lceil \tfrac{\max(0, u-v)}{g} \right\rceil \right] \right\}.$$

Case 2: $\bar{u} < \bar{c} - 1$ *or* $u < |X| - c$. Without loss of generality, assume $\bar{u} < \bar{c} - 1$. (In the other case, one can reverse all strings to obtain this situation.) Then, the period g of $\exp(\bar{X}_\ell) = X[\bar{c}..c]$ breaks at position $X[\bar{c} - 1 - \bar{u}]$. The corresponding position within any occurrence of X in T must also break the period; whenever $X[\bar{c}..c]$ is aligned within $T[\bar{a}..a]$, the only candidate position is $T[\bar{a} - 1 - \bar{v}]$. Thus, the only candidate for an occurrence is $T[\bar{a} - \bar{c} + \bar{u} - \bar{v} + 1..\bar{a} - \bar{c} + \bar{u} - \bar{v} + |X|]$, corresponding to $i = \tfrac{\bar{u} - \bar{v}}{g}$; it can be verified with a single LCE query.

Summary. We were able to construct the set of occurrences of X in Y with up to five LCE queries for each of the $\mathcal{O}(1)$ arithmetic progressions representing $\mathrm{Occ}(\bar{X}_\ell, Y'_\ell)$. Each of these queries can be answered in $\mathcal{O}(r)$ time using Proposition 13. Therefore, our verification algorithm has a runtime of $\mathcal{O}(r)$.

Lemma 19. *Given* $\mathsf{Occ}(\bar{X}_\ell, Y'_\ell)$ *as a set of* $\mathcal{O}(1)$ *arithmetic progressions with differences at most* $|\bar{X}_\ell|$, *we can compute* $\mathsf{Occ}(X, Y)$ *in time* $\mathcal{O}(r)$.

Lemma 19 is similar to [19, Lemma 1.14 (b)]. Both answer a set of LCE queries in a periodic string while only interested in the indices of the largest results. Since the LCE queries in our compressed setting are slower than in theirs, we get a larger runtime in total.

References

1. Alstrup, S., Brodal, G.S., Rauhe, T.: Pattern matching in dynamic texts. In: Shmoys, D.B. (ed.) 11th Annual ACM-SIAM Symposium on Discrete Algorithms, SODA 2000, pp. 819–828. ACM/SIAM (2000). http://dl.acm.org/citation.cfm?id=338219.338645
2. Amir, A., Benson, G.: Efficient two-dimensional compressed matching. In: Storer, J.A., Cohn, M. (eds.) IEEE Data Compression Conference, DCC 1992, pp. 279–288. IEEE Computer Society (1992). https://doi.org/10.1109/DCC.1992.227453
3. Amir, A., Landau, G.M., Vishkin, U.: Efficient pattern matching with scaling. J. Algorithms-cognition Inform. Logic **13**(1), 2–32 (1992). https://doi.org/10.1016/0196-6774(92)90003-U
4. Charalampopoulos, P., et al.: Approximate circular pattern matching. In: Chechik, S., Navarro, G., Rotenberg, E., Herman, G. (eds.) 30th Annual European Symposium on Algorithms, ESA 2022,. LIPIcs, vol. 244, pp. 35:1–35:19. Schloss Dagstuhl–Leibniz-Zentrum für Informatik (2022). https://doi.org/10.4230/LIPICS.ESA.2022.35
5. Charalampopoulos, P., Kociumaka, T., Wellnitz, P.: Faster approximate pattern matching: A unified approach. In: Irani, S. (ed.) 61st IEEE Annual Symposium on Foundations of Computer Science, FOCS 2020, pp. 978–989. IEEE Computer Society (2020). https://doi.org/10.1109/FOCS46700.2020.00095
6. Charalampopoulos, P., Kociumaka, T., Wellnitz, P.: Faster pattern matching under edit distance: A reduction to dynamic puzzle matching and the seaweed monoid of permutation matrices. In: 63rd IEEE Annual Symposium on Foundations of Computer Science, FOCS 2022, pp. 698–707. IEEE (2022). https://doi.org/10.1109/FOCS54457.2022.00072
7. Charalampopoulos, P., Pissis, S.P., Radoszewski, J., Rytter, W., Waleń, T., Zuba, W.: Approximate circular pattern matching under edit distance. In: Beyersdorff, O., Kanté, M.M., Kupferman, O., Lokshtanov, D. (eds.) 41st International Symposium on Theoretical Aspects of Computer Science, STACS 2024. LIPIcs, vol. 289, pp. 24:1–24:22. Schloss Dagstuhl–Leibniz-Zentrum für Informatik (2024). https://doi.org/10.4230/LIPICS.STACS.2024.24
8. Chung, K.L.: Fast string matching algorithms for run-length coded strings. Computing **54**(2), 119–125 (1995). https://doi.org/10.1007/BF02238127
9. Clifford, R., Gawrychowski, P., Kociumaka, T., Martin, D.P., Uznański, P.: The dynamic k-mismatch problem. In: Bannai, H., Holub, J. (eds.) 33rd Annual Symposium on Combinatorial Pattern Matching, CPM 2022. LIPIcs, vol. 223, pp. 18:1–18:15. Schloss Dagstuhl–Leibniz-Zentrum für Informatik (2022). https://doi.org/10.4230/LIPICS.CPM.2022.18

10. Gawrychowski, P., Karczmarz, A., Kociumaka, T., Łącki, J., Sankowski, P.: Optimal dynamic strings. In: Czumaj, A. (ed.) 29th Annual ACM-SIAM Symposium on Discrete Algorithms, SODA 2018, pp. 1509–1528. SIAM (2018). https://doi.org/10.1137/1.9781611975031.99
11. I, T.: Longest common extensions with recompression. In: Kärkkäinen, J., Radoszewski, J., Rytter, W. (eds.) 28th Annual Symposium on Combinatorial Pattern Matching, CPM 2017. LIPIcs, vol. 78, pp. 18:1–18:15. Schloss Dagstuhl–Leibniz-Zentrum für Informatik (2017). https://doi.org/10.4230/LIPIcs.CPM.2017.18
12. Jeż, A.: Faster fully compressed pattern matching by recompression. ACM Trans. Algorithms **11**(3), 20:1–20:43 (2015). https://doi.org/10.1145/2631920
13. Jeż, A.: Recompression: A simple and powerful technique for word equations. J. ACM **63**(1), 4:1–4:51 (2016). https://doi.org/10.1145/2743014
14. Kempa, D., Kociumaka, T.: Resolution of the Burrows-Wheeler Transform conjecture. In: Irani, S. (ed.) 61st IEEE Annual Symposium on Foundations of Computer Science, FOCS 2020, pp. 1002–1013. IEEE Computer Society (2020). https://doi.org/10.1109/FOCS46700.2020.00097
15. Kempa, D., Kociumaka, T.: Dynamic suffix array with polylogarithmic queries and updates. In: Leonardi, S., Gupta, A. (eds.) 54th Annual ACM SIGACT Symposium on Theory of Computing, STOC 2022, pp. 1657–1670. ACM (2022). https://doi.org/10.1145/3519935.3520061
16. Kempa, D., Kociumaka, T.: Collapsing the hierarchy of compressed data structures: suffix arrays in optimal compressed space. In: 64th IEEE Annual Symposium on Foundations of Computer Science, FOCS 2023, pp. 1877–1886. IEEE (2023). https://doi.org/10.1109/FOCS57990.2023.00114
17. Kociumaka, T., Navarro, G., Prezza, N.: Towards a definitive compressibility measure for repetitive sequences. IEEE Trans. Inf. Theory **69**(4), 2074–2092 (2023). https://doi.org/10.1109/TIT.2022.3224382
18. Kociumaka, T., Radoszewski, J., Rytter, W., Waleń, T.: Internal pattern matching queries in a text and applications. In: Indyk, P. (ed.) 26th Annual ACM-SIAM Symposium on Discrete Algorithms, SODA 2015, pp. 532–551. SIAM (2015). https://doi.org/10.1137/1.9781611973730.36
19. Kociumaka, T., Radoszewski, J., Rytter, W., Waleń, T.: Internal pattern matching queries in a text and applications. Accepted SIAM J. Comput. (2024). https://arxiv.org/abs/1311.6235v5
20. Landau, G.M., Vishkin, U.: Fast string matching with k differences. J. Comput. Syst. Sci. **37**(1), 63–78 (1988). https://doi.org/10.1016/0022-0000(88)90045-1
21. Mehlhorn, K., Sanders, P.: Algorithms and data structures: the basic toolbox. Springer (2008). https://doi.org/10.1007/978-3-540-77978-0

Bounded-Ratio Gapped String Indexing

Arnab Ganguly[1], Daniel Gibney[2(✉)], Paul MacNichol[3], and Sharma V. Thankachan[3]

[1] University of Wisconsin - Whitewater, Whitewater, WI, USA
gangulya@uww.edu
[2] University of Texas at Dallas, Dallas, TX, USA
daniel.gibney@utdallas.edu
[3] North Carolina State University, Raleigh, NC, USA
{pemacnic,svalliy}@ncsu.edu

Abstract. In the *gapped string indexing* problem, one is given a text $T[1 \mathinner{.\,.} n]$ to preprocess. At query time, a gapped pattern $P = P_1[\alpha \mathinner{.\,.} \beta]P_2$ and an integer range $[\alpha \mathinner{.\,.} \beta]$ are provided, where P_1 and P_2 are strings of total length m. The goal of the query is to report all pairs of occurrences of P_1 and P_2 with a gap falling within $[\alpha \mathinner{.\,.} \beta]$. An existing (conditional) lower bound reveals that any index with query time $\widetilde{\mathcal{O}}(m + occ)$ must occupy almost quadratic space, where occ is the output size. However, there are interesting special cases where more efficient solutions are possible. For example, queries with a bounded gap, i.e., $\beta \leq G$ (fixed at construction) can be answered optimally using an $\widetilde{\mathcal{O}}(nG)$ space structure. In this paper, we bring out an interesting version of the problem where rather than having a fixed upper bound on β, we fix γ and allow any $\beta \leq \gamma \cdot m$ (i.e., allow longer gaps for longer patterns; *gap-to-pattern ratio* is bounded). We show that such queries can be answered optimally using an $\widetilde{\mathcal{O}}(n\gamma)$ space structure.

1 Introduction

Let $T[1 \mathinner{.\,.} n]$ be a string (called the text) over an alphabet Σ and $P = P_1[\alpha \mathinner{.\,.} \beta]P_2$ be a gapped pattern, where P_1 and P_2 are strings over Σ and $[\alpha \mathinner{.\,.} \beta]$ is an integer interval called the *gap range*. Also, let $m = |P_1| + |P_2|$. An occurrence of P in T is a pair (i, j) such that $T[i \mathinner{.\,.} i + |P_1|) = P_1$, $T[j \mathinner{.\,.} j + |P_2|) = P_2$ and the gap $g = j - (i + |P_1|) \in [\alpha \mathinner{.\,.} \beta]$. In gapped-pattern matching, we are interested in finding all those occurrences efficiently. The algorithmic version of this problem can be solved in $\widetilde{\mathcal{O}}(n + m + occ)$ time [6,7,10,15,16], where $\widetilde{\mathcal{O}}(\cdot)$ suppresses polylogarithmic factors. We focus on the indexing version of this problem, where text is initially given for preprocessing and the pattern comes as a query.

Conditioned on the Strong Set-Disjointness Conjecture [11], Bille et al. [5] showed a lower bound that any structure that can answer queries in $\widetilde{\mathcal{O}}(m+occ)$

This research is supported in part by the U.S. National Science Foundation (NSF) award CCF-2315822.

time takes $n^{2-o(1)}$ space.[1] Recently, Bille et al. [4] showed that subquadratic space of $\min\{\widetilde{\mathcal{O}}(n^{2-\delta/3}), \mathcal{O}(n^{3-2\delta})\}$ is possible with higher query time of $\widetilde{\mathcal{O}}(m + n^\delta(occ+1))$, where $\delta \in (0,1)$ is a parameter fixed at construction; note that the time per occurrence is sublinear. See Aronov et al. [2] for a similar result.

There are also several existing results based on restrictions on the pattern and gap lengths. For a version where the lengths $|P_1|, |P_2|$ and the gap of length $\alpha = \beta = g$ are given at the preprocessing time, Peterlongo et al. [17] give an $\mathcal{O}(n)$ space solution with optimal $\mathcal{O}(m + occ)$ query time. For the version where only the gap length g is known at preprocessing, a solution was presented by Iliopoulos and Rahman [12] having $\mathcal{O}(m + \log\log n + occ)$ query time and $\mathcal{O}(n \log^{1+\epsilon} n)$ space, where $\epsilon > 0$ is an arbitrarily small constant. Again, for the case where g is known in advance, Bille and Gørtz [3] presented an improved solution with optimal query time and $\mathcal{O}(n \log^\epsilon n)$ space. The key technique behind all these solutions is to reduce the problem to an equivalent 2D range searching.

For $k+1$ patterns and k gap ranges, $P_1[\alpha_1 .. \beta_1]P_2 \ldots P_k[\alpha_k .. \beta_k]P_{k+1}$, but all $\beta_i \leq G$, Lewenstein [14] presented an $\mathcal{O}(nG^{2k} \log^k n)$ space solution with $\mathcal{O}(m + 2^k \log\log n)$ query time where $m = \sum_{i=1}^{k+1} |P_i|$. This can be viewed as a generalization of the classic result by Cole et al. [8]. For $k = 1$, the space can be easily improved to $\widetilde{\mathcal{O}}(Gn)$ by reducing the problem to an equivalent 3D range searching; we shall refer to this version as *bounded* gapped string indexing.

In this paper, we consider a novel variant of gapped string indexing where rather than having a fixed upper bound G on β at preprocessing time, the upper bound on β is at most some fixed multiple γ of the pattern length, i.e., $\beta \leq \gamma \cdot m$. We assume that γ is known at the time of preprocessing of T. We call this version of the problem *bounded-ratio* gapped string indexing. In contrast with the *bounded* gapped string indexing, we allow longer gaps for longer patterns. Our main result is stated in Theorem 1.

Theorem 1. *By maintaining an index of size $\mathcal{O}(\gamma \cdot n \cdot \log^{4+\epsilon} n)$, we can find all occ occurrences of any gapped pattern $P = P_1[\alpha .. \beta]P_2$ of length $m = |P_1| + |P_2|$ in $T[1..n]$, such that $\beta \leq \gamma \cdot m$, in optimal $\mathcal{O}(m + occ)$ time. Here γ is fixed while index construction, and $\epsilon > 0$ is an arbitrarily small constant.*

We highlight that within a fixed space budget (i.e., when nG and $n\gamma$ are close within polylogarithmic factors), the queries that can be optimally handled by our solution is a *"superset"* of the queries that can be handled by the $\widetilde{\mathcal{O}}(Gn)$ space solution for the bounded gapped string indexing.

2 Preliminaries

We consider strings over a polynomially-sized integer alphabet Σ. For a string T, let $|T|$ be its length and $T[i]$ be its ith character. We use $T[i..j]$ to denote the substring $T[i] \cdot \ldots \cdot T[j]$, where $\cdot$ denotes concatenation. We also use $T[i..j)$

---

[1] Although stated for reporting consecutive occurrences with gap constraints, their proof easily extends to gapped string indexing.

to denote the substring $T[i] \cdot \ldots \cdot T[j-1]$, $T(i..j]$ to denote the substring $T[i+1] \cdot \ldots \cdot T[j]$, $T[i..]$ to denote the suffix $T[i..|T|]$, and $T[..j]$ to denote the prefix $T[1..j]$. The reverse of T is written as $\overleftarrow{T}$, and we write $\overleftarrow{T[i..j]}$ for the reverse of the substring $T[i..j]$.

2.1 Suffix Trees and Suffix Arrays

Let $\$ \notin \Sigma$ be a character lexicographically smaller than all symbols in Σ. The suffix tree [18] of T, written $\text{ST}(T)$ is a compacted trie of all suffixes of $T \cdot \$$. We henceforth let $n = |T \cdot \$|$. For a node v in $\text{ST}(T)$, the string obtained by concatenating edge labels on the root-to-v path is denoted as $\text{path}(v)$. The leaves of $\text{ST}(T)$, $\ell_1, \ell_2, \ldots$, are ordered such that $\text{path}(\ell_i)$ is i^{th} suffix in lexicographically ascending order. The suffix array $\text{SA}[1..n]$ is defined such that $\text{SA}[i]$ is the starting position in T of the ith suffix in lexicographically ascending order. For a node u, the *suffix link* is the node v, such that $\text{path}(v) = \text{path}(u)[2..]$.

The locus of a pattern P is defined as the highest node u (if it exists), such that P is a prefix of $\text{path}(u)$, and it can be found in $\mathcal{O}(|P|)$ time. Let $[L..R]$ be the range of leaves such that $\ell_L, \ell_{L+1}, \ldots, \ell_R$ are leaves in the subtree of the locus of P. The range $[L..R]$ is called the *suffix range* of P. The starting positions of all occurrences of P in T is $\{\text{SA}[i] \mid i \in [L..R]\}$. Given the locus of $P[k..]$ (a suffix of the pattern), we can compute the locus of $P[k+1..]$ using a suffix link in $\mathcal{O}(1)$ time. The suffix tree (including suffix links) and suffix array can be constructed in linear time for polynomially-sized integer alphabets [9].

2.2 Range Reporting Data Structures

The range reporting problem is to store a set of n d-dimensional points $\mathcal{P}$ in a data structure so that for a query rectangle Q, all points in $Q \cap \mathcal{P}$ can be reported. The following results will be utilized in our solution.

Lemma 1 (Karpinski and Nekrich [13].) *Orthogonal range reporting of a set of n d-dimensional points ($d > 3$) can be performed in $\mathcal{O}(\frac{\log^{d-3} n}{(\log \log n)^{d-6}} + occ)$ time by maintaining an $\mathcal{O}(n \log^{d-2+\epsilon} n)$ space index, where occ represents the size of the output and $\epsilon > 0$ is an arbitrarily small constant.*

Lemma 2 (Alstrup et al [1].) *Orthogonal range reporting of a set of n 1-dimensional points can be performed in $\mathcal{O}(occ)$ time by maintaining an $\mathcal{O}(n)$ space index, where occ represents the size of the output.*

3 Our Solution

We describe the solution for the case where $|P_2| \geq |P_1|$. The case where $|P_1| \geq |P_2|$ can be handled by building our index also over $\overleftarrow{T}$ and modifying the query pattern $P_1[\alpha..\beta]P_2$ to $\overleftarrow{P_2}[\alpha..\beta]\overleftarrow{P_1}$. Note that $|P_1| \leq |P_2|$ and $\beta \leq \gamma(|P_1|+|P_2|)$ together imply $\beta \leq 2\gamma |P_2|$.

The overall idea of the solution is to sample a set of positions for each potential ending position i of P_1 such that some sampled position should lie inside P_2 in any occurrence. Since the length of P_2 must increase as the allowable gap between P_1 and P_2 increases, the farther from i we are, the fewer samples have to be taken. This is formalized by defining a set of blocks for each position i and sample positions within each block.

3.1 Data Structure

We fix a parameter $r \geq 2\gamma$ at construction time. The exact value of r will be specified in the analysis in Sect. 3.4. For $i \in [1..n]$, we define *blocks* B_1^i, B_2^i .... The block $B_b^i = [s_b^i .. s_{b+1}^i)$ where $s_1^i = i+1$ and $s_b^i = i + 2 + (2^{b-1} - 1)r$ for $b > 1$. Note that the first block ($b = 1$) has length $r + 1$, the second block ($b = 2$) has length $2r$, the third block has length $4r$, and so on, with the length of the blocks changing by a multiplicative factor of 2. For every i, we consider the above blocking scheme as starting at $i + 1$ and extending until the end of the string T.

Within each block B_b^i, we define a set of *sample positions*. In block B_1^i we sample all $r + 1$ positions. For blocks B_b^i with $b > 1$, we sample exactly r evenly spaced positions. In particular, for $j = 1, \ldots, r$, the jth sample position in block B_b^i is at the position $s_b^i + j2^{b-1} - 1$. See Fig. 1.

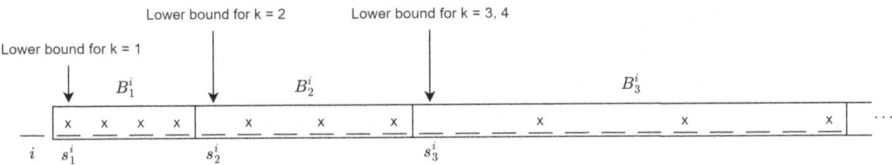

Fig. 1. Blocks B_1^i, B_2^i, and B_3^i for $r = 3$. Sample positions are indicated with an 'x'. The relative lower bounds for different query iterations over k are shown as well.

To create our data structure, for each $i \in [1..n]$ and sampled position j in some block B_b^i, we compute:

- The leaf ℓ_x in $\mathrm{ST}(\overleftarrow{T})$ such that $\overleftarrow{T[..i]} = \mathrm{path}(\ell_x)$.
- The leaf ℓ_y in $\mathrm{ST}(\overleftarrow{T})$ such that $\overleftarrow{T[..j-1]} = \mathrm{path}(\ell_y)$.
- The leaf ℓ_z in $\mathrm{ST}(T)$ such that $T[j..] = \mathrm{path}(\ell_z)$.

Based on i and j, and the x, y, and z values of the suffix tree leaves computed above, we create the 4D point $(x, y, z, j - i)$. All points are added to the 4D range reporting data structure of Karpinski and Nekrich.

3.2 Querying

Our querying procedure iterates over $k \in [1 .. \lceil \beta/r \rceil]$. Our algorithm first finds the suffix range $[L_1 .. R_1]$ of $\overleftarrow{P_1}$ in $\mathrm{ST}(\overleftarrow{T})$. Next, for each $k \in [1 .. \lceil \beta/r \rceil]$, we find the suffix ranges $[L_2^k .. R_2^k]$ of $\overleftarrow{P_2[.. k-1]}$ in $\mathrm{ST}(\overleftarrow{T})$ as well as $[L_3^k .. R_3^k]$ of $P_2[k..]$ in $\mathrm{ST}(T)$. This is accomplished in $\mathcal{O}(|P_2| + \lceil \beta/r \rceil)$ time by finding the locus of $\overleftarrow{P_2[1 .. \lceil \beta/r \rceil]}$ and then applying suffix links.

Having obtained these suffix ranges, for $k \in [1 .. \lceil \beta/r \rceil]$, we make the following range reporting query

$$[L_1 .. R_1] \times [L_2^k .. R_2^k] \times [L_3^k .. R_3^k] \times [\max(s^1_{\lceil \log k \rceil + 1} - 1, \alpha + k) .. \beta + k]$$

For each point $(x, y, z, j - i)$ reported for a given k, we output the pair $(\mathrm{SA}[z] - (j - i) - |P_1| + 1, \mathrm{SA}[z] - k + 1)$ as the starting positions of P_1 and P_2, respectively.

3.3 Correctness

We say that a sampled position j in T *anchors* an occurrence of $P_1 = T(i - |P_1| .. i]$ and $P_2 = T[h .. h + |P_2|)$ if $j \in [h .. h + |P_2|)$. Note that for an anchor at index j in T aligning to index k in P_2, the starting position of P_2 is $j - k + 1$, and the gap between P_1 and P_2 is $g = j - k - i$. Based on the gap range $[\alpha .. \beta]$, we require that $\alpha \leq g = j - k - i \leq \beta$ or $j - i \in [\alpha + k .. \beta + k]$. See Fig. 2.

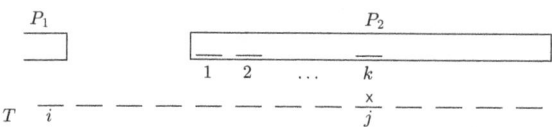

Fig. 2. Index positions of anchor j in T aligning to index k in P_2.

We call a valid gapped match of P_1 and P_2 a *type-k occurrence* if its first anchor in P_2 aligns with an index k in P_2; we report such occurrences via kth range query.

Note that we adjust the lower bound on $j - i$ in the range query based on iteration, k. For $k = 1$, we wish to consider anchors in blocks B_1^i, B_2^i, ... (with respect to i). For $k = 2$, we wish to consider anchors in blocks B_2^i, B_3^i, ... and so on. Subsequently, we wish to change the lower bound from B_b^i to B_{b+1}^i after 2^{b-2} iterations of k with the lower bound starting at B_b^i. See Fig. 1. This can be equivalently described by wanting to capture anchors at position j in T satisfying $s^i_{\lceil \log k \rceil + 1} \leq j$. Since $s^i_b = s^1_b - 1 + i$, this can equivalently be expressed as $s^1_{\lceil \log k \rceil + 1} - 1 \leq j - i$. In the following lemmas, we will prove that all occurrences are reported and that this adjustment guarantees each occurrence is reported exactly once.

Lemma 3. *For any valid occurrence of P_1 and P_2 there exists some $k \in [1\mathinner{.\,.}\lceil \beta/r \rceil]$ such that the occurrence is a type-k occurrence.*

Proof. The case where the gap $g = 0$ will result in a type-1 occurrence. For gap $g > 0$, suppose P_2 starts in block B_b^i. Then the gap g satisfies $g \geq (2^{b-1}-1)r+1$. Since $g \leq \beta \leq 2\gamma|P_2|$, we have $|P_2| \geq \lceil g/(2\gamma) \rceil \geq \lceil g/r \rceil \geq \lceil 2^{b-1} - 1 + 1/r \rceil = 2^{b-1}$. In block B_b^i, every substring of length at least 2^{b-1} must contain an anchor, hence P_2 is anchored. Furthermore, since $2^{b-1} \leq \lceil g/r \rceil \leq \lceil \beta/r \rceil$, we have that an anchor occurs in the first $\lceil \beta/r \rceil$ positions of P_2. □

Lemma 4. *For each $k \in [1\mathinner{.\,.}\lceil \beta/r \rceil]$, all type-$k$ occurrences are reported.*

Proof. All type-1 occurrences are reported in the first iteration. Now consider a type-k occurrence ($k > 1$) of P_1 and P_2 that has P_1 ending at i in T and P_2 anchored at j in T. See Fig. 2. It suffices to show that $s^1_{\lceil \log k \rceil + 1} - 1 \leq j - i$, or equivalently $s^1_{\lceil \log k \rceil + 1} - 1 + i = s^i_{\lceil \log k \rceil + 1} \leq j$. For the sake of contradiction, suppose that j is in block B_b^i with $b < \lceil \log k \rceil + 1$. That is, $b \leq \lceil \log k \rceil$. The occurrence of P_2 starts at position $j - k + 1$. Observe also that there exists a sampled position at $j - 2^{b-1}$. This sampled position is also an anchor for P_2, i.e., $j - k + 1 \leq j - 2^{b-1}$ or equivalently,

$$2^{b-1} \leq k - 1.$$

Note that the inequality $2^{\lceil \log k \rceil - 1} \leq k - 1$ holds for all $k \geq 2$. Hence,

$$2^{b-1} \leq 2^{\lceil \log k \rceil - 1} \leq k - 1$$

contradicting that this is a type-k occurrence. □

Lemma 5. *No occurrence is reported more than once.*

Proof. Consider a type-k occurrence of P_1 and P_2 with P_1 ending at i in T and P_2 anchored first at j in T and anchored again at some $j' > j$ in T. First, suppose j is in the block B_b^i and j' is also in B_b^i. Then, $j' - j = 2^{b-1}$. However, after 2^{b-2} (or 1 if $b = 1$) iterations we modify the lower bound on $j - i$ from $s_b^1 - 1$ to $s_{b+1}^1 - 1$, excluding the possibility of the occurrence being reported for j'. In particular, by the iteration where we are considering the suffix ranges for $\overleftarrow{P_2[\mathinner{.\,.} k + (j' - j)]}$ and $P_2[k + (j' - j) + 1\mathinner{.\,.}]$, the procedure has already shifted the lower bound past the block B_b^i. Similarly, if j' is in a block $B_{b'}^i$ where $b' > b$, then $j' - j \geq 2^{b'-1}$. Observe that after at most $2^{b-2} + 2^{b-1} + \ldots + 2^{b'-2} < 2^{b'-1}$ iterations, the lower bound on $j - i$ is updated to $s^1_{b'+1} - 1$. Therefore, by the iteration where we are considering the suffix ranges for $\overleftarrow{P_2[\mathinner{.\,.} k + (j' - j)]}$ and $P_2[k + (j' - j) + 1\mathinner{.\,.}]$, the block whose anchors are being considered has already passed $B_{b'}^i$. This prevents the occurrence from being reported for anchor j'. □

3.4 Complexity Analysis

Query Time: We find the suffix range for $\overleftarrow{P_1}$ in $\mathcal{O}(|P_1|)$ time and obtain the desired suffix ranges for P_2 in time $\mathcal{O}(|P_2| + \lceil \beta/r \rceil)$. The time for this step is therefore $\mathcal{O}(|P_2| + \lceil \gamma |P_2|/r \rceil)$. We also make $\lceil \beta/r \rceil = \mathcal{O}(\lceil \gamma |P_2|/r \rceil)$ range queries, each of which reports distinct occurrences by Lemma 5. Each range query takes $\mathcal{O}(\log n / (\log \log n)^2)$ time. The total time taken is therefore

$$\mathcal{O}\left(|P_1| + |P_2| + \left\lceil \frac{\gamma |P_2|}{r} \right\rceil \frac{\log n}{(\log \log n)^2} + occ \right)$$

We now make $r = 2\gamma \cdot \lceil \log n / (\log \log n)^2 \rceil$. The total query time becomes,

$$\mathcal{O}\left(|P_1| + |P_2| + \frac{\log n}{(\log \log n)^2} + occ \right)$$

We show in Sect. 4 how to improve the query time to $\mathcal{O}(m + occ)$.

Space Complexity: The total number of 4D points we create for a given i is $\mathcal{O}(r \log n)$. Thus, the total number of 4D points overall is $\mathcal{O}(rn \log n)$. The 4D range reporting structure built over these points requires $\mathcal{O}(rn \log^{3+\epsilon} n)$ space. Using our choice of r, the total space is $\mathcal{O}(\gamma n \frac{\log^{4+\epsilon} n}{(\log \log n)^2})$, which is $\mathcal{O}(\gamma n \log^{4+\epsilon} n)$.

4 Obtaining Optimal Query Time

In the case where $m = \omega(\log n / (\log \log n)^2)$, we achieve optimal query time with the above algorithm. Here, we consider an algorithm that can be applied whenever $m \leq \lceil \log n \rceil$.

Data Structure: For $\ell_1 \in [1..\lceil \log n \rceil]$, $\ell_2 \in [1..\lceil \log n \rceil]$, and $g \in [1..\gamma(\ell_1 + \ell_2)]$, we consider starting locations $i \in [1..n - (\ell_1 + g + \ell_2) + 1]$. Let $T_1 := T[i..i+\ell_1)$ and $T_2 := T[i+\ell_1+g..i+\ell_1+g+\ell_2)$. We store all tuples (i, T_1, T_2, g).

Now, for all distinct pairs of T_1 and T_2 found above, we create a 1D range reporting structure of Alstrup, say $R_{(T_1,T_2)}$ over the tuples (i, T_1, T_2, g), where the 1D points are over the g values and the corresponding i values are stored as a list for a given g. Moreover, we can associate $R_{(T_1,T_2)}$ with a leaf of a suffix tree built over all distinct concatenations $T_1\$_1 T_2\$_2$ so that given T_1 and T_2 we can access $R_{(T_1,T_2)}$ in $\mathcal{O}(|T_1| + |T_2|)$ time.

Querying: Given $P_1[\alpha..\beta]P_2$ we first check whether $m \geq \lceil \log n \rceil$ and if so, use the data structure described in Sect. 3.1. If instead $m \leq \lceil \log n \rceil$, we locate the range reporting structure $R_{(P_1,P_2)}$ in $\mathcal{O}(m)$ time. We then perform the 1D range reporting query $[\alpha..\beta]$ on $R_{(P_1,P_2)}$. This requires $\mathcal{O}(occ)$ time. For each point (g) returned, we access its list of i values and output as an occurrence starting locations $(i, i + |P_1| + g)$.

Space Complexity: Observe that the total number of tuples is $n \cdot \log n \cdot \log n \cdot 2\gamma \log n = \mathcal{O}(\gamma n \log^3 n)$. The range reporting structures are built over disjoint sets of tuples. Using the structure of Alstrup et al. [1], the total space taken over all range reporting structures remains $\mathcal{O}(\gamma n \log^3 n)$. Since there are $\mathcal{O}(\gamma n \log^3 n)$ distinct pairs T_1, T_2 and each $T_1\$_1T_2\$_2$ has length $\mathcal{O}(\log n)$, the total space needed for the suffix tree is $\mathcal{O}(\gamma n \log^4 n)$.

References

1. Alstrup, S., Brodal, G.S., Rauhe, T.: Optimal static range reporting in one dimension. In: Vitter, J.S., Spirakis, P.G., Yannakakis, M. (eds.) Proceedings on 33rd Annual ACM Symposium on Theory of Computing, 6-8 July 2001, Heraklion, Crete, Greece, pp. 476–482. ACM (2001). https://doi.org/10.1145/380752.380842
2. Aronov, B., Cardinal, J., Dallant, J., Iacono, J.: A general technique for searching in implicit sets via function inversion. In: 2024 Symposium on Simplicity in Algorithms (SOSA), pp. 215–223. SIAM (2024)
3. Bille, P., Gørtz, I.L.: Substring range reporting. Algorithmica **69**(2), 384–396 (2014). https://doi.org/10.1007/S00453-012-9733-4
4. Bille, P., Gørtz, I.L., Lewenstein, M., Pissis, S.P., Rotenberg, E., Steiner, T.A.: Gapped string indexing in subquadratic space and sublinear query time. CoRR abs/ arXiv: 2211.16860 (2022)
5. Bille, P., Gørtz, I.L., Pedersen, M.R., Steiner, T.A.: Gapped indexing for consecutive occurrences. Algorithmica **85**(4), 879–901 (2023). https://doi.org/10.1007/S00453-022-01051-6
6. Bille, P., Gørtz, I.L., Vildhøj, H.W., Wind, D.K.: String matching with variable length gaps. Theor. Comput. Sci. **443**, 25–34 (2012). https://doi.org/10.1016/J.TCS.2012.03.029
7. Bille, P., Thorup, M.: Regular expression matching with multi-strings and intervals. In: Charikar, M. (ed.) Proceedings of the Twenty-First Annual ACM-SIAM Symposium on Discrete Algorithms, SODA 2010, Austin, Texas, USA, 17-19 January 2010, pp. 1297–1308. SIAM (2010). https://doi.org/10.1137/1.9781611973075.104
8. Cole, R., Gottlieb, L., Lewenstein, M.: Dictionary matching and indexing with errors and don't cares. In: Babai, L. (ed.) Proceedings of the 36th Annual ACM Symposium on Theory of Computing, Chicago, IL, USA, 13-16 June 2004, pp. 91–100. ACM (2004). https://doi.org/10.1145/1007352.1007374
9. Farach, M.: Optimal suffix tree construction with large alphabets. In: 38th Annual Symposium on Foundations of Computer Science, FOCS 1997, Miami Beach, Florida, USA, 19-22 October 1997, pp. 137–143. IEEE Computer Society (1997). https://doi.org/10.1109/SFCS.1997.646102
10. Fredriksson, K., Grabowski, S.: Efficient algorithms for pattern matching with general gaps, character classes, and transposition invariance. Inf. Retr. **11**(4), 335–357 (2008). https://doi.org/10.1007/S10791-008-9054-Z
11. Goldstein, I., Kopelowitz, T., Lewenstein, M., Porat, E.: Conditional lower bounds for space/time tradeoffs. In: Ellen, F., Kolokolova, A., Sack, J. (eds.) WADS 2017, LNCS, vol. 10389, pp. 421–436. Springer (2017). https://doi.org/10.1007/978-3-319-62127-2_36
12. Iliopoulos, C.S., Rahman, M.S.: Indexing factors with gaps. Algorithmica **55**(1), 60–70 (2009). https://doi.org/10.1007/S00453-007-9141-3

13. Karpinski, M., Nekrich, Y.: Space efficient multi-dimensional range reporting. In: Ngo, H.Q. (ed COCOON 2009. LNCS, vol. 5609, pp. 215–224. Springer (2009). https://doi.org/10.1007/978-3-642-02882-3_22
14. Lewenstein, M.: Indexing with gaps. In: Grossi, R., Sebastiani, F., Silvestri, F. (eds.) SPIRE 2011. LNCS, vol. 7024, pp. 135–143. Springer (2011). https://doi.org/10.1007/978-3-642-24583-1_14
15. Morgante, M., Policriti, A., Vitacolonna, N., Zuccolo, A.: Structured motifs search. J. Comput. Biol. **12**(8), 1065–1082 (2005). https://doi.org/10.1089/CMB.2005.12.1065
16. Navarro, G., Raffinot, M.: Fast and simple character classes and bounded gaps pattern matching, with applications to protein searching. J. Comput. Biol. **10**(6), 903–923 (2003). https://doi.org/10.1089/106652703322756140
17. Peterlongo, P., Allali, J., Sagot, M.: Indexing gapped-factors using a tree. Int. J. Found. Comput. Sci. **19**(1), 71–87 (2008). https://doi.org/10.1142/S0129054108005541
18. Weiner, P.: Linear pattern matching algorithms. In: 14th Annual Symposium on Switching and Automata Theory, Iowa City, Iowa, USA, 15-17 October 1973. pp. 1–11. IEEE Computer Society (1973). https://doi.org/10.1109/SWAT.1973.13

Simultaneously Building and Reconciling a Synteny Tree

Mathieu Gascon, Mattéo Delabre, and Nadia El-Mabrouk

Département d'informatique et de recherche opérationnelle, Université de Montréal, Québec, Canada
{mathieu.gascon.1,matteo.delabre}@umontreal.ca, mabrouk@iro.umontreal.ca

Abstract. We present *FullSynesth*, a tree reconciliation algorithm predicting the evolution of a set of homologous genomic regions or *syntenies*, inside a species tree. The considered evolutionary model involves *segmental events* (i.e. acting on multiple genes) including duplications (D), losses (L), synteny fissions and transfers possibly going through unsampled or extinct species. Formally, given a set of syntenies in a set of genomes and a set $\mathcal{G}$ of consistent gene trees for the gene families composing the syntenies, the problem is to infer a most parsimonious evolutionary history explaining the observed gene trees and syntenies given a species tree. The problem is NP-hard for the DL distance. FullSynesth is based on *Synesth* explicating the evolution of a set of syntenies given a single *synteny tree*, which can be obtained from $\mathcal{G}$ by selecting an "optimal" supertree. Rather than trying each supertree in turn, FullSynesth is based on a two-in-one approach simultaneously building and reconciling a *synteny supertree*. The running time of this algorithm is exponential in the number of gene trees rather than in the size of gene trees. We show on simulated datasets that FullSynesth significantly improves the running time of Synesth applied to each possible supertree. An implementation of the algorithm is available at: http://www.iro.umontreal.ca/~mabrouk/.

1 Introduction

A *gene/species trees reconciliation* is an embedding of a gene tree into a species tree explaining the difference between the two trees through a sequence of events shaping the gene family inside the species tree. Reconciliation has been widely studied [1], first focusing on duplications (D) and losses (L), then extending to horizontal gene transfers (HGT) and other events [2,3]. Various software tools have been developed (e.g. Ranger-DTL [4], Eucalypt [5], ecceTERA [6]). However, a major drawback of classical reconciliation is that gene families are considered separately from one another, which is not appropriate for genes organized in *syntenies*, i.e. colocalized genes evolving together through segmental events. This is the case of several operons such that the one containing the Cas genes of the CRISPR-Cas systems allowing prokaryotes to defend against invading viruses and plasmids [7]. Various methods have been developed to infer the evolution of adjacencies [8], group individual events into segmental ones [9], or minimize duplication episodes [10,11], but none of them are intended to explicitly look for evolutionary scenarios with segmental events.

We presented the first algorithm generalizing reconciliation to segmental events and synteny trees (i.e. with leaves representing syntenies rather than single genes) for the Duplication-Loss (DL) distance in [12], for the Duplication-Transfer-Loss (DTL) distance in [13], and more recently for a more flexible evolutionary model also involving cuts (fissions) and the possibility of transient events going through unsampled or extinct species [14]. This latter algorithm is called *Synesth* (for *SYNteny Evolution in SegmenTal Histories*).

For any tree reconciliation method, obtaining accurate input trees is crucial. This is particularly challenging in the case of a reconciliation model requiring a synteny tree as input, as phylogenetic studies usually lead to gene trees. If the individual gene trees are consistent, then a supertree (a tree displaying them all) can be obtained [15–17]. We call a *synteny supertree* such a supertree leaf-labeled with the corresponding syntenies. However, the number of such synteny supertrees can be exponential in the number of syntenies.

In this paper, given a species tree S, a set $\mathcal{F}$ of gene families, a set $\mathcal{X}$ of syntenies containing genes from $\mathcal{F}$, and a set of consistent gene trees $\mathcal{G} = \{G_1, G_2, \ldots, G_k\}$ for each gene family of $\mathcal{F}$, we seek a synteny supertree for $\mathcal{G}$ leading to a most parsimonious reconciliation with S explaining the evolution of $\mathcal{X}$. The problems of finding a minimum supertree [18] or synteny supertree [12], respectively for single and segmental events, have been shown NP-hard for the simplest DL distance model[1]. A naive approach consists in generating all supertrees (respec. synteny supertrees), reconcile each of them with the species tree, and keep one leading to the minimum cost. Here, we present FullSynesth, an algorithm for simultaneously building and reconciling an optimal synteny supertree, which combines the dynamic programming approach of Synesth with the supertree reconstruction strategy of [19]. In contrast to the naive approach whose running time is exponential in the sum of sizes of all gene trees, FullSynesth is exponential in the number of gene families composing the syntenies (i.e. size of operons or paralogons), which is usually a much smaller parameter. By running the algorithms on simulated datasets, we show that FullSynesth significantly improves the running time of Synesth applied to each possible supertree.

2 Notations on Trees and Syntenies

All trees considered in this paper are rooted. Given a tree T, we denote by $r(T)$ its root, by $V(T)$ its node set and by $L(T) \subseteq V(T)$ its leafset. We say that T is *a tree on* L if $L(T) \subseteq L$. A node v' is an *ancestor* of v if v' is on the (inclusive) path between v and the root, and we then call v a *descendant* of v'. The node $v' = p(v)$ immediately preceding $v \neq r(T)$ on this path is the *parent*

---

[1] Note that in [12] the model accounts for gene orders in the syntenies, yet the NP-hardness proof can easily be adapted to our unordered definition of syntenies.

of v, and then v is a *child* of v'. The ancestor–descendant relation is denoted $\leq$ and forms a partial order on nodes, in which the root is minimal and the leaves are maximal. Any pair of nodes v and v' not ordered by this relation are said to be *separated*, which we denote $v \parallel v'$. The set of children of any node v is denoted by ch(v). If $|\text{ch}(v)| = 1$, then v is said to be *unary* and we denote its only child by v_c. If $|\text{ch}(v)| = 2$, then it is said to be *binary* and, unless specified otherwise, we denote its children by v_ℓ and v_r in no particular order. We denote by $E(T)$ the edge set of T, where each edge is represented by a pair of nodes $(p(v), v)$. For any two nodes v and v' of T, there exists a unique path from v to v' that we denote $\text{P}_T(v, v') \subseteq \text{E}(T)$.

We denote by T_v the subtree of T rooted at a node v, i.e. obtained from T by removing all the nodes which are not descendants of v. A *binary tree* is a tree where all internal (non-leaf) nodes are binary. For a binary node v, we call $(L(T_{v_l}), L(T_{v_r}))$ the bipartition of v. If all internal nodes are unary or binary, then the tree is *partially binary*. For two trees T_1 and T_2, we write $T_1 = T_2$ iff there is an isomorphism preserving leaf labels between T_1 and T_2.

A tree T' is said to be an *extension* of a tree T if T' can be obtained from T by a sequence of operations among: (1) *subdividing* an edge (u, w) by adding a new node v and replacing (u, w) by two edges (u, v) and (v, w); (2) *grafting* a new node v below an existing node u by adding the edge (u, v); (3) *rerooting* the tree to a new node u by adding the edge $(u, \text{r}(T))$. If T' is an extension of T and v is a node of T, then we denote by $\Psi_{T'}(v)$ the node of T' corresponding to v.

The *lowest common ancestor* (lca) of a subset L' of $L(T)$, denoted $lca_T(L')$, is the ancestor common to all nodes in L' most distant from the root. The restriction $T|_{L'}$ of T to L' is the tree with leafset L' obtained from the subtree of T rooted at $lca_T(L')$ by removing all leaves not in L', and removing all internal nodes with a single child. We generalize this notation to a set of trees: For a set $\mathbb{T}$ of trees, $\mathbb{T}_{L'} = \{T|_{L'} : T \in \mathbb{T}\}$. Let T' be a tree such that $L(T') = L' \subseteq L(T)$. We say that T *displays* T' if $T|_{L'} = T'$. A set $\mathbb{T}$ of trees is said to be *consistent* if there is a tree T displaying each tree of $\mathbb{T}$. A *supertree* for a set $\mathbb{T}$ of consistent trees is a tree displaying them all. See Fig. 1 for an example.

Species, Genes and Synteny Trees: A *species tree* on a set Σ of species is a tree S with a bijection s between $\text{L}(S)$ and Σ, while a *gene tree* G for a gene family Γ is a tree with a bijection between $\text{L}(G)$ and Γ. A set of syntenies is a set of genomic segments in a set of genomes (potentially with multiple segments per genome). In this paper, each synteny contains a set of genes from a set of gene families $\mathcal{F}$, where the genes of a given synteny all belong to different gene families (i.e. repeated gene copies inside a synteny are ignored). Therefore, from now on, a gene is simply identified by the family $\Gamma \in \mathcal{F}$ it belongs to. Consider a set $\mathcal{X}$ of syntenies in Σ. A *synteny tree* for $\mathcal{X}$ is a tree T with a bijection between $L(T)$ and $\mathcal{X}$. Let $\Gamma \in \mathcal{F}$ be a gene family with a one-to-one function between Γ and $\mathcal{X}$. We will identify the leaves of the gene tree for Γ by the synteny to which the corresponding gene belongs. Conceptually, the leaves of a synteny tree T represent syntenies, while the leaves of a gene tree G represent genes identified by the synteny they belong to. Let $\mathcal{G}$ be a set of consistent gene

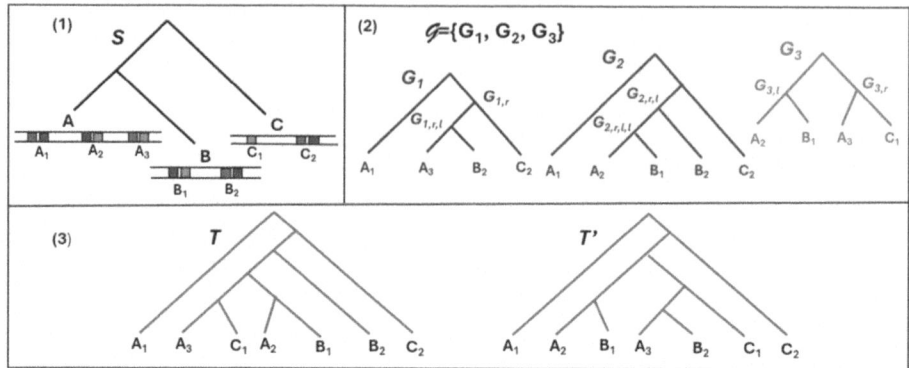

Fig. 1. (1) A species tree S for the set of species $\Sigma = \{A, B, C\}$. Each species is represented as a linear genome with syntenies on the set $\mathcal{F}$ containing the blue, red and green gene families. The set of syntenies is $\mathcal{X} = \{A_1, A_2, A_3, B_1, B_2, C_1, C_2\}$, each X_i belonging to the species X. (2) The set $\mathcal{G} = \{G_1, G_2, G_3\}$ of the gene trees for the blue, red and green gene families. Gene copies are identified (on leaves) by the synteny each one belongs to. This set of trees is consistent. (3) Two synteny supertrees for $\mathcal{G}$. (Colour figure online)

trees, one for each family of $\mathcal{F}$. With the above notation (leaves of gene trees identified by the syntenies they belong to), we can consider them all as trees on $\mathcal{X}$. We call *synteny supertree* a supertree for $\mathcal{G}$. See an example in Fig. 1.

For a synteny or gene tree T, we define $x : L(T) \to \mathcal{P}(\mathcal{F})$ the function mapping each synteny to its content and $s : L(T) \to \Sigma$ the function mapping each synteny to its corresponding species. We say that a set $\mathcal{G}$ of trees *is on* $\mathcal{X}$ if $\cup_{G \in \mathcal{G}} L(G) = \mathcal{X}$. For a set $\mathcal{G}$ of trees on $\mathcal{X}$, we denote $x(\mathcal{G}) = \cup_{\xi \in \mathcal{X}} x(\xi)$.

Finally, the restriction of any function f to a subset A of its domain is denoted by $f|_A$. Moreover, we use the Iverson bracket notation $[\![P]\!]$, where P is a boolean statement, and $[\![P]\!] = 1$ if P is true and $[\![P]\!] = 0$ otherwise.

3 Evolutionary Histories for Syntenies

We consider the evolutionary model for syntenies defined in [14], which includes divergence following speciation ("Spe"), duplication of a synteny segment ("Dup"), fission of a synteny ("Cut"), transfer of a duplicated or cut segment (resp. "TrDup" or "TrCut"), and gain or loss of a segment ("Gain", "Loss"). Losses can be partial in the sense that only a segment of a synteny can be lost. Evolutionary histories are sequences of such events, as formally defined below. See Fig. 2 for an example.

Definition 1 (Evolutionary History for Syntenies). *A history $\mathcal{H}$ on a binary species tree S is a tuple $\langle H, e, x, s \rangle$, where H is a partially binary tree. Each node $v \in V(H)$ is labeled with a species $s(v) \in V(S)$ and a synteny content $x(v)$. Each internal node is additionally labeled with an event $e(v) \in \mathcal{E} = \{$Spe,*

Dup, Cut, TrDup, TrCut, Gain, Loss} *acting on $x(v)$ and $s(v)$. These labelings satisfy the following conditions:*

1. *If $e(v) = $ Spe, then v is binary; let $\mathrm{ch}(v) = \{v', v''\}$ and $\sigma = s(v)$, then $x(v) = x(v') = x(v'')$, $s(v') = \sigma_\ell$, and $s(v'') = \sigma_r$.*
2. *If $e(v) \in \{\text{Dup}, \text{Cut}, \text{TrDup}, \text{TrCut}\}$, then v is binary; let $\mathrm{ch}(v) = \{v_t, v_k\}$:*
   1. *if $e(v) \in \{\text{Dup}, \text{TrDup}\}$, then $x(v_t) \subseteq x(v) = x(v_k)$;*
   2. *if $e(v) \in \{\text{Cut}, \text{TrCut}\}$, then $x(v_t) \cup x(v_k) = x(v)$ and $x(v_t) \cap x(v_k) = \emptyset$ (but $x(v_t) = \emptyset$ or $x(v_k) = \emptyset$ is allowed);*
   3. *if $e(v) \in \{\text{Dup}, \text{Cut}\}$, then $s(v) = s(v_t) = s(v_k)$;*
   4. *if $e(v) \in \{\text{TrDup}, \text{TrCut}\}$, then $s(v) \parallel s(v_t)$, and $s(v) = s(v_k)$.*
3. *If $e(v) \in \{\text{Gain}, \text{Loss}\}$, then v is unary; let $\mathrm{ch}(v) = \{v_c\}$, then $s(v_c) = s(v)$ and:*
   1. *if $e(v) = $ Gain, then $x(v_c) \supsetneq x(v)$.*
   2. *if $e(v) = $ Loss, then $x(v_c) \subsetneq x(v)$ (a loss is full if $x(v_c) = \emptyset$, and partial otherwise);*
4. *For each gene family Γ, exactly one Gain event in H involves Γ.*
5. *The only nodes v of H which can have an empty synteny ($x(v) = \emptyset$) are its root and leaves.*

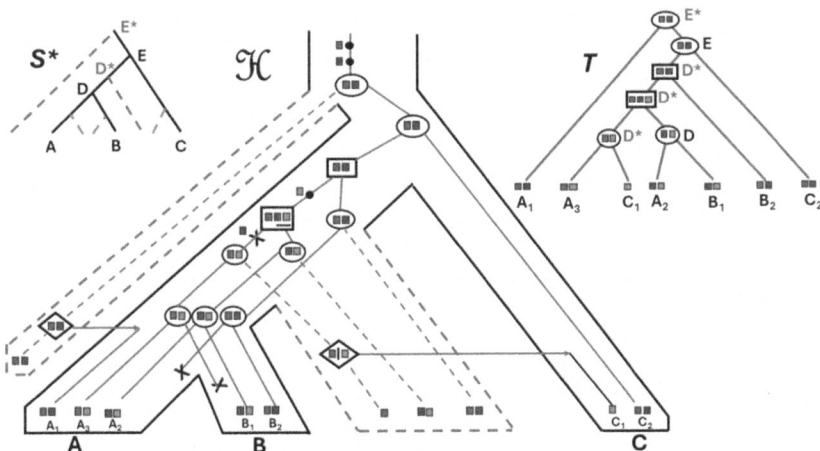

Fig. 2. (Left): The augmented species tree S^* obtained from species tree S in Fig. 1.(1); The blue dotted branches of S^* represent the edges leading to unsampled leaves which are used in $\mathcal{H}$. (Center): A history $\mathcal{H}$ explicating the synteny supertree T in Fig. 1.(3), the history tree is represented with thin lines inside the species. Events are represented as follows: "Spe" by ovals, "Dup" and "Cut" by rectangles, "TrDup" and "TrCut" by diamonds, "Loss" by crosses and "Gain" by circles. The synteny contents are represented inside of each event, while the associated species is represented implicitly by the position of each event on top of the species tree. Duplicated or transferred genes are underlined, while fissions are represented by a separation in the synteny. (Right): The synteny supertree T with internal nodes labeled as its corresponding nodes in $\mathcal{H}$.

Condition (4) excludes *convergent gains*, i.e. histories where the same gene family is gained separately in different parts of the history. Moreover, as in [14], we allow for transfers from unsampled species by augmenting the species tree.

Definition 2 (Augmented Species Tree). *A tree S can be augmented into S^* by adding unsampled leaves as follows: (1) subdivide each edge (v, v') of $\mathrm{E}(S)$ into two edges (v, z) and (z, v') through a new node z, (2) connect each z to a new unsampled leaf, and (3) create a new root $\mathrm{r}(S^*)$ whose two children are $\mathrm{r}(S)$ and a new unsampled leaf. A node $\sigma \in V(S^*)$ which is not an unsampled leaf is said to be* sampled.

Definition 3 (Visible Leaves). *A leaf l of a history $\mathcal{H} = \langle H, e, x, s \rangle$ is said to be* visible *if $x(l) \neq \emptyset$ and $s(l)$ is sampled, and* invisible *otherwise. The set of visible leaves of $\mathcal{H}$ is denoted $\mathrm{L}_V(\mathcal{H})$.*

Definition 4 (Explicatory Histories). *For a species tree S, a history $\mathcal{H} = \langle H, e, x', s' \rangle$ on S^* is said to* explicate *S and $\langle T, x, s \rangle$, a synteny tree T with its associated mappings x and s, if:*

1. H *is an extension of T.*
2. $\forall\ v \in L(T)$, $x'(\Psi_H(v)) = x(v)$ *and* $s'(\Psi_H(v)) = s(v)$.
3. $\mathrm{L}_V(\mathcal{H}) = \mathrm{L}(T)$.
4. *No $(u,v) \in \mathrm{E}(T)$ is such that $s(\Psi_H(v)) < s(\Psi_H(u))$.*

The set of all such histories is denoted by $\mathbb{H}(T, S)$.

Condition (4) excludes assignments of species which create cycles between adjacent nodes of the synteny tree. This condition is a necessary, but not sufficient, condition for acyclicity. Imposing a full acyclicity condition would lead to computationally-intractable problems [3].

If T is a synteny supertree for a set of consistent gene trees $\mathcal{G}$, then any node v of a tree in $\mathcal{G}$ corresponds to a node of T which itself corresponds to a node of H denoted $\Psi_H(v)$.

4 Problem Statement

The cost of a history will depend on the number of events of each type. As in [14], we forbid segmental gains. Thus, the number of gains simply corresponds to the number of gene families, and can be ignored.

Definition 5 (Event Vector). *Let $\mathcal{H} = \langle H, e, x, s \rangle$ be a history. We define $\mathrm{ev}(\mathcal{H}) = (c_{\mathrm{Spe}}, c_{\mathrm{Dup}}, c_{\mathrm{Cut}}, c_{\mathrm{TrDup}}, c_{\mathrm{TrCut}}, c_{\mathrm{Loss}}) \in \mathbb{N}^6$ as the vector such that c_{ev} is the number of events of type $ev \in \mathcal{E}$ in $\mathcal{H}$.*

Given an event cost vector $\delta = (\delta_{\mathrm{Spe}}, \delta_{\mathrm{Dup}}, \delta_{\mathrm{Cut}}, \delta_{\mathrm{TrDup}}, \delta_{\mathrm{TrCut}}, \delta_{\mathrm{Loss}})$ such that $\delta \in (\mathbb{R}_{\geq 0} \cup \{\infty\})^6$, we can associate an overall scalar cost $c(\mathcal{H}) = \delta \cdot \mathrm{ev}(\mathcal{H})$ to each history. As usual for reconciliation, we will ignore the number of speciations (i.e. we assume $\delta_{\text{Spe}} = 0$), since they do not allow to meaningfully distinguish histories.

For a set $\mathcal{G} = \{G_1, G_2, \ldots, G_k\}$ of consistent gene trees on $\mathcal{X}$, we denote by $\mathbb{T}(\mathcal{G})$ the set of all binary synteny supertrees of $\mathcal{G}$ on $\mathcal{X}$. Moreover, for a synteny supertree T and a species tree S, we denote $c^{\min}(T, S) = \min\{c(\mathcal{H}) \mid \mathcal{H} \in \mathbb{H}(T, S)\}$. We are now ready to set our problem which consists in simultaneously building and reconciling a synteny supertree.

MINIMUM SYNTENY SUPERTREE PROBLEM (MINSYNSUPERTREE):
Input: A species tree S on Σ, a set $\mathcal{X}$ of syntenies in Σ, a set $\mathcal{G} = \{G_1, \ldots, G_k\}$ of consistent gene trees on $\mathcal{X}$ and a cost vector δ.
Output: $\min_{T \in \mathbb{T}(\mathcal{G})} \{c^{\min}(T, S)\}$.

5 Method

We successively describe, in this section, the different parts of our dynamic programming algorithm *FullSynesth* for solving MINSYNSUPERTREE. For conciseness, the described algorithm simply outputs the minimum cost. However, it can easily be modified to also output the corresponding synteny supertree T and a corresponding history $\mathcal{H}$ by keeping track of the synteny tree and history corresponding to the minimum cost at each step. The full pseudocode is given in Sect. 5.5. The correctness of the pseudocode follows from a series of theorems (Theorems 2 to 6) that we successively state (the proofs are omitted in this extended abstract). They lead to the following time complexity.

Theorem 1. *Let $k = |\mathcal{G}|$, $m = |\mathcal{X}|$ and $n = |V(S^*)|$. FullSynesth solves* MIN-SYNSUPERTREE *in $O(m^k \times 8^k \times (k^2 + n))$ time.*

In the following, we consider as input a species tree S for a set of species Σ, an augmented species tree S^*, a set $\mathcal{F}$ of gene families, a set $\mathcal{X}^{input}$ of syntenies in Σ and a set $\mathcal{G}^{input} = \{G_1^{input}, \ldots, G_k^{input}\}$ on $\mathcal{X}^{input}$ of consistent gene trees such that $x(\mathcal{G}^{input}) = \mathcal{F}$. Moreover, we will need a dummy gene family $\Gamma^* \notin \mathcal{F}$.

5.1 Bipartitions

We take advantage of the strategy considered in [19] for the classical reconciliation model with DL distance, the difference being the evolutionary model and the way the cost is recursively computed. An optimal synteny supertree T is constructed from the root to the leaves. At each step, i.e. for each internal node v being constructed in T, each possible bipartition $(L(T_{v_l}), L(T_{v_r}))$ that could be induced by v is tried, and the algorithm continues on each of $L(T_{v_l})$ and $L(T_{v_r})$. For example, at the root, the goal is to find a best bipartition of $\mathcal{X}^{input}$, i.e. one leading to the minimum cost. Figure 3.(3) illustrates the recurrence for reconstructing a synteny supertree. Each step of the recurrence (each triangle in Fig. 3.(3)) involves a tree set $\mathcal{G} = \mathcal{G}^{input}|_{\mathcal{X}}$ for a set of leaves $\mathcal{X} \subseteq \mathcal{X}^{input}$. We call such a set *a restriction of* $\mathcal{G}^{input}$. From now on, all considered tree sets are restrictions of $\mathcal{G}^{input}$.

One of the key ideas behind the proof of Theorem 1 is that the constraint of displaying the input gene trees induces a strong constraint on the bipartitions. In fact, only *compatible bipartitions* defined below have to be tested, reducing their number from $\Theta(2^{|\mathcal{X}|})$ to $O(4^{|\mathcal{G}|})$.

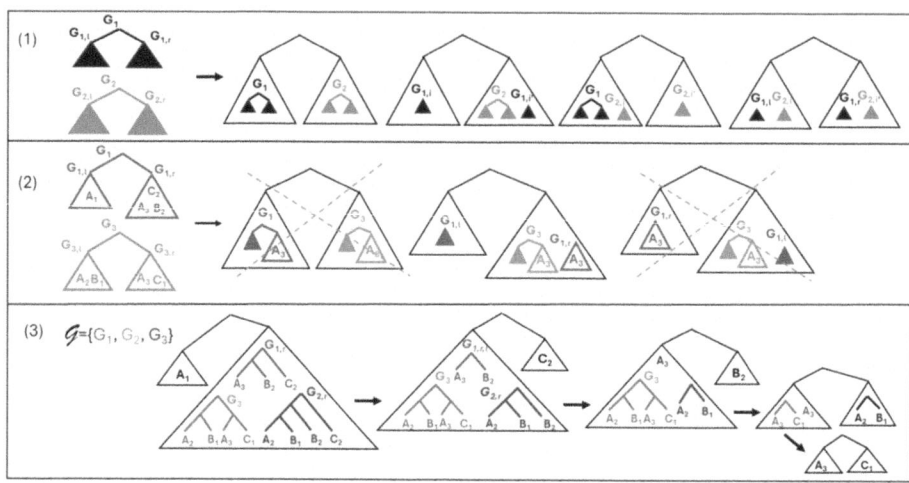

Fig. 3. (1) The procedure for displaying all the compatible bipartitions for two consistent trees G_1 and G_2. Each bipartition is obtained by "sending" $\mathcal{X}_1 \in \{L(G_1), L(G_{1,l}), L(G_{1,r}), L(G_{1,r}), \emptyset\}$ in the left part, and the complement $L(G_1) \setminus \mathcal{X}_1$ in the right part. The same process is then applied to G_2. In the figure, the index $(i,i') \in \{(l,r),(r,l)\}$, leading to seven possible splits. (2) Splits do not necessarily lead to bipartitions as illustrated on the trees G_1 and G_3 of Fig. 1. (3) A sequence of bipartitions for the set $\mathcal{G} = \{G_1, G_2, G_3\}$ leading to the synteny supertree T in Fig. 1.

Definition 6 (Compatible Bipartition). *Let $\mathcal{G} = \{G_1, \ldots, G_k\}$ be a tree set on $\mathcal{X}$. A bipartition $(\mathcal{X}_\ell, \mathcal{X}_r)$ of $\mathcal{X}$ is compatible with $\mathcal{G}$ (or simply "compatible" if no ambiguity on the set of leaves) if there exists a supertree of $\mathcal{G}$ such that the bipartition of its root is $(\mathcal{X}_\ell, \mathcal{X}_r)$.*

We denote by $\mathcal{B}(\mathcal{G})$ the set of all compatible bipartitions.

We define below the cost of a history corresponding to $\mathcal{G}$. The boolean b in this definition will be required later for testing ancestral syntenies.

Definition 7 (Cost of History). *Let $\mathcal{G} = \{G_1, G_2, \ldots, G_k\}$ be a tree set on $\mathcal{X}$ ($|\mathcal{X}| \geq 2$). Let $(\mathcal{X}_\ell, \mathcal{X}_r)$ be a compatible bipartition of $\mathcal{X}$, b be a boolean, and $\sigma \in V(S^*)$. Consider the following conditions, where $\mathcal{H} = \langle H, e, x, s \rangle$ is a history explicating a synteny supertree T of $\mathcal{G}$, and $v = \Psi_H(r(T))$:*

1. $s(v) = \sigma$

2. $\begin{cases} x(v) \subseteq x(\mathcal{G}) & \text{if } b = \text{False} \\ x(v) \not\subseteq x(\mathcal{G}) & \text{otherwise} \end{cases}$

3. $\left((\cup_{\xi \in (\mathcal{X}^{input} - \mathcal{X})} x(\xi)) \cap x(\mathcal{G}) \right) \subseteq x(v)$
4. The bipartition of $r(T)$ is $(\mathcal{X}_\ell, \mathcal{X}_r)$

We define $c(\mathcal{G}, b, \sigma, (\mathcal{X}_\ell, \mathcal{X}_r))$ (respec. $c(\mathcal{G}, b, \sigma)$) as the minimum cost of any history such $\mathcal{H}$ explicating any synteny supertree T, both satisfying conditions 1 to 4 (respec. conditions 1 to 3). Note that $c(\mathcal{G}, b, \sigma)$ is also defined in the same way for $|\mathcal{X}| = 1$.

Condition 3 ensures that each gene family is gained exactly once in the history. We are now ready to formulate the main step of the recurrence algorithm, which is depicted in Fig. 3.(1).

Theorem 2 (Testing Bipartitions). Let $\mathcal{G} = \{G_1, G_2, \ldots, G_k\}$ be a tree set on $\mathcal{X}$. Let $\sigma \in S^*$ and b be a boolean. Then,

If $|\mathcal{X}| = 1$, $\quad c(\mathcal{G}, b, \sigma) = \begin{cases} 0 & \text{if } \sigma = s(G_1) \text{ and } b \text{ is False} \\ \infty & \text{otherwise} \end{cases}$

Otherwise, $\quad c(\mathcal{G}, b, \sigma) = \min_{(\mathcal{X}_\ell, \mathcal{X}_r) \in \mathcal{B}(\mathcal{G})} \{c(\mathcal{G}, b, \sigma, (\mathcal{X}_\ell, \mathcal{X}_r))\}$

5.2 Syntenies

It is shown in [14] that, for computing the minimum cost of a history H explicating a given synteny supertree T for a set $\{G_1^{input}, G_2^{input}, \ldots, G_k^{input}\}$ of consistent gene trees, it is sufficient to place the gain event for each gene family Γ_i at the lowest common ancestor of the leaves they appear in, which corresponds to $\Psi_H(r(G_i^{input}))$. From now on, we will assume that this is the case for all histories. The next lemma, following from Lemma 2 in [14], states that only two synteny contents (i.e. one for each possible value of b) have to be tested at each internal node of the synteny supertree being constructed.

Lemma 1 (Minimum Synteny Contents). Let $\mathcal{G} = \{G_1, G_2, \ldots, G_k\}$ be a tree set on $\mathcal{X}$ $(|\mathcal{X}| \geq 2)$. Let $(\mathcal{X}_\ell, \mathcal{X}_r)$ be a compatible bipartition of $\mathcal{X}$ and b be a boolean. Let $gain(\mathcal{G}, \mathcal{X}') = \{\Gamma_i \mid (i \in \{1, \ldots, k\}) \land (r(G_i^{input}) \in V(G_i|_{\mathcal{X}'}))\}$ and

$$x^{syn}(\mathcal{G}, b, \mathcal{X}_\ell, \mathcal{X}_r) = \begin{cases} x(\mathcal{G}) - gain(\mathcal{G}, \mathcal{X}_\ell) - gain(\mathcal{G}, \mathcal{X}_r) \cup \{\Gamma^*\} & \text{if } b \\ x(\mathcal{G}) - gain(\mathcal{G}, \mathcal{X}_\ell) - gain(\mathcal{G}, \mathcal{X}_r) & \text{otherwise} \end{cases}$$

There exists a history $\langle H, e, x, s \rangle$ of cost $c(\mathcal{G}, b, \sigma, (\mathcal{X}_\ell, \mathcal{X}_r))$ explicating a synteny supertree T of $\mathcal{G}$ and satisfying conditions 1 to 4 of Definition 7 such that $x(\Psi_H(r(T))) = x^{syn}(\mathcal{G}, b, \mathcal{X}_\ell, \mathcal{X}_r)$.

We next introduce the notations allowing to decompose a history according to the root of the corresponding synteny tree, as illustrated in Fig. 4. Notice that, in Definition 8, the booleans y and z will be used later (Theorem 5) to define the syntenies Y and Z, respectively.

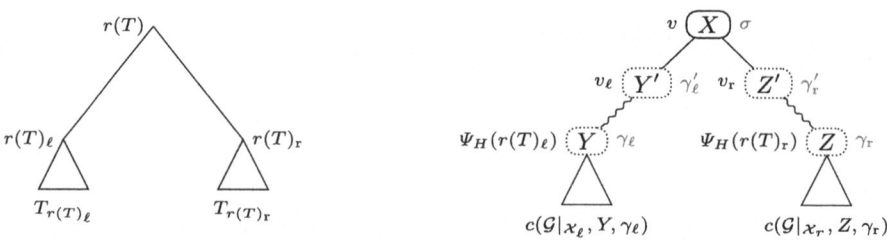

Fig. 4. A decomposition for a history H (on the right) explicating a synteny tree T on the left. The notations illustrate those introduced in Definition 8. The wavy lines correspond to paths

Definition 8. Let $\mathcal{G} = \{G_1, G_2, \ldots, G_k\}$ be a tree set on $\mathcal{X}$ ($|\mathcal{X}| \geq 2$). Let event $\in \{\text{Spe, Dup, Cut, TrDup, TrCut}\}$, $\sigma, \gamma_\ell, \gamma'_\ell, \gamma_r, \gamma'_r \in V(S^*)$, $(\mathcal{X}_\ell, \mathcal{X}_r)$ be a compatible bipartition of $\mathcal{X}$, y, z be two booleans and $X, Y', Z' \subseteq \mathcal{F}$. Consider the following conditions, where $\langle H, e, x, s \rangle$ is a history explicating a synteny supertree T of $\mathcal{G}$, and $v = \Psi_H(r(T))$:

1. $e(v) = event$;
2. The bipartition of $r(T)$ is $(\mathcal{X}_\ell, \mathcal{X}_r)$;
3. $s(v) = \sigma$, $x(v) = X$;
4. $\begin{cases} x(\Psi_H(r(T)_\ell)) \subseteq x(\mathcal{G}|_{\mathcal{X}_\ell}) & \text{if } y = \text{False} \\ x(\Psi_H(r(T)_\ell)) \not\subseteq x(\mathcal{G}|_{\mathcal{X}_\ell}) & \text{otherwise}; \end{cases}$
5. $\begin{cases} x(\Psi_H(r(T)_r)) \subseteq x(\mathcal{G}|_{\mathcal{X}_r}) & \text{if } z = \text{False} \\ x(\Psi_H(r(T)_r)) \not\subseteq x(\mathcal{G}|_{\mathcal{X}_r}) & \text{otherwise}; \end{cases}$
6. $((\cup_{\xi \in (\mathcal{X}^{input} - \mathcal{X})} x(\xi)) \cap x(\mathcal{G})) \subseteq x(v)$;
7. $s(\Psi_H(r(T)_\ell)) = \gamma_\ell$, $s(\Psi_H(r(T)_r)) = \gamma_r$;
8. $s(v_\ell) = \gamma'_\ell$, $x(v_\ell) = Y'$;
9. $s(v_r) = \gamma'_r$, $x(v_r) = Z'$.

We define $c(\mathcal{G}, \sigma, (\mathcal{X}_\ell, \mathcal{X}_r), event, X, ([Y', \gamma'_\ell], [y, \gamma_\ell]), ([Z', \gamma'_r], [z, \gamma_r]))$ (respec. $c(\mathcal{G}, \sigma, (\mathcal{X}_\ell, \mathcal{X}_r), event, X, y, z))$ as the minimum cost of a history satisfying conditions 1 to 9 (respec. 1 to 6) if such history exists, and ∞ otherwise.

Theorem 3 (Testing all Possible Syntenies). Let $\mathcal{G} = \{G_1, G_2, \ldots, G_k\}$ be a tree set on $\mathcal{X}$ ($|\mathcal{X}| \geq 2$). Let $(\mathcal{X}_\ell, \mathcal{X}_r)$ be a compatible bipartition of $\mathcal{X}$, $\sigma \in S^*$ and b be a boolean. Then,

$$c(\mathcal{G}, b, \sigma, (\mathcal{X}_\ell, \mathcal{X}_r)) = \min_{\substack{X = x^{syn}(\mathcal{G}, b, \mathcal{X}_\ell, \mathcal{X}_r) \\ y \in \{False, True\} \\ z \in \{False, True\} \\ event \in \{\text{Spe,Dup,Cut,TrDup,TrCut}\}}} \{c(\mathcal{G}, \sigma, (\mathcal{X}_\ell, \mathcal{X}_r), event, X, y, z)\}$$

5.3 Events

For each $event$, the cost $c(\mathcal{G}, \sigma, (\mathcal{X}_\ell, \mathcal{X}_r), event, X, y, z)$ is given in the next theorem, which follows from Definition 8 and Theorem 1 in [14].

Theorem 4 (Testing all Possible Events). *Let $\mathcal{G} = \{G_1, G_2, \ldots, G_k\}$ be a tree set on $\mathcal{X}$ ($|\mathcal{X}| \geq 2$). Let $B = (\mathcal{X}_\ell, \mathcal{X}_r)$ be a compatible bipartition of $\mathcal{X}$, $\sigma \in S^*$, y, z be booleans and $X \subseteq \mathcal{F}$. Then,*

Spe: $c(\mathcal{G}, \sigma, (\mathcal{X}_\ell, \mathcal{X}_r), \text{Spe}, X, y, z) =$

$$\begin{cases} \infty & \text{if } \sigma \in L(S^*) \\ \min_{\gamma_\ell, \gamma_r \not< \sigma} \{c(\mathcal{G}, \sigma, B, \text{Spe}, X, ([X, \sigma_\ell], [y, \gamma_\ell]), ([X, \sigma_r], [z, \gamma_r])), & \text{otherwise} \\ \qquad c(\mathcal{G}, \sigma, B, \text{Spe}, X, ([X, \sigma_r], [y, \gamma_\ell]), ([X, \sigma_\ell], [z, \gamma_r]))\} \end{cases}$$

Dup: $c(\mathcal{G}, \sigma, (\mathcal{X}_\ell, \mathcal{X}_r), \text{Dup}, X, y, z) =$

$$\min_{\gamma_\ell, \gamma_r \not< \sigma} \{c(\mathcal{G}, \sigma, B, \text{Dup}, X, ([X \cap x(\mathcal{G}|_{\mathcal{X}_\ell}), \sigma], [y, \gamma_\ell]), ([X, \sigma], [z, \gamma_r])),$$
$$c(\mathcal{G}, \sigma, B, \text{Dup}, X, ([X, \sigma], [y, \gamma_\ell]), ([X \cap x(\mathcal{G}|_{\mathcal{X}_r}), \sigma], [z, \gamma_r]))\}$$

Cut: $c(\mathcal{G}, \sigma, (\mathcal{X}_\ell, \mathcal{X}_r), \text{Cut}, X, y, z) =$

$$\begin{cases} \infty & \text{if } x(\mathcal{G}|_{\mathcal{X}_\ell}) \cap x(\mathcal{G}|_{\mathcal{X}_r}) \neq \emptyset \\ \min_{\gamma_\ell, \gamma_r \not< \sigma} & \text{otherwise} \\ \{c(\mathcal{G}, \sigma, B, \text{Cut}, X, ([X \cap x(\mathcal{G}|_{\mathcal{X}_\ell}), \sigma], [y, \gamma_\ell]), ([X - x(\mathcal{G}|_{\mathcal{X}_\ell}), \sigma], [z, \gamma_r])), \\ \quad c(\mathcal{G}, \sigma, B, \text{Cut}, X, ([X - x(\mathcal{G}|_{\mathcal{X}_r}), \sigma], [y, \gamma_\ell]), ([X \cap x(\mathcal{G}|_{\mathcal{X}_r}), \sigma], [z, \gamma_r]))\} \end{cases}$$

TrDup: $c(\mathcal{G}, \sigma, (\mathcal{X}_\ell, \mathcal{X}_r), \text{TrDup}, X, y, z) =$

$$\begin{cases} \infty & \text{if } \sigma = r(S^*) \\ \min_{\substack{\gamma_i \| \sigma; \gamma_t \not< \gamma_i, \sigma \\ \gamma_k \not< \sigma}} & \text{otherwise} \\ \{c(\mathcal{G}, \sigma, B, \text{TrDup}, X, ([X \cap x(\mathcal{G}|_{\mathcal{X}_\ell}), \gamma_i], [y, \gamma_t]), ([X, \sigma], [z, \gamma_k])), \\ \quad c(\mathcal{G}, \sigma, B, \text{TrDup}, X, ([X, \sigma], [y, \gamma_k]), ([X \cap x(\mathcal{G}|_{\mathcal{X}_r}), \gamma_i], [z, \gamma_t]))\} \end{cases}$$

TrCut: $c(\mathcal{G}, \sigma, (\mathcal{X}_\ell, \mathcal{X}_r), \text{TrCut}, X, y, z)\}) =$

$$\begin{cases} \infty & \text{if } \sigma = r(S^*) \text{ or } x(\mathcal{G}|_{\mathcal{X}_\ell}) \cap x(\mathcal{G}|_{\mathcal{X}_r}) \neq \emptyset \\ \min_{\substack{\gamma_i \| \sigma; \gamma_t \not< \gamma_i, \sigma \\ \gamma_k \not< \sigma}} & \text{otherwise} \\ \{c(\mathcal{G}, \sigma, B, \text{TrCut}, X, ([X \cap x(\mathcal{G}|_{\mathcal{X}_\ell}), \gamma_i], [y, \gamma_t]), ([X - x(\mathcal{G}|_{\mathcal{X}_\ell}), \sigma], [z, \gamma_k])), \\ \quad c(\mathcal{G}, \sigma, B, \text{TrCut}, X, ([X - x(\mathcal{G}|_{\mathcal{X}_r}), \gamma_i], [y, \gamma_t]), ([X \cap x(\mathcal{G}|_{\mathcal{X}_r}), \sigma], [z, \gamma_k])), \\ \quad c(\mathcal{G}, \sigma, B, \text{TrCut}, X, ([X \cap x(\mathcal{G}|_{\mathcal{X}_\ell}), \sigma], [y, \gamma_k]), ([X - x(\mathcal{G}|_{\mathcal{X}_\ell}), \gamma_i], [z, \gamma_t])), \\ \quad c(\mathcal{G}, \sigma, B, \text{TrCut}, X, ([X - x(\mathcal{G}|_{\mathcal{X}_r}), \sigma], [y, \gamma_k]), ([X \cap x(\mathcal{G}|_{\mathcal{X}_r}), \gamma_i], [z, \gamma_t]))\} \end{cases}$$

5.4 History Decomposition

We next explain how the cost of a history verifying the conditions of Definition 8, i.e. decomposed into paths and sub-histories as shown in Fig. 4, is computed.

We first need some preliminary notations. A node v of a history such that $x(v) = X$ and $s(v) = \sigma$ is denoted $[X, \sigma]$. Moreover, define a path from a node

v to w of a history as an acyclic history whose root is v and whose only visible leaf is w, and define cPath($[X, \sigma], [Y, \gamma]$) as the minimum cost of any path from $v = [X, \sigma]$ to $w = [Y, \gamma]$.

Let $\mathcal{G} = \{G_1, G_2, \ldots, G_k\}$ be a tree set on $\mathcal{X}$, b be a boolean, $\mathcal{X}' \subseteq \mathcal{X}$ and $X \subseteq \mathcal{F}$. We define

$$syn(\mathcal{G}, X, b, \mathcal{X}') = \begin{cases} X & \text{if } b, \\ x(\mathcal{G}|_{\mathcal{X}'}) & \text{otherwise.} \end{cases}$$

Theorem 5 (Cost of History Decomposition). *Let $\mathcal{G} = \{G_1, G_2, \ldots, G_k\}$ be a tree set on $\mathcal{X}$ ($|\mathcal{X}| \geq 2$). Let event $\in \{\text{Spe}, \text{Dup}, \text{Cut}, \text{TrDup}, \text{TrCut}\}$, $\sigma, \gamma_\ell, \gamma'_\ell, \gamma_r, \gamma'_r \in V(S^*)$, $(\mathcal{X}_\ell, \mathcal{X}_r)$ be a compatible bipartition of $\mathcal{X}$, y, z be two booleans, $X, Y', Z' \subseteq \mathcal{F}$, $Y = syn(\mathcal{G}, Y', y, \mathcal{X}_\ell)$ and $Z = syn(\mathcal{G}, Z', z, \mathcal{X}_r)$.*

If there exists a history verifying conditions 1 to 8 of Definition 8, then:

$$\begin{aligned} c(\mathcal{G}, &\sigma, (\mathcal{X}_\ell, \mathcal{X}_r), event, X, ([Y', \gamma'_\ell], [y, \gamma_\ell]), ([Z', \gamma'_r], [z, \gamma_r])) \\ &= \delta_{event} + (\text{cPath}([Y', \gamma'_\ell], [Y, \gamma_\ell]) + c(\mathcal{G}|_{\mathcal{X}_\ell}, y, \gamma_\ell)) \\ &\quad + (\text{cPath}([Z', \gamma'_r], [Z, \gamma_r]) + c(\mathcal{G}|_{\mathcal{X}_r}, z, \gamma_r)) \end{aligned}$$

We finally show how the cost of a minimum path is computed.

Theorem 6 (Minimum Cost of Paths). *Let X and Y be two syntenies and σ and γ be two species. If $\sigma \leq \gamma$, let $\Omega(\sigma, \gamma) = \sum_{(u,v) \in P_{S^*}(\sigma, \gamma)} [\![v \notin V(S)]\!]$. If $\sigma < \gamma$, let $chSep(\sigma, \gamma)$ be the child of σ that is separated from γ and let $chAnc(\sigma, \gamma)$ be the child of σ that is an ancestor of γ. If γ is a strict ancestor of σ (i.e. γ is an ancestor of σ and $\gamma \neq \sigma$), then $\text{cPath}([X, \sigma], [Y, \gamma]) = \infty$. Otherwise,*

$\text{cPath}([X, \sigma], [Y, \gamma]) = \min\{$

$\delta_{\text{Loss}} \cdot \Omega(\sigma, \gamma) + \delta_{\text{Loss}} \cdot [\![X \not\subseteq Y]\!] + \infty \cdot [\![\sigma \parallel \gamma]\!],$

$\delta_{\text{Loss}} \cdot \Omega(\sigma, \gamma) + \min\{\delta_{\text{Dup}}, \delta_{\text{Cut}}\} \cdot [\![X \not\subseteq Y]\!] + \infty \cdot [\![(\gamma \text{ sampled }) \vee (\sigma \parallel \gamma)]\!],$

$\delta_{\text{TrDup}} + [\![\sigma \text{ sampled }]\!] \cdot \delta_{\text{Loss}} + \infty \cdot [\![\sigma \leq \gamma]\!],$

$\delta_{\text{TrCut}} + [\![(X \not\subseteq Y) \wedge (\sigma \text{ sampled })]\!] \cdot \delta_{\text{Loss}} + \infty \cdot [\![\sigma \leq \gamma]\!]$

$\delta_{\text{TrDup}} + \delta_{\text{Loss}} \cdot [\![chAnc(\sigma, \gamma) \text{ sampled }]\!]$
$\quad + \delta_{\text{Loss}} \cdot [\![chSep(\sigma, \gamma) \text{ sampled }]\!] + \infty \cdot [\![\sigma \parallel \gamma]\!],$

$\delta_{\text{TrCut}} + \delta_{\text{Loss}} \cdot [\![chAnc(\sigma, \gamma) \text{ sampled }]\!]$
$\quad + \delta_{\text{Loss}} \cdot [\![(X \not\subseteq Y) \wedge (chSep(\sigma, \gamma) \text{ sampled })]\!] + \infty \cdot [\![\sigma \parallel \gamma]\!],$

$2\delta_{\text{TrDup}} + \delta_{\text{Loss}} \cdot [\![\sigma \text{ sampled }]\!] + \infty \cdot [\![\sigma = r(S^*)]\!],$

$\min\{\delta_{\text{TrDup}}, \delta_{\text{TrCut}}\} + \delta_{\text{TrCut}} + \infty \cdot [\![\sigma = r(S^*)]\!]\}$

5.5 Pseudocode

Algorithm 1 with input $(\mathcal{G}^{input}, S^*, \delta)$ solves MINSYNSUPERTREE.

Algorithm 1: FullSynesth$(\mathcal{G}, S^*, \delta)$

---
1 Compute $c(\mathcal{G}, b, \sigma)$ for all $\sigma \in V(S^*)$ and $b \in \{False, True\}$ by calling FullSynesthRecursion$(\mathcal{G}, S^*, \delta)$
2 Return $\min_{\sigma \in V(S^*)} \{c(\mathcal{G}, False, \sigma)\}$

Algorithm 2: FullSynesthRecursion$(\mathcal{G}, S^*, \delta)$

---
1 **if** $|\cup_{G \in \mathcal{G}} L(G)| = 1$ **then**
2   **for** *each* $\sigma \in V(S^*)$ **do**
3   Compute $c(\mathcal{G}, False, \sigma)$ and $c(\mathcal{G}, True, \sigma)$ according to Theorem 2
4   **end**
5 **else**
6   **for** *each* $(\mathcal{X}_\ell, \mathcal{X}_r) \in \mathcal{B}(\mathcal{G})$ **do**
7   Compute $c(\mathcal{G}|_{\mathcal{X}_\ell}, b, \sigma)$ for all $\sigma \in V(S^*)$ and $b \in \{False, True\}$ by calling FullSynesthRecursion$(\mathcal{G}|_{\mathcal{X}_\ell}, S^*, \delta)$
8   Compute $c(\mathcal{G}|_{\mathcal{X}_r}, b, \sigma)$ for all $\sigma \in V(S^*)$ and $b \in \{False, True\}$ by calling FullSynesthRecursion$(\mathcal{G}|_{\mathcal{X}_r}, S^*, \delta)$
9   **end**
10   **for** *each* $\sigma \in V(S^*)$ *and* $b \in \{False, True\}$ **do**
11   Compute $c(\mathcal{G}, b, \sigma)$ according to Theorems 2 to 6.
12   **end**
13 **end**
14 Return $c(\mathcal{G}, b, \sigma)$ for all $\sigma \in V(S^*)$ and $b \in \{False, True\}$

---

6 Results

We developed *Syntesim*, a simulator generating evolutionary histories of species, gene families and syntenies. Starting from a synteny with a single gene belonging to an ancestral species, the simulation chooses, at each step, an event among Speciation, Extinction, Duplication, Cut, Transfer, Gain and Loss according to given rates on operations. The output of one simulation is a full history including a species tree S, a set $\mathcal{F}$ of gene families, a set $\mathcal{X}$ of syntenies, a set $\mathcal{G}$ of consistent gene trees for the gene families in $\mathcal{F}$, and a synteny tree T for $\mathcal{X}$ which is a supertree of the gene trees of $\mathcal{G}$. We used our simulator with the stopping criterion being to reach $|\Sigma| = 10$ species. We retained the 2300 datasets with $20 \leq |\mathcal{X}| \leq 30$. We then run FullSynesth on each of the obtained datasets on a desktop computer equipped with an AMD Ryzen 9 5900x processor and 64GB of RAM. The scatter plot left of Fig. 5 reflects the increase in running time of FullSynesth, from 0.3 s for a single gene tree to about 10 min for 15 gene trees.

We then compared FullSynesth with *NaiveSynesth* which consists in solving MINSYNSUPERTREE by applying Synesth to each supertree for $\mathcal{G}$. As Naive-Synesth is directly proportional to the number of supertrees, it is clearly slower than FullSynesth if $|\mathbb{T}(\mathcal{G})|$ is large enough (NaiveSynesth would take more than 10 min to run if $|\mathbb{T}(\mathcal{G})| > 2000$). We thus restricted the comparison to low values of $|\mathbb{T}(\mathcal{G})|$. To this end, we implemented AllTrees [17] allowing to output the full set of binary supertrees for a given set of consistent trees. We then grouped our 2300 datasets according to the number of supertrees obtained from the set $\mathcal{G}$ of gene trees for each dataset. From the obtained results shown in the right of Fig. 5, we see that while FullSynesth is slightly slower for large sets of supertrees (from an average of 0.3 s for less than 10 supertrees, to an average of 2 min for 500 supertrees or more), it is always faster on average than NaiveSynesth even for less than 10 supertrees.

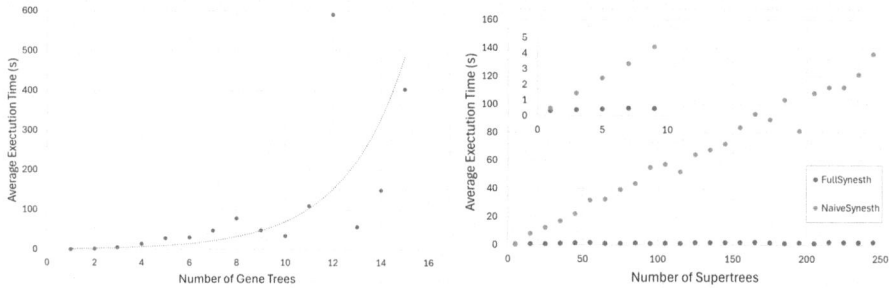

Fig. 5. (Left): Average running time of FullSynesth on simulated datasets grouped according to the number of gene trees, keeping only samples containing more than 10 simulations. (Right): Average running time of FullSynesth and NaiveSynesth on simulated datasets grouped according to the number of possible supertrees. Each dot represents the average for a number of supertrees in an interval of size 10 around the dot. The scatter plot on top is a zoom on the first interval of the main scatter plot.

7 Conclusion

FullSynesth adds a key piece to the Synesth architecture, since with FullSynesth, the synteny tree does not need to be known in advance. We show that Full-Synesth is several orders of magnitude faster than Synesth applied to each possible supertree. As the two algorithms are exact, comparing their output is not required as they both return the same predictions.

There are a number of possible future directions for improvements, probably the most important being to deal with the case of a set $\mathcal{G} = \{G_1, G_2, \ldots, G_k\}$ of gene trees that are inconsistent, and thus not included in a supertree. For a given tree distance, one can be interested in finding a median tree, i.e. a tree on the leafset of $\mathcal{G}$ minimizing the sum of distances with the gene trees. This problem has been shown NP-hard for the D and DL reconciliation distance [20–22], and heuristics have been developed [23,24]. Another way would be to find a consensus of gene trees in form of a sufficiently constrained phylogenetic network, such as a tree-child network, and design a strategy for building and reconciling a synteny tree from this network [25].

Acknowledgements. We would like to thank Aïda Ouangraoua for helpful discussions on this topic.

Funding. This research was funded by the *Fonds de recherche du Québec - Nature et technologies* [grant numbers 335135 (MG) and 335893 (MD)] and the Natural Sciences and Engineering Research Council of Canada [grant number RN000743 (NEM)].

Disclosure of Interests. The authors declare no conflicts of interest.

References

1. Goodman, M., Czelusniak, J., Moore, G.W., Romero-Herrera, A.E., Matsuda, G.: Fitting the gene lineage into its species lineage, a parsimony strategy illustrated by cladograms constructed from globin sequences. Syst. Zool. **28**, 132–163 (1979)
2. Bansal, M.S., Alm, E.J., Kellis, M.: Efficient algorithms for the reconciliation problem with gene duplication, horizontal transfer and loss. Bioinformatics **28**(12), 283–291 (2012)
3. Tofigh, A., Hallett, M., Lagergren, J.: Simultaneous identification of duplications and lateral gene transfers. IEEE/ACM Trans. Comput. BiolBioinform. **8**(2), 517–535 (2011)
4. Bansal, M.S., Kellis, M., Kordi, M., Kundu, S.: RANGER-DTL 2.0: rigorous reconstruction of gene-family evolution by duplication, transfer and loss. Bioinformatics **34**, 3214–3216 (2018)
5. Donati, B., Baudet, C., Sinaimeri, B., Crescenzi, P., Sagot, M.F.: EUCALYPT: efficient tree reconciliation enumerator. Algorithms Mol. Biol. **10**(3) (2015)
6. Jacox, E., Chauve, C., Szöllősi, G.J., Ponty, Y., Scornavacca, C.: ecceTERA: comprehensive gene tree-species tree reconciliation using parsimony. Bioinformatics **32**(13), 2056–2058 (2016)
7. Makarova, K.S., et al.: Evolutionary classification of CRISPR-Cas systems: a burst of class 2 and derived variants. Nat. Rev. Microbiol. **18**(2), 67–83 (2020)
8. Duchemin, W., et al.: DeCoSTAR: reconstructing the ancestral organization of genes or genomes using reconciled phylogenies. Genome Biol. Evol. **9**(5), 1312–1319 (2017)
9. Duchemin, W.: Phylogeny of dependencies and dependencies of phylogenies in genes and genomes. Université de Lyon, Theses (2017)
10. Dondi, R., Lafond, M., Scornavacca, C.: Reconciling multiple genes trees via segmental duplications and losses. Algorithms Molecular Biol. **14**(1), 03 (2019)
11. Paszek, J., Gorecki, P.: Efficient algorithms for genomic duplication models. IEEE/ACM Trans Comput. Biol. Bioinform. (2017)
12. Delabre, M., et al.: Evolution through segmental duplications and losses: a super-reconciliation approach. Algorithms Molecular Biol. **15**, 12 (2020)
13. Anselmetti, Y., Delabre, M., El-Mabrouk, N.: Reconciliation with segmental duplication, transfer, loss and gain. In: Jin, L., Durand, D (eds.) Comparative Genomics, Cham, pp. 124–145 (2022)
14. Delabre, M., El-Mabrouk, N.: Synesth: comprehensive syntenic reconciliation with unsampled lineages. Algorithms **17**(5) (2024)

15. Aho, A.V., Yehoshua, S., Szymanski, T.G., Ullman, J.D.: Inferring a tree from lowest common ancestors with an application to the optimization of relational expressions. SIAM J. Comput. **10**(3), 405–421 (1981)
16. Constantinescu, M., Sankoff, D.: An efficient algorithm for supertrees. J. Classif. **12**, 101–112 (1995)
17. Ng, M.P., Wormald, N.C.: Reconstruction of rooted trees from subtrees. Discrete Appl. Math **69**, 19–31 (1996)
18. Lafond, M., Ouangraoua, A., El-Mabrouk, N.: Reconstructing a supergenetree minimizing reconciliation. BMC-Genomics **16**, S4 (2015), Special issue of RECOMB-CG 2015
19. Lafond, M., Chauve, C., El-Mabrouk, N., Ouangraoua, A.: Gene tree construction and correction using supertree and reconciliation. IEEE/ACM Trans. Comput. Biol. Bioinf. **15**(5), 1560–1570 (2018)
20. Ma, B., Li, M., Zhang, L.: From gene trees to species trees. SIAM J. Comput. **30**, 729–752 (2000)
21. Bansal, M.S., Shamir, R.: A note on the fixed parameter tractability of the gene-duplication problem. IEEE/ACM Trans. Comput. Biol. Bioinform. **8**(3), 848–850 (2011)
22. Blin, G., Bonizzoni, P., Dondi, R., Rizzi, R., Sikora, F.: Complexity insights of the minimum duplication problem. Theoret. Comput. Sci. **530**, 66–79 (2014)
23. Bansal, M.S., Burleigh, J.G., Eulenstein, O., Fernández-Baca, D.: Robinson-foulds supertrees. Algorithms Molecular Biol. **5**(1), 18 (2010)
24. Bayzid, M.S., Mirarab, S., Warnow, T.: Inferring optimal species trees under gene duplication and loss. Pac Symp. Biocomput. 250–261 (2013)
25. Wu, Y., Zhang, L.: Computing the bounds of the number of reticulations in a tree-child network that displays a set of trees. J. Comput. Biol. **31**(4), 345–359 (2024)

Open Access This chapter is licensed under the terms of the Creative Commons Attribution 4.0 International License (http://creativecommons.org/licenses/by/4.0/), which permits use, sharing, adaptation, distribution and reproduction in any medium or format, as long as you give appropriate credit to the original author(s) and the source, provide a link to the Creative Commons license and indicate if changes were made.

The images or other third party material in this chapter are included in the chapter's Creative Commons license, unless indicated otherwise in a credit line to the material. If material is not included in the chapter's Creative Commons license and your intended use is not permitted by statutory regulation or exceeds the permitted use, you will need to obtain permission directly from the copyright holder.

Quantum Algorithms for Longest Common Substring with a Gap

Daniel Gibney[✉][iD] and Md Helal Hossen[iD]

University of Texas at Dallas, Texas 75080, USA
{daniel.gibney,mdhelal.hossen}@utdallas.edu

Abstract. Recent breakthroughs have provided a sublinear time quantum algorithm for the Longest Common Substring Problem running in $\widetilde{\mathcal{O}}(n^{2/3}/d^{1/6})$ time for two strings of length at most n, where d is the length of the solution. At the same time, no subquadratic time quantum algorithm for the Longest Common Subsequence Problem is known, implying increasing difficulty as gaps are allowed within the solution. In this work, we consider the problem of finding two ordered matching substrings such that their total length is maximized. We present a strongly sublinear-time quantum algorithm.

Keywords: Longest Common Subsequence · Longest Common Substring · Quantum Algorithms

1 Introduction

Longest Common Substring and Longest Common Subsequence are two of the most fundamental problems in the field of string algorithms. Finding the longest common substring of two strings of length at most n was conjectured by Knuth to require $\mathcal{O}(n \log n)$ time, and the refutation of this conjecture with a linear time algorithm by Wiener in 1973 [28] led to the design of the suffix tree, a data structure that is now one of the pillars of string algorithms. The Longest Common Subsequence Problem, and its apparent required quadratic time complexity has been a significant motivator for innovations in fine-grained complexity [1,2], parameterized algorithms [17], and approximation algorithms [4,25].

The problem we study here can be viewed as a bridge between these two theoretically significant problems, with a parameter that transitions between the two. Given two strings T_1 and T_2 of length at most n and an integer $k \geq 1$, the problem is to find up to k ordered matching substrings so that the sum of the lengths of these substrings is maximized. Formally, we say a set of k matching substrings $(T_1[i_1 \mathinner{.\,.} i'_1], T_2[j_1 \mathinner{.\,.} j'_1]), (T_1[i_2 \mathinner{.\,.} i'_2], T_2[j_2 \mathinner{.\,.} j'_2]), \ldots, (T_1[i_k \mathinner{.\,.} i'_k], T_2[j_k \mathinner{.\,.} j'_k])$ is ordered if $i'_h < i_{h+1}$ and $j'_h < j_{h+1}$ for $h \in [1 \mathinner{.\,.} k-1]$. The problem we study here is a special case of:

Problem 1 (Longest Common Factor with Gaps (k-LCFg)). *Given strings $T_1[1 \mathinner{.\,.} n]$, $T_2[1 \mathinner{.\,.} m]$ ($n \geq m$) and an integer k, find an ordered set of*

at most k matching substrings such that the total combined length of the substrings is maximized.

For $k=1$, k-LCFg is equivalent to the Longest Common Substring Problem, and for $k=m$, it is equivalent to the Longest Common Subsequence Problem. k-LCFg can be solved in $\mathcal{O}(kn^2)$ time classically through dynamic programming. This problem was also consider by Li et al. for multiple strings under the name Longest k Tuple Common Substring [21].

Our work considers the special case of 2-LCFg. For this problem, existing results on suffix-longest common substring queries [6] imply that there exists a quasilinear, $\mathcal{O}(n \log^2 n)$ time classical algorithm. Our main result is a sublinear time quantum algorithm with the running time stated in Theorem 1.

Theorem 1 *There exists a quantum algorithm solving 2-LCFg in $\widetilde{\mathcal{O}}(n^{15/16+o(1)})$ quantum time.*[1]

In the full version of this work, we also provide a simpler quantum algorithm with a query complexity of $\widetilde{\mathcal{O}}(n^{9/10+o(1)})$, albeit with an increased time complexity. As we discuss in the conclusion, this suggest further improvement in time complexity is likely possible.

Related Work: Several other variants of Longest Common Substring have been studied in the classical setting. The related problem of finding the longest common substring between two strings with at most k-mismatches has been studied previously. Babenko and Starikovskaya considered the case with 1 mismatch and provided an $\mathcal{O}(nm)$ time algorithm [7]. Flouri et al. later provided a quadratic time algorithm for arbitrary k [12]. Kociumaka et al. proved that, for $k = \omega(\log n)$, a strongly subquadratic time algorithm is not possible under the Strong Exponential Time Hypothesis and considered approximate solutions [20]. Work by Thankachan et al. demonstrates that the problem can also be solved in $\mathcal{O}(n \log^k n)$ time [26]. The same problem, but with the further restriction that each match has length at least ℓ, was studied by Charalampopoulos et al., who provided a solution running in time $\mathcal{O}(n + (n \log^{k+1} n)/\sqrt{\ell})$ [10]. The longest common substring that is a concatenation of matches all having length at least k has also been considered [24,27]. The known algorithms all have worst-case quadratic time complexity.

Quantum algorithms for string problems have also been a very active area of research over recent years. This line of research was initiated by Ramesh and Vinay who provided a quantum algorithm for determining whether a pattern $P[1..m]$ occurs in a text $T[1..n]$ in $\widetilde{\mathcal{O}}(\sqrt{n} + \sqrt{m})$ time [16]. Additional past results include determining the LZ77 factorization [29] of a text in $\widetilde{\mathcal{O}}(\sqrt{zn})$ time where z is the number of LZ77 factors and edit distance in $\widetilde{\mathcal{O}}(\sqrt{kn} + k^2)$ time where k is the number of edits [14]. Most relevant to this work, Gall and Seddighin presented the first sublinear time quantum algorithm for longest common substring running in time $\widetilde{\mathcal{O}}(n^{5/6})$ and an $\Omega(n^{2/3})$ lower bound based

---

[1] $\widetilde{\mathcal{O}}(\cdot)$ suppresses polylogarithmic factors.

on a reduction from element distinctness [13]. This result was later improved to $\widetilde{\mathcal{O}}(n^{2/3})$ by Akmal and Jin [3] and then to $\widetilde{\mathcal{O}}(n^{2/3}/d^{1/6})$ by Jin and Nogler [18] where d is the length of the longest common string. Conditional $\Omega(n^{1.5})$ lower bounds on the quantum time complexity of Longest Common Subsequence based on a quantum version of the Strong Exponential Time Hypothesis have also been established by Buhrman et al. [9].

1.1 Technical Preliminaries

We use $[i\mathrel{..}j]$ to denote the set of integers $\{i, i+1, \ldots, j-1, j\}$. A string $T[1\mathrel{..}n]$ is a sequence of n symbols over some alphabet Σ. We use $T[i]$ to denote the ith symbol in T and $T[i\mathrel{..}j]$ to denote the substring $T[i] \cdot T[i+1] \cdot \ldots \cdot T[j]$, and $T[i\mathrel{..}j)$ to denote $T[i] \cdot T[i+1] \cdot \ldots \cdot T[j-1]$, where $\cdot$ means concatenation. A suffix of T is a substring of the form $T[i\mathrel{..}n]$ for some index $i \in [1\mathrel{..}n]$ and a prefix is a substring of the form $T[1\mathrel{..}i]$ for some index $i \in [1\mathrel{..}n]$. We use $\overline{T}$ to denote the reverse of the string T. For two strings $T_1[1\mathrel{..}n]$ and $T_2[1\mathrel{..}m]$, we say that two equal length substrings $T_1[i_1\mathrel{..}i_2]$ and $T_2[j_1\mathrel{..}j_2]$ form a match $(T_1[i_1\mathrel{..}i_2], T_2[j_1\mathrel{..}j_2])$ if $T_1[i_1+h] = T_2[j_1+h]$ for all $h \in [0\mathrel{..}i_2-i_1]$. The match $(T_1[i_1\mathrel{..}i_2], T_2[j_1\mathrel{..}j_2])$ is called *left maximal* if $i_1 = 1$ or $j_1 = 1$ or $T_1[i_1-1] \neq T_2[j_1-1]$. The match $(T_1[i_1\mathrel{..}i_2], T_2[j_1\mathrel{..}j_2])$ is called *right maximal* if $i_2 = n$ or $j_2 = m$ or $T_1[i_2+1] \neq T_2[j_2+1]$.

Suffix Trees, LCP, and LCE. The suffix tree $\mathcal{T}$ of a string $T[1\mathrel{..}n]$ is a compact trie of all suffixes of $T \cdot \$$ where $\$$ is the lexicographically smallest symbol and does not appear in T. The ith leaf of the suffix tree read from left to right corresponds to the ith lexicographically largest suffix. Each node v in the suffix tree is associated with string, $\mathrm{str}(v)$, that equals the concatenation of string labels on the path from the root to v. The string length of a node v in $\mathcal{T}$, is defined as $\mathrm{strlen}(v) = |\mathrm{str}(v)|$. The *longest common extension* between two suffixes $T[i\mathrel{..}n]$ and $T[j\mathrel{..}n]$, denoted $\mathrm{LCE}(i,j)$, is equal to the length of the longest common prefix of $T[i\mathrel{..}n]$ and $T[j\mathrel{..}n]$. We also use $\overline{\mathrm{LCE}}(i,j)$ to denote the longest common suffix of $T[1\mathrel{..}i]$ and $T[1\mathrel{..}j]$. The suffix array $\mathrm{SA}[1\mathrel{..}n]$ is defined such that $\mathrm{SA}[i]$ is the starting position in T of the ith suffix in lexicographically ascending order. The $\mathrm{LCP}[2\mathrel{..}n]$ array, is define as $\mathrm{LCP}[i] = \mathrm{LCE}(\mathrm{SA}[i-1], \mathrm{SA}[i])$. A generalized suffix tree of a set of strings $T_1, \ldots, T_k$ is the suffix tree for the string $T = T_1\$_1 T_2\$_2 \ldots T_s\$_s$ where each $\$_i$ symbol only occurs only once in T. The suffix tree and LCP array of a string of length n can be built in $\mathcal{O}(n)$ time for polynomial sized alphabets [11]. After constructing the suffix tree, $\mathrm{LCE}(i,j)$ for arbitrary i and j can be computed in $\mathcal{O}(1)$ time.

Dynamic Orthogonal Range Queries. Given a set of n d-dimensional points with assigned values, each point is defined by a tuple $(x_1, x_2, \ldots, x_d, value)$. We can construct a classical data structure in $\mathcal{O}(n \log^d n)$ time that can report the largest value of any point in a query orthogonal range $[a_1\mathrel{..}b_1] \times [a_2\mathrel{..}b_2] \times \ldots \times [a_d\mathrel{..}b_d]$ and insert, delete, or modify the value of any specified point, each in $\mathcal{O}(\log^d n)$ time [8]. We use RMQ($[a_1 .. b_1] \times [a_2 .. b_2] \times \ldots \times [a_d .. b_d]$) to denote the result of a range maximum query. Additional details needed for using RMQ structures in our quantum algorithm are provided in Sect. 2.2.

Quantum Algorithms. We assume that the input strings are provided as an input oracle. In particular, we assume access to a unitary operator that performs the transformation $|i\rangle|y\rangle \to |i\rangle|y \oplus T_1 \cdot T_2[i]\rangle$ for $i \in [1..n+m]$. We refer the interested reader to [23] for more preliminaries on quantum computing. In this work, we assume quantum random access memory (QRAM), which allows quantum routines, such as the quantum walk algorithm and Grover's search, to invoke classical algorithms with only a constant factor overhead [3]. We are interested in two complexity measures: the *query complexity*, which is the number of input queries made, and the *quantum time complexity*, which is the sum of the number of input queries, the number of elementary quantum gates, the number of classical operations, and the number classical RAM and QRAM accesses. Since quantum algorithms are probabilistic, we say a quantum algorithm solves a problem if it outputs the correct solution with a constant probability greater than 1/2. The probability can be made arbitrarily small with logarithmic factor overhead using standard techniques.

Grover's Search. Given a function $f : [1..n] \to \{0,1\}$ which can be evaluated for a given index in time t, Grover's search is a quantum algorithm that allows one to determine the existence and (if it exists) an index $i \in [1..n]$ such that $f(i) = 1$ in time $\widetilde{\mathcal{O}}(\sqrt{n} \cdot t)$ [15].

Quantum Walks. A quantum walk is an algorithm for finding marked vertices in a graph. The technique was first employed by Ambainis et al. to provide an optimal quantum algorithm for element distinctness [5]. It has since been further applied to several other problems, including Longest Common Substring [13,18]. Similar to these works, we will consider quantum walks on *Johnson graphs*. In the Johnson graphs we consider, the vertices consist of all subsets of r elements of a domain $\{1, 2, \ldots, N\}$ and two vertices are adjacent if they contain $r-1$ common elements. To perform a quantum walk, we must define what it means for a vertex to be marked (usually indicating the subset contains a solution to the problem of interest) and specify dynamic data structures that support checking if a given vertex is marked. Let S denote the initial cost of setting up one of these data structures with r elements, U the cost of adding or removing an element from the set, and C the cost of checking whether a given vertex is marked. Then the total cost for finding a marked vertex is logarithmic factors from

$$S + \frac{1}{\sqrt{\epsilon}}(\sqrt{r}U + C)$$

where ϵ is a lower bound on the probability that a randomly selected vertex is marked, given some marked vertices exist [22]. The case where no vertices are marked can also be handled without increasing the asymptotic time complexity. For the Johnson graphs we consider, where the subsets are of size r, if a vertex is marked when it contains a special subset of k elements, then $\varepsilon \geq \binom{N-k}{r-k}/\binom{N}{r} = \Omega(r^k/N^k)$.

The data structures used in a quantum walk must also be *history-independent* in the sense that they can only depend on the current subset of r elements and not the order in which elements were inserted or deleted to arrive at the current subset.

2 Our Algorithm

The following lemma will allow us to only consider finding an optimal solution where the left match is both left maximal and right maximal.

Lemma 1. *There exists an optimal solution to the 2-LCFg problem of the form $(T_1[i_1..i_2], T_2[j_1..j_2]), (T_1[i_3..i_4], T_2[j_3..j_4])$ with $i_2 < i_3$ and $j_2 < j_3$ where the left match, i.e. $(T_1[i_1..i_2], T_2[j_1..j_2])$, is both left maximal and right maximal.*

Proof. Clearly, the left match in an optimal solution is left maximal. Suppose the left match $T_1[i_1..i_2] = T_2[j_1..j_2]$ and right match $T_1[i_3..i_4] = T_2[j_3..j_4]$ form an optimal 2-LCFg solution where $T_1[i_1..i_2] = T_2[j_1..j_2]$ is not right maximal, i.e., $T_1[i_2+1] = T_2[j_2+1]$. If $i_2 < i_3-1$ and $j_2 < j_3-1$, then we can extend the left match, contradicting the solution's optimality. If $i_2 = i_3 - 1$, then the common substrings $T_1[i_1..i_2 + 1] = T_2[j_1..j_2 + 1]$ and $T_1[i_3 + 1..i_4] = T_2[j_3 + 1..j_4]$ provide a solution with equal objective value. A similar argument holds in the case where $j_2 = j_3 - 1$. Repeatedly extending the left match to the right, we eventually obtain a solution of the same value where the left match is right maximal.

2.1 Classical Algorithm for 2-LCFg

It is a straightforward consequence of existing results by Amir et al. [6] that 2-LCFg can be solved classically in $\mathcal{O}(n \log^2 n)$ time. As a starting point, we provide this algorithm. It makes use of the data structure described in the following lemma.

Lemma 2 (Suffix-LCS Queries [6]). *After $\mathcal{O}(n \log^2 n)$ time preprocessing, the longest common substrings of any two suffixes of T_1 and T_2 can be computed in $\mathcal{O}(\log n)$ time.*

2-LCFg Classical Algorithm Description: We start by constructing the generalized suffix tree $\mathcal{T}$ for T_1 and T_2. We traverse the nodes in $\mathcal{T}$ in a bottom-up fashion, and for each node v, we obtain the leftmost occurrences of str(v) in T_1 and T_2 if they exist. Due to Lemma 1, we can assume an optimal solution where the left match must correspond to str(v) for some node v in $\mathcal{T}$, otherwise, the left match would not be right maximal.

Iterating through the nodes of $\mathcal{T}$, suppose for a node v, the right boundary of the leftmost occurrence of str(v) in T_1 is i_2 and the right boundary of the leftmost occurrence of str(v) in T_2 is j_2. We then apply the data structure of Lemma 2 on the suffixes $T_1[i_2+1..n]$ and $T_2[j_2+1..m]$ and sum the length of the result with strlen(v). Taking the pair of matches with the maximum gives the solution.

2.2 Quantum Algorithm for 2-LCFg

Overview: If the suffix tree and suffix-LCS query data structures used in the classical algorithm described above were already constructed, a natural approach to a quantum algorithm for 2-LCFg would be to apply Grover's search to the nodes of the suffix tree to search for the left match and perform a subsequent suffix-LCS query. However, the time required to construct these data structures precludes a sublinear time quantum algorithm. Instead, we apply a sampling strategy combined with sparse suffix tree constructions to simulate the aforementioned approach. The exact sampling strategy will differ based on the length of the right match, with the case of the smaller right match being handled with a quantum walk looking for four indices that define the boundaries of a solution. The case of the longer right match is handled using *anchor sets* (defined in Sect. 2.4) and a two-level Grover's search.

For the remainder of this section, we will always assume that the left match is both left and right maximal and that the smaller of the two matches is on the right. By repeating the same overall algorithm with the reversed strings, we solve the case where the left match is smaller. The first idea is to break the problem into two subproblems based on the range of lengths for the right match. To accomplish this, we define the problem 2-LCFg(L, U) as follows.

Problem 2 (2-LCFg(L, U)). *Given two strings $T_1[1..n]$ and $T_2[1..m]$, find substrings $T_1[i_1..i_2]$, $T_2[j_1..j_2]$, $T_1[i_3..i_4]$, $T_2[j_3..j_4]$ such that*

- $T_1[i_1..i_2] = T_2[j_1..j_2]$ and $T_1[i_3..i_4] = T_2[j_3..j_4]$
- $i_2 < i_3$ and $j_2 < j_3$
- *the left match is at least as long as the right match, i.e., $i_2 - i_1 \geq i_4 - i_3$*
- *the left match is right maximal, i.e., $T_1[i_2+1] \neq T_2[j_2+1]$, or $i_2 = n$, or $j_2 = m$.*
- *the length of the second match in the range $[L..U]$, i.e., $L \leq i_4 - i_3 + 1 \leq U$*
- *the combined length $(i_2 - i_1 + 1) + (i_4 - i_3 + 1)$ is maximized while respecting the above constrains.*

We can also assume both matches to be non-empty (if one is empty, the solution can be obtained with the quantum algorithm for Longest Common Substring). We present different solutions to 2-LCFg($1, \kappa$) and 2-LCFg($\kappa+1, n$), where κ is some threshold value that we optimize later. By doing so, we solve for all possible lengths of the right match and can then take the best solution.

In the quantum setting, Jin and Nogler provide the following result, which we apply often.

Lemma 3 (Quantum LCE-Queries [18]). Given a string $T \in \Sigma^n$ and integer $1 \leq \tau \leq n/2$, there is a quantum preprocessing algorithm that outputs in $T_{prep} = \widetilde{\mathcal{O}}(n/\tau^{1/2-o(1)})$ time a classical data structure D (with high success probability), such that: given any $i, j \in [1 \mathinner{.\,.} n]$, one can compute $\mathrm{LCE}(i,j)$ in $T_{ans} = \widetilde{\mathcal{O}}(\sqrt{\tau})$ quantum time, given access to T and D.

As a preprocessing step, we construct LCE and $\overline{\mathrm{LCE}}$ data structures for $T_1 \$_1 T_2 \$_2$. This requires $\widetilde{\mathcal{O}}(n/\tau^{1/2-o(1)})$ time where τ is a parameter we specify later. After this, each LCE and $\overline{\mathrm{LCE}}$ query can be answered in $\widetilde{\mathcal{O}}(\sqrt{\tau})$ time.

2.3 Algorithm: Short Right Match

In this section, we present the solution for 2-LCFg$(1, \kappa)$. We utilize a quantum walk algorithm discussed in Sect. 1.1. We define a threshold α that we do a binary search over, looking for the combined lengths of the matches to be at least α. Each step in this binary search constitutes its own quantum walk, but this only increases the total complexity by at most a logarithmic factor. Within a subset of r indices, we wish to know if four indices are contained in the subset, two defining the right boundaries of the left match and two defining left boundaries of the right match, such that the sum of the match lengths is at least α.

Data Structures: We will use RMQ data structures to identify the longest right match in our given set of r sampled boundaries. The following is slight modification of the 2D Range Sum data structure used in by Akmal and Jin [3] to be a 2D-RMQ structure.

Lemma 4 (2D Range Maximum). *Let integer $N \leq n^{O(1)}$. There is a history-independent data structure of size $\widetilde{\mathcal{O}}(r)$ that maintains a set of at most r points $\{(x_1, y_1, v_1) \ldots, (x_r, y_r, v_r)\}$ with integer coordinates $x_i, y_i, v_i \in [1 \mathinner{.\,.} N]$, and supports the following operations with worst-case $\widetilde{\mathcal{O}}(1)$ time complexity and with high probability of success:*

- *Insertion: Add a new point (x, y, v) into the set.*
- *Deletion: Delete the point (x, y, v) from the set.*
- *Range maximum: Given $1 \leq x_1 \leq x_2 \leq N$, $1 \leq y_1 \leq y_2 \leq N$, return a point and value (x, y, v) such that $(x, y) \in [x_1, x_2] \times [y_1, y_2]$ and v is the maximum among all such points.*

Proof (Sketch). The construction is nearly identical to the segment tree based construction presented in [3] for 2D-range sum queries, only differing in that for the Cartesian product of rectangles stored in the structure we maintain the maximum value and the corresponding point rather than the sum of points in that range. Given a query range, we decompose it to $\mathcal{O}(\log^2 N)$ many query rectangles and take the maximum over them rather summing their individual values.

We will also require a history independent dynamic array data structure.

Lemma 5 (Dynamic Arrays [3]). There is a history-independent data structure of size $\widetilde{\mathcal{O}}(r)$ that maintains an array of items $(key_1, value_1)$, $(key_2, value_2)$, ..., $(key_r, value_r)$ with distinct keys (neither the keys nor the values are necessarily sorted in increasing order), and supports the following operations with worst-case $\widetilde{\mathcal{O}}(1)$ time complexity and high success probability:

- Indexing: Given an index $1 \leq i \leq r$, return the ith item $(key_i, value_i)$.
- Insertion: Given an index $1 \leq i \leq r+1$ and a new item, insert it into the array between the $(i-1)^{st}$ item and the ith item (shifting later items to the right).
- Deletion: Given an index $1 \leq i \leq r$, delete the ith item from the array (shifting later items to the left).
- Location: Given a key, return its position i in the array.
- Range Minimum Query: Given $1 \leq a \leq b \leq r$, return $\min_{a \leq i \leq b}\{value_i\}$.

We are now ready to describe our quantum walk algorithm. We call elements in the current subset of size r *sampled*. The data structures maintained for the quantum walk are the following:

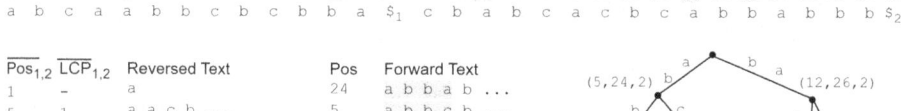

Fig. 1. (Top) The concatenated string $T_1\$_1 T_2\$_2$ with the $r = 6$ samples indicated. Here $\kappa = 4$. (Bottom left) The arrays for the co-lexicographically sorted prefix indices and $\overline{LCP}$ are maintained. We also maintain the arrays $\overline{Pos}_1 = [1, 5, 12]$ and $\overline{Pos}_2 = [17, 24, 26]$. (Bottom middle) The lexicographically sorted suffix positions. These are explicitly not stored. (Bottom right) A compact trie for the truncated suffixes with length $\kappa = 4$ is maintained. Note that edge labels are shown for illustration, but not explicitly stored as strings. The internal nodes are associated with two-dimensional points put into a 2D-RMQ structure.

- We maintain a compact trie denoted $\mathcal{T}_s$ for all r indexes in the sampled set, using only the κ long prefixes of suffixes starting at those indices (if the suffix has a length less than κ we use the entire suffix). For convenience, each truncated suffix is given a unique # symbol at its end. We maintain for every node v in $\mathcal{T}_s$ the rightmost sampled occurrence of $str(v)$ in T_1 and T_2 (if it exists).

- We maintain a 2D-RMQ structure for the nodes of $\mathcal{T}_s$. For every node v whose subtree has leaves corresponding to positions in both substrings, we associate coordinates (x, y) where x is the rightmost sampled starting position of str(v) in T_1, y is the rightmost sampled starting position of str(v) in T_2. The value assigned to the point is strlen(v). See Fig. 1.
- We maintain three dynamic arrays for the prefixes belonging to the current r sized subset. The first array $\overline{\text{Pos}_{1,2}}$ maintains the ending positions for each sampled prefix in sorted co-lexicographic order, with indices as keys and the $\overline{LCP}$ between adjacent prefixes as values. An array $\overline{\text{Pos}_1}$ contains the indices of sampled prefixes for only T_1 in co-lexicographically sorted order as both values and as keys. An array $\overline{\text{Pos}_2}$ containing the indices of sampled prefixes for only T_2 in co-lexicographically sorted order as both values and as keys. These arrays are used to simulate a sparse suffix tree used for the left match.

Setup: The initialization cost for these structures is $S = \widetilde{\mathcal{O}}(r\sqrt{\tau})$. This is because we can use LCE queries to lexicographic sort all of the truncated suffixes and then by processing them in sorted order add each new node, leaf, and edge to $\mathcal{T}_s$, each in $\widetilde{\mathcal{O}}(1)$ time. We do one bottom-up pass over the nodes $\mathcal{T}_s$ to compute the information stored in the suffix tree nodes and add the points to the associated 2D-RMQ structure. We can use $\overline{LCE}$ queries and binary search to find the correct location in $\overline{\text{Pos}_{1,2}}$, $\overline{\text{Pos}_1}$, and $\overline{\text{Pos}_2}$ for each sample and get the $\overline{LCP}$ value for $\overline{\text{Pos}_{1,2}}$.

Updating: To insert a new sample, we use binary search and $\overline{LCE}$ queries to find the correct sorted position and update the $\overline{\text{Pos}_{1,2}}$, $\overline{\text{Pos}_1}$, and $\overline{\text{Pos}_2}$ arrays. We use LCE queries and binary search to find the position to insert the new leaf into the trie $\mathcal{T}_s$ for the length κ prefix of the suffix at the sample position. Knowing the position and the string depth with which to insert this leaf, it can be inserted by traversing the path starting an adjacent leaf and splitting appropriate edge in $\widetilde{\mathcal{O}}(\kappa)$ time. We then traverse a path in $\mathcal{T}_s$ from the inserted leaf, updating at most κ nodes in the 2D-RMQ structure. Doing so makes the update time for insertion $\widetilde{\mathcal{O}}(\sqrt{\tau} + \kappa)$. For deleting a particular index, we use the deletion functionality of dynamic arrays. Removing a leaf may again require traversing a path in $\mathcal{T}_s$ of at most κ nodes to update the corresponding points in the 2D-RMQ query structure. This takes $\widetilde{\mathcal{O}}(\kappa)$ time, making $U = \widetilde{\mathcal{O}}(\sqrt{\tau} + \kappa)$.

Checking: To check if the given subset of indices contains two matches with a combined length of at least α, we use Grover's search on the keys of $\overline{\text{Pos}_{1,2}}$. Suppose Grover's search is evaluating an index i corresponding to the string $T_1[i - \overline{LCP}[i] + 1 .. i]$ (this could instead be a substring of T_2). We first find the ranges in $\overline{\text{Pos}_1}$ and $\overline{\text{Pos}_2}$ corresponding to prefixes that have the suffix $T_1[i - \overline{LCP}[i]+1 .. i]$. Since the positions are sorted according to co-lexicographic order, this can be accomplished using $\overline{LCE}$ queries in $\widetilde{\mathcal{O}}(\sqrt{\tau})$ time with binary search. After this, we perform range minimum queries on these ranges in $\overline{\text{Pos}_1}$ and $\overline{\text{Pos}_2}$, which provide us with the leftmost sampled occurrences of $T_1[i - \overline{LCP}[i]+1 .. i]$ in T_1 and T_2 respectively (if they exist, otherwise Grover's search fails on i).

Suppose the leftmost occurrence in T_1 has a right boundary of i_2 and the leftmost occurrence in T_2 has a right boundary of j_2. Next, using the 2D-RMQ structure built over the nodes of $\mathcal{T}_s$ with the query range $[i_2+1\mathinner{.\,.} n] \times [j_2+1\mathinner{.\,.} m]$, we find the maximum sampled right match for the remaining suffixes. We finally check whether the combined length of these two matches is at least α and the left match is at least as long as the right match, otherwise the Grover's search fails on i. Overall, this requires $C = \widetilde{\mathcal{O}}(\sqrt{r} \cdot \sqrt{\tau})$ time due to Grover's search over $\overline{\mathrm{Pos}_{1,2}}$ and the LCE queries to used within to identify the corresponding ranges in $\overline{\mathrm{Pos}_1}$ and $\overline{\mathrm{Pos}_2}$. The additional RMQ query contributes only logarithmic factors to the time complexity.

Time Complexity: If there exists a solution for the given α, we have four indices corresponding to a solution in our subset with probability at least $\epsilon \geq \binom{n-4}{r-4}/\binom{n}{r} = \Omega(r^4/n^4)$. The overall time complexity for solving 2-LCFg$(1, \kappa)$ is logarithmic factors from

$$\frac{n}{\tau^{1/2-o(1)}} + S + \frac{1}{\sqrt{\epsilon}}(\sqrt{r}U + C) = \frac{n}{\tau^{1/2-o(1)}} + r\sqrt{\tau} + \frac{n^2}{r^2}(\sqrt{r}(\sqrt{\tau}+\kappa) + \sqrt{\tau r})$$

2.4 Algorithm: Long Right Match

Next, we solve 2-LCFg$(\kappa+1, n)$. Instead of a quantum walk, we use a two-level Grover's search and *anchor sets*. We first state the definition of a left-anchor set.

Definition 1 (Left-Anchor Set). *A set of indices A_L is a left-anchor set for length ℓ if for any length $\ell' \geq \ell$ matching substrings $T_1[i\mathinner{.\,.} i+\ell') = T_2[j\mathinner{.\,.} j+\ell')$, with minimized $i+j$, there exist a shift $s \in \{0, 1, \ldots, \ell-1\}$ such that $i+s, j+s \in A_L$.*

It is also useful for us to consider right-anchor sets, defined as follows:

Definition 2 (Right-Anchor Set). *A set of indices A_R is a right-anchor set for length ℓ if for any length $\ell' \geq \ell$ matching substrings $T_1[i\mathinner{.\,.} i+\ell') = T_2[j\mathinner{.\,.} j+\ell')$, with maximized $i+j$, there exist a shift $s \in \{0, 1, \ldots, \ell-1\}$ such that $i+\ell'-1-s, j+\ell'-1-s \in A_R$.*

Anchor sets are constructed by combining synchronizing sets [19] with the first occurrences of Lyndon words of repetitive substrings [10]. In the quantum setting, we have the following result by Jin and Nogler.

Lemma 6 (Quantum Algorithm for Left-Anchor Sets [18]). *Given a string $T \in \Sigma^n$ and integer $1 \leq d \leq n$, there is a quantum algorithm that computes a left-anchor set for length d of size $\widetilde{\mathcal{O}}(n/d)$ in $\widetilde{\mathcal{O}}(n/d^{1/2-o(1)})$ time.*

We first apply Lemma 6 to compute left-anchor and right-anchor sets for length κ. The sizes of these sets are both $\widetilde{\mathcal{O}}(n/\kappa)$ and we can compute both in $\widetilde{\mathcal{O}}(n/\kappa^{1/2-o(1)})$ time. Note that the right-anchor set for length κ is obtained by computing the left-anchor set of the reversed strings. Let $A = A_L \cup A_R$, which we simply refer to now as the anchor set. We have that $|A| = \widetilde{\mathcal{O}}(n/\kappa)$.

Like before, we do a binary search over a target total length α. We again construct the LCE and $\overline{\text{LCE}}$ data structures using parameter τ, in $\widetilde{\mathcal{O}}(n/\tau^{1/2-o(1)})$ time to answer queries in $\widetilde{\mathcal{O}}(\sqrt{\tau})$ time. For each α tested, we do an **outer level Grover's search** for an offset $o \in \{0, 1, \ldots, \kappa - 1\}$. The offset o represents the difference from right boundary of the right match from the anchor pair in A. This works since we know the target right match has length at least κ. We build a compact trie $\mathcal{T}_l$ for the reversed prefixes resulting from the anchor set plus offset o. Specifically, we construct $\mathcal{T}_l$ from the reversed prefixes ending at $i + o$ for each $i \in A$. We call the set $\{i + o \mid i \in A\}$ the set of sampled positions. Each insertion in $\mathcal{T}_l$ requires $\widetilde{\mathcal{O}}(\sqrt{\tau})$ time for finding the position in the co-lexicographically sorted order using an $\overline{\text{LCE}}$ query. After constructing $\mathcal{T}_l$, we construct three RMQ structures over all the nodes in $\mathcal{T}_l$. These capture the different ways in which the left match can intersect with a potential right match and the resulting total length.

Suppose for a node v in $\mathcal{T}_l$, the rightmost sampled position for $\overline{\text{str}(v)}$ is x in T_1 and rightmost sampled position for $\overline{\text{str}(v)}$ is y in T_2. Let $s = \text{strlen}(v)$.

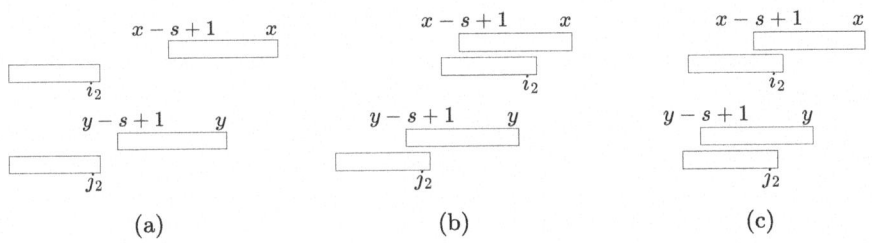

Fig. 2. The three cases considered for RMQ structures in the outer level of Grover's search. We consider T_1 as being on the top T_2 on the bottom. Matches are shown as rectangles. Relevant indices for RMQ_1, RMQ_2, and RMQ_3 are indicated.

(a) RMQ_1 gets point $(x - s + 1, y - s + 1)$ with value s. This will capture the case where the potential right match ends at x and y in T_1 and T_2 (respectively) and the left match, $(T_1[i_1 .. i_2], T_2[j_1 .. j_2])$, does not intersect with the rightmost sampled occurrence of $\overline{\text{str}(v)}$. In particular, our query to RMQ_1 will capture when $i_2 < x - s + 1$ and $j_2 < y - s + 1$. See Fig. 2a.

(b) RMQ_2 gets point $(x - s + 1, x, x - y)$ with value x. This will capture the case where the potential right match ends at x and y in T_1 and T_2 (respectively) and the left match, $(T_1[i_1 .. i_2], T_2[j_1 .. j_2])$, intersects with the rightmost sampled occurrence of $\overline{\text{str}(v)}$ and the overlap in T_1 is at least the overlap in T_2. In particular, our query to RMQ_2 will capture when $x - s + 1 \leq i_2 \leq x$ and $x - i_2 \leq y - j_2$. See Fig. 2b. Note that this includes the possibility of $j_2 < y - s + 1$.

(c) RMQ$_3$ gets point $(y-s+1, y, x-y)$ with value y. This will capture the case where the potential right match ends at x and y in T_1 and T_2 (respectively) and the left match, $(T_1[i_1..i_2], T_2[j_1..j_2])$, intersects with the rightmost sampled occurrence of str(v), and the overlap is greater in T_2. In particular, our query to RMQ$_3$ will capture when $y-s+1 \le j_2 \le y$ and $x-i_2 > y-j_2$. See Fig. 2c. Note that this includes the possibility of $i_2 < x-s+1$.

Observe that all possible ways that the left match in a solution can overlap with the MEM obtained by extending the right match are captured with the three cases above. The time for this step is dominated by the initial construction of T_l, which is $\widetilde{\mathcal{O}}(n\sqrt{\tau}/\kappa)$.

Next, we do an **inner level Grover's search** over the pairs of anchors. This is to pick out the anchor pair for the left match. We know there must be a pair of indices in A that anchor the left match because we have that the left match is at least as long as the right match, which has length at least κ. Applying that the left match is both left and right maximal from Lemma 1, for a given pair indices from A, one in T_1 and one in T_2, we take the left and right most extension starting at the anchor pair using $\overline{\text{LCE}}$ and LCE queries, respectively. We then use the RMQ structure built over the nodes of T_l to find the largest right match possible with the sampled suffixes. Let i_2 be the right boundary of the left match in T_1 and j_2 the right boundary of the left match in T_2.

- We query RMQ$_1$ with $[i_2+1..n] \times [j_2+1..m]$ and take the value of the point return.
- We query RMQ$_2$ with $[0..i_2] \times [i_2..n] \times [-n..i_2-j_2]$ and subtract from the value of the point returned $(i_2 - 1)$. This is since, the overlap in T_1 being at least the overlap in T_2, i.e., $x-i_2 \le y-j_2$, implies we can only use at most $x-i_2+1$ of the potential right match.
- We query RMQ$_3$ with $[0..j_2] \times [j_2..n] \times [i_2-j_2+1..n]$ and subtract from the value of the point returned $(j_2 - 1)$. This is since the overlap in T_2 be greater than the overlap in T_1 implies we can only use at most $y-j_2+1$ of the potential right match.

A maximum is taken over the sum of the length of the first match and the values computed above. If the combined length is at least α, then Grover's search succeeds.

Time Complexity: The inner level Grover's search requires $\widetilde{\mathcal{O}}(\sqrt{|A|^2} \cdot \sqrt{\tau})$, or equivalently, $\widetilde{\mathcal{O}}(n\sqrt{\tau}/\kappa)$ time. Combining the time for constructing the LCE data structures, the anchor set, and the outer and inner level Grover's search, the time complexity is logarithmic factors from

$$\frac{n}{\tau^{1/2-o(1)}} + \frac{n}{\kappa^{1/2-o(1)}} + \sqrt{\kappa}\left(\frac{n\sqrt{\tau}}{\kappa}\right) = \widetilde{\mathcal{O}}\left(\frac{n}{\tau^{1/2-o(1)}} + \frac{n\sqrt{\tau}}{\sqrt{\kappa}}\right) \qquad (1)$$

Note that the omission of $n/\kappa^{1/2-o(1)}$ from the right-hand side will be justified by our choice of τ, which is polynomial in n.

2.5 Combining and Optimizing

We seek to determine the parameters r, τ, and κ to minimize the overall time complexity of finding solutions to 2-LCFg$(1, \kappa)$ and 2-LCFg$(\kappa+1, n)$. The overall time complexity is logarithmic factors from

$$\left(\frac{n}{\tau^{1/2-o(1)}} + r\sqrt{\tau} + \frac{n^2\sqrt{\tau}}{r^{3/2}} + \frac{n^2\kappa}{r^{3/2}}\right) + \frac{n\sqrt{\tau}}{\sqrt{\kappa}}$$

To optimize, we assume $r = n^a$, $\tau = n^b$, and $\kappa = n^c$ for some unknowns a, b, c. Ignoring the little $o(1)$ terms, the exponents on n of the terms in the above expression are $1 - b/2$, $a + b/2$, $2 + b/2 - 3a/2$, $2 + c - 3a/2$, and $1 + b/2 - c/2$. We set up and solve the corresponding linear program with variables $0 \leq a, b, c \leq 1$ and new variable E:

$$\begin{aligned}
\text{minimize} \quad & E \quad \text{s.t.} \\
& 1 - b/2 \leq E \\
& a + b/2 \leq E \\
& 2 + b/2 - 3a/2 \leq E \\
& 2 + c - 3a/2 \leq E \\
& 1 + b/2 - c/2 \leq E
\end{aligned}$$

Solving this linear program yields $a = 7/8$, $b = 1/8$, $c = 1/4$, and $E = 15/16$, implying $r = n^{7/8}$, $\tau = n^{1/8}$, $r = n^{1/4}$, which are all valid choices for their respective parameters. The overall time complexity of the algorithm is $\widetilde{\mathcal{O}}(n^{15/16+o(1)})$. This completes the proof of Theorem 1.

3 Discussion

Our goal in this work was to demonstrate that 2-LCFg, a natural variant of the Longest Common Substring Problem, can be solved in sublinear time by a quantum algorithm. We suspect that the time complexity presented in this work can be further improved, perhaps by breaking the problem down in to further sub-cases. In the full version of this work, we demonstrate that 2-LCFg can be solved with a simpler quantum algorithm having a query complexity of $\widetilde{\mathcal{O}}(n^{9/10+o(1)})$, although this approach currently has a worse time complexity. This is the result of using a simpler data structure and a more brute force approach to checking if a solution is contained in the current subset during the quantum walk.

Other future research directions include providing sub-linear time quantum algorithms for any constant k. Proving lower bounds that are higher than those known for Longest Common Substring for either 2-LCFg, or more generally, k-LCFg, would also advance our understanding of this problem's computational complexity.

References

1. Abboud, A., Backurs, A., Williams, V.V.: Tight hardness results for LCS and other sequence similarity measures. In: Guruswami, V. (ed.) IEEE 56th Annual Symposium on Foundations of Computer Science, FOCS 2015, Berkeley, CA, USA, 17-20 October, 2015. pp. 59–78. IEEE Computer Society (2015). https://doi.org/10.1109/FOCS.2015.14
2. Abboud, A., Hansen, T.D., Williams, V.V., Williams, R.: Simulating branching programs with edit distance and friends: or: a polylog shaved is a lower bound made. In: Wichs, D., Mansour, Y. (eds.) Proceedings of the 48th Annual ACM SIGACT Symposium on Theory of Computing, STOC 2016, Cambridge, MA, USA, 18-21 June 2016. pp. 375–388. ACM (2016). https://doi.org/10.1145/2897518.2897653
3. Akmal, S., Jin, C.: Near-optimal quantum algorithms for string problems. In: Naor, J.S., Buchbinder, N. (eds.) Proceedings of the 2022 ACM-SIAM Symposium on Discrete Algorithms, SODA 2022, Virtual Conference / Alexandria, VA, USA, 9 - 12 January 2022, pp. 2791–2832. SIAM (2022). https://doi.org/10.1137/1.9781611977073.109
4. Akmal, S., Williams, V.V.: Improved approximation for longest common subsequence over small alphabets. In: Bansal, N., Merelli, E., Worrell, J. (eds.) 48th International Colloquium on Automata, Languages, and Programming, ICALP 2021, 12-16 July 2021, Glasgow, Scotland (Virtual Conference). LIPIcs, vol. 198, pp. 13:1–13:18. Schloss Dagstuhl - Leibniz-Zentrum für Informatik (2021). https://doi.org/10.4230/LIPIcs.ICALP.2021.13
5. Ambainis, A.: Quantum walk algorithm for element distinctness. SIAM J. Comput. **37**(1), 210–239 (2007). https://doi.org/10.1137/S0097539705447311
6. Amir, A., Charalampopoulos, P., Pissis, S.P., Radoszewski, J.: Dynamic and internal longest common substring. Algorithmica **82**(12), 3707–3743 (2020). https://doi.org/10.1007/s00453-020-00744-0
7. Babenko, M.A., Starikovskaya, T.: Computing the longest common substring with one mismatch. Probl. Inf. Transm. **47**(1), 28–33 (2011). https://doi.org/10.1134/S0032946011010030
8. de Berg, M., Cheong, O., van Kreveld, M.J., Overmars, M.H.: Computational geometry: algorithms and applications, 3rd Edition. Springer (2008). https://www.worldcat.org/oclc/227584184
9. Buhrman, H., Patro, S., Speelman, F.: A framework of quantum strong exponential-time hypotheses. In: Bläser, M., Monmege, B. (eds.) 38th International Symposium on Theoretical Aspects of Computer Science, STACS 2021, 16-19 March 2021, Saarbrücken, Germany (Virtual Conference). LIPIcs, vol. 187, pp. 19:1–19:19. Schloss Dagstuhl - Leibniz-Zentrum für Informatik (2021). https://doi.org/10.4230/LIPIcs.STACS.2021.19
10. Charalampopoulos, P., et al.: Linear-time algorithm for long LCF with k mismatches. In: Navarro, G., Sankoff, D., Zhu, B. (eds.) Annual Symposium on Combinatorial Pattern Matching, CPM 2018, 2-4 July 2018 - Qingdao, China. LIPIcs, vol. 105, pp. 23:1–23:16. Schloss Dagstuhl - Leibniz-Zentrum für Informatik (2018). https://doi.org/10.4230/LIPIcs.CPM.2018.23
11. Farach, M.: Optimal suffix tree construction with large alphabets. In: 38th Annual Symposium on Foundations of Computer Science, FOCS '97, Miami Beach, Florida, USA, 19-22 October 1997, pp. 137–143. IEEE Computer Society (1997). https://doi.org/10.1109/SFCS.1997.646102

12. Flouri, T., Giaquinta, E., Kobert, K., Ukkonen, E.: Longest common substrings with k mismatches. Inf. Process. Lett. **115**(6–8), 643–647 (2015). https://doi.org/10.1016/J.IPL.2015.03.006
13. Gall, F.L., Seddighin, S.: Quantum meets fine-grained complexity: sublinear time quantum algorithms for string problems. Algorithmica **85**(5), 1251–1286 (2023). https://doi.org/10.1007/s00453-022-01066-z
14. Gibney, D., Jin, C., Kociumaka, T., Thankachan, S.V.: Near-optimal quantum algorithms for bounded edit distance and lempel-ziv factorization. In: Woodruff, D.P. (ed.) Proceedings of the 2024 ACM-SIAM Symposium on Discrete Algorithms, SODA 2024, Alexandria, VA, USA, 7-10 January 2024, pp. 3302–3332. SIAM (2024). https://doi.org/10.1137/1.9781611977912.118
15. Grover, L.K.: A fast quantum mechanical algorithm for database search. In: Proceedings of the 28th Annual ACM Symposium on the Theory of Computing (STOC 1996), pp. 212–219 (1996). https://doi.org/10.1145/237814.237866
16. Hariharan, R., Vinay, V.: String matching in õ(sqrt(n)+sqrt(m)) quantum time. J. Dis. Algorithms **1**(1), 103–110 (2003). https://doi.org/10.1016/S1570-8667(03)00010-8
17. Heeger, K., Nichterlein, A., Niedermeier, R.: Parameterized lower bounds for problems in P via fine-grained cross-compositions. In: Berenbrink, P., Bouyer, P., Dawar, A., Kanté, M.M. (eds.) 40th International Symposium on Theoretical Aspects of Computer Science, STACS 2023, 7-9 March 2023, Hamburg, Germany. LIPIcs, vol. 254, pp. 35:1–35:19. Schloss Dagstuhl - Leibniz-Zentrum für Informatik (2023). https://doi.org/10.4230/LIPIcs.STACS.2023.35
18. Jin, C., Nogler, J.: Quantum speed-ups for string synchronizing sets, longest common substring, and k-mismatch matching. In: Bansal, N., Nagarajan, V. (eds.) Proceedings of the 2023 ACM-SIAM Symposium on Discrete Algorithms, SODA 2023, Florence, Italy, 22-25 January 2023, pp. 5090–5121. SIAM (2023). https://doi.org/10.1137/1.9781611977554.ch186
19. Kempa, D., Kociumaka, T.: String synchronizing sets: sublinear-time BWT construction and optimal LCE data structure. In: Charikar, M., Cohen, E. (eds.) Proceedings of the 51st Annual ACM SIGACT Symposium on Theory of Computing, STOC 2019, Phoenix, AZ, USA, 23-26 June 2019, pp. 756–767. ACM (2019). https://doi.org/10.1145/3313276.3316368
20. Kociumaka, T., Radoszewski, J., Starikovskaya, T.: Publisher correction: longest common substring with approximately k mismatches. Algorithmica **85**(10), 3323 (2023). https://doi.org/10.1007/S00453-023-01119-X
21. Li, T., Zhu, D., Jiang, H., Feng, H., Cui, X.: Longest k-tuple common sub-strings. In: Adjeroh, D.A., et al. (eds.) IEEE International Conference on Bioinformatics and Biomedicine, BIBM 2022, Las Vegas, NV, USA, 6-8 December 2022, pp. 63–66. IEEE (2022). https://doi.org/10.1109/BIBM55620.2022.9995199
22. Magniez, F., Nayak, A., Roland, J., Santha, M.: Search via quantum walk. In: Johnson, D.S., Feige, U. (eds.) Proceedings of the 39th Annual ACM Symposium on Theory of Computing, San Diego, California, USA, 11-13 June 2007, pp. 575–584. ACM (2007). https://doi.org/10.1145/1250790.1250874
23. Nielsen, M.A., Chuang, I.L.: Quantum Computation and Quantum Information (10th Anniversary edition). Cambridge University Press (2016). https://www.cambridge.org/de/academic/subjects/physics/quantum-physics-quantum-information-and-quantum-computation/quantum-computation-and-quantum-information-10th-anniversary-edition?format=HB
24. Pavetic, F., Zuzic, G., Sikic, M.: *lcsk++*: Practical similarity metric for long strings. CoRR arXiv: abs/1407.2407 (2014)

25. Rubinstein, A., Seddighin, S., Song, Z., Sun, X.: Approximation algorithms for LCS and LIS with truly improved running times. In: Zuckerman, D. (ed.) 60th IEEE Annual Symposium on Foundations of Computer Science, FOCS 2019, Baltimore, Maryland, USA, 9-2 November 2019, pp. 1121–1145. IEEE Computer Society (2019). https://doi.org/10.1109/FOCS.2019.00071
26. Thankachan, S.V., Apostolico, A., Aluru, S.: A provably efficient algorithm for the k-mismatch average common substring problem. J. Comput. Biol. **23**(6), 472–482 (2016). https://doi.org/10.1089/cmb.2015.0235
27. Ueki, Y., et al.: Longest common subsequence in at least k length order-isomorphic substrings. In: Steffen, B., Baier, C., van den Brand, M., Eder, J., Hinchey, M., Margaria, T. (eds.) SOFSEM 2017. LNCS, vol. 10139, pp. 363–374. Springer (2017). https://doi.org/10.1007/978-3-319-51963-0_28
28. Weiner, P.: Linear pattern matching algorithms. In: 14th Annual Symposium on Switching and Automata Theory, Iowa City, Iowa, USA, 15-17 October 1973, pp. 1–11. IEEE Computer Society (1973). https://doi.org/10.1109/SWAT.1973.13
29. Ziv, J., Lempel, A.: A universal algorithm for sequential data compression. IEEE Trans. Inf. Theory **23**(3), 337–343 (1977). https://doi.org/10.1109/TIT.1977.1055714

Online Computation of String Net Frequency

Peaker Guo[1](✉)[iD], Seeun William Umboh[1,3](✉)[iD], Anthony Wirth[1,2](✉)[iD], and Justin Zobel[1](✉)[iD]

[1] School of Computing and Information Systems, The University of Melbourne, Parkville, Australia
zifengg@student.unimelb.edu.au,
{william.umboh,awirth,jzobel@unimelb.edu.au
[2] School of Computer Science, The University of Sydney, Sydney, Australia
anthony.wirth@sydney.edu.au
[3] ARC Training Centre in Optimisation Technologies, Integrated Methodologies, and Applications (OPTIMA), Parkville, Australia

Abstract. The *net frequency* (NF) of a string, of length m, in a text, of length n, is the number of occurrences of the string in the text with unique left and right *extensions*. Recently, Guo et al. [CPM 2024] showed that NF is combinatorially interesting and how two key questions can be computed efficiently in the *offline* setting. First, SINGLE-NF: reporting the NF of a query string in an input text. Second, ALL-NF: reporting an occurrence and the NF of each string of positive NF in an input text. For many applications, however, facilitating these computations in an *online* manner is highly desirable. We are the first to solve the above two problems in the online setting, and we do so in optimal time, assuming, as is common, a constant-size alphabet: SINGLE-NF in $\mathcal{O}(m)$ time and ALL-NF in $\mathcal{O}(n)$ time. Our results are achieved by first designing new and simpler offline algorithms using suffix trees, proving additional properties of NF, and exploiting Ukkonen's online suffix tree construction algorithm and results on *implicit node* maintenance in an *implicit suffix tree* by Breslauer and Italiano.

Keywords: Suffix trees · Implicit nodes · Suffix links · Weiner links

1 Introduction

The *net frequency* (NF) of a string S, of length m, in a text T, of length n, is the number of occurrences of S in T with unique left and right *extensions*. For example, let T be rstk<u>st</u>castarstast\$. Among the five occurrences of st, only the underlined occurrence has a unique left extension kst and a unique right extension stc. In contrast, the other occurrences of st either have a repeated left extension (rst and ast), or a repeated right extension (sta). Thus the NF of st is 1 in T. Introduced by Lin and Yu [14], NF has been demonstrated to be useful for Chinese phoneme-to-character (and character-to-phoneme) conversion and other NLP tasks [14,15].

Recently, Guo et al. [10] reconceptualised NF and simplified the original definition. They thus identified strings with positive NF, including in Fibonacci words. They also showed that there could be at most n distinct strings in T with positive NF. They then bounded the sum of lengths of strings with positive NF in T of length n between $\Omega(n)$ and $\mathcal{O}(n \log \delta)$, where $\delta := \max\{S(k)/k : k \in [n]\}$ and $S(k)$ is the number of distinct strings of length k in T.

Although NF can be efficiently computed in the offline setting [10], it is also useful to determine NF in an online setting in which a text is being dynamically extended. In this setting, at all times we have read the first n characters from a stream and further characters are to be read in turn, that is, n is being incremented. As each character arrives, the data structure is updated and a query on NF computation can be answered. Throughout our inspection of the online setting, T is an input text of which we have read the first n characters, while S is a length-m query string. We consider the following two problems in the online setting. SINGLE-NF: report the NF of S in $T[1 \ldots n]$. ALL-NF: for each string of positive NF in $T[1 \ldots n]$, report one occurrence and its NF.

We have found that adapting the existing offline approach based on a suffix array to the online setting is not trivial. In the offline setting, suppose we have read n characters of an input text T so far, after building the structure for $T[1 \ldots n]$, should we append a single character to T, we would have to build the structure for $T[1 \ldots n+1]$ from scratch, without reusing the structure for $T[1 \ldots n]$. In contrast, in this paper, the structure for $T[1 \ldots n+1]$ derives from the structure for $T[1 \ldots n]$. This setting aligns with the *online* suffix tree construction algorithm by Ukkonen [21], which is widely used due to its simplicity and online nature, cited in numerous follow-up works [6,7,12,13,20].

Our Contribution. In this work, we first introduce a new characteristic of NF (Theorem 2), specifically designed for NF computation with a suffix tree. With this characteristic and Weiner links, assuming a constant-size alphabet, we present our optimal-time offline SINGLE-NF algorithm (Algorithm 1). Applying this characteristic in an asynchronous fashion, our optimal-time offline ALL-NF algorithm (Algorithm 2) using suffix links is arguably simpler than the state-of-the-art suffix array-based solution. We then adapt our offline algorithms to the online setting – to our knowledge online algorithms have not previously been reported. With additional properties of NF and prior results on implicit nodes in an implicit suffix tree, we obtain optimal $\mathcal{O}(m)$-time online SINGLE-NF and $\mathcal{O}(n)$-time online ALL-NF algorithms (Theorem 3 and Theorem 4).

2 Preliminaries

Strings. Let Σ be a constant-size alphabet throughout this paper. The constant-size assumption follows previous work on suffix trees [6,13]. Let $SP \equiv S \cdot P$ be the concatenation of two strings, S and P. Let $[n]$ denote the set

$\{1, 2, \ldots, n\}$. A substring of string T with starting position $i \in [n]$ and end position $j \in [n]$ is written as $T[i \ldots j]$. A substring $T[1 \ldots j]$ is called a *prefix* of T, while $T[i \ldots n]$ is called a *suffix* of T. Let T_i denote the ith suffix of T, $T[i \ldots n]$. An *occurrence* in the text T is a pair of starting and ending positions $(i, j) \in [n] \times [n]$. We say (i, j) is an *occurrence of string* S if $S = T[i \ldots j]$, and i is an occurrence of S if $S = T[i \ldots i + |S| - 1]$. An occurrence (i', j') is a *sub-occurrence* of (i, j) if $i \leq i'$, while $j' \leq j$. The *frequency* of S (in T), denoted by $f(S)$, is the number of occurrences of S in T.

Net Frequency. The NF of a string was originally defined by Lin and Yu [14], and reconceptualised and simplified by Guo et al. [10]. The NF of a unique string is defined to be zero; the NF of a repeated string S, written as $\phi(S)$, is the number of occurrences of S with unique left and right extensions.

Definition 1 (Net occurrence [10]). *An occurrence (i, j) is a net occurrence if $f(T[i \ldots j]) \geq 2$, $f(T[i - 1 \ldots j]) = 1$, and $f(T[i \ldots j + 1]) = 1$. When $i = 1$, $f(T[i - 1 \ldots j]) = 1$ is assumed to be true; when $j = n$, $f(T[i \ldots j + 1]) = 1$ is assumed to be true.*

NF can be also formulated in terms of symbols from Σ, rather than occurrences.

Lemma 1 (NF characteristic [10]). *Given a repeated string S,*

$$\phi(S) = |\{ (x, y) \in \Sigma \times \Sigma : f(xS) = 1 \text{ and } f(Sy) = 1 \text{ and } f(xSy) = 1 \}|.$$

Here, strings xS, Sy, and xSy are the *left*, *right*, and *bidirectional extensions* of S, respectively. Note that in Definition 1, the condition on the bidirectional extension $f(T[i - 1 \ldots j + 1]) = 1$ is not needed because it is implied by $f(T[i - 1 \ldots j]) = 1$ and $f(T[i \ldots j + 1]) = 1$. A string S is *branching* in T if S is the longest common prefix of two distinct suffixes of T. The following result says that branching strings are the only strings that could have positive NF.

Lemma 2 ([10]). *If S is not a branching string, then $\phi(S) = 0$.*

Suffix Trees. Introduced by Weiner [22], the suffix tree is arguably one of the most significant and versatile data structures in string processing with a wide range of applications [11]. The *suffix tree* of T is a rooted directed tree whose edges are labelled with substrings of T. It contains n leaf nodes, each labelled from 1 to n. For each $i \in [n]$, the concatenation of the edge labels on the path from the root to leaf i forms the suffix T_i. Each non-root and non-leaf node, known as a *branching node*, has at least two children, and the first character of the label on the edge to each child is distinct.

It is commonly assumed that a unique sentinel character $\$ \notin \Sigma$ is appended to the text. This ensures that no suffix is a prefix of another, thereby ensuring that each suffix is represented as a distinct leaf in the suffix tree. Given a node u, the *path label* of u, denoted by $\text{str}(u)$, is the concatenation of the edge labels on the path from the root to u. We call $|\text{str}(u)|$ the *string depth* of u, denoted by $\text{depth}(u)$.

In suffix-tree construction algorithms [16,21,22], a type of pointer called a *suffix link*[4] helps traverse the suffix tree efficiently; it is a key ingredient in achieving linear time. In our algorithms, besides suffix links, we also use *Weiner links*, also known as *reverse suffix links*. Given a branching node, u, the *suffix link* of u points from u to another branching node, v, where $\text{str}(v)$ is the string obtained by removing the first character of $\text{str}(u)$, that is, the longest proper suffix of $\text{str}(u)$. Whenever there is a suffix link from u to v, there is a *Weiner link* from v to u. Suffix links and Weiner links are usually only needed among branching nodes, but some applications also require them among leaf nodes [19].

We now define the following notations in a suffix tree used in subsequent sections. Consider a node u. Let $\text{start}(u)$ be the starting position of some occurrence of $\text{str}(u)$ in the text. Let $\text{parent}(u)$ be the parent node of u. Let $\text{child}(u, y)$ be the child node v of u such that the label of the edge uv starts with symbol y, and $\bot$ if such v does not exist. Let $\text{children}(u) := \{(v, y) : v = \text{child}(u, y) \neq \bot\}$. Let $\text{slink}(u)$ be the node that receives the suffix link from u. Let $\text{wlink}(u, x)$ be the node v that receives a Weiner link from u and whose path label is $\text{str}(v) = x \cdot \text{str}(u)$, and $\bot$ if such v does not exist. Let $\text{wlinks}(u) := \{(v, x) : v = \text{wlink}(u, x) \neq \bot\}$.

A suffix tree of T can be also defined as a *trie* for all the suffixes of T where each non-branching path is compressed into a single edge. A *locus* in a suffix tree corresponds to a node in the uncompressed trie. More precisely, it is a location in the suffix tree specified as a pair (u, d) where u is a node and d is an integer that satisfies $\text{depth}(\text{parent}(u)) < d \leq \text{depth}(u)$.

Given a substring S of T, after traversing the suffix tree from the root following the characters in S, the traversal always ends at a unique locus, (u, d). We say the traversal ends *within* an edge if $d < \text{depth}(u)$ and ends *at* a node if $d = \text{depth}(u)$. Observe that the frequency of S equals the number of leaf nodes in the subtree rooted at u. So S is unique if u is a leaf node and is repeated if u is a branching node. Also note that $S = \text{str}(u)[1 \ldots d]$.

Implicit Suffix Trees. The implicit suffix tree of a text can be defined by modifying its suffix tree as follows. First, every edge with only the sentinel character, \$, as its label is removed. Then, for each node u with only one child, v, we perform these operations: let $w := \text{parent}(u)$; remove node u, together with edges wu and uv; and create a new edge from w to v with edge label $T[\text{start}(v)+\text{depth}(w) \ldots \text{start}(v)+\text{depth}(v)]$, the concatenation of the edge labels of wu and uv. The tree obtained is called an *implicit suffix tree*. In an implicit suffix tree, branching nodes and leaf nodes are referred to as *explicit nodes*. Note that in a suffix tree, each suffix is unique and corresponds to a distinct leaf node. However, in an implicit suffix tree, repeated suffixes do not correspond

---

[4] A suffix link is also referred to as a *vine pointer* in the Prediction by Partial Matching algorithm [17] or as a *failure link* in the Aho–Corasick algorithm [2].

to leaf nodes. For example, the repeated suffixes of the text aabaabababaa are the strings a, aa, baa, and abaa.

Definition 2 (Implicit node). *An implicit node (u, d) is a locus such that $\text{str}(u)[1\ldots d]$ is a repeated suffix of the text.*

Each implicit node in an implicit suffix tree corresponds to a branching node in the suffix tree with only two children and one of them is $. Note that if an implicit node satisfies $d = \text{depth}(u)$, then it coincides with a branching node in the implicit suffix tree. Figure 1 demonstrates the major aspects of the suffix tree and implicit suffix tree. On its right, among the implicit nodes numbered 10–12, node 12 coincides with a branching node.

The key idea of Ukkonen's algorithm is successive building of implicit suffix trees for each prefix of the text, then adding the $ at the end. The algorithm maintains a pointer called the *active point* to the locus of the longest repeated suffix of the text. One limitation of Ukkonen's algorithm is that the locations of the implicit nodes are not maintained during suffix tree construction. Breslauer and Italiano [6] provide techniques for such task and extend Ukkonen's algorithm. They also further classify edges and nodes in an implicit suffix tree.

Definition 3. *An edge uv is called an* external edge *if v is a leaf node and an* internal edge *otherwise. An implicit node u is called an* internal implicit node *if u is within an internal edge and an* external implicit node *otherwise.*

If the locus each suffix corresponds to has an outgoing suffix link (including leaf and implicit nodes), then these suffix links form a path, which is called the *suffix chain* by Breslauer and Italiano. The path starts from the leaf labelled by T and ends at the root, going through each suffix and satisfying the following.

Lemma 3 ([6]). *The suffix chain can be partitioned into the following consecutive segments: leaves, external implicit nodes, internal implicit nodes, and implicit nodes that coincide with branching nodes.*

On the right of Fig. 1, these segments are: nodes 1–8, 9–10, 11, and 12.

The main result by Breslauer and Italiano that we use together with Ukkonen's online suffix tree construction algorithm is summarised as follows.

Theorem 1 ([6,21]). *An implicit suffix tree on the first n characters of T, together with the suffix links and Weiner links can be built in $\mathcal{O}(n)$ amortized time. A query returning the implicit nodes within a specific suffix tree edge takes worst-case $\mathcal{O}(1)$ time.*

Breslauer and Italiano showed that there is at most one implicit node within an internal edge. Although each external edge may contain multiple implicit nodes, their representation as an arithmetic progression can be returned in $\mathcal{O}(1)$ time [6]. This result also supports other queries in the online setting [13].

3 Offline NF Computation with Suffix Trees

We present our suffix-tree based approaches for offline NF computation, which are arguably simpler than the suffix-array based approaches [10]. We adapt them from the offline to the online setting in the following section.

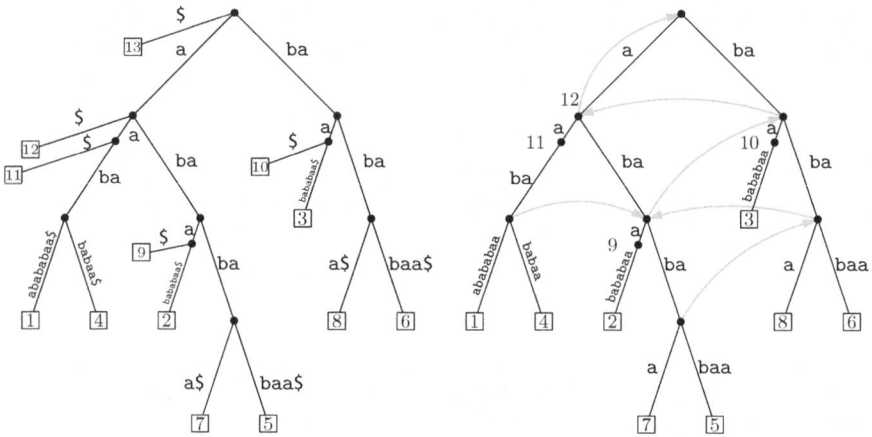

Fig. 1. The suffix tree (left) and implicit suffix tree (right) for text **aabaabababaa**. Leaves (squares) and implicit nodes (red dots) are numbered; green arrows are suffix links coming from branching nodes. (Color figure online)

3.1 Offline SINGLE-NF Algorithms

The NF characteristic for suffix array might lead to explicit character matching in a suffix tree. Moreover, when leaf nodes do not have incoming Weiner links, we are unable to enumerate unique left extensions of a string; so Lemma 1 is unhelpful with a suffix tree. We therefore introduce a new characteristic, which is more suitable for computing NF in a suffix tree. Essentially, it bypasses enumerating unique left extensions of the string.

Theorem 2 (Suffix tree NF characteristic). *Given a repeated string S, let $L(S) := \{x \in \Sigma : f(xS) \geq 2\}$ and $r(S) := \{y \in \Sigma : f(Sy) = 1\}$, then*

$$\phi(S) = |r(S)| - \sum_{x \in L(S)} |r(xS) \cap r(S)|.$$

Proof. We first define the following three sets:

$$L_{=1} := \{(x,y) \in \Sigma \times \Sigma : f(xS) = 1 \text{ and } f(Sy) = 1 \text{ and } f(xSy) = 1\},$$
$$L_{\geq 1} := \{(x,y) \in \Sigma \times \Sigma : f(xS) \geq 1 \text{ and } f(Sy) = 1 \text{ and } f(xSy) = 1\}, \text{ and}$$
$$L_{\geq 2} := \{(x,y) \in \Sigma \times \Sigma : f(xS) \geq 2 \text{ and } f(Sy) = 1 \text{ and } f(xSy) = 1\}.$$

Observe that $|L_{=1}| = |L_{\geq 1}| - |L_{\geq 2}|$ and $\phi(S) = |L_{=1}|$. Next, note that for each $y \in \Sigma$, if $f(Sy) = 1$, then $f(xSy) \leq 1$ holds for each $x \in \Sigma$. Further, if $f(xSy) = 1$ then $f(xS) \geq 1$; otherwise, $f(xSy) = 0$ and $f(xS) = 0$. Thus, we have $|r(S)| = |L_{\geq 1}|$. Finally, we derive that $\sum_{x \in L(S)} |r(xS) \cap r(S)| = \left|\bigcup_{x \in L(S)} r(xS) \cap r(S)\right| = |L_{\geq 2}|$. Therefore, we have proved the desired result. □

Take $T = \mathtt{rstkstcastarstast\$}$ and $S = \mathtt{st}$ as an example. We have $r(S) = \{\$, \mathtt{c}, \mathtt{k}\}$ and $L(S) = \{\mathtt{r}, \mathtt{a}\}$. When $x = \mathtt{r}, r(xS) = \{\mathtt{a}, \mathtt{k}\}$ and $r(xS) \cap r(S) = \{\mathtt{k}\}$. When $x = \mathtt{a}, r(xS) = \{\$, \mathtt{a}\}$ and $r(xS) \cap r(S) = \{\$\}$. Thus, $\phi(S) = 3 - (1+1) = 1$. The following informs checking of whether a right extension is unique.

Proposition 1. *Consider a branching node u and $y \in \Sigma$. Let $S := \mathrm{str}(u)$, then, Sy is unique if $\mathrm{child}(u, y)$ is a leaf.*

Theorem 2 immediately suggests a criterion for early termination of the algorithm when it is determined that the NF of the query string is zero.

Corollary 1. *If $|r(S)| = 0$, then $\phi(S) = 0$.*

Lemma 2 was introduced to narrow down candidates with a potentially positive NF when solving ALL-NF. However, due to certain limitations of a suffix array, the result could not be efficiently applied to SINGLE-NF. Now, with a suffix tree, we can utilise this result to detect a zero NF input string for SINGLE-NF.

Our SINGLE-NF algorithm using a suffix tree is presented in Algorithm 1. It uses Lemma 2 and Corollary 1 for zero NF detection. The algorithm computes the two terms in Theorem 2 separately and also utilises Weiner links for left extension enumeration. The correctness of the algorithm follows from Theorem 2.

Given a string S, Algorithm 1 runs in $\mathcal{O}(|S|)$ time in the worst case. Locating the locus of S in the tree takes $\mathcal{O}(|S|)$ time. Computing $|r(S)|$ takes at most $\mathcal{O}(|\Sigma|)$ time, a constant in our analysis. Then the time to compute $\sum_{x \in L(S)} |r(xS) \cap r(S)|$ is bounded by $|\Sigma|^2$, also a constant.

3.2 Offline ALL-NF Algorithm

From Lemma 2, only branching strings could have positive NF. In this section, we present an offline ALL-NF algorithm that extracts and stores the NF of each branching string in its corresponding branching node in the suffix tree. The stored positive NF values can be reported afterwards.

In Algorithm 1, the two terms in in Theorem 2 are computed in separate steps of the algorithm. For our ALL-NF algorithm, we compute these two terms in an *asynchronous* fashion: before we finish computing the NF of one branching string, we might start computing the NF of another branching string. In other words, the NF of each branching string is partially computed and updated as the algorithm progresses. Specifically, suppose we are visiting a branching node v, let $u := \mathrm{slink}(v)$, $xS := \mathrm{str}(v)$, and $S := \mathrm{str}(u)$. We update the NF for both xS and S as follows. We compute $|r(xS)|$ to update $\phi(xS)$ and compute $|r(xS) \cap r(S)|$

Algorithm 1: for offline SINGLE-NF

1 $(u, d) \leftarrow$ the locus of a query string S;
   // *unique or non-branching strings have zero NF (Lemma 2)*
2 **if** u *is a leaf or* $d <$ depth(u) **then return** 0;
   // *initialise the NF of S by counting $|r(S)|$ in Theorem 2*
3 $\phi \leftarrow$ the number of leaf child nodes of u;
   // *if $|r(S)| = 0$, then $\phi(S) = 0$ (Corollary 1)*
4 **if** $\phi = 0$ **then return** 0;
   // *compute $\sum_{x \in L(S)} |r(xS) \cap r(S)|$ in Theorem 2*
5 **foreach** $(v, x) \in$ wlinks(u) **do**
6 $\quad$ **foreach** $(w, y) \in$ children(v) **do**
   $\quad\quad$ // *xSy and Sy are both unique (Proposition 1)*
7 $\quad\quad$ **if** w *is a leaf and* child(u, y) *is a leaf* **then** $\phi \leftarrow \phi - 1$;
8 **return** ϕ;

Algorithm 2: for offline ALL-NF

// *assume $\phi(\text{str}(v))$ is initialised to 0 for each branching node v in the suffix tree*
1 **foreach** *branching node v in the suffix tree* **do**
2 $\quad u \leftarrow$ slink(v); $xS \leftarrow$ str(v); $S \leftarrow$ str(u);
3 $\quad$ **foreach** $(w, y) \in$ children(v) **do**
4 $\quad\quad$ **if** w *is a leaf* **then**
   $\quad\quad\quad$ // *xSy is unique*
5 $\quad\quad\quad \phi(xS) \leftarrow \phi(xS) + 1$;
6 $\quad\quad\quad$ **if** child(u, y) *is a leaf* **then**
   $\quad\quad\quad\quad$ // *Sy is unique*
7 $\quad\quad\quad\quad \phi(S) \leftarrow \phi(S) - 1$;

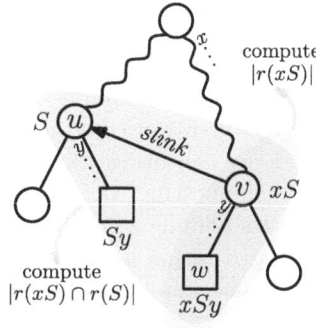

Fig. 2. Algorithm 2 illustration.

to update $\phi(S)$. Note that, after visiting node v, it is possible that neither $\phi(xS)$ nor $\phi(S)$ has the correct value, as they may have only been partially computed. But at the end of the algorithm, each NF value will be correct.

Our offline ALL-NF algorithm is listed in Algorithm 2 and illustrated in Fig. 2. It does not require traversing the branching nodes in a particular order, and runs in $\mathcal{O}(n)$ time in the worst case. Since checking whether a node is a leaf takes $\mathcal{O}(1)$ time, the overall time usage is bounded by the number of nodes visited throughout execution. Given a branching node, u, each of its child nodes, w, is visited exactly once; each node in the suffix tree is the child node of exactly one branching node. Moreover, there are at most n branching nodes in the tree. Thus, the total number of nodes visited is bounded by $\mathcal{O}(n)$.

4 Online NF Computation with Implicit Suffix Trees

With our offline algorithms, we first introduce additional properties of NF then present our online algorithms based on these results.

4.1 Online SINGLE-NF Algorithm

Our offline SINGLE-NF algorithm computes the two terms in Theorem 2 for a query string whose locus is a branching node. In the online setting, our online approach is still based on Theorem 2, but also taking implicit nodes into account in the implicit suffix tree. We first reconsider Proposition 1, as Proposition 2.

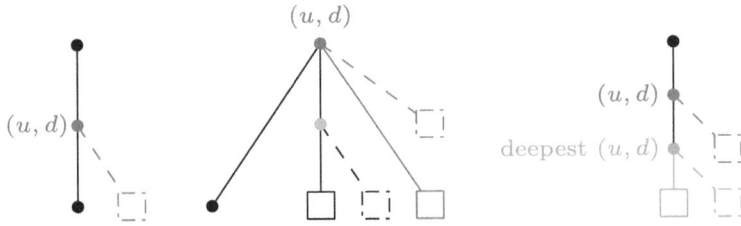

Fig. 3. Illustration of Lemma 4: Case 1 (left), Case 2 (middle), and Case 3 (right). Black dots, coloured dots, and squares represent branching nodes, implicit nodes, and leaves, respectively. Each dashed (non-existent) edge has label $ and leads to a dashed (non-existent) leaf node. In each case, each implicit node has its own colour. The implicit node (u, d) is also labelled, and the leaf nodes corresponding to its unique right extensions share the same colour as (u, d). (Color figure online)

Proposition 2. *Consider a branching node, u, and $y \in \Sigma$. Let $S := \text{str}(u)$, then, Sy is unique if $v := \text{child}(u, y)$ is a leaf and edge uv has no implicit node.*

Let (u, d) be the locus of the query string S. In the offline case, $|r(S)|$ in Theorem 2 is computed by simply counting the number of leaf child nodes of u. But it is more involved in the online setting.

Definition 4. *Consider an implicit node (u, d) and let $\rho(u, d)$ denote the number of unique right extensions of $\text{str}(u)[1 \ldots d]$.*

Lemma 4. *Consider an implicit node (u, d) and let $S := \text{str}(u)[1 \ldots d]$. The following cases are illustrated in Fig. 3.*

Case 1 If (u, d) is an internal implicit node, then $\rho(u, d) = 1$.
Case 2 If (u, d) is an internal implicit node that coincides with a branching node, then $\rho(u, d)$ equals one plus the number of leaf child nodes of u whose leading edge does not contain an implicit node.

Case 3 If (u,d) is an external implicit node, then $\rho(u,d) = 2$ if (u,d) is the deepest such on the edge, and $\rho(u,d) = 1$ otherwise.

Proof. In Case 1, the only unique right extension character is the \$. Case 2 follows from Proposition 2, and we add one for the \$. In Case 3, the deepest such has the \$ and a leaf child, while the others only has the \$. □

The NF of the longest repeated suffix is given as follows.

Lemma 5. *Consider an implicit node (u,d) and let $S := \text{str}(u)[1\ldots d]$. If S is the longest repeated suffix of T, then $\phi(S) = \rho(u,d)$.*

Proof. Observe that each occurrence of S has a unique left extension character as otherwise S would not be the longest repeated suffix. So, $\phi(S) = \rho(u,d)$. □

The following result provides another zero NF detection mechanism.

Lemma 6. *Consider an implicit node (u,d) and let $S := \text{str}(u)[1\ldots d]$. If S is not the longest repeated suffix and (u,d) does not coincide with a branching node, then $\phi(S) = 0$.*

Proof. First consider each occurrence of S that is not a sub-occurrence of an occurrence of the longest repeated suffix: the left extension character of such occurrence is always $T[n-d]$. Next, consider each occurrence of S except the rightmost one: since (u,d) does not coincide with a branching node, the right extension character of such occurrence is always $T[i+d]$ for any occurrence i of S with $i < n-d$. Thus, no occurrence of S is a net occurrence and $\phi(S) = 0$. □

For the remaining case, not covered by Lemma 5 or Lemma 6, the NF of the query string cannot be easily deduced. Further computation, assisted by Weiner links, is required. In Algorithm 1, the recipient of each Weiner link is a branching node, however, for online SINGLE-NF, *implicit* Weiner links, whose recipients are implicit nodes, are also needed. We present the following result on how to compute an implicit Weiner link. The result is illustrated in Fig. 4a.

Lemma 7. *Consider an implicit node (u,d) that coincides with a branching node. Let (q,ℓ) be another implicit node such that $\text{str}(q)[1\ldots\ell] = x \cdot \text{str}(u)[1\ldots d]$ for some $x \in \Sigma$. Define $w := \text{parent}(q)$. If there exists an ancestor v of u such that $\text{wlink}(v,x)$ exists, then $w = \text{wlink}(v^*,x)$ where v^* is the lowest such ancestor of u; otherwise, w is the root.*

Proof. The proof is illustrated in Fig. 4b.

If v^* exists, let $w' := \text{wlink}(v^*, x)$ and assume, by contradiction, that $w' \neq w$. Let $P := \text{str}(v^*)$, then $\text{str}(w') = xP$. Since xP is a prefix of xS, w' is an ancestor of w. Let $P' := \text{str}(w)[2\ldots\text{depth}(w)]$, then $xP' = \text{str}(w)$. Since w is a branching node, there exists a branching node v' with $\text{str}(v') = P'$. This implies that $\text{wlink}(v',x) = w$. Since v' is an ancestor of u, this contradicts that v^* is the lowest such ancestor. Thus, our assumption is false and $w = w'$.

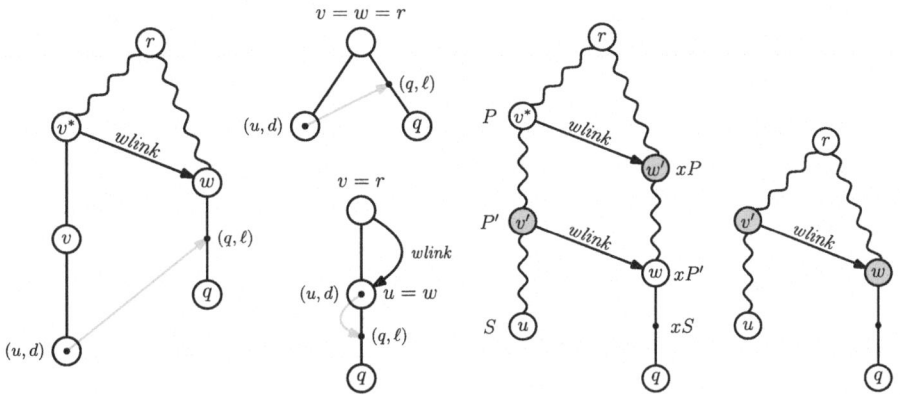

(a) Illustration of Lemma 7. (b) Illustration of proof of Lemma 7.

Fig. 4. Let r be the root node of the implicit suffix tree. An edge is shown straight; a path is shown squiggly. In Fig. 4a, some possible locations of nodes u, w, and $v :=$ parent(u) are shown: $v \neq r \neq w$ (left), $v = r = w$ (top right), $v = r \neq u = w$ (bottom right). Each green arrow indicates an implicit Weiner link from (u,d) to (q,ℓ). In Fig. 4b, compare scenarios when v^* exists (left) and when v^* does not exist (right). A node is coloured grey to indicate that it exists only under false assumption. Next to several nodes are corresponding path labels.

If v^* does not exist, assume, by contradiction, that w is not the root. Similarly, since w is a branching node, there exists a branching node v' such that wlink$(v', x) = w$. Since v' is an ancestor of u, this contradicts that v^* does not exist. Thus, our assumption is false and w is the root. □

Note that the concept of implicit Weiner links has also been described before [4, 9, 18], but with either a different definition or way of computing them. We next introduce a corollary of Theorem 2 that splits the set $L(S)$ into two disjoint sets.

Corollary 2. *Consider a locus* (u, d) *with* $d = $ depth(u). *Let* $S := $ str(u),

$$L_{exp}(S) := \{x \in \Sigma : f(xS) \geq 2 \text{ and the locus of } xS \text{ is an explicit node}\},$$
$$L_{imp}(S) := \{x \in \Sigma : f(xS) \geq 2 \text{ and the locus of } xS \text{ is an implicit node}\},$$

and let $\lambda(x) := |r(xS) \cap r(S)|$, *then*

$$\phi(S) = |\rho(u,d)| - \sum_{x \in L_{exp}(S)} \lambda(x) - \sum_{x \in L_{imp}(S)} \lambda(x).$$

Using Lemma 7 and Corollary 2, together with the results introduced in this section earlier, we present our online SINGLE-NF algorithm in Algorithm 3. Computing implicit Weiner links takes $\mathcal{O}(|S| \cdot |\Sigma|^2) \subseteq \mathcal{O}(|S|)$ time, assuming a constant-size alphabet. The rest of the analysis follows from the analysis of Algorithm 1 and we have the following result.

Theorem 3. *Online* SINGLE-NF *can be solved in worst-case* $\mathcal{O}(|S|)$ *time.*

Algorithm 3: for online SINGLE-NF
---
1 $(u,d) \leftarrow$ the locus of a query string S;
2 **if** (u,d) *is the active point* **then return** $\rho(u,d)$; // Lemma 5
3 **if** u *is a leaf or* $d < \text{depth}(u)$ **then return** 0; // Lemma 2 and Lemma 6
4 $\phi \leftarrow \rho(u,d)$; // initialise the NF of S by computing $\rho(u,d)$ using Lemma 4
5 **if** $\phi = 0$ **then return** 0; // Corollary 1
   // compute $\sum_{x \in L_{exp}(S)} \lambda(x)$ in Corollary 2
6 **foreach** $(w,x) \in \text{wlinks}(u)$ **do**
7 | **foreach** $(q,y) \in \text{children}(v)$ **do**
8 | | **if** q *is a leaf and edge* vq *does not contain an implicit node and*
     | | *child*(u,y) *is a leaf* **then** $\phi \leftarrow \phi - 1$;
9 **if** (u,d) *is not an implicit node* **then return** ϕ;
   // compute Λ, the set of implicit Weiner links from (u,d) using Lemma 7
10 $\Lambda \leftarrow \emptyset$; $r \leftarrow$ the root; $p \leftarrow u$;
11 **while** $p \neq r$ **do**
12 | $p \leftarrow \text{parent}(p)$;
13 | **foreach** $(w,\_) \in \text{wlinks}(p)$ **do** // "_" is used when the character is unused
14 | | **foreach** $(q,\_) \in \text{children}(w)$ **do**
15 | | | **if** *there exists implicit node* (q,ℓ) *with* $\ell = d+1$ **then**
       | | | $\Lambda \leftarrow \Lambda \cup \{(q,\ell)\}$;
16 **foreach** $(q,\_) \in \text{children}(r)$ **do**
17 | **if** *there exists implicit node* (q,ℓ) *with* $\ell = d+1$ **then** $\Lambda \leftarrow \Lambda \cup \{(q,\ell)\}$
   // compute $\sum_{x \in L_{imp}(S)} \lambda(x)$ in Corollary 2; note that each $\text{str}(q)[1\ldots\ell]$ has exactly two
   right extension characters, $\$$ (assumed to be unique) and $T[\text{start}(q) + \ell + 1]$
18 **foreach** $(q,\ell) \in \Lambda$ **do**
19 | $\phi \leftarrow \phi - 1$; // for $\$$
20 | $y \leftarrow T[\text{start}(q) + \ell + 1]$;
21 | **if** q *is a leaf and child*(u,y) *is a leaf* **then** $\phi \leftarrow \phi - 1$;
22 **return** ϕ;

4.2 Online ALL-NF Algorithm

Our arguably very simple offline ALL-NF algorithm forms the basis of our online approach. Similar to the online SINGLE-NF algorithm, we need to deal with left extensions whose loci are implicit nodes. We make an additional observation.

Lemma 8. *Consider a repeated suffix S and its left extension, xS, which is also a repeated suffix. If the locus of xS coincides with a branching node, then the locus of S also does. If the locus of S does not coincide with a branching node, then the locus of xS also does not.*

Proof. Consider the suffix chain in Lemma 3 in reverse and we have the desired result. □

With Lemma 8, observe that at most one string might require implicit Weiner links: namely, the longest repeated suffix whose locus coincides with a branching node. The other repeated suffixes whose locus coincides with a branching node do not have implicit Weiner links. Our online ALL-NF algorithm is adapted from our offline ALL-NF algorithm as follows. We first compute the NF of the longest repeated suffix. We then invoke offline ALL-NF, but use Proposition 2 instead of Proposition 1. While traversing each branching node, we keep track of string depth and find τ, the longest repeated suffix whose locus coincides with a branching node. Once we do, we invoke online SINGLE-NF on τ. Overall, the cost is bounded by $\mathcal{O}(n + |\tau|) \subseteq \mathcal{O}(n)$ where $\mathcal{O}(n)$ is the cost for offline ALL-NF and $\mathcal{O}(|\tau|)$ is the cost for online SINGLE-NF on τ.

Theorem 4. *Online* ALL-NF *can be solved in worst-case* $\mathcal{O}(n)$ *time.*

5 Conclusion and Future Work

In this work, we present, to our knowledge, the first and, indeed, optimal-runtime online algorithms for both SINGLE-NF and ALL-NF. Having unsuccessfully investigated online approaches based on suffix arrays – the basis of previous offline methods – we turned our attention to suffix trees, using which we found offline and online algorithms that are runtime-optimal. The results are based on new characteristics and properties of net frequency and prior results on auxiliary pointers in suffix trees and on implicit node maintenance.

An avenue of future work is design and engineering an efficient implementation of our solution, in particular the structures required for implicit nodes such as structures for dynamic nearest marked ancestors in trees [3,8] and structures for ancestor relationships [5].

Acknowledgements. The authors thank Patrick Eades for insightful discussions during the early stage of this work. The authors also thank the anonymous reviewers for their suggestions. This work was supported by the Australian Research Council, grant number DP190102078, and an Australian Government Research Training Program Scholarship.

References

1. Kulikov, A.S., Kuznetsov, S.O., Pevzner, P. (eds.): CPM 2014. LNCS, vol. 8486. Springer, Cham (2014). https://doi.org/10.1007/978-3-319-07566-2
2. Aho, A.V., Corasick, M.J.: Efficient string matching: an aid to bibliographic search. Commun. ACM **18**(6), 333–340 (1975). https://doi.org/10.1145/360825.360855
3. Amir, A., Farach, M., Idury, R.M., Poutré, J.A.L., Schäffer, A.A.: Improved dynamic dictionary matching. Inf. Comput. **119**(2), 258–282 (1995). https://doi.org/10.1006/INCO.1995.1090
4. Belazzougui, D., Cunial, F.: Fully-functional bidirectional Burrows-Wheeler indexes and infinite-order de Bruijn graphs. In: 30th Annual Symposium on Combinatorial Pattern Matching, CPM 2019, June 18-20, 2019, Pisa, Italy. LIPIcs, vol. 128, pp. 10:1–10:15. Schloss Dagstuhl - Leibniz-Zentrum für Informatik (2019). https://doi.org/10.4230/LIPICS.CPM.2019.10

5. Bender, M.A., Cole, R., Demaine, E.D., Farach-Colton, M., Zito, J.: Two simplified algorithms for maintaining order in a list. In: Möhring, R., Raman, R. (eds.) ESA 2002. LNCS, vol. 2461, pp. 152–164. Springer, Heidelberg (2002). https://doi.org/10.1007/3-540-45749-6_17
6. Breslauer, D., Italiano, G.F.: On suffix extensions in suffix trees. Theoret. Comput. Sci. **457**, 27–34 (2012). https://doi.org/10.1016/J.TCS.2012.07.018
7. Breslauer, D., Italiano, G.F.: Near real-time suffix tree construction via the fringe marked ancestor problem. J. Discrete Algorithms **18**, 32–48 (2013). https://doi.org/10.1016/J.JDA.2012.07.003
8. Feigenblat, G., Porat, E., Shiftan, A.: An improved query time for succinct dynamic dictionary matching. In: 25th Annual Symposium on Combinatorial Pattern Matching, CPM 2014, June 16-18, Moscow, Russia [1], pp. 120–129 (2014). https://doi.org/10.1007/978-3-319-07566-2_13
9. Fujishige, Y., Tsujimaru, Y., Inenaga, S., Bannai, H., Takeda, M.: Linear-time computation of DAWGs, symmetric indexing structures, and MAWs for integer alphabets. Theoret. Comput. Sci. **973**, 114093 (2023). https://doi.org/10.1016/J.TCS.2023.114093
10. Guo, P., Eades, P., Wirth, A., Zobel, J.: Exploiting new properties of string net frequency for efficient computation. In: 35th Annual Symposium on Combinatorial Pattern Matching, CPM 2024, June 25-27 2024, Fukuoka, Japan. LIPIcs, vol. 296, pp. 16:1–16:16. Schloss Dagstuhl - Leibniz-Zentrum für Informatik (2024). https://doi.org/10.4230/LIPICS.CPM.2024.16
11. Gusfield, D.: Algorithms on Strings, Trees, and Sequences - Computer Science and Computational Biology. Cambridge University Press, Cambridge (1997). https://doi.org/10.1017/CBO9780511574931
12. Inenaga, S., Takeda, M.: On-line linear-time construction of word suffix trees. In: Lewenstein, M., Valiente, G. (eds.) CPM 2006. LNCS, vol. 4009, pp. 60–71. Springer, Heidelberg (2006). https://doi.org/10.1007/11780441_7
13. Larsson, N.J.: Most recent match queries in on-line suffix trees. In: 25th Annual Symposium on Combinatorial Pattern Matching, CPM 2014, June 16-18, 2014. Moscow, Russia [1], pp. 252–261 (2014). https://doi.org/10.1007/978-3-319-07566-2_26
14. Lin, Y., Yu, M.: Extracting Chinese frequent strings without dictionary from a Chinese corpus and its applications. J. Inf. Sci. Eng. **17**(5), 805–824 (2001), https://jise.iis.sinica.edu.tw/JISESearch/pages/View/PaperView.jsf?keyId=86_1308
15. Lin, Y., Yu, M.: The properties and further applications of Chinese frequent strings. Int. J. Comput. Linguist. Chin. Lang. Process. **9**(1), 113–128 (2004). http://www.aclclp.org.tw/clclp/v9n1/v9n1a7.pdf
16. McCreight, E.M.: A space-economical suffix tree construction algorithm. J. ACM **23**(2), 262–272 (1976). https://doi.org/10.1145/321941.321946
17. Moffat, A.: Implementing the PPM data compression scheme. IEEE Trans. Commun. **38**(11), 1917–1921 (1990). https://doi.org/10.1109/26.61469
18. Nakashima, K., et al.: Parameterized DAWGs: efficient constructions and bidirectional pattern searches. Theoret. Comput. Sci. **933**, 21–42 (2022). https://doi.org/10.1016/J.TCS.2022.09.008
19. Starikovskaya, T., Vildhøj, H.W.: A suffix tree or not a suffix tree? J. Discrete Algorithms **32**, 14–23 (2015). https://doi.org/10.1016/J.JDA.2015.01.005
20. Takagi, T., Inenaga, S., Arimura, H.: Fully-online construction of suffix trees for multiple texts. In: 27th Annual Symposium on Combinatorial Pattern Matching, CPM 2016, June 27-29, 2016, Tel Aviv, Israel. LIPIcs, vol. 54, pp. 22:1–22:13.

Schloss Dagstuhl - Leibniz-Zentrum für Informatik (2016). https://doi.org/10.4230/LIPICS.CPM.2016.22
21. Ukkonen, E.: On-line construction of suffix trees. Algorithmica **14**(3), 249–260 (1995). https://doi.org/10.1007/BF01206331
22. Weiner, P.: Linear pattern matching algorithms. In: 14th Annual Symposium on Switching and Automata Theory, Iowa City, Iowa, USA, October 15–17, 1973, pp. 1–11. IEEE Computer Society (1973). https://doi.org/10.1109/SWAT.1973.13

On the Number of Non-equivalent Parameterized Squares in a String

Rikuya Hamai[1], Kazushi Taketsugu[1], Yuto Nakashima[2](✉)[iD],
Shunsuke Inenaga[2][iD], and Hideo Bannai[3][iD]

[1] Department of Information Science and Technology, Kyushu University,
Fukuoka, Japan
hamai.rikuya.226@s.kyushu-u.ac.jp
[2] Department of Informatics, Kyushu University, Fukuoka, Japan
{nakashima.yuto.003,inenaga.shunsuke.380}@m.kyushu-u.ac.jp
[3] M&D Data Science Center, Tokyo Medical and Dental University, Tokyo, Japan
hdbn.dsc@tmd.ac.jp

Abstract. A string s is called a parameterized square when $s = xy$ for strings x, y and x and y are parameterized equivalent. Kociumaka et al. showed the number of *parameterized squares*, which are non-equivalent in parameterized equivalence, in a string of length n that contains σ distinct characters is at most $2\sigma!n$ [TCS 2016]. In this paper, we show that the maximum number of non-equivalent parameterized squares is less than σn, which significantly improves the best-known upper bound by Kociumaka et al.

Keywords: Parameterized equivalence of strings · Squares · Periodicity

1 Introduction

In combinatorics on words, properties of repetitive structures (e.g., maximal repetitions, squares) are well-studied topics. For instance, a string w is said to be a square if w can be represented as $w = xx$ for some string x. Let $Sq(s)$ be the number of distinct squares in a string s. Fraenkel and Simpson [7] showed that $Sq(s) < 2n$ holds for any string s of length n, and conjectured that $Sq(s) < n$, which subsequently became a long-standing and well-known open question. This conjecture was recently proved by Brlek and Li [4]. In fact, they showed that $Sq(s) \leq n - |\Sigma_s| + 1$, where Σ_s denotes the set of distinct characters in s.

On the other hand, variants of this problem on parameterized equivalence [3], order-preserving equivalence [14], and Abelian equivalence were considered by Kociumaka et al. [15]. For each of these equivalence models, they introduced two types of the distinctness: one is counting non-standard squares which are distinct as strings, and the other is counting non-standard squares which are non-equivalent in the equivalence model.

Our focus in this paper is the parameterized equivalence model: For any two length-k strings x and y over an alphabet Σ, x and y are said to be parameterized equivalent (denoted by $x \approx y$) if there exists a bijection $f : \Sigma \to \Sigma$ such that $f(x[i]) = y[i]$ for any position $i \leq k$. Parameterized matching was first introduced by Baker [3] with motivations to software maintenance, and various algorithms and data structures have been proposed for pattern matching and other string processing under the parameterized equivalence model (see [1,5,8, 9,11–13,16,19,20] and references therein).

We hereby say that a string $w = xy$ is a *parameterized square* iff $x \approx y$. Kociumaka et al. [15] showed that any string of length n over an alphabet of size σ can contain at most $2(\sigma!)^2 n$ *distinct* parameterized squares (which are distinct as strings), and at most $2\sigma! n$ *non-equivalent* parameterized squares. We note that these bounds do not count standard squares (on the exact equality).

In this paper, we present a new upper bound on the latter: any string of length n over an alphabet of size σ can contain less than σn non-equivalent parameterized squares. Our result significantly improves the previous upper bound $2\sigma! n$ by Kociumaka et al. [15].

The *periodicity lemmas* [6,17] are the main tools in the analysis of periodic properties of strings under the exact equivalence model. Apostolico and Giancarlo [2] presented a parameterized version of the periodicity lemma for character bijections which are commutative. Ideguchi et al. [11] proposed a variant of the parameterized periodicity lemma that does not use the commutativity of the bijections. We present improved (i.e. tighter) versions of these two parameterized periodicity lemmas, which are used for showing our σn bound for the number of non-equivalent parameterized squares in a string.

2 Preliminaries

2.1 Strings

Let Σ be an *alphabet*. An element of Σ^* is called a *string*. The length of a string s is denoted by $|s|$. The empty string ε is the string of length 0. Let Σ^+ be the set of non-empty strings, i.e., $\Sigma^+ = \Sigma^* \setminus \{\varepsilon\}$. For any strings x and y, let $x \cdot y$ (or sometimes xy) denote the concatenation of the two strings. For a string $s = xyz$, x, y and z are called a *prefix*, *substring*, and *suffix* of s, respectively. Let $\mathsf{Substr}(s)$ denote the set of substrings of s. The i-th symbol of a string w is denoted by $w[i]$, where $1 \leq i \leq |w|$. For a string w and two integers $1 \leq i \leq j \leq |w|$, let $w[i..j]$ denote the substring of w that begins at position i and ends at position j. For convenience, let $w[i..j] = \varepsilon$ when $i > j$. Also, let $w[..i] = w[1..i]$, $w[i..] = w[i..|w|]$, and $w[i..j] = w(i-1..j] = w[i..j+1)$. For any string w, let $w^1 = w$ and let $w^k = ww^{k-1}$ for any integer $k \geq 2$. A string w is said to be *primitive* if w cannot be written as x^k for any $x \in \Sigma^+$ and integer $k \geq 2$.

2.2 Parameterized Squares

Two strings x and y of length k each are said to be *parameterized equivalent* iff there is a bijection f on Σ such that $f(x[i]) = y[i]$ for all $1 \leq i \leq k$. For instance, let $\Sigma = \{a, b, c, d\}$, and consider two strings $x = $ aabcacbbdad and $y = $ bbcabaccdbd. These two strings are parameterized equivalent, since x can be transformed to y by applying a bijection f such that $f(a) = b$, $f(b) = c$, $f(c) = a$, and $f(d) = d$ to the characters in x. We write $x \approx y$ iff two strings x and y are parameterized equivalent.

A string w is called a parameterized square when $w = xy$ for strings x, y such that x and y are parameterized equivalent. We say that two strings w and z of equal length are *non-equivalent* (under the parameterized equivalence) iff $w \not\approx z$. Let $\mathsf{PS}(s)$ denote the set of all parameterized squares occurring in a string s. Let $PS(s)$ denote the number of equivalence classes on $\mathsf{PS}(s)$ with respect to $\approx$. We call $PS(s)$ the number of non-equivalent parameterized squares in s.

Let $s = $ aabbac. For example, s contains

$$\text{aa, bb}$$

as standard squares,

$$\text{aa, ab, ac, ba, bb, aabb, abba}$$

as parameterized squares, and

$$\text{aa} \approx \text{bb, ab} \approx \text{ac} \approx \text{ba, aabb, abba}$$

as non-equivalent parameterized squares, and $PS(s) = 4$.

Due to the definition of parameterized squares, any standard square is also a parameterized square. In the previous study [15], they do not count standard squares (of form ww) as parameterized squares. More precisely, they considered the set $\mathsf{PS}'(s)$ of all parameterized squares ww' ($w \approx w'$) occurring in a string s that satisfies $w \neq w'$. Let $PS'(s)$ denote the number of equivalence classes on $\mathsf{PS}'(s)$ with respect to $\approx$. It is clear from the definitions that $\mathsf{PS}'(s) \subseteq \mathsf{PS}(s)$ and $PS'(s) \leq PS(s)$ hold for any string s. In the above example, $PS'(s) = 3$ since aa $\approx$ bb is not contained in $\mathsf{PS}'(s)$. In this paper, we consider an upper bound for $PS(s)$. By the above definitions, any upper bound for $PS(s)$ can also apply to $PS'(s)$.

A bijection can be seen as a permutation of Σ. A cyclic permutation is called a transposition if the length is 2.

Fact 1. *Any permutation can be represented by a product of transpositions. The parity (odd or even) of the number of transpositions is uniquely determined for any permutation.*

2.3 Parameterized Periods

An integer p is said to be a *parameterized period* (*p-period*) of string s if $s[1..|s| - p] \approx s[p+1..|s|]$ holds. For any p-period p of s, we write $p \parallel_f s$ if $s[1..|s| - p] \approx$

$s[p+1..|s|]$ holds by a bijection f. We sometimes drop the subscript f (i.e., we just write $p \parallel s$) when no confusions occur. The smallest p-period of s is denoted by $\mathrm{p}(s)$. By the definition of p-periods, we obtain the following fact which we will use in our proof.

Fact 2 (cf. [18]). *Let s be a string that satisfies $p \parallel_f s$. For any position i satisfying $1 \leq i \leq |s| - p$, $f(s[i]) = s[i+p]$.*

Apostolico and Giancarlo [2] showed a variant of the periodicity lemma on the parameterized equivalence model.

Lemma 1 (Lemma 3 of [2]). *Let s be a string that satisfies $p \parallel_f s$ and $q \parallel_g s$. If $p + q \leq |s|$ and $f \circ g = g \circ f$, then $\gcd(p,q) \parallel s$.*

The above lemma uses the commutativity of the two bijections. Recently, Ideguchi et al. [11] proposed a variant of the parameterized periodicity lemma which does not use the commutativity of the bijections as follows:

Lemma 2 (Lemma 5 of [11]). *Let s be a string that satisfies $s \in \Sigma^*$, $p \parallel s$, $q \parallel s$. If $p + q + \min(p,q) \cdot (|\Sigma_s| - 1) \leq |s|$, then $\gcd(p,q) \parallel s$.*

The number of distinct characters in a string may play an important role in bounding the parameterized periodicity, as can be seen in the previous lemma. In our result here we will extensively show more of such relations. The next lemma, shown by Ideguchi et al. [11], also gives a useful relation between them.

Lemma 3 (Lemma 4 of [11]). *Let s be a string. For any substring s' of s, If $|s'| \geq \mathrm{p}(s) \cdot (|\Sigma_s| - 1)$, then $|\Sigma_{s'}| \geq |\Sigma_s| - 1$ holds.*

In the next section, we present a tighter version for each of Lemmas 2 and 3.

3 Upper Bound of Non-equivalent Parameterized Squares

We show the following upper bound of the maximum number $PS(s)$ of non-equivalent parameterized squares in a string s.

Theorem 1. *For any string s of length n that contains σ distinct characters, $PS(s) < \sigma n$ holds.*

To prove Theorem 1, it suffices for us to prove the following theorem: Since any (parameterized) square cannot begin at the last position in s, the following theorem immediately implies our upper bound.

Theorem 2. *For any string s that contains σ distinct characters, there can be at most σ prefixes of s that are parameterized squares and have no other parameterized occurrence in s.*

In order to prove Theorem 2, we first prove Lemma 4. This lemma explains that any sufficiently long substring w.r.t. the shortest p-period of the whole string contains many distinct characters. Lemma 4 is a generalized property of a tighter version of the previous statement (Lemma 3).

Lemma 4. *Let s be a string that satisfies $p \parallel s$. For any substring s' of s and any integer k satisfying $2 \leq k \leq |\Sigma_s|+1$, if $|s'| \geq p \cdot (k-2)+1$, then $|\Sigma_{s'}| \geq k-1$ holds.*

Proof. We prove this lemma by induction on k. It is clear that the statement holds for $k = 2$. Suppose that the statement holds for $k = m$ for some integer m satisfying $2 \leq m \leq |\Sigma_s|$. Namely, if $|s'| \geq p \cdot (m-2) + 1$, then $|\Sigma_{s'}| \geq m - 1$ holds. We show that the statement holds for $k = m+1$. Since $|s'| \geq p \cdot (m - 1) + 1 \geq p \cdot (m - 2) + 1$, $|\Sigma_{s'}| \geq m - 1$ holds by the induction hypothesis. Assume on the contrary that $|\Sigma_{s'}| < m$. This implies that $|\Sigma_{s'}| = m - 1$. Let us consider the prefix $s'[1..i]$ of length $i = p \cdot (m-2) + 1$ of s'. By the induction hypothesis, $|\Sigma_{s'[1..i]}| \geq m - 1$. Moreover, since $s'[1..i]$ is a substring of s', $|\Sigma_{s'[1..i]}| = m-1$ holds. Due to the p-period $p \parallel_f s$, $f(s'[1..i]) = s'[1+p..i+p]$ holds. This implies that $\Sigma_{s'[1..i]} = \Sigma_{s'[1+p..i+p]}$ (if not, there exists a character $c \in \Sigma_{s'[1+p..i+p]} \setminus \Sigma_{s'[1..i]}$ and then it contradicts to $|\Sigma_{s'}| = m - 1$). We can also see that $\Sigma_{f^\ell(s'[1..i])} = \Sigma_{s'[1..i]}$ for any integer ℓ since f is a bijection. In a similar way, s should be covered by $f^\ell(s')$ over an alphabet $\Sigma_{s'[1..i]}$ of size $m - 1$. This contradicts the fact that $m \leq |\Sigma_s|$. □

When we set $k = |\Sigma_s|$, we obtain the following statement.

Corollary 1. *Let s be a string that satisfies $p \parallel s$. For any substring s' of s, if $|s'| \geq p \cdot (|\Sigma_s| - 2) + 1$, then $|\Sigma_{s'}| \geq |\Sigma_s| - 1$ holds.*

By using this corollary and the next lemma, we can also obtain a tighter version of the periodicity lemma (Lemma 2) for parameterized strings over an alphabet of size more than one. To prove it, we introduce a relation between the number of distinct characters of a substring and the commutativity of any two permutations of the alphabet.

Lemma 5. *Assume that $|\Sigma| \geq 2$. Let Σ' be a subset of Σ satisfying $|\Sigma'| = |\Sigma| - 2$. For any permutations f and g of Σ, if $(f \circ g)(a) = (g \circ f)(a)$ for all $a \in \Sigma'$, then f and g commute.*

Proof. Let f and g be permutations of $\Sigma = \{c_1, c_2, \ldots, c_\sigma\}$. Also, let $\Sigma' = \{c_1, c_2, \ldots, c_{\sigma-2}\}$ be a subset of Σ such that $(f \circ g)(a) = (g \circ f)(a)$ holds for every $a \in \Sigma'$. Assume on the contrary that $(f \circ g)(c_{\sigma-1}) \neq (g \circ f)(c_{\sigma-1})$. Let $(f \circ g)(c_{\sigma-1}) = c_i$ and $(g \circ f)(c_{\sigma-1}) = c_j$ for some $i \neq j$. Then, $(f \circ g)(c_\sigma) = c_j$ and $(g \circ f)(c_\sigma) = c_i$. Thus, $g \circ f = (c_i, c_j) \circ (f \circ g)$. This implies that $g \circ f$ and $(c_i, c_j) \circ (f \circ g)$ are products of transposition with different parity. This fact contradicts Fact 1. □

Lemma 6. *Let s be a string that satisfies $p \parallel s$, $q \parallel s$, and $|\Sigma_s| \geq 2$. If $p + q + \min(p, q) \cdot (|\Sigma_s| - 2) \leq |s|$, $\gcd(p, q) \parallel s$ holds.*

Proof. Let s be a string that satisfies $p \parallel_f s$, $q \parallel_g s$, and $|\Sigma_s| \geq 2$. Also, let $s' = s[1..\min(p,q) \cdot (|\Sigma_s|-2)]$. Then, $(g \circ f)(s') = s[1+p+q..\min(p,q) \cdot (|\Sigma_s|-2)+p+q]$ and $(f \circ g)(s') = s[1+p+q..\min(p,q) \cdot (|\Sigma_s|-2)+p+q]$. Therefore, for any $a \in \Sigma_{s'}$, $(f \circ g)(a) = (g \circ f)(a)$. By Lemma 4, $|\Sigma_{s'}| \geq |\Sigma_s|-2$, thus, by Lemma 5, $f \circ g = g \circ f$. Hence, by Lemma 1, $\gcd(p,q) \parallel s$ holds. □

In the proof of Theorem 2, we need to discuss structures of overlapping parameterized squares. If there exist two strings that are overlapping each other and have the same standard period (in the exact matching model), we can see that the string covered by the two strings also has the same period. This property does not always hold for the parameterized equivalence, since p-periods in the two overlapping strings may come from different bijections. In the next lemma, we show that a similar property holds if the overlapping part is sufficiently long (i.e., the overlapping has many distinct characters due to the above lemmas).

Lemma 7. *For any non-empty strings $x, y,$ and z, if $p \parallel xy$, $p \parallel yz$, and $|y| \geq p \cdot (|\Sigma_s| - 1) + 1$, then $p \parallel xyz$.*

Proof. Let f (resp. g) be a bijection of $p \parallel xy$ (resp. $p \parallel yz$), and $\sigma = |\Sigma_s|$. Since $p \parallel_f xy$ and $p \parallel_g yz$ and $|y| \geq p + 1$, then, $p \parallel_f y$ and $p \parallel_g y$. This implies that $f(y[1..(\sigma-2)p+1]) = y[p+1..(\sigma-1)p+1]$, and $g(y[1..(\sigma-2)p+1]) = y[p+1..(\sigma-1)p+1]$. Thus, $f(c) = g(c)$ for any character $c \in \Sigma_{y[1..(\sigma-2)p+1]}$. By Lemma 4, $|\Sigma_{y[1..(\sigma-2)p+1]}| \geq \sigma - 1$ holds. These facts implies that $f = g$. Thus, $f((xyz)[1..|xyz|-p]) = (xyz)[p+1..|xyz|]$ holds and the lemma holds. □

The next lemma explains that a p-period of a substring which is sufficiently long can extend to the whole string, if the whole string has a p-period that is a multiple of the substring's p-period.

Lemma 8. *Let t be a string that satisfies $q \parallel_g t$, and s be a prefix of t that satisfies $p \parallel_f s$. If $p(|\Sigma_t| - 2) + q + 1 \leq |s|$ and $q = kp$ for some integer k, then $p \parallel_f t$.*

Proof. Let $s' = s[1..p(|\Sigma_t|-2)+1]$. Then, $g(s') = g(s[1..p(|\Sigma_t|-2)+1]) = s[1+q..p(|\Sigma_t|-2)+1+q] = s[1+kp..p(|\Sigma_t|-2)+1+kp]$ and $f^k(s') = f^k(s[1..p(|\Sigma_t|-2)+1]) = s[1+kp..p(|\Sigma_t|-2)+1+kp]$. Therefore, for any $a \in \Sigma_{s'}$, $g(a) = f^k(a)$.

If $|\Sigma_{s'}| \leq |\Sigma_t| - 2$, s should be covered by $f^{\ell_1}(s')$ for any integer satisfying $0 \leq \ell_1 \leq k$ over an alphabet $\Sigma_{s'}$ by a similar discussion to Lemma 4. Since for any $a \in \Sigma_{s'}$, $g(a) = f^k(a)$, t should be covered by $g^{\ell_2}(s)$ for any integer satisfying $0 \leq \ell_2 \leq \lceil \frac{|t|-|s|}{q} \rceil$ over an alphabet $|\Sigma_s| = |\Sigma_{s'}| \leq |\Sigma_t| - 2$ by a similar discussion to Lemma 4, which is a contradiction. Thus, $|\Sigma_{s'}| \geq |\Sigma_t| - 1$. Therefore, $g = f^k$.

Let $v = s[1..q]$ and $v' = v\left[1..q - \left\lfloor \frac{|t|}{q} \right\rfloor \cdot q\right]$, then,

$$t = v \cdot g(v) \cdot g^2(v) \cdots g^{\lfloor \frac{|t|}{q} \rfloor - 1}(v) \cdot g^{\lfloor \frac{|t|}{q} \rfloor}(v').$$

Let $u = v[1..p]$, then, $v = uf(u)\cdots f^{k-1}(u)$ and

$$t = v \cdot g(v) \cdot g^2(v) \cdots g^{\lfloor\frac{|t|}{q}\rfloor-1}(v) \cdot g^{\lfloor\frac{|t|}{q}\rfloor}(v')$$
$$= uf(u)\cdots f^{k-1}(u)f^k(u)f^{k+1}(u)\cdots f^{2k-1}(u)\cdots f^{\lfloor\frac{|t|}{p}\rfloor}(u'),$$

where $u' = u[1..|t| - \lfloor\frac{|t|}{p}\rfloor \cdot p]$. Therefore, $p \parallel_f t$. □

Now, we are ready to prove the following main theorem.

Theorem 2. *For any string s that contains σ distinct characters, there can be at most σ prefixes of s that are parameterized squares and have no other parameterized occurrence in s.*

Proof. Suppose that there are $\sigma+1$ prefixes of s that are parameterized squares and have no other parameterized occurrence in s. Let $x_1 x_1', \ldots, x_{\sigma+1} x_{\sigma+1}'$ denote the $\sigma+1$ parameterized square prefixes that satisfies $|x_1| < \cdots < |x_{\sigma+1}|$ and $x_i \approx x_i'$ for every $i \in [1, \sigma+1]$. It is clear from the definition that $|x_{\sigma+1}|/2 < |x_1|$ holds (if not, $x_{\sigma+1}'[1..|x_1 x_1'|] \approx x_1 x_1'$). We also consider the length $r_k = |x_{k+1}| - |x_k|$ for every integer k satisfying $1 \le k \le \sigma$ (see also Fig. 1). Since $x_k \approx x_k' \approx x_{k+1}[1..|x_k|] \approx x_{k+1}'[1..|x_k|]$, then $r_k \parallel x_k$ for any k. By $r_k \parallel x_k$, it is clear that $r_k \parallel x_1$ for any k.

From now on, we prove $r_\sigma = \ell \cdot p(x_1)$ for some integer $\ell \ge 1$. Assume on the contrary that the statement does not hold. From the definitions of $r_1, \ldots, r_\sigma$,

$$p(x_1) \cdot (|\sigma| - 1) + r_\sigma \le r_1 + \cdots + r_\sigma \le |x_1|$$

holds. This implies that $p(x_1) + r_\sigma + p(x_1) \cdot (|\sigma| - 2) \le |x_1|$. Thus, $\gcd(p(x_1), r_\sigma) \parallel x_1$ also holds by Lemma 6. By the assumption, $\gcd(p(x_1), r_\sigma) < p(x_1)$, which is a contradiction. In a similar way, we can see that

$$p(x_1) \cdot (|\sigma| - 2) + r_\sigma + 1 \le r_1 + \cdots + r_\sigma \le |x_1|$$

also holds. Hence, this fact and the condition $r_\sigma = \ell \cdot p(x_1)$ for some integer $\ell \ge 1$ imply that $p(x_1) \parallel x_\sigma$ by Lemma 8.

Next, we prove $|x_1| = m \cdot p(x_1)$ for some integer $m \ge 1$. Assume on the contrary that the statement does not hold. It is clear that $|x_1| \parallel x_\sigma$. From the definitions of $r_1, \ldots, r_\sigma$,

$$p(x_1) \cdot (|\sigma| - 1) + |x_1| \le |x_1| + r_1 + \cdots + r_{\sigma-1} = |x_\sigma|$$

holds. This implies that $p(x_1) + |x_1| + p(x_1) \cdot (|\sigma| - 2) \le |x_\sigma|$. Thus, by Lemma 6, $\gcd(p(x_1), |x_1|) \parallel x_\sigma$ also holds. By the assumption, $\gcd(p(x_1), |x_1|) < p(x_1)$ holds. Moreover, by the definitions of x_1 and x_σ, $\gcd(p(x_1), |x_1|) \parallel x_1$ also holds, which is a contradiction. It is clear from the definitions that $|x_1| \parallel x_{\sigma+1}$. In a similar way, we can see that

$$p(x_1) \cdot (|\sigma| - 2) + |x_1| + 1 \le |x_1| + r_1 + \cdots + r_{\sigma-1} = |x_\sigma|$$

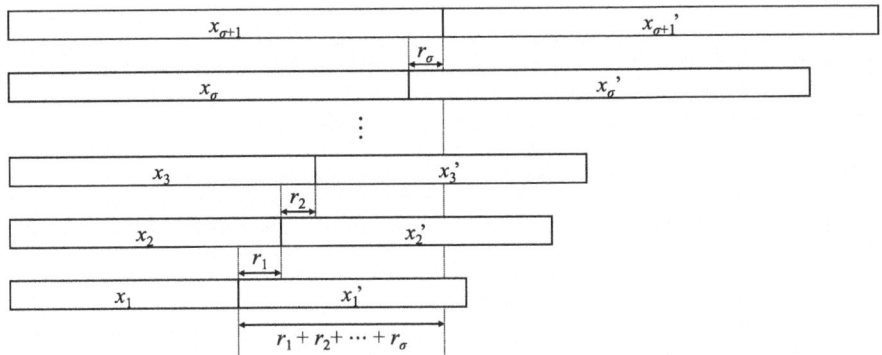

Fig. 1. Illustration for the proof of Theorem 2: $\sigma + 1$ non-equivalent parameterized squares cannot begin at the same position.

also holds. Hence, this fact and the condition $|x_1| = m \cdot p(x_1)$ for some integer $m \geq 1$ imply that $p(x_1) \parallel x_{\sigma+1}$ by Lemma 8.

We consider $x_{\sigma+1}$ and x_1' in s. We can see that they overlap each other and the length of the overlap is $r_1 + \cdots + r_\sigma$. Since $r_1 + \cdots + r_\sigma \geq p(x_1) \cdot (\sigma - 1) + 1$ and both strings have a p-period $p(x_1)$, $p(x_1) \parallel x_1 x_1'$ by applying Lemma 7. If we apply a similar argument to $x_1 x_1'$ and x_2' that overlap each other and have the same p-period $p(x_1)$, then we can obtain $p(x_1) \parallel x_2 x_2'$. Therefore, $x_2 x_2'[1 + p(x_1)..|x_1 x_1'| + p(x_1)] \approx x_1 x_1'$, which is a contradiction. □

Therefore, all proofs are done for our upper bound.

4 Open Questions

To the best of our knowledge, no non-trivial lower bound is known at all for this problem of counting non-equivalent parameterized squares occurring in a string. If we count the parameterized squares which are *distinct as strings*, there is an $\Omega(\sigma n)$ lower bound which comes from the lower bound for the number of order-preserving squares in a string which are distinct as strings [10].

Kociumaka et al. [15] conjectured $PS'(s) \in \Theta(n)$ (Conjecture 7.1). We conjecture that even a stronger statement would hold:

Conjecture 1. For any string s of length n, $PS(s) < n$.

On the other hand, we have a tight lower bound for Theorem 2 which is given as follows:

Lemma 9. *For any σ, there exists a string with alphabet size σ such that there are σ prefixes that are parameterized squares and have no other parameterized occurrence.*

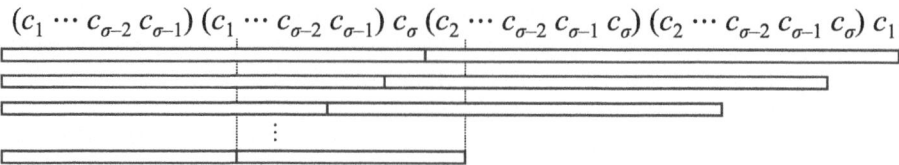

Fig. 2. Illustration for Lemma 9: There are σ non-equivalent parameterized square prefixes.

Proof. Let $\{c_1, \ldots, c_\sigma\}$ be an alphabet of size σ. We can see that the string

$$(c_1 \cdots c_{\sigma-1})^2 c_\sigma (c_2 \cdots c_\sigma)^2 c_1$$

is such a string (see also Fig. 2). □

Acknowledgments. We would like to thank the anonymous reviewers for their suggestions to improve the presentation of this paper. This work was supported by JSPS KAKENHI Grant Numbers JP21K17705 (YN), JP20H05964, JP23K18466, JP23K24808 (SI), and JP24K02899 (HB).

References

1. Amir, A., Farach, M., Muthukrishnan, S.: Alphabet dependence in parameterized matching. Inf. Process. Lett. **49**(3), 111–115 (1994)
2. Apostolico, A., Giancarlo, R.: Periodicity and repetitions in parameterized strings. Discrete Appl. Math. **156**(9), 1389–1398 (2008). general Theory of Information Transfer and Combinatorics
3. Baker, B.S.: Parameterized pattern matching: algorithms and applications. J. Comput. Syst. Sci. **52**(1), 28–42 (1996)
4. Brlek, S., Li, S.: On the number of squares in a finite word (2022). arXiv preprint arXiv:2204.10204
5. Deguchi, S., Higashijima, F., Bannai, H., Inenaga, S., Takeda, M.: Parameterized suffix arrays for binary strings. In: PSC 2008, pp. 84–94 (2008). http://www.stringology.org/event/2008/p08.html
6. Fine, N.J., Wilf, H.S.: Uniqueness theorems for periodic functions. Proc. Am. Math. Soc. **16**(1), 109–114 (1965)
7. Fraenkel, A.S., Simpson, J.: How many squares can a string contain? J. Comb. Theor. Ser. A **82**(1), 112–120 (1998)
8. Fujisato, N., Nakashima, Y., Inenaga, S., Bannai, H., Takeda, M.: Direct linear time construction of parameterized suffix and LCP arrays for constant alphabets. In: Brisaboa, N.R., Puglisi, S.J. (eds.) SPIRE 2019. LNCS, vol. 11811, pp. 382–391. Springer, Cham (2019). https://doi.org/10.1007/978-3-030-32686-9_27
9. Ganguly, A., Shah, R., Thankachan, S.V.: pBWT: achieving succinct data structures for parameterized pattern matching and related problems. In: SODA 2017, pp. 397–407 (2017). https://doi.org/10.1137/1.9781611974782.25
10. Gawrychowski, P., Ghazawi, S., Landau, G.M.: Order-preserving squares in strings (2023). arXiv preprint arXiv:2302.00724

11. Ideguchi, H., Hendrian, D., Yoshinaka, R., Shinohara, A.: Efficient parameterized pattern matching in sublinear space. In: International Symposium on String Processing and Information Retrieval, pp. 1389–1398 (2023)
12. Idury, R.M., Schäffer, A.A.: Multiple matching of parametrized patterns. Theor. Comput. Sci. **154**(2), 203–224 (1996). https://doi.org/10.1016/0304-3975(94)00270-3
13. Iseri, K., I, T., Hendrian, D., Köppl, D., Yoshinaka, R., Shinohara, A.: Breaking a barrier in constructing compact indexes for parameterized pattern matching. In: ICALP 2024. LIPIcs, vol. 297, pp. 89:1–89:19 (2024). https://doi.org/10.4230/LIPICS.ICALP.2024.89
14. Kim, J., et al.: Order-preserving matching. Theoret. Comput. Sci. **525**, 68–79 (2014)
15. Kociumaka, T., Radoszewski, J., Rytter, W., Waleń, T.: Maximum number of distinct and nonequivalent nonstandard squares in a word. Theoret. Comput. Sci. **648**, 84–95 (2016)
16. Lewenstein, M.: Parameterized pattern matching. In: Encyclopedia of Algorithms, pp. 1525–1530. Springer (2016). https://doi.org/10.1007/978-1-4939-2864-4_282
17. Lyndon, R.C., Schützenberger, M.P., et al.: The equation $a^m = b^n c^p$ in a free group. Michigan Math. J **9**(4), 289–298 (1962)
18. Matsuoka, Y., Aoki, T., Inenaga, S., Bannai, H., Takeda, M.: Generalized pattern matching and periodicity under substring consistent equivalence relations. Theor. Comput. Sci. **656**, 225–233 (2016). https://doi.org/10.1016/J.TCS.2016.02.017
19. Nakashima, K., et al.: Parameterized dawgs: efficient constructions and bidirectional pattern searches. Theor. Comput. Sci. **933**, 21–42 (2022). https://doi.org/10.1016/J.TCS.2022.09.008
20. Osterkamp, E.M., Köppl, D.: Extending the parameterized burrows-wheeler transform. In: DCC 2024, pp. 143–152 (2024). https://doi.org/10.1109/DCC58796.2024.00022

Another Virtue of Wavelet Forests

Aaron Hong[1](✉), Christina Boucher[1](✉), Travis Gagie[2](✉),
Yansong Li[2](✉), and Norbert Zeh[2](✉)

[1] Department of Computer and Information Science and Engineering,
University of Florida, Gainesville, FL, USA
yu.hong@ufl.edu, cboucher@cise.ufl.edu
[2] Department of Computer Science, Dalhousie University, Halifax, NS, Canada
travis.gagie@gmail.com, Yansong.Li@dal.ca, nzeh@cs.dal.ca

Abstract. A wavelet forest for a text $T[1..n]$ over an alphabet σ takes $nH_0(T) + o(n \log \sigma)$ bits of space and supports access and rank on T in $O(\log \sigma)$ time. Kärkkäinen and Puglisi (2011) implicitly introduced wavelet forests and showed that when T is the Burrows-Wheeler Transform (BWT) of a string S, then a wavelet forest for T occupies space bounded in terms of higher-order empirical entropies of S even when the forest is implemented with uncompressed bitvectors. In this paper we show experimentally that wavelet forests also have better access locality than wavelet trees and are thus interesting even when higher-order compression is not effective on S, or when T is not a BWT at all.

1 Introduction

A wavelet tree [10,16] for a text $T[1..n]$ over an alphabet of size σ stores T in $nH_0(T) + o(n \log \sigma)$ bits and supports access, rank and select on T in $O(\log \sigma)$ time, where $H_k(T)$ is the kth-order empirical entropy of T, $T.\mathrm{rank}_c(i)$ is the number of copies of c in $T[1..i]$ and $T.\mathrm{select}_c(i)$ is the position of the ith copy of c in T. One important application of wavelet trees is in FM-indexes [7], for which we use rank queries over the Burrows-Wheeler Transform (BWT) [2] of the indexed string.

Ferragina et al. [6] showed that when T is the BWT of a string S and we carefully split T into blocks of varying lengths, then storing a wavelet tree for each block takes a total of $nH_k(S) + o(n \log \sigma)$ bits for $k = o(\log_\sigma n)$. Mäkinen and Navarro [15] then showed that a single wavelet tree for T implemented with RRR-compressed bitvectors [17] achieves the same space bound. Finally, Kärkkäinen and Puglisi [11] showed that splitting T into the right number of fixed-length blocks and storing a wavelet tree for each block together with the rank of each distinct character at the beginning of each block, achieves the same space bound even when the wavelet trees are implemented with uncompressed bitvectors. Uncompressed bitvectors are faster than compressed ones, so Kärkkäinen and Puglisi's scheme is used in some of the most competitive implementations of FM-indexes [9]. Kärkkäinen and Puglisi called their scheme *fixed-block compression boosting* but, because we think it may be useful even in cases when higher-order compression is not effective, we use the name *wavelet forest*. One such case is storing a single human genome for DNA alignment, which is the main practical application of FM-indexes; for example, Bowtie [12,13] and BWA [14] store their BWTs in $\lg \sigma = 2$ bits per character. Can wavelet forests help even here?

One significant weakness of FM-indexes is their poor access locality [4]. Searching for a pattern $P[1..m]$ requires a sequence of $\Theta(m)$ rank queries at positions that are usually scattered throughout the BWT, and these queries must be performed in a specific order. Consequently, searching for a pattern $P[1..m]$ may result in $\Theta(m)$ cache misses, especially when n is large. In fact, the situation can be even worse. If $H_0(T) \approx \log \sigma$, then a wavelet tree for T supports an average rank query on T as rank queries on approximately $\log \sigma$ bitvectors, each of length n. Therefore, searching for P can potentially cause $\Theta(m \log \sigma)$ cache misses on average. This was recently demonstrated by Ceregini et al. [3], who showed that using quaternary wavelet trees—which double the alphabet size from binary to quaternary, double the degree, and roughly halve the height—reduces the cache misses during a search in an FM-index by about half, thereby halving the search time. However, implementing wavelet trees with very high degrees in a compact space is challenging [1]. Thus, exploring other methods to reduce cache misses remains an interesting area of research.

Suppose we store a wavelet forest for T with block length b. Then a rank query on T is again supported as rank queries on about $\log \sigma$ bitvectors, but with standard implementations all the bitvectors in the wavelet tree for a block are stored together in $O(b \log \sigma)$ bits. As long as b is not too big, intuitively it is likely that searching for P may now actually cause only $\Theta(m)$ cache misses. This idea, illustrated in Fig. 1, is intuitive but has not been evaluated in practice. For example, neither Kärkkäinen and Puglisi [11] or Gog et al. [9] evaluated whether fixed-block compression boosting could benefit from access locality. In this paper, we investigate this idea by implementing the wavelet forest and benchmarking its performance against the conventional wavelet matrix [5] provided by SDSL [8]. In particular, we measure performance by determining the number of cache misses for random access queries in both a simulated and real computing environment. All source code is publicly available at https://github.com/AaronHong1024/wavelet_forest. We show that, as we conjectured, wavelet forests have better access locality than wavelet trees, particularly for larger alphabets.

2 Methods

We built several wavelet forests and measured how efficiently they supported random access in comparison to a wavelet matrix. Although implementing random access compactly with wavelet forests is simpler than implementing rank and select, it provides a straightforward manner to determine whether the various block sizes increases the locality and leads to improved efficiency. Importantly, we used uncompressible data so any increase in the time efficiency cannot be attributed to compression.

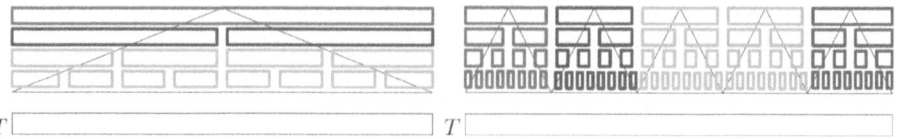

Fig. 1. A wavelet tree (**left**) and a wavelet forest (**right**) for T over an alphabet of size 16. Coloured blocks indicate bitvectors. If T is between 4 and 5 times too big for cache, then querying bitvectors of different colours will usually cause a cache miss, but querying bitvectors of the same colour (which are stored together) should not. A root-to-leaf descent to answer an access or rank query will usually cause 4 cache misses in the wavelet tree but (perhaps) only one in the wavelet forest.

Datasets. We used the random number generation utilities provided by the C++ standard library to create an 8 GB file of random digits between 0 and 255, which is stored in memory. Next, we converted this file to 64 billion binary digits, 32 billion quaternary digits, 24 billion octal digits, 16 billion hexadecimal digits, and lastly, 8 billion ASCII characters by representing every digit between 0 and 255 as a digit of the corresponding base. For example, 200 would be 11001000 in binary, 3020 in quaternary, 310 in octal, and c8 in hexadecimal. Lastly, we stored the file to hard disk using sdsl-lite's store-to-file.

Implementation. We implemented the following data structures for each file from our 5 alphabet sizes: (a) we built a single wavelet matrix with uncompressed bitvectors over the dataset; and (b) we built a forest of wavelet matrices with various block sizes with uncompressed bitvectors. In particular, we built forests with 0.5 MB blocks, 1 MB blocks, 5 MB blocks, 10 MB blocks, and 100 MB blocks for each dataset. We used sdsl-lite [8] for construction for both the wavelet tree and wavelet forest. We used wavelet matrix because sdsl-lite's balanced wavelet trees seem to have height of 8 regardless of the alphabet size, whereas wavelet matrix will have expected depth at most about $\log \sigma + 1$, where σ is the effective alphabet size. We built the wavelet matrices and wavelet forests with g++ version 12.2.0 on a server with an AMD EPYC 75F3 CPU running Red Hat Enterprise Linux 7.7 then saved them to file to hard disk using sdsl-lite's store-to-file.

Experimental Set-up. We used a machine equipped with 4 GB of RAM, 500 GB of swap space, and an Intel Core i7-11700 CPU for all experiments. Accurately measuring cache performance is nontrivial as all methods have trade-offs. Therefore, we used two different methods to measure performance to balance the shortcomings. For the first set of performance measurements, we used valgrind. We note that the shortcoming of valgrind is that it will only *simulate* the performance of the CPU rather than provide the actual CPU performance. The benefit to this simulation is that it is not sensitive to other processes that are presently running on the machine.

Next, we used perf to measure performance. This method performance tool is part of the Linux kernel tools which uses performance counters provided by modern CPUs to collect data about the program's execution, including both kernel and user space. It can trace function calls, monitor hardware events, and more, with a much smaller performance impact compared to `valgrind`. It provides the actual performance measurement on the CPU but it can be sensitive to other processes that are currently running. For this reason, when running the experiments, we shut down all possible processes currently running; but mention that some processes (such as those from the operating system) cannot be halted.

For each wavelet matrix and forest, we pseudo-randomly selected 10,000 characters from the corresponding dataset and measured both the access time, the average L3 access time, and the L3 hit rate.

3 Results and Discussion

We first evaluated the performance using the access time reported by `perf`. In particular, we compared the access time for each of the various block sizes and alphabet sizes. The heatmap depicted in Fig. 2 illustrates the comparative performance between wavelet matrices and forests, showing a consistent increase in speed for wavelet forests across a variety of alphabets and block sizes. Notably, a uniform pattern is observed for each alphabet across all datasets, demonstrating the impact of block size on access time. To explore further into these performance trends and assess the consistency of access times over different block sizes, we present the line graphs in Fig. 2. The graphs reveal that wavelet forests were faster—presumably because of better access locality—for the octal, hexadecimal, and ASCII datasets for all block lengths. Especially, at the optimal block size, wavelet forests demonstrate a performance gain of 2.5 times over the standard wavelet matrix.

To verify the factors contributing to our observed speedup, we also investigated the cache miss rates. While `valgrind` offers a simulated view of CPU activity, providing estimates of access time and cache misses, it is important to note that `perf` operates directly on the hardware, yielding a more precise measurement. Thus, we used both tools to ensure the validation of our results. Our analysis particularly focused on the third level of cache (L3). Due to its larger size and position in the memory hierarchy, it is most-likely to significantly influence CPU time by mitigating the delay caused by cache misses. Thus, in our analysis, we calculated the average access time for the L3 cache. The average access time was determined by using the following formula:

$$\text{Average L3 Access Time} = \alpha_{L3} \times \beta_{L3} + \alpha_{mem} \times (1 - \beta_{L3}),$$

where α_{L3} is the L3 access time, i.e., the time required to retrieve data from the L3 cache upon a cache hit occurrence. α_{L3} was determined by measuring the number of CPU clock cycles required for a series of memory access operations to complete. We measured this using the `rdtsc` instruction to capture precise start and end times before and after accessing the memory allocated to simulate cache hits. The difference in these timestamps provided a direct measurement of the cache access time, allowing for an accurate estimation of the L3 cache's efficiency in data retrieval. Next, α_{mem} is the memory access time indicating the time to retrieve data directly from the system memory when the L3 cache does not contain the requested data, resulting in a cache miss. The presence of a cache miss means the system must go through a longer process to access the required data, directly affecting the performance and efficiency of data retrieval operations. This measurement was obtained by Intel Memory Latency Checker (MLC), which is a tool developed by Intel to measure various latency and bandwidth characteristics of a computer's memory system. Among other metrics, it can accurately determine the α_{mem} by simulating scenarios where the CPU must fetch data from the main memory rather than the cache. And lastly, β_{L3} is the L3 hit rate, which is the proportion of access attempts that successfully retrieve data from the L3 cache out of the total number of access attempts. Essentially, it measures how often the CPU can obtain data from the L3 cache instead of resorting to the slower main system memory. Both `perf` and `valgrind` were used to obtain β_{L3}. Hence, this formula yields a weighted average access time that measures the effectiveness of the L3 cache in handling data requests and the consequences of resorting to the slower main memory due to cache misses.

We illustrate this measurement for all block sizes in Fig. 3, showing a consistent correlation between L3 cache misses and access times across most cases, despite occasional anomalies in other datasets. The findings from `valgrind` align with those obtained through `perf`, reinforcing the validity of our observations. This parallelism in results between `valgrind` and `perf` not only confirms our initial findings but also highlights the reliability of our analysis methods in capturing the performance dynamics of our system.

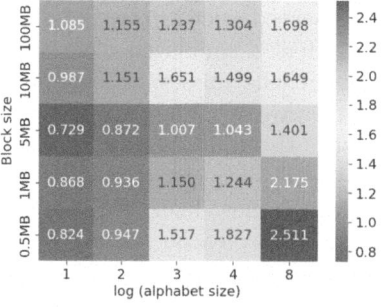 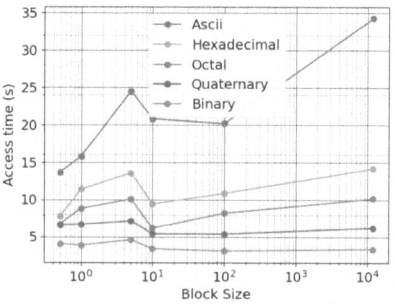

Fig. 2. The figure on the left illustrates the ratio of access times obtained within the source code through using standard C++ `chrono` library packages for wavelet matrices and forests across block sizes of 0.5 MB, 1 MB, 5 MB, 10 MB, and 100 MB, when querying 10,000 characters in their respective alphabets. The right-most node represent the access time usage for wavelet matrix. Each heatmap cell indicates the speed-up of the wavelet forest relative to the baseline wavelet matrix operations. The figure on the right illustrates the total access time for 10,000 characters.

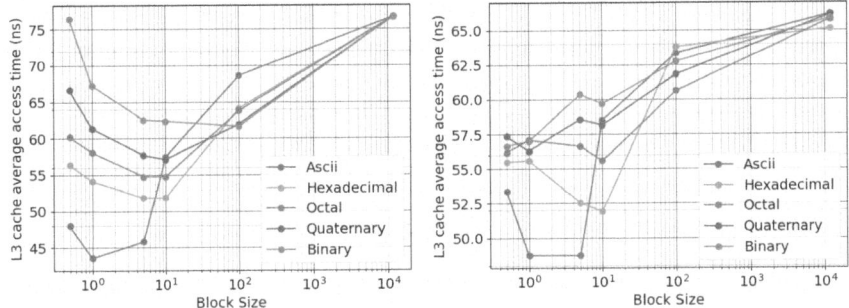

Fig. 3. Comparison of the average L3 access time estimated for the wavelet matrix and the wavelet forest using both the L3 hit rate provided by `valgrind` (left) and `perf` (right)

4 Discussion

Using different performance analysis tools, like `valgrind` and `perf`, may lead to contradictory results due to the inherent differences in their operational methods and performance metrics measurement. `valgrind`, which uses dynamic binary instrumentation, provides highly detailed information about memory usage and cache behavior by inserting additional code into the program. However, this method incurs significant overhead, slowing down the program considerably, and potentially altering its normal execution patterns, including cache behaviors. In contrast, `perf`, a sampling profiler, leverages hardware performance counters to collect data with minimal overhead, thus capturing performance metrics that are more reflective of real-world conditions. `perf`'s broad coverage allows it to monitor various system-wide and per-process events accurately. These differences can lead to discrepancies in results; for instance, `valgrind`'s detailed but perturbing instrumentation might show different cache usage patterns compared to `perf`'s low-interference, reflective sampling.

In the specific case of comparing wavelet forests and wavelet matrices, both tools confirmed efficiency improvements due to reduced L3 cache misses. However, the observation that wavelet forest with 5MB yields similar overall access times to wavelet matrix despite having lower average L3 access times highlights the complexity of performance metrics and their interactions. While L3 cache access time is a significant factor, it is not the sole determinant of overall access time. Other factors, such as the number of memory accesses, the efficiency of cache line utilization, and the specific access patterns, play crucial roles. In the case of wavelet forest with 5MB, it is possible that the improved locality and reduced cache misses contribute to a lower average L3 access time. However, if the total number of accesses or other overheads are higher, this could offset the benefits, resulting in similar overall access times as wavelet matrix. Additionally, the impact of prefetching mechanisms and the specific architecture of the

CPU can influence these results. Therefore, the seemingly contradictory results between Figs. 2 and 3 can be attributed to the multifaceted nature of performance factors, where L3 access time is one of many contributing elements. Further investigation into the access patterns and detailed profiling could provide more insights into these dynamics. Accurately determining access to all levels of cache is challenging, especially when focusing on a single specific process.

5 Conclusion and Future Work

We experimentally demonstrated that wavelet forests, with an optimized block size of 500 KB and data interpreted as ASCII characters, are about 2.5 times faster than standard wavelet matrices. This efficiency, as confirmed by using both `perf` and `valgrind`, is largely due to reduced L3 cache misses, enhancing locality. In summary, we demonstrated that improved locality and increased efficiency can be obtained by simple insights in the implementation—insights that should be incorporated into the design of future data structures. Furthermore, additional methods to improve locality in random access, rank, and select operations deserve exploration. Initially, we hypothesized that clustering BWT blocks could minimize long-distance LF steps. Our preliminary tests involved segmenting the Douglas Fir genome's BWT into blocks and clustering them using Metis. We found that most LF steps occurred within the same cluster for practical block and cluster sizes. Further experiments involved creating a virtual machine with 1 GB of RAM to test substring extraction from an FM-index using both standard and clustered BWTs of the Douglas Fir genome. Results showed a notable speedup, attributed primarily to blocking rather than clustering, which seemed effective in theory but not in practice. This led to the insights presented in this paper, though the practical benefits of clustering remain a point of interest.

Acknowledgments. Many thanks to Professor Alin Dobra (at the University of Florida), Simon Gog, and Professor Prabhat Mishra (at the University of Florida) for helpful discussions. This research was funded by NIH/NIAID grant R01AI14180, NSF/BIO grant DBI-2029552 and NSF/SCH grant INT-2013998 to Christina Boucher, NSERC grant RGPIN-07185-2020 to Travis Gagie and NIH/NHGRI grant R01HG011392 to Ben Langmead.

References

1. Bowe, A.: Multiary wavelet trees in practice. Bachelor's thesis, RMIT University, Melbourne, Australia (2010)
2. Burrows, M., Wheeler, D.J.: A block-sorting lossless data compression algorithm. Tech. Rep. 124, Digital Equipment Corporation (1994)
3. Ceregini, M., Kurpicz, F., Venturini, R.: Faster wavelet trees with quad vectors. arXiv preprint arXiv:2302.09239 (2023)
4. Chien, Y.F., Hon, W.K., Shah, R., Thankachan, S.V., Vitter, J.S.: Geometric BWT: compressed text indexing via sparse suffixes and range searching. Algorithmica **71**(2), 258–278 (2015)

5. Claude, F., Navarro, G., Ordóñez, A.: The wavelet matrix: an efficient wavelet tree for large alphabets. Inf. Syst. **47**, 15–32 (2015)
6. Ferragina, P., Giancarlo, R., Manzini, G., Sciortino, M.: Boosting textual compression in optimal linear time. J. ACM **52**(4), 688–713 (2005)
7. Ferragina, P., Manzini, G.: Indexing compressed text. J. ACM **52**(4), 552–581 (2005)
8. Gog, S., Beller, T., Moffat, A., Petri, M.: From theory to practice: plug and play with succinct data structures. In: Proceedings of the 13th Symposium on Experimental Algorithms (SEA), pp. 326–337 (2014)
9. Gog, S., Kärkkäinen, J., Kempa, D., Petri, M., Puglisi, S.J.: Fixed block compression boosting in FM-indexes: theory and practice. Algorithmica **81**, 1370–1391 (2019)
10. Grossi, R., Gupta, A., Vitter, J.S.: High-order entropy-compressed text indexes. In: Proceedings of the 14th Symposium on Discrete Algorithms (SODA), pp. 841–850 (2003)
11. Kärkkäinen, J., Puglisi, S.J.: Fixed block compression boosting in FM-indexes. In: Proceedings of the 18th Symposium on String Processing and Information Retrieval (SPIRE), pp. 174–184 (2011)
12. Langmead, B., Salzberg, S.L.: Fast gapped-read alignment with bowtie 2. Nat. Methods **9**(4), 357–359 (2012)
13. Langmead, B., Trapnell, C., Pop, M., Salzberg, S.L.: Ultrafast and memory-efficient alignment of short DNA sequences to the human genome. Genome Biol. **10**(3), 1–10 (2009)
14. Li, H., Durbin, R.: Fast and accurate short read alignment with Burrows-Wheeler transform. Bioinformatics **25**(14), 1754–1760 (2009)
15. Mäkinen, V., Navarro, G.: Implicit compression boosting with applications to self-indexing. In: Proceedings of the 14th Symposium on String Processing and Information Retrieval (SPIRE), pp. 229–241 (2007)
16. Navarro, G.: Wavelet trees for all. J. Discrete Algorithms **25**, 2–20 (2014)
17. Raman, R., Raman, V., Satti, S.R.: Succinct indexable dictionaries with applications to encoding k-ary trees, prefix sums and multisets. ACM Trans. Algorithms **3**(4), 43 (2007)

All-Pairs Suffix-Prefix on Dynamic Set of Strings

Masaru Kikuchi[1] and Shunsuke Inenaga[2]()

[1] Department of Information Science and Technology, Kyushu University, Fukuoka, Japan
kikuchi.masaru.484@s.kyushu-u.ac.jp
[2] Department of Informatics, Kyushu University, Fukuoka, Japan
inenaga.shunsuke.380@m.kyushu-u.ac.jp

Abstract. The *all-pairs suffix-prefix* (*APSP*) problem is a classical problem in string processing which has important applications in bioinformatics. Given a set $\mathcal{S} = \{S_1, \ldots, S_k\}$ of k strings, the APSP problem asks one to compute the longest suffix of S_i that is a prefix of S_j for all k^2 ordered pairs $\langle S_i, S_j \rangle$ of strings in $\mathcal{S}$. In this paper, we consider the *dynamic* version of the APSP problem that allows for insertions of new strings to the set of strings. Our objective is, each time a new string S_i arrives to the current set $\mathcal{S}_{i-1} = \{S_1, \ldots, S_{i-1}\}$ of $i-1$ strings, to compute (1) the longest suffix of S_i that is a prefix of S_j and (2) the longest prefix of S_i that is a suffix of S_j for all $1 \leq j \leq i$. We propose an $O(n)$-space data structure which computes (1) and (2) in $O(|S_i| \log \sigma + i)$ time for each new given string S_i, where n is the total length of the strings.

Keywords: all-pairs suffix-prefix problem · Aho-Corasick automata · directed acyclic word graphs (DAWGs)

1 Introduction

The *all-pairs suffix-prefix* (*APSP*) problem is a classical problem in string processing which has important applications in bioinformatics, since it is the first step of genome assembly [9]. Given a set $\mathcal{S} = \{S_1, \ldots, S_k\}$ of k strings, the APSP problem asks one to compute the longest suffix of S_i that is a prefix of S_j for all k^2 ordered pairs $\langle S_i, S_j \rangle$ of strings in $\mathcal{S}$.

A straightforward solution, mentioned in [10], is to use a variant of the KMP pattern matching algorithm [13] for each pair of two strings S_i, S_j in $O(|S_i|+|S_j|)$ time. This however leads to an inefficient $O(kn)$-time complexity, where $n = \|\mathcal{S}\|$ is the total length of the strings in $\mathcal{S}$.

Gusfield et al. [10] proposed the first efficient solution to the APSP problem that takes $O(n)$ space, which is based on the *generalized suffix tree* [5,10] for the set $\mathcal{S}$ of strings. After building the generalized suffix tree in $O(n \log \sigma)$ time in the case of a general ordered alphabet of size σ [22] or in $O(n)$ time in the case of an integer alphabets of size $\sigma = n^{O(1)}$ [8], Gusfield et al.'s algorithm [10] works in $O(n + k^2)$ optimal time. Ohlebusch and Gog [18] proposed an alternative algorithm for solving the APSP problem with the same complexities as above, using (enhanced) suffix arrays [1,17]. Tustumi et a. [21] gave an improved suffix-array based algorithm that is fast and memory efficient in practice. All these approaches share the common concepts of building the generalized suffix tree/array for $\mathcal{S}$.

Another data structure that can be used to solve the APSP problem is the *Aho-Corasick automata (AC-automata)* [2]. This is intuitive since the AC-automaton is a generalization of the KMP-automaton for multiple strings. The use of the AC-automaton for solving APSP was suggested by Lim and Park [14], but they did not give any details of an algorithm nor the complexity. Recently, Loukides and Pissis [15] proposed an AC-automaton based algorithm for the APSP problem. After building the AC-automaton in $O(n \log \sigma)$ time in the case of a general ordered alphabet of size σ [2] or in $O(n)$ time in the case of an integer alphabets of size $\sigma = n^{O(1)}$ [7], the algorithm of Loukides and Pissis [15] runs in optimal $O(n + k^2)$ time. Their algorithm can also solve a length-threshold version of the problem in optimal time. Recently, Loukides et al. [16] considered a query-version of the APSP problem. They presented a data structure of $O(n)$-space that can report the longest suffix-prefix match between one string S_i and all the other strings in $O(k)$ time. Their data structure is based on the AC-automaton and the micro-macro decomposition of trees [3].

In this paper, we consider the *dynamic* version of the APSP problem that allows for insertions of new strings to the set of strings. Our objective is, each time a new string S_i arrives to the current set $\mathcal{S}_i = \{S_1, \ldots, S_{i-1}\}$ of strings, to compute the following longest suffix-prefix matches between a new string S_i and all the other strings $S_1, \ldots, S_{i-1}$:

- the longest suffix of S_i that is a prefix of S_j for $1 \leq j \leq i$, and
- the longest prefix of S_i that is a suffix of S_j for $1 \leq j \leq i$.

We propose an $O(n)$-space data structure based on the *directed acyclic word graph (DAWG)* for a growing set of strings, which is able to compute the suffix-prefix matches for a given new string S_i in $O(|S_i| \log \sigma + i)$ time. After iterating this for all k strings $S_1, \ldots, S_k$ arriving to $\mathcal{S}$, the total running time becomes $O(n \log \sigma + k^2)$, which is as fast as the state-of-the-art algorithms [10,15] for a static set of strings in the case of general ordered alphabets of size σ. We note that the suffix-tree based algorithm of Gusfield et al. [10] performs a DFS, and thus, does not seem to extend directly to the dynamic case.

Before describing our APSP algorithm for the dynamic case (Sect. 5), we present a new AC-automata based APSP algorithm for a static set of strings (Sect. 4). While the algorithm of Loukides and Pissis [15] is based on a reduction of the APSP problem to the ULIT (union of labeled intervals on a tree) problem, our new APSP algorithm does not use this reduction. Instead, we use simple failure link traversal of the AC-automaton and marking operations on a compact prefix trie that is a compact version of the AC-trie. The simple nature of our algorithm makes it possible to work also on the DAWG structure in the dynamic case.

The *hierarchical overlap graphs* (*HOGs*) and their variants have a close relationship to the APSP problem. There are AC-automata oriented approaches for computing these graphs [6,12,19]. We remark that all of their approaches are for the static case, and do not seem to immediately extend to the dynamic setting.

2 Preliminaries

2.1 Strings

Let Σ denote an ordered *alphabet* of size σ. An element of Σ^* is called a *string*. The length of a string $S \in \Sigma^*$ is denoted by $|S|$. The *empty string* ε is the string of length 0. Let $\Sigma^+ = \Sigma^* \setminus \{\varepsilon\}$. For string $S = xyz$, x, y, and z are called the *prefix*, *substring*, and *suffix* of S, respectively. Let $\mathsf{Prefix}(S)$, $\mathsf{Substr}(S)$, and $\mathsf{Suffix}(S)$ denote the sets of prefixes, substrings, and suffixes of S, respectively. The elements of $\mathsf{Prefix}(S) \setminus \{S\}$, $\mathsf{Substr}(S) \setminus \{S\}$, and $\mathsf{Suffix}(S) \setminus \{S\}$ are called the *proper prefixes*, *proper substrings*, and *proper suffixes* of S, respectively. For a string S of length n, $S[i]$ denotes the i-th symbol of S and $S[i..j] = S[i] \cdots S[j]$ denotes the substring of S that begins at position i and ends at position j for $1 \leq i \leq j \leq n$. For convenience, let $S[i..j] = \varepsilon$ for $i > j$. The *reversed string* of a string S is denoted by S^R, that is, $S^R = S[|S|] \cdots S[1]$.

For a set $\mathcal{S} = \{S_1, \ldots, S_k\}$ of strings, let $\mathsf{Substr}(\mathcal{S}) = \bigcup_{i=1}^{k} \mathsf{Substr}(S_i)$.

2.2 All-Pairs Suffix-Prefix Overlap (APSP) Problems

In this section, we introduce the problems that we tackle in this paper.

Let $\mathcal{S} = \{S_1, \ldots, S_k\}$ be a set of non-empty strings, where $S_i \in \Sigma^+$ for $1 \leq i \leq k$. For each ordered pair $\langle S_i, S_j \rangle$ of two strings in $\mathcal{S}$, the longest string in the set $\mathsf{Suffix}(S_i) \cap \mathsf{Prefix}(S_j)$ is called the *longest suffix-prefix overlap* of S_i and S_j, and is denoted by $\mathsf{lspo}(i,j)$. We encode each $\mathsf{lspo}(i,j)$ by a tuple $(i, j, |\mathsf{lspo}(i,j)|)$ using $O(1)$ space.

APSP Problems on Static Sets

Problem 1 (static-APSP). The *static all-pairs suffix-prefix overlaps* (*static-APSP*) problem is, given a static set $\mathcal{S} = \{S_1, \ldots, S_k\}$ of k non-empty strings, to compute

$$\mathcal{L}_\mathcal{S} = \{\mathsf{lspo}(i,j) \mid 1 \leq i \leq k, 1 \leq j \leq k\}.$$

Note that the output size of Problem 1 is $|\mathcal{L}_\mathcal{S}| = \Theta(k^2)$ as we compute $\mathsf{lspo}(i,j)$ for all ordered pairs $\langle S_i, S_j \rangle$ of strings in $\mathcal{S}$.

Problem 2 (length-bounded static-APSP). The *length-bounded* static-APSP problem is, given a static set $\mathcal{S} = \{S_1, \ldots, S_k\}$ of k non-empty strings and an integer threshold $\ell \geq 0$, compute

$$\mathcal{L}_\mathcal{S}^{\geq \ell} = \{\mathsf{lspo}(i,j) \mid 1 \leq i \leq k, 1 \leq j \leq k, |\mathsf{lspo}(i,j)| \geq \ell\}.$$

APSP Problems on Dynamic Sets with Insertions. Let $\mathcal{S}_0 = \emptyset$ and for $1 \leq i \leq k$, let $\mathcal{S}_i = \{S_1, \ldots, S_i\}$ denote the subset of $\mathcal{S}$ consisting of the first i strings. We consider a *dynamic* variant of APSP allowing for insertions of strings, named *dynamic-APSP*, that is defined as follows:

Problem 3 (*dynamic*-APSP). The *dynamic*-APSP problem is, each time a new string S_i is added to the current set $\mathcal{S}_{i-1} = \{S_1, \ldots, S_{i-1}\}$, to compute the two following sets

$$\mathcal{F}_i = \{\mathsf{lspo}(i,j) \mid 1 \leq j \leq i\},$$
$$\mathcal{B}_i = \{\mathsf{lspo}(j,i) \mid 1 \leq j \leq i\}.$$

We remark that $\mathcal{L}_\mathcal{S} = \bigcup_{i=1}^{k} (\mathcal{F}_i \cup \mathcal{B}_i)$ holds for $\mathcal{S} = \{S_1, \ldots, S_k\}$. Thus, a solution to Problem 3 is also a solution to Problem 1.

Problem 4 (*length-bounded dynamic*-APSP). Let $\ell > 0$ be a length threshold. The *dynamic*-APSP problem is, each time a new string S_i is added to the current set $\mathcal{S}_{i-1} = \{S_1, \ldots, S_{i-1}\}$, to compute the two following sets

$$\mathcal{F}_i^{\geq \ell} = \{\mathsf{lspo}(i,j) \mid 1 \leq j \leq i, |\mathsf{lspo}(i,j)| \geq \ell\},$$
$$\mathcal{B}_i^{\geq \ell} = \{\mathsf{lspo}(j,i) \mid 1 \leq j \leq i, |\mathsf{lspo}(i,j)| \geq \ell\}.$$

Note that a solution to Problem 4 is also a solution to Problem 2 as well.

3 Tools

In this section we list the data structures that are used as building blocks of our static-APSP and dynamic-APSP algorithms.

3.1 Tries and Compact Tries

A *trie* T is a rooted tree (V, E) such that

- each edge in E is labeled by a single character from Σ and
- the character labels of the out-going edges of each node begin with distinct characters.

For a pair of nodes u, v in a trie where u is an ancestor of v, let $\langle u, v \rangle$ denote the path from u to v. The *path label* of $\langle u, v \rangle$ is the concatenation of the edge labels from u to v. The *string label* of a node v in a trie, denoted $\mathsf{str}(v)$, is the path label from the root to v. The *string depth* of a node v in a trie is the length of the string label of v, that is $|\mathsf{str}(v)|$.

For a set $\mathcal{S} = \{S_1, \ldots, S_k\}$ of k strings, let $\mathsf{Trie}(\mathcal{S})$ denote the trie that represents the strings in $\mathcal{S}$. We assume that each of the k nodes in $\mathsf{Trie}(\mathcal{S})$ representing S_i stores a unique integer i called the *id*.

We build and maintain a compacted trie $\mathsf{ComTrie}(\mathcal{S})$ that is obtained from $\mathsf{Trie}(\mathcal{S})$ by keeping

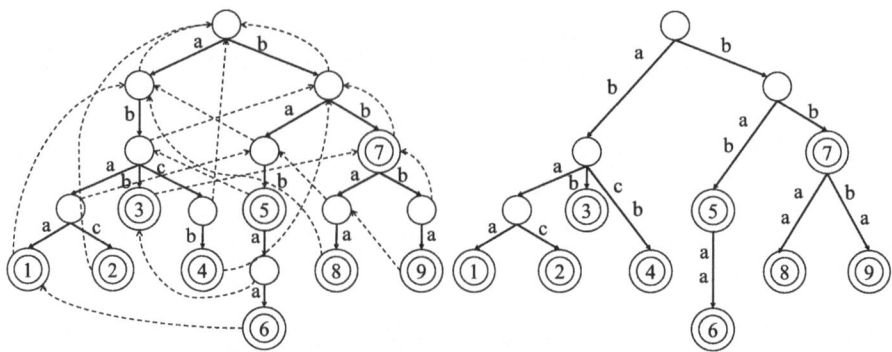

Fig. 1. Illustrations of AC($\mathcal{S}$) (left) and ComTrie($\mathcal{S}$) (right) for the set $\mathcal{S}$ = {abaa, abac, abb, abcb, bab, babaa, bb, bbaa, bbba} of strings. The bold solid arcs represent trie edges and the dashed arcs represent failure links. The nodes representing the strings in $\mathcal{S}$ are depicted by double-lined circles with the string id's.

- all the leaves and the (possibly non-branching) internal nodes that store id's $1, \ldots, k$ for the k strings $S_1, \ldots, S_k$, and
- the root and all the branching nodes.

All other nodes, each having a single child, are removed and their corresponding edges are contracted into paths. For each node v in Trie($\mathcal{S}$), let des(v) denote the shallowest descendant of v that exists in ComTrie($\mathcal{S}$). Note that the number of nodes in ComTrie($\mathcal{S}$) is $O(k)$.

3.2 Aho-Corasick Automata

For each non-root node v of Trie($\mathcal{S}$), define its *failure link* by flink(v) = u iff str(u) is the longest proper suffix of str(v) that can be spelled out from the root of Trie($\mathcal{S}$). In other words, str(u) is the longest prefix of some string(s) represented by Trie($\mathcal{S}$) that is also a proper suffix of str(v). The failure links also form a (reversed) tree, and let us denote it by FLTree($\mathcal{S}$).

The *Aho-Corasick automaton* [2] for a set $\mathcal{S}$ of strings, denoted AC($\mathcal{S}$), is a finite automaton with two kinds of transitions[1]: the goto function that is represented by the edges of the trie Trie($\mathcal{S}$), and the failure function that is represented by the edges of the reversed trie FLTree($\mathcal{S}$). It is clear that the total number of nodes and edges of AC($\mathcal{S}$) is $O(n)$, where $n = \|\mathcal{S}\|$. See Fig. 1 for concrete examples of the AC-automaton and its trie, compact trie, and failure links.

Theorem 1 ([2,7]). *For a set $\mathcal{S}$ of strings of total length n, AC($\mathcal{S}$) can be built*

- *in $O(n)$ time and space for an integer alphabet of size $\sigma = n^{O(1)}$;*
- *in $O(n \log \sigma)$ time and $O(n)$ space for a general ordered alphabet of size σ.*

---

[1] We do not use the output function in our algorithms.

3.3 Directed Acyclic Word Graphs (DAWGs)

For a set $\mathcal{S}$ of strings, let $\mathsf{End\_Pos}_\mathcal{S}(w)$ denote the set of pairs of string ids and ending positions of all occurrences of a substring $w \in \mathsf{Substr}(\mathcal{S})$, that is,

$$\mathsf{End\_Pos}_\mathcal{S}(w) = \{(i,j) \mid S_i[j-|w|+1..j] = w, 1 \le i \le k, 1 \le j \le |S_i|\}.$$

We consider an equivalence relation $\equiv_\mathcal{S}$ of strings over Σ w.r.t. $\mathcal{S}$ such that, for any two strings w and u, $w \equiv_\mathcal{S} u$ iff $\mathsf{End\_Pos}_\mathcal{S}(w) = \mathsf{End\_Pos}_\mathcal{S}(u)$. For any string $x \in \Sigma^*$, let $[x]_\mathcal{S}$ denote the equivalence class for x w.r.t. $\equiv_\mathcal{S}$.

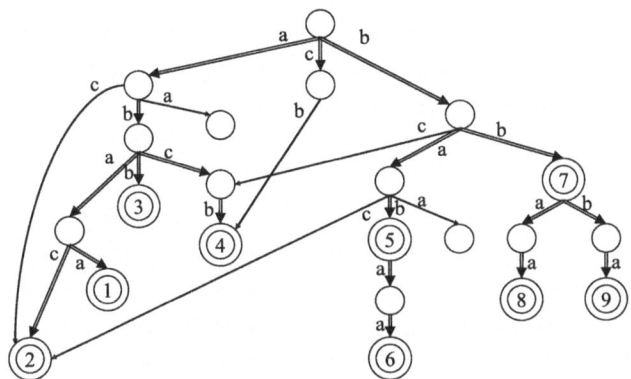

Fig. 2. Illustration of $\mathsf{DAWG}(\mathcal{S})$ for the same set $\mathcal{S} = \{\mathsf{abaa}, \mathsf{abac}, \mathsf{abb}, \mathsf{abcb}, \mathsf{bab}, \mathsf{babaa}, \mathsf{bb}, \mathsf{bbaa}, \mathsf{bbba}\}$ of strings as in Fig. 1. The induced tree consisting only of the double-lined arcs is $\mathsf{Trie}(\mathcal{S})$.

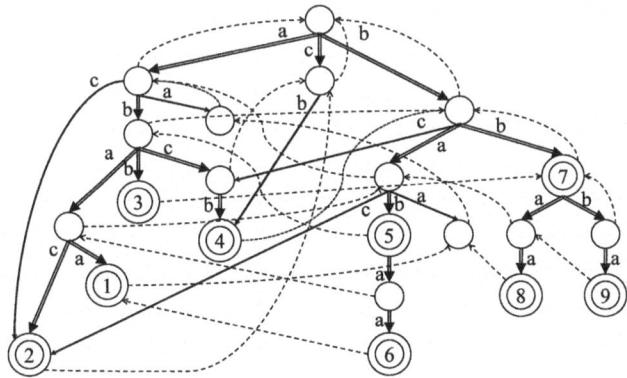

Fig. 3. Illustration of the suffix links of $\mathsf{DAWG}(\mathcal{S})$ for the same set $\mathcal{S}$ of strings as in Fig. 2.

The DAWG of a set $\mathcal{S}$ of strings, denoted $\mathsf{DAWG}(\mathcal{S})$, is an edge-labeled DAG (V, E) such that

$$V = \{[x]_\mathcal{S} \mid x \in \mathsf{Substr}(\mathcal{S})\},$$
$$E = \{([x]_\mathcal{S}, b, [xb]_\mathcal{S}) \mid x, xb \in \mathsf{Substr}(\mathcal{S}), b \in \Sigma\}.$$

We also define the set L of *suffix links* of $\mathsf{DAWG}(\mathcal{S})$ by

$$L = \{([ax]_\mathcal{S}, a, [x]_\mathcal{S}) \mid x, ax \in \mathsf{Substr}(\mathcal{S}), a \in \Sigma, [ax]_\mathcal{S} \neq [x]_\mathcal{S}\}.$$

Namely, two substrings x and y in $\mathsf{Substr}(\mathcal{S})$ are represented by the same node of $\mathsf{DAWG}(\mathcal{S})$ iff the ending positions of x and y in the strings of $\mathcal{S}$ are equal. See Fig. 2 and Fig. 3 for examples.

For a node $v \in V$ of $\mathsf{DAWG}(\mathcal{S})$, let $\mathsf{long}(v)$ denote the longest string represented by v (i.e., $|\mathsf{long}(v)|$ is the length of the longest path from the source to v). Also, let $\mathsf{slink}(v) = u$ denote the suffix link from node u to node v.

An edge $(u, a, v) \in E$ of $\mathsf{DAWG}(\mathcal{S})$ is called an *primary edge* if $|\mathsf{long}(u)| + 1 = |\mathsf{long}(v)|$, and it is called a *secondary edge* otherwise ($|\mathsf{long}(u)| + 1 < |\mathsf{long}(v)|$). The primary edges form a spanning tree of $\mathsf{DAWG}(\mathcal{S})$ which consists of the longest paths from the source to all the nodes, and let us denote this spanning tree by $\mathsf{LPTree}(\mathcal{S})$. By the definition of the equivalence class $[\cdot]_\mathcal{S}$, for each $S_i \in \mathcal{S}$ it holds that $S_i = \mathsf{long}([S_i]_\mathcal{S})$. Thus we have the following:

Observation 1. *For each $S_i \in \mathcal{S}$, the path that spells out S_i from the source of $\mathsf{DAWG}(\mathcal{S})$ is a path of $\mathsf{LPTree}(\mathcal{S})$.*

$\mathsf{DAWG}(\mathcal{S})$ is a partial DFA of size $O(n)$ that accepts $\mathsf{Substr}(\mathcal{S})$ [4,5]. The following lemma permits us to efficiently update the DAWG each time a new string is inserted to the current set of strings:

Theorem 2 ([5]). *Suppose that $\mathsf{DAWG}(\mathcal{S}_{i-1})$ has been built for a set $\mathcal{S}_{i-1} = \{S_1, \ldots, S_{i-1}\}$ of $i - 1$ strings. Given a new string S_i to insert, one can update $\mathsf{DAWG}(\mathcal{S}_{i-1})$ to $\mathsf{DAWG}(\mathcal{S}_i)$ in $O(|S_i| \log \sigma)$ time.*

4 Algorithm for Static-APSP

In this section, we present a new, very simple algorithm for the static-APSP problem for a static set $\mathcal{S} = \{S_1, \ldots, S_k\}$ of k strings of total length $\|\mathcal{S}\| = n$.

Our main data structure is the AC-automaton $\mathsf{AC}(\mathcal{S})$. For each string $S_i \in \mathcal{S}$ (in an arbitrary order), set node v to be the node that represents S_i, i.e. initially $\mathsf{str}(v) = S_i$. All the nodes of $\mathsf{ComTrie}(\mathcal{S})$ are initially unmarked. Our algorithm is shown in Algorithm 1. See also Fig. 4 for illustration.

Algorithm 1: Compute lspo(i, j) for all $1 \le j \le k$

(1) Set v to be the node that represents S_i (i.e. str(v) = S_i).
(2) – Set $u \leftarrow$ des(v).
 – Traverse the *induced* subtree T_u of ComTrie($\mathcal{S}$) that is rooted at u, and consists only of unmarked nodes (and thus we do not traverse the subtrees under the marked nodes in T_u).
 – Report lspo(i, j) = |str(v)| for all id's j found in the induced subtree T_u.
 – Mark node u on ComTrie($\mathcal{S}$).
(3) If v is the root, finish the procedure for the given string S_i. Unmark all marked nodes.
(4) If v is not the root, take the failure link from v on Trie($\mathcal{S}$), and set $v \leftarrow$ flink(v). Continue Step (2).

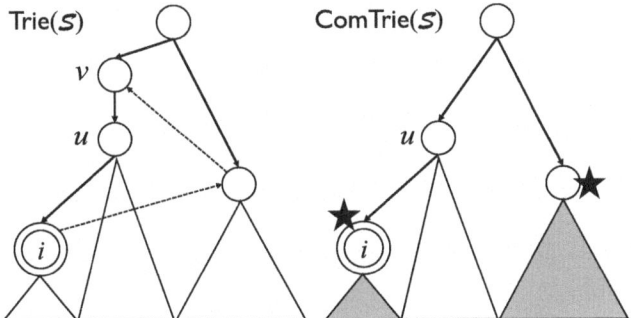

Fig. 4. Illustration for Algorithm 1 that solves the static-APSP problem, in which we have climbed up the failure links from the node with id i to node v. We access $u =$ des(v) in ComTrie($\mathcal{S}$), and traverse only the white subtree below u.

Theorem 3. *For a static set $\mathcal{S} = \{S_1, \ldots, S_k\}$ of k strings of total length n, our algorithm solves the static-APSP problem with $O(n)$ working space and*

– *in $O(n + k^2)$ time for an integer alphabet of size $\sigma = n^{O(1)}$;*
– *in $O(n \log \sigma + k^2)$ time for a general alphabet of size σ.*

Proof. The correctness of our algorithm follows from the definitions of the lspo($\cdot, \cdot$) function and the failure links of AC($\mathcal{S}$), and from our failure link traversal starting from the node v representing S_i, in decreasing order of the node depths towards the root r.

For each string S_i, Step (1) clearly takes $O(1)$ time. The total number l of nodes v, flink(v), flink^2(v), ..., flink^{l-1}(v) = r visited in the chain of failure links starting from node v with str(v) = S_i is at most $|S_i|+1$. The size of ComTrie($\mathcal{S}$) is $O(k)$. As we process the nodes v, flink(v), flink^2(v), ..., flink^{l-1}(v) = r in decreasing order of their depths, any marked node is visited at most once. This is because, if a marked node w is visited twice, then there must be two distinct ancestors p, q of w in ComTrie($\mathcal{S}$), such that p is an ancestor of q. This however is a contradiction, since q was marked before p, and our algorithm does not visit the subtree under q when processing the induced tree T_p. Therefore, each node is marked at most once, and each marked node is visited at most once. Thus, the traversals on ComTrie($\mathcal{S}$) with our marking operations takes time linear in its size $O(k)$. Overall, for each $S_i \in \mathcal{S}$, we can compute $\mathsf{lspo}(i,j)$ for all $1 \leq i \leq k$ in optimal $O(|S_i| + i)$ time, which leads to an overall $O(n + k^2)$ time for all the k strings in $\mathcal{S}$.

We build AC($\mathcal{S}$) by Theorem 1. We can compute ComTrie($\mathcal{S}$) and $\mathsf{des}(u) = v$ for all nodes v in Trie($\mathcal{S}$) in time linear in the size of Trie($\mathcal{S}$), which is $O(n)$. □

For the length-bounded APSP problem, we have the following:

Corollary 1. *For a static set $\mathcal{S} = \{S_1, \ldots, S_k\}$ of k strings of total length n, our algorithm solves the length-ℓ static-APSP problem with $O(n)$ working space and*

- *in $O(n + |\mathcal{L}_{\mathcal{S}}^{\geq \ell}|)$ time for an integer alphabet of size $\sigma = n^{O(1)}$;*
- *in $O(n \log \sigma + |\mathcal{L}_{\mathcal{S}}^{\geq \ell}|)$ time for a general alphabet of size σ.*

Our data structure can be reused for different values of the length-threshold ℓ.

Proof. It suffices for us to change Step (3) and Step (4) in Algorithm 1 as follows:

(3) If $\underline{|\mathsf{str}(v)| \leq \ell}$, finish the procedure for the given string S_i. Unmark all marked nodes.
(4) If $\underline{|\mathsf{str}(v)| > \ell}$, take the failure link from v on Trie($\mathcal{S}$), and set $v \leftarrow \mathsf{flink}(v)$. Continue Step (2).

We can reuse our data structure for different values of ℓ, as we only need to change the underlined conditions in the above. □

5 Algorithm for Dynamic-APSP

In this section, we present our algorithm for solving the dynamic-APSP problem: Given a new string S_i that is inserted to the current set $\mathcal{S}_{i-1} = \{S_1, \ldots, S_{i-1}\}$ of $i-1$ strings, our goal is to compute the outputs $\mathcal{F}_i$ and $\mathcal{B}_i$ of Problem 3.

In what follows, we will show the following:

Theorem 4. *There exists a data structure of $O(n_i)$ space, which solves the dynamic-APSP problem of computing $\mathcal{F}_i$ and $\mathcal{B}_i$ for a given new string S_i in $O(|S_i| \log \sigma + i)$ time, where $n_i = \|\mathcal{S}_i\| = \sum_{j=1}^{i} |S_i|$.*

Corollary 2. *Let $\ell > 0$ be a length threshold. There exists a data structure of $O(n_i)$ space, which solves the length-bounded dynamic-APSP problem of computing $\mathcal{F}_i^{\geq \ell}$ and $\mathcal{B}_i^{\geq \ell}$ for a given new string S_i in $O(|S_i| \log \sigma + i)$ time, where $n_i = \|\mathcal{S}_i\| = \sum_{j=1}^{i} |S_i|$.*

We will focus on proving Theorem 4. Corollary 2 can be obtained from Theorem 4, in a similar manner to the static case in Sect. 4.

5.1 Computing $\mathcal{F}_i$

We use the DAWG data structure for the dynamic case. Consider a set $\mathcal{S} = \{S_1, \ldots, S_{i-1}\}$ of $i-1$ strings. By Observation 1, $\mathsf{Trie}(\mathcal{S}_{i-1})$ is an induced tree of $\mathsf{LPTree}(\mathcal{S}_{i-1})$ that consists only of the paths spelling out $S_1, \ldots, S_{i-1}$ from the source of $\mathsf{DAWG}(\mathcal{S}_{i-1})$. Also, if $\mathsf{flink}(u) = v$ for a node u of $\mathsf{Trie}(\mathcal{S}_{i-1})$, then there is a chain of suffix links $u, \mathsf{slink}(u), \ldots, v$ of length $k \geq 1$ from the node u to the node v in $\mathsf{DAWG}(\mathcal{S}_{i-1})$ (i.e. $\mathsf{slink}^k(u) = v$). It is known that the suffix links of $\mathsf{DAWG}(\mathcal{S}_{i-1})$ form an edge-reversed tree that is equivalent to the suffix tree [22] of the *reversed* strings [5,20]. We denote this suffix link tree by $\mathsf{SLTree}(\mathcal{S})$, and will use $\mathsf{SLTree}(\mathcal{S})$ in the dynamic case, instead of the tree $\mathsf{FLTree}(\mathcal{S})$ of failure links in the static case.

Now, we design our algorithm for the dynamic case with the DAWG structure, using the idea from Sect. 4 for the static case with the AC-automaton (see also Algorithm 1). Suppose that we have built and maintained the aforementioned data structures for the dynamic set $\mathcal{S}_{i-1} = \{S_1, \ldots, S_{i-1}\}$ of $i-1$ strings. Given a new string S_i to insert, we update $\mathsf{DAWG}(\mathcal{S}_{i-1})$ to $\mathsf{DAWG}(\mathcal{S}_i)$ in $O(|S_i| \log \sigma)$ time with Theorem 2. We have now updated $\mathsf{SLTree}(\mathcal{S}_{i-1})$ to $\mathsf{SLTree}(\mathcal{S}_i)$ as well.

To perform equivalent procedures as in Algorithm 1, we climb up the suffix link path from the node v with $\mathsf{str}(v) = S_i$ toward the root on $\mathsf{SLTree}(\mathcal{S}_i)$. Namely, we use slink instead of flink. Each time we visit a node v that is contained in $\mathsf{Trie}(\mathcal{S}_i)$, we access $u = \mathsf{des}(v)$ on $\mathsf{ComTrie}(\mathcal{S}_i)$ and perform the same procedures as in Algorithm 1. This gives us $\mathsf{lspo}(i, j)$ for all $1 \leq j \leq i$ in a total of $O(|S_i| \log \sigma + i)$ time, since $\mathsf{ComTrie}(\mathcal{S}_i)$ is of size $O(i)$. We can determine whether a node v is in $\mathsf{Trie}(\mathcal{S}_i)$ or not in $O(1)$ time after a simple $O(|S_i| \log \sigma)$-time preprocessing per new string S_i: After inserting S_i to the DAWG, we simply trace the path that spells out S_i from the source on the DAWG.

What remains is how to update $\mathsf{ComTrie}(\mathcal{S}_{i-1})$ to $\mathsf{ComTrie}(\mathcal{S}_i)$, and update the des function. We do this by naïvely inserting the new string S_i to $\mathsf{ComTrie}(\mathcal{S}_{i-1})$, traversing the tree with S_i from the root in $O(|S_i| \log \sigma)$ time. The des function needs to be updated when an edge (v, u) on $\mathsf{ComTrie}(\mathcal{S}_{i-1})$ is split into two edges (v, w) and (w, u), where w is the parent of the new leaf representing S_i on $\mathsf{ComTrie}(\mathcal{S}_i)$. Let $v = v_1, \ldots, v_x = w$ be the nodes in the uncompacted trie $\mathsf{Trie}(\mathcal{S}_i)$ that are implicit on the edge (v, w) in the compacted trie $\mathsf{ComTrie}(\mathcal{S}_i)$. We update $\mathsf{des}(v_1) \leftarrow w, \ldots, \mathsf{des}(v_x) \leftarrow w$. We set $\mathsf{des}(y) \leftarrow s$ for all new nodes y of $\mathsf{Trie}(\mathcal{S}_i)$, where s is the new leaf representing S_i. These updates can be done in $O(|S_i|)$ time.

To summarize this subsection, we have proven the following:

Lemma 1. *There is a data structure of $O(n_i)$ space, which computes $\mathcal{F}_i$ in $O(|S_i| \log \sigma + i)$ time per new string S_i inserted.*

Remark 1. Diptarama et al. [11] showed how to simulate the failure link tree of the AC-automaton with the suffix link tree of the DAWG that is augmented with a nearest marked ancestor (NMA) data structure [23]. It should be noted that NMA data structures are not required in our algorithm.

5.2 Computing $\mathcal{B}_i$

We can easily reduce the problem of computing $\mathcal{B}_i$ to computing $\mathcal{F}_i$ for the set $\mathcal{S}_i^R = \{S_1^R, \ldots, S_i^R\}$ of *reversed strings*, where the roles of suffixes and prefixes are swapped. This immediately gives us the following:

Corollary 3. *There is a data structure of $O(n_i)$ space, which computes $\mathcal{B}_i$ in $O(|S_i| \log \sigma + i)$ time per new string S_i inserted.*

6 Conclusions and Future Work

In this paper, we presented an efficient algorithm for the APSP problem for a dynamic set of strings $\mathcal{S} = \{S_1, \ldots, S_k\}$ of total length n. Our algorithm occupies $O(n)$ space and takes $O(|S_i| \log \sigma + i)$ time for each string S_i inserted, for $1 \leq i \leq k$. The running time is near-optimal except for the $\log \sigma$ factor, that is indeed required if one maintains $\mathcal{S}$ in a tree or DAG based data structure for a general ordered alphabet of size σ.

An intriguing question is whether our method can be modified to handle deletions of strings from $\mathcal{S}$. Diptarama et al. [11] showed how to update the failure links of the AC-automaton in $O(\sigma|S| + u_f)$ time when a string S is deleted from $\mathcal{S}$, where u_f is the number of failure links that need to be updated. It is interesting to see whether our DAWG-based method can be modified for the *fully* dynamic set of strings allowing for both insertions and deletions.

It is also interesting to see whether our dynamic algorithm can be extended to efficient construction/update of the HOG [6,12,19] for a dynamic set of strings.

Acknowledgments. This work was supported by JSPS KAKENHI Grant Numbers and JP20H05964, JP23K24808, JP23K18466 (SI).

References

1. Abouelhoda, M.I., Kurtz, S., Ohlebusch, E.: Replacing suffix trees with enhanced suffix arrays. J. Discrete Algorithms **2**(1), 53–86 (2004)
2. Aho, A.V., Corasick, M.J.: Efficient string matching: an aid to bibliographic search. Commun. ACM **18**, 333–340 (1975)
3. Alstrup, S., Holm, J., de Lichtenberg, K., Thorup, M.: Minimizing diameters of dynamic trees. In: Degano, P., Gorrieri, R., Marchetti-Spaccamela, A. (eds.) ICALP 1997. LNCS, vol. 1256, pp. 270–280. Springer, Heidelberg (1997). https://doi.org/10.1007/3-540-63165-8_184
4. Blumer, A., Blumer, J., Haussler, D., Ehrenfeucht, A., Chen, M.T., Seiferas, J.I.: The smallest automaton recognizing the subwords of a text. Theoret. Comput. Sci. **40**, 31–55 (1985)
5. Blumer, A., Blumer, J., Haussler, D., McConnell, R., Ehrenfeucht, A.: Complete inverted files for efficient text retrieval and analysis. J. ACM **34**(3), 578–595 (1987). https://doi.org/10.1145/28869.28873
6. Cánovas, R., Cazaux, B., Rivals, E.: The compressed overlap index. CoRR arXiv:1707.05613 (2017)

7. Dori, S., Landau, G.M.: Construction of Aho Corasick automaton in linear time for integer alphabets. Inf. Process. Lett. **98**(2), 66–72 (2006)
8. Farach-Colton, M., Ferragina, P., Muthukrishnan, S.: On the sorting-complexity of suffix tree construction. J. ACM **47**(6), 987–1011 (2000)
9. Gusfield, D.: Algorithms on Strings, Trees, and Sequences - Computer Science and Computational Biology. Cambridge University Press (1997)
10. Gusfield, D., Landau, G.M., Schieber, B.: An efficient algorithm for the all pairs suffix-prefix problem. Inf. Process. Lett. **41**(4), 181–185 (1992)
11. Hendrian, D., Inenaga, S., Yoshinaka, R., Shinohara, A.: Efficient dynamic dictionary matching with DAWGs and AC-automata. Theor. Comput. Sci. **792**, 161–172 (2019)
12. Khan, S.: Optimal construction of hierarchical overlap graphs. In: CPM 2021. LIPIcs, vol. 191, pp. 17:1–17:11 (2021)
13. Knuth, D.E., Morris, J.H., Jr., Pratt, V.R.: Fast pattern matching in strings. SIAM J. Comput. **6**(2), 323–350 (1977)
14. Lim, J., Park, K.: A fast algorithm for the all-pairs suffix-prefix problem. Theor. Comput. Sci. **698**, 14–24 (2017)
15. Loukides, G., Pissis, S.P.: All-pairs suffix/prefix in optimal time using Aho-Corasick space. Inf. Process. Lett. **178**, 106275 (2022)
16. Loukides, G., Pissis, S.P., Thankachan, S.V., Zuba, W.: Suffix-prefix queries on a dictionary. In: CPM 2023. LIPIcs, vol. 259, pp. 21:1–21:20 (2023)
17. Manber, U., Myers, E.W.: Suffix arrays: a new method for on-line string searches. SIAM J. Comput. **22**(5), 935–948 (1993)
18. Ohlebusch, E., Gog, S.: Efficient algorithms for the all-pairs suffix-prefix problem and the all-pairs substring-prefix problem. Inf. Process. Lett. **110**(3), 123–128 (2010)
19. Park, S., Park, S.G., Cazaux, B., Park, K., Rivals, E.: A linear time algorithm for constructing hierarchical overlap graphs. In: CPM 2021. LIPIcs, vol. 191, pp. 22:1–22:9 (2021)
20. Takagi, T., Inenaga, S., Arimura, H., Breslauer, D., Hendrian, D.: Fully-online suffix tree and directed acyclic word graph construction for multiple texts. Algorithmica **82**(5), 1346–1377 (2020)
21. Tustumi, W.H.A., Gog, S., Telles, G.P., Louza, F.A.: An improved algorithm for the all-pairs suffix-prefix problem. J. Discrete Algorithms **37**, 34–43 (2016)
22. Weiner, P.: Linear pattern matching algorithms. In: 14th Annual Symposium on Switching and Automata Theory, pp. 1–11 (1973)
23. Westbrook, J.R.: Fast incremental planarity testing. In: ICALP 1992. Lecture Notes in Computer Science, vol. 623, pp. 342–353 (1992)

Adaptive Dynamic Bitvectors

Gonzalo Navarro[1,2,3]

[1] Department of Computer Science, University of Chile, Santiago, Chile
gnavarro@uchile.cl
[2] CeBiB—Center for Biotechnology and Bioengineering, Santiago, Chile
[3] IMFD—Millennium Institute for Foundational Research on Data, Santiago, Chile

Abstract. While operations *rank* and *select* on static bitvectors can be supported in constant time, lower bounds show that supporting updates raises the cost per operation to $\Theta(\log n / \log \log n)$. This is a shame in scenarios where updates are possible but uncommon. We develop a representation of bitvectors that, if there are q queries per update, supports all the operations in $O(\log(n/q))$ amortized time. Our experimental results support the theoretical findings, displaying speedups of orders of magnitude compared to standard dynamic implementations.

1 Introduction

Bitvectors are the basic bricks of most compact data structures [11]. Apart from the basic query access(B, i), which retrieves the bit $B[i]$ of the bitvector $B[1..n]$, they support two fundamental queries: rank$_b(B, i)$, which tells the number of times the bit $b \in \{0, 1\}$ occurs in $B[1..i]$, and select$_b(B, j)$, which gives the position of the jth occurrence of $b \in \{0, 1\}$ in B. It is well known since the nineties that those operations can be supported in $O(1)$ time with a bitvector representation that uses $n + o(n)$ bits of space [4,10]. Things are considerably different, however, if we aim to allow updates to the bitvector: just supporting rank and bit flips requires $\Omega(\log n / \log \log n)$ time [7]. Indeed, one can incorporate in $O(\log n / \log \log n)$ time, and still $n + o(n)$ bits of space, the operations write(B, i, v), which sets $B[i] = v$, insert(B, i, v), which inserts the bit value v at position i in B, and delete(B, i), which removes the bit $B[i]$ from B [12].

This almost logarithmic gap between static and dynamic bitvectors permeates through most compact data structures that build on them, making dynamic compact data structures considerably slower than their static counterparts, and not as competitive with classic data structures. Although this price is in principle unavoidable, one may wonder whether it must be so high in cases where updates are sparse compared to queries, as is the case in many applications. As an extreme example, since the static data structures can be built in linear time, one could have $O(1)$ amortized time if queries were $\Omega(n)$ times more frequent than updates, by just rebuilding the static structure upon each update. The idea

---

Funded by ANID, Chile, via Basal Funds FB0001, Millennium Science Initiative Program – Code ICN17_002, and Fondecyt Grant 1-230755.

© The Author(s), under exclusive license to Springer Nature Switzerland AG 2025
Z. Lipták et al. (Eds.): SPIRE 2024, LNCS 14899, pp. 204–217, 2025.
https://doi.org/10.1007/978-3-031-72200-4_16

degrades quickly, however: If queries are q times more frequent than updates, this technique yields $O(n/q)$ amortized times.

In this paper we introduce a representation of dynamic bitvectors $B[1..n]$ that uses at most $4n+o(n)$ bits and offers $O(\log(n/q))$ amortized time for all the operations, if queries are q times more frequent than updates. We modify classic dynamic bitvector representations [3,9]. Our structure is a binary tree whose leaves may either be "dynamic", storing $O(\log^2 n)$ bits and supporting updates, or long "static" bitvectors handling only queries. A whole subtree is converted into static—which we call "flattening"—when it has received sufficient queries to amortize the cost of building the static structures (i.e., linear in the number of bits it represents). When an update falls in a static leaf, the leaf is recursively halved into static leaves of decreasing lengths until producing a (short) dynamic leaf where the update is executed—a process we call "splitting". For maintaining balance in the tree we resort to weight-balancing [1,13], which interacts well with our new operations of flattening and splitting.

2 Our Work in Context

Our problem is an instance of the so-called "dynamic bitvector with indels" problem, which as said requires $\Omega(\log n/ \log \log n)$ time per operation even if we support only rank and write [7]. Several solutions have matched this lower bound, or been close to. Hon et al. [8] store a dynamic bitvector $B[1..n]$ in $n+o(n)$ bits of space, handling queries in time $O(\log_b n)$ and updates in time $O(b)$, for any $b = \Omega((\log n/\log \log n)^2)$. Their main structure is a weight-balanced B-tree [5,15]. Chan et al. [3] use balanced binary trees with leaves containing $\Theta(\log n)$ bits, obtaining $O(n)$ bits of space and $O(\log n)$ time for all the operations. Mäkinen and Navarro [9] still use balanced binary trees, but use leaves of $\Theta(\log^2 n)$ bits, retaining their $O(\log n)$ times but reducing the space to $n + o(n)$ bits. Finally, Navarro and Sadakane [12] replace binary trees by structures closer to B-trees, retaining the $n + o(n)$ bits of space and supporting all the operations in the optimal time $O(\log n/\log \log n)$. In those terms the problem is considered closed.

In this paper we aim at obtaining $O(\log(n/q))$ times under a regime where there are, on average, q queries per update. Our results are amortized, as we rely on converting whole subtrees into static structures (which answer queries in constant time) when they have received sufficient queries to pay for that conversion. The conversion needs to temporarily double the space for the bits stored in the converted tree, thus we cannot aim at using $n + o(n)$ bits of space. For simplicity, we will aim at using $4n + o(n)$ bits—though we could reduce it to $(3+\epsilon)n$ for any $\epsilon > 0$. To reach such a concrete constant, we use the bigger leaves of $\Theta(\log^2 n)$ bits, but we allow them to be partly filled, which considerably simplifies matters in comparison with the solutions using $n + o(n)$ bits [9,12].

There has been work to store the bitvectors within entropy space, which means Hn bits with $H = \frac{m}{n} \log_2 \frac{n}{m} + \frac{n-m}{n} \log_2 \frac{n}{n-m}$, m being the number of 1s in the bitvector. Assuming $m < n$, Blandford and Blelloch [2] obtain $O(nH + \log n)$ bits of space while supporting all operations in $O(\log n)$ time, using a balanced binary tree where the distances between consecutive 1 s are gap-encoded in the leaves. Mäkinen and Navarro [9] improve the space to $nH + o(n)$ bits, while retaining $O(\log n)$ time for the operations. Navarro and Sadakane [12] retain this space and reduce the time to the optimal $O(\log n/\log \log n)$. We discuss in the Conclusions how our results can be extended to use entropy-bounded space.

3 Adaptive Dynamic Bitvectors

We use the transdichotomous RAM model of computation, with computer words of $w = \Theta(\log N)$ bits, N being the maximum size of a bitvector that fits in memory. Pointers use w bits. We call $n \leq N$ is the current size of bitvector B.

3.1 Structure

As anticipated, our data structure is a binary tree. Its leaves are of two types:

- A "dynamic leaf", which allocates space for $b = \Theta(w \log n)$ bits and no pre-computed answers. A dynamic leaf answers access queries in $O(1)$ time and rank/select queries in $O(\log n)$ time, via word-wise scanning [9,11].[1]
- A "static leaf", which stores arbitrarily large bitvectors with their corresponding precomputation to solve access/rank/select in $O(1)$ time [4,10].

The internal tree nodes v record, apart from their two children v.left and v.right, the following fields:

v.size : total number of bits represented in the subtree rooted at v.
v.ones : total number of 1-bits represented in the subtree rooted at v.
v.leaves : number of leaves below v (static leaves count as many, see later).
v.queries : number of queries that traversed v since the last update that traversed v, or since the creation of v.

Our binary tree is maintained with balanced weight [1,13]: given a parameter $1/2 < \alpha < 1$, for every node v, v.left.size $\leq \alpha \cdot v$.size and v.right.size $\leq \alpha \cdot v$.size. This implies that the tree has height at most $\log_{\frac{1}{1-\alpha}} n = O(\log n)$. Balancing will be maintained by reconstructing biased trees as perfectly balanced [1]. Similarly, we will ensure that v.leaves $= O(v.\text{size}/(w \log n))$, by converting nodes into static when the leaves below them are too empty. Since each tree node uses $O(w)$ bits of space, this ensures that all the tree nodes use together $O(n/\log n)$ bits.

---

[1] Mäkinen and Navarro [9] show how to maintain b when $\lceil \log n \rceil$ changes without affecting any complexity. Our implementation uses a fixed value of b for simplicity.

3.2 Queries

The queries use in principle the standard mechanism for dynamic bitvectors [9]: access(B, i) traverses the tree from the root, going to the left child if $i \le$ v.size and to the right otherwise (subtracting v.size from i in that case). When arriving at a leaf, the query completes in $O(1)$ additional time. Since our trees are balanced, access takes time $O(\log n)$ in the worst case.

The procedure for $\mathsf{rank}_1(B, i)$ is analogous, adding up v.left.ones whenever we descend to v.right, and adding at the end $\mathsf{rank}_1(i)$ on the leaf. For $\mathsf{rank}_0(B, i)$ we compute $i - \mathsf{rank}_1(B, i)$. For $\mathsf{select}_1(B, j)$, we descend guided by v.ones instead of by v.size, subtract v.left.ones from j when going to v.right, and accumulate v.left.size instead of v.left.ones; at the end we add $\mathsf{select}_1(j)$ on the leaf. Finally, $\mathsf{select}_0(B, j)$ is analogous to $\mathsf{select}_1(B, j)$, using v.left.size $-$ v.left.ones instead of v.left.ones. Both rank and select take time $O(\log n)$ plus the time spent on the leaf, which is $O(\log n)$ on dynamic leaves—by counting the 1 s in w-bit words in constant time [11, Sect. 4.2]—and $O(1)$ on static ones.

Flattening. The novelty in our adaptive scheme is that, every time we traverse an internal node v for any of the three queries, we increment v.queries, and if v.queries $\ge \theta \cdot v$.size, for an appropriate tuning constant θ, we convert the whole subtree of v into a static leaf, which we call "flattening" v. Flattening is done in time $O(v.\text{size})$, by traversing and deleting the subtree of v, while writing the bits of all the leaves onto a new bitvector, which is finally preprocessed for constant-time queries and converted into the static leaf corresponding to v. We show later, however, that its amortized cost is absorbed by the preceding $\theta \cdot v$.size queries.

Note that flattening temporarily increases the space used by v.size, which may be as much as n if v is the root. Note also that flattening does not change v.size, and thus it does not affect the balancing.

3.3 Updates

Updates are handled, in principle, as in previous work [9]. To perform write(B, i), we traverse the tree as for access, modify the corresponding bit in the (dynamic) leaf we arrive at (we consider soon the case where we arrive at a static leaf), and update v.ones as we return from the recursion. This takes time $O(\log n)$ because our trees are balanced. Note that write has no effect on the tree balancing.

Insertions and deletions are analogous, yet at the end they insert or delete a bit in a dynamic leaf and must update v.size and v.ones along the path. Apart from costing $O(\log n)$ for shifting the needed bits wordwise, one must handle overflows and underflows in leaves. An overflow occurs when we insert a bit in a leaf v containing b bits already, and it is handled by splitting the leaf into two holding $b/2$ bits and making v the parent of both new leaves. An underflow occurs when we delete the only bit in a leaf v, in which case v is eliminated together with its parent node. We also act when the leaf v that receives the deletion is the sibling of another (dynamic) leaf, so that after the deletion both leaves add up to at most γb bits, where $1/2 \le \gamma < 1$ is the desired fill ratio

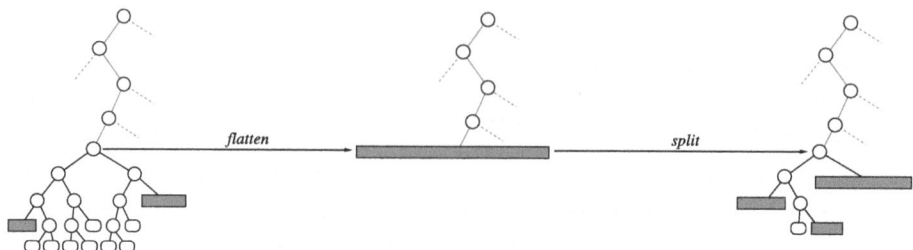

Fig. 1. Flattening and posterior splitting of a node, the former after receiving over $\theta \cdot v.\text{size}$ consecutive queries, and the latter when receiving an update at a static leaf; the leaf is recursively halved until the update falls in a dynamic leaf. Circles are internal nodes, round rectangles are dynamic leaves, and gray rectangles are static leaves. (Color figure online)

of newly created leaves. In this case both leaves are concatenated into one and their parent is eliminated. As we return from the recursion, we keep trying to merge the two children of the current node if both are dynamic leaves. As a consequence, every internal node v has $v.\text{size} \geq \gamma b$ (and can be flattened).

Splitting. The interesting part of updates occurs when we arrive at a static leaf v. In this case we halve the bitvector stored at v, and make v an internal node with one child holding each half (halving is slightly corrected to make the left half contain a multiple of γb bits, so that dynamic leaves can be created of size γb). The bitvector half that does not contain i is converted into a leaf again (static if it is of size over γb and dynamic if not). The half containing i is recursively halved, until it contains at most γb bits and so it becomes a dynamic leaf. Note that splitting v does not change $v.\text{size}$, and thus has no effect on the tree balance.

Splitting takes total time $O(v.\text{size})$, as we create leaves in time proportional to their bit length. Though the worst-case cost of updates can be $\Theta(n)$, we prove logarithmic amortized bounds later. Figure 1 illustrates flattening and splitting.

Balancing. We maintain the weight balance of the tree by checking, at every internal node v in the path from the root to the leaf where the insertion or deletion will take place, that the weight-balancing conditions will be satisfied once we insert or delete the bit on $v.\text{left}$ or $v.\text{right}$, depending on the case. For example, if the insertion must continue by the left, and it holds that $v.\text{left.size} + 1 > \alpha \cdot (v.\text{size} + 1)$, then node v will become unbalanced after the insertion.

When we detect that v will become unbalanced in our top-down traversal to insert or delete at position i, we (i) flatten v, and (ii) split v at position i. This will correct the imbalance of v, ensuring that all the nodes in the path towards position i are perfectly balanced until the final dynamic leaf. The cost is $O(v.\text{size})$, of the same order of the tree balancing performed on weight balanced trees, in the variant that rebuilds whole subtrees within low amortized time [1].

Because our splitting by half can be slightly shifted to ensure that lengths are multiples of γb, we avoid balancing when this shifting would leave the tree unbalanced anyway. This can be the case when the subtree is very small (e.g., if $v.\text{size} = 3\gamma b$, splitting will produce a child of size $2\gamma b$ and another of size γb, which is considered unbalanced if $\gamma > 1/3$), or if γ is too close to $1/2$. This correction does not affect the asymptotically logarithmic height of the tree.

As anticipated, despite our attempts to maintain leaves as full as possible, a constant fill ratio cannot be guaranteed. We then monitor, just like we do for bias in the left versus right subtree sizes, the ratio between $v.\text{leaves}$ and $v.\text{size}$. This cannot be predicted in the top-down traversal as the bias, so we check instead, when returning from the recursion, that $v.\text{size} \geq (b/3) \cdot v.\text{leaves}$, and otherwise flatten v (there is no point in eagerly splitting v after flattening, as the deletion already took place). Note that leaves are created with fill ratio $1/2$ when a leaf overflows, and with fill ratio $\gamma \geq 1/2$ when a static leaf is split; we rebuild when the fill ratio falls below $1/3$, so this flattening cost will be amortized by the deletions. We assume $v.\text{leaves} = v.\text{size}/(\gamma b)$ for static leaves v, as γb is the fill ratio of the leaves it will create if split (thus splitting v does not alter $v.\text{leaves}$).

4 Amortized Analysis

We first show that all the operations have an amortized cost of $O(\log n)$; later we analyze the case where the frequency of updates is $1/q$. Because our trees are balanced, the actual cost of all the operations is always $O(\log n)$, except for flattening and splitting, and their use for balancing.

Our amortized analysis will define three potential functions, ϕ_v, ψ_v, and β_v, for every node v of the tree, and the global potential will be $\Phi = \sum_v (\phi_v + \psi_v + \beta_v)$. We then have that $\Phi = 0$ when the tree is empty. The potentials ϕ_v, ψ_v, and β_v will accumulate work to later pay for the operations of flattening, splitting, and balancing, respectively. They are defined operationally, as follows:

1. Dynamic leaves and internal nodes v are always created with $\phi_v = \beta_v = 0$, but those potentials can increase later. Instead, they always have $\psi_v = 0$.
2. Static leaves v created via flattening start with $\psi_v = 0$ and with $\phi_v = \sum_{u \sqsubseteq v} \phi_u$, where $u \sqsubseteq v$ means that u is v or descends from v. Some of this ϕ_v is later transferred to the potentials ψ_u of static leaves u created upon splitting, as seen later. Note that transfers of potentials do not alter the sum Φ. Static leaves may inherit β_v from former descendants, as seen later.
3. Query operations increment ϕ_v on every visited internal node and dynamic leaf v. If they arrive at a static leaf v, they increase ϕ_v by $\log_{1/\alpha}(v.\text{size}/b)$.
4. Update operations increase β_v by a constant τ, to be defined later, on every visited internal node and dynamic leaf v. They can also trigger splittings, which are analyzed separately.

Consider query operations. Their actual cost is $O(\log n)$. In addition, they increase Φ by $O(\log n)$, in part by increasing ϕ_v by $O(1)$ on the internal nodes and dynamic leaves v they traverse, and in part by increasing ϕ_v on the static leaves they reach by $O(\log n)$. Update operations cost $O(\log n)$ and also increase Φ by $O(\log n)$. Their amortized cost is then also $O(\log n)$. We analyze flattening, splitting, and balancing as separate operations.

Flattening. Now consider the flattening operation. Recall that v is flattened whenever $v.\text{queries} \geq \theta \cdot v.\text{size}$, and that the actual cost of flattening is $v.\text{size}$. Since the updates that traverse v reset $v.\text{queries}$ to zero, it follows that the last $v.\text{queries}$ operations that traversed v have been queries. Each of those incremented the potential ϕ_v, and also increased the potentials ϕ_u of descendants u of v by at least $\log_{1/\alpha}(v.\text{size}/b)$ in total:

- If the query ended in a dynamic leaf, $\log_{1/\alpha}(v.\text{size}/b)$ is the minimum possible distance from v to a dynamic leaf (i.e., when the path to the leaf is as short as possible in the α-balanced tree and that the leaves are all full).
- In case the query arrived at a static leaf u, it increased ϕ_u by $\log_{1/\alpha}(u.\text{size}/b)$. Since the distance between v and u is at least $\log_{1/\alpha}(v.\text{size}/u.\text{size})$, the query also incremented ϕ at that many internal nodes between v and u, and summed with the increase it produced in ϕ_u adds up to $\log_{1/\alpha}(v.\text{size}/b)$.

The potential ϕ_u of the descendants u of v is then $\geq \theta \cdot v.\text{size} \cdot \log_{1/\alpha}(v.\text{size}/b)$. When flattening v, its descendants u disappear and we add their ϕ_u to ϕ_v, which then becomes $\phi_v \geq \theta \cdot v.\text{size} \cdot (1 + \log_{1/\alpha}(v.\text{size}/b))$. Assume $\theta \geq 2$, so $\phi_v \geq v.\text{size} \cdot (2 + \theta \log_{1/\alpha}(v.\text{size}/b))$. From ϕ_v we spend $v.\text{size}$ to pay for the flattening and transfer the rest to ψ_v (which does not alter Φ). Flattening has then zero amortized cost and, after it, it holds $\phi_v = 0$ and $\psi_v \geq v.\text{size} \cdot (1 + \theta \log_{1/\alpha}(v.\text{size}/b))$. The potential ψ_v will be used to pay for future splittings, and will be lost if, instead, some ancestor of v is flattened and makes v disappear.

Splitting. Now assume a static leaf v is split by an update operation. Assuming for simplicity that $v.\text{size}$ is of the form $2^k \cdot \gamma b$ (otherwise only constants change), splitting creates a sequence of static leaves $u_1, u_2, \ldots$ of lengths $u_1.\text{size} = v.\text{size}/2$, $u_2.\text{size} = v.\text{size}/4, \ldots$, until $u_k.\text{size} = u_{k+1}.\text{size} = \gamma b$, where u_1 to u_{k-1} are static and the last two are dynamic. Let the actual cost of splitting be $v.\text{size}$. This cost will be paid from ψ_v, so that splitting has zero amortized cost. The remaining potential in ψ_v, $v.\text{size} \cdot \theta \log_{1/\alpha}(v.\text{size}/b)$, will be transferred to the potentials ψ_{u_i} of the static leaves just created. Concretely, the static leaves u_i are created with $\phi_{u_i} = 0$ and $\psi_{u_i} = u_i.\text{size} \cdot (1 + \theta \log_{1/\alpha}(u_i.\text{size}/b))$, which is what those static leaves need to face their own possible future splittings. There remains enough potential in ψ_v to feed all the new potentials ψ_{u_i} because

$$\sum_{i=1}^{k-1} \psi_{u_i} = \sum_{i=1}^{k-1} u_i.\text{size} \cdot (1 + \theta \log_{1/\alpha}(u_i.\text{size}/b))$$

$$= \sum_{i=1}^{k-1} \frac{v.\text{size}}{2^i} \cdot \left(1 + \theta \log_{1/\alpha}\left(\frac{v.\text{size}/b}{2^i}\right)\right)$$

$$= v.\text{size} \cdot \sum_{i=1}^{k-1} \frac{1}{2^i} \left(1 + \theta(\log_{1/\alpha}(v.\text{size}/b) - i \log_{1/\alpha} 2)\right)$$

$$< v.\text{size} \cdot \theta \log_{1/\alpha}(v.\text{size}/b) + v.\text{size} \cdot \left(1 - \frac{1}{2^{k-1}} - \theta \left(2 - \frac{k+1}{2^{k-1}}\right) \log_{1/\alpha} 2\right).$$

The first term is what we have available in ψ_v to distribute across the potentials ψ_{u_i}. It suffices that $\theta \geq \log_2(1/\alpha)$ for the second term to be nonpositive for all $k \geq 1$. As we have assumed $\theta \geq 2$ and $\log_2(1/\alpha) \geq 1$, we define $\theta = 1 + \log_2(1/\alpha)$.

Balancing. Balancing on v invokes flattening plus splitting when either $v.\text{left.size} > \alpha \cdot v.\text{size}$ or $v.\text{right.size} > \alpha \cdot v.\text{size}$. Note that, when creating a dynamic node v by splitting an overflowing leaf, child sizes differ by 1, and when creating v by splitting a static node, they differ by at most γb. We avoid balancing nodes with $v.\text{size} < 5\gamma b$, so as to ensure that $\max(v.\text{left.size}, v.\text{right.size})/v.\text{size} \leq 3/5$ and thus nodes created by splitting do not immediately need balancing (this adds just $O(1)$ to the maximum tree height). We can then use any $3/5 < \alpha < 1$.

It follows that, once a (balanceable) dynamic node v is created, it must undergo i insertions on the larger child (whose initial size can be up to $3/5 \cdot v.\text{size}$) to become unbalanced, because $3/5 \cdot v.\text{size} + i > \alpha(v.\text{size} + i)$. Thus, more than $\frac{\alpha - 3/5}{1 - \alpha} \cdot v.\text{size}$ insertions must occur in v before balancing takes place. Deletions on the smaller child pose a more stringent condition, as it suffices that d deletions occur, with $3/5 \cdot v.\text{size} > \alpha \cdot (v.\text{size} - d)$, that is, $d > \frac{\alpha - 3/5}{\alpha} \cdot v.\text{size}$.

We then set $\tau = \theta / \frac{\alpha - 3/5}{\alpha}$ as the constant by which updates increase the potential β_v of the traversed nodes v. Because updates always reach leaves, an update on v increases the potentials β on v and its descendants by at least $\tau(1 + \log_{1/\alpha}(v.\text{size}/b))$. If a dynamic node v must be balanced, at least $\frac{\alpha - 3/5}{\alpha} \cdot v.\text{size}$ updates have traversed it, and thus they have increased the potentials β_u, for all $u \sqsubseteq v$, by at least $\frac{\alpha - 3/5}{\alpha} \cdot v.\text{size}$ times $\tau(1 + \log_{1/\alpha}(v.\text{size}/b))$, which is $v.\text{size} \cdot \theta(1 + \log_{1/\alpha}(v.\text{size}/b))$. This is exactly the potential needed for flattening (and later splitting) v at zero amortized cost.

The potentials β_v of nodes are maintained upon creation and destruction: merged leaves add up their potentials β, leaves that are split also split their potential β (does not matter how), static leaves created by flattening inherit all the potentials β_u of their destroyed descendants u, and when splitting v we may leave β_v in one of the (one or two) dynamic leaves that are created. The goal is that ancestors of v that may be later balanced preserve the needed potentials β below them. Static leaves v do not need to store β_v for themselves, because when split they will be created as (at least) 2/5–3/5 balanced trees.

Flattening to Maintain Fill Ratios. We also flatten v when $v.\text{size} < (b/3) \cdot v.\text{leaves}$. Note that leaves are created by overflowing, upon an insertion, with fill ratio at least $1/2$; only deletions can drive the fill ratio below that fraction. Potential flattening in a subtree of v only improves the average fill ratio, to $\gamma \geq 1/2$ in that subtree. Therefore, a node v must undergo at least $v.\text{size}/6$ deletions before it must be flattened to maintain fill ratios. Just as for balancing, it suffices to assign deletions (i.e., to τ) an additional amortized cost of $\theta/(1/6) = 6\theta$ they pay for the future flattenings (plus possible later splittings) they may trigger.

Since this ensures, in particular, that there are at most $3n/b$ leaves, each using b bits of space, the total space allocated in the leaves is at most $3n$ bits. We may need n additional bits of temporary space when flattening, which sets the maximum usage to $4n$ bits, plus the $o(n)$ bits needed by the tree nodes and static rank/select data structures. Splitting also requires temporary space, but this is less stringent because the v.size bits are already packed in a static array.

Theorem 1. *An adaptive dynamic bitvector starting empty can be maintained within $4n+o(n)$ bits of space, where n is the current number of bits it represents, so that any operation on it has $O(\log n)$ amortized cost.*

We can reduce the space to $(3+\epsilon)n$ for any constant $\epsilon > 0$, by flattening when $v.\text{size} < (b/(2+\epsilon/2)) \cdot v.\text{leaves}$ and not allocating the b bits for the leaves. Instead, we maintain space for only $(1+\epsilon/2) \cdot v.\text{size}$ bits, and reallocate as necessary.

4.1 Adaptive Analysis

We now show that, if only a fraction $1/q$ of the operations are updates, then the amortized cost per operation is $O(\log(n/q))$. This is clearly true for the updates: though each one costs $O(\log n)$, they are only a fraction $1/q$ of the total, thus their contribution to the global amortized cost is $O(\log(n)/q) \subseteq O(\log(n/q))$.

For the queries, the intuition is that nodes v with $v.\text{size} = \Theta(q)$ are in general static leaves, so the query traverses $O(\log n - \log q)$ nodes to finish. To show that this is the case, we start with a particular regime: consider a sequence that starts on a static leaf of length $n = 2^k \cdot \gamma b$, with one write$(B, 1)$ and then $q-1$ access$(B, 1)$ queries, for q to be determined soon. After the split, the first $\theta \cdot 2\gamma b$ accesses will cost $\log_2(n/(\gamma b))$, at which point the parent of the leaf (of $2\gamma b$ bits) will be flattened; the accesses will cost $\log_2(n/(\gamma b)) - 1$ from now. The grandparent needs other $\theta \cdot 2\gamma b$ accesses to reach $\theta \cdot 4\gamma b$ and be flattened in turn; now the access costs will be $\log_2(n/(\gamma b)) - 2$ for other $\theta \cdot 4\gamma b$ accesses, at which point the grandgrandparent reaches $\theta \cdot 8\gamma b$ accesses and is flattened, and so on. After ℓ rounds, the total number of queries is $(1 + \sum_{i=0}^{\ell-1} 2^i) \cdot \theta \gamma b = q-1$, so let us set $q = 1 + 2^\ell \theta \gamma b$, or $\ell = \log_2((q-1)/(\theta \gamma b))$. The total cost of accesses is

$$\theta \gamma b \cdot \log_2(n/(\theta\gamma)) + \sum_{i=0}^{\ell-1} 2^i \cdot \theta \gamma b \cdot (\log_2(n/(\theta b)) - i)$$
$$= \theta \gamma b \cdot \left(2^\ell \log_2(n/(\gamma b)) - 2^\ell(\ell-2) - 2\right) \leq (q-1)\left(\log_2(n/(q-1)) + 2\right).$$

That is, the cost per query is $O(\log(n/q))$. The cost of the $\ell - 1$ flattenings adds up to $O(2^\ell) = O(q/(\theta\gamma b)) = o(q)$ in total, thus adding $o(1)$ amortized time. Note that we have assumed that $q \geq \theta \gamma b$ for this analysis to hold.

Interestingly, the general case cannot be worse than this particular case. In general, for any q, each update can, via splitting, create a path of new nodes up to depth $\log_2(n/(\gamma b))$. Repeatedly accessing the deepest node in the path can produce a cost over $\log_2(n/q)$ only to the next $q-1$ queries, after which the path is flattened at height $\ell = \log_2((q-1)/(\theta\gamma b))$. Per our analysis of the particular case, the amortized cost incurred by each of those queries is still $O(\log(n/q))$.

After this flattening, the extra cost induced by the update is canceled, as further queries will traverse $O(\log(n/q))$ nodes. Deviations from this regime only decrease costs: (i) the updates are most effective in increasing the cost if they open other paths (starting on nodes of size $\Theta(n)$, if possible), disjoint from the current one, so we can assume that no other updates fall in the path during the next $q-1$ queries; (ii) the worst case is that we access the deepest node in the path, because accessing higher leaves does not postpone their flattening.

We still need to consider the costs of flattening, splitting, and balancing, which required amortized analysis. Charging 1 on ϕ_v and τ on β_v for the traversed nodes v does not change the actual cost $O(\log(n/q))$, and as we have seen, suffices to pay for flattening and balancing, but not for splittings. For those, we had charged $\log_{1/\alpha}(v.\text{size}/b) = O(\log n)$ cost to the queries (to transfer them to ψ_v when needed), independently of the depth of the leaf they reach. This now exceeds our budget, so we analyze splittings in another way, without using ψ_v.

Consider the following model. There is a bag of static leaves that evolves over time, and the tree always has n nodes for simplicity. At any moment, with $\theta \ell$ queries we can create a new static leaf of length ℓ. Along m queries, we have created in total r static leaves of lengths $\ell_1, \ell_2, \ldots, \ell_r$ so that $\sum_{i=1}^{r} \ell_i = m/\theta$. Over each static leaf i, we have applied u_i updates, each producing splits in the most costly way, so that we applied $\sum_{i=1}^{r} u_i = m/q$ updates in total.

To measure how much can $u > 0$ updates over a static leaf v may cost, let $v.\text{size} = \ell = 2^k \cdot \gamma b$ for $k \geq 1$ and $u = 2^d$ for $d \geq 0$; the general case has the same order. The first update costs ℓ, and creates static leaves of lengths $\ell/2, \ell/4$, etc. The second chooses the leaf of size $\ell/2$, costing $\ell/2$ and creating leaves of size $\ell/4, \ell/8$, etc. Now there are two leaves of size $\ell/4$, which are the next ones chosen, and so on. Let $L(d)$ be the number of leaves of length $\ell/2^d$ created in the process. Because leaves of size $\ell/2^d$ are created by leaves of size $\ell, \ell/2, \ldots, \ell/2^{d-1}$, the recurrence is $L(d) = \sum_{j=0}^{d-1} L(j)$, which solves to $L(0) = 1$ and $L(d) = 2^{d-1}$ for $d > 0$. If we apply updates to all the leaves up to size $\ell/2^d$, we will perform $\sum_{j=0}^{d} L(j) = 2^d$ updates. The cost incurred by those $u = 2^d$ updates is $\sum_{j=0}^{d} (\ell/2^j) \cdot L(j) = \ell(1 + d/2) = \ell(1 + \frac{1}{2}\log_2 u)$.

The maximum possible cost we can produce is then $\sum_{i=1}^{r} \ell_i(1 + \frac{1}{2}\log_2 u_i)$, where $\sum_{i=1}^{r} \ell_i = m/\theta$ and $\sum_{i=1}^{r} u_i = m/q$ (if $u_i = 0$, we assume $1 + \frac{1}{2}\log_2 u_i = 0$). The maximum is obtained when we create $r = m/(\theta n)$ leaves of maximum size $\ell_i = n$ and distribute the updates uniformly on them, $u_i = (m/q)/r$, to split the longest possible leaves. This yields total cost $(m/\theta)(1 + \frac{1}{2}\log_2(\theta n/q)) = O(m\log(n/q))$, or $O(\log(n/q))$ extra amortized cost per operation. If $m < \theta n$, we create only $r = 1$ leaf of length $\ell_1 = m/\theta$ and apply the $u_1 = m/q$ updates on it, yielding maximum cost $O(m\log(m/q)) \subseteq O(m\log(n/q))$. The other border case is $m/q < r$, where the best is to apply $u_i = 1$ updates to m/q leaves of maximum length n, which yields total cost $O((m/q)\log n) \subseteq O(m\log(n/q))$.

Theorem 2. *An adaptive dynamic bitvector starting empty can be maintained in $4n + o(n)$ bits, where n is the current number of bits it represents, so that* if the fraction of updates over total operations so far is $1/q$, then the bitvector operations have $O(\log(n/q))$ amortized cost.

5 Implementation and Experiments

We made a proof-of-concept implementation in C, to illustrate the gain in performance that our data structure obtains as the frequency of updates decreases.

5.1 Implementation

Our machine word has $w = 64$ bits. We strive for a lightweight data structure to answer rank/select, as the time to build it impacts in our overall times as well. We implement static rank with blocks and superblocks [4,10] of $4w$ and 2^{16} bits, respectively, storing block counters in 16 bits and superblock counters in 64 bits. The space overhead over the bit data is 6.35%, solving rank with 2 accesses to counters plus at most 4 consecutive access to the bitvector array (plus the popcount operations). Operation select performs interpolation search in the superblocks, then on the blocks, and ends with a linear scan in the 4 words.

Dynamic leaves allocate 32 w-bit words for the bit data (i.e., 2048 bits) and implement all the operations sequentially and word-wise, using bit-parallelism. All the operations are then handled by scanning 32 words, each in constant time.

We use $\alpha = 0.65$ as the balance factor, and $\gamma = 3/4$ as the initial fill ratio of leaves after splitting. We try to avoid overflows by transferring bits to a sibling leaf when possible, leaving both leaves with the same fill ratio, yet we avoid it if the transfer would be less than $1/8$ of the leaf space (this avoids a cascade of small transfers as the leaf size approaches the maximum).

A difficult parameterization has been the value of θ, which rules the frequency of flattening. A too large value retains the tree in dynamic form for too long, thereby inducing higher query times. A too low value builds the static subtree too eagerly, just to be soon split again by updates. No single constant worked well for all update frequencies, especially for large n. We opted for finding reasonable values by hand, with the aim of showing the best that can be achieved with proper parameterization. All the rest is implemented as described in the paper.

As a sanity check, we compared with a recent highly optimized dynamic bitvector implementation we call DPR after its authors [6], with $O(\log n)$ time complexity for all the operations, independently of q. The well-known DYNAMIC library [14] was considerably slower, so we omit it in the comparisons.

5.2 Experiments

Our machine is a 64-bit 12th Gen Intel Core i7-1260P at 4.7 GHz, with 16 CPUs and 16GB RAM, running Ubuntu 22.04.4 LTS. We compiled with gcc -O3.

We generated random bitvectors of sizes $n = 2^{20}$ to $n = 2^{28}$, built from those a flattened leaf, and carried out $m = n$ operations on them. The operations are insertions of random bits at random positions, $m/(2q)$ times, deletions at

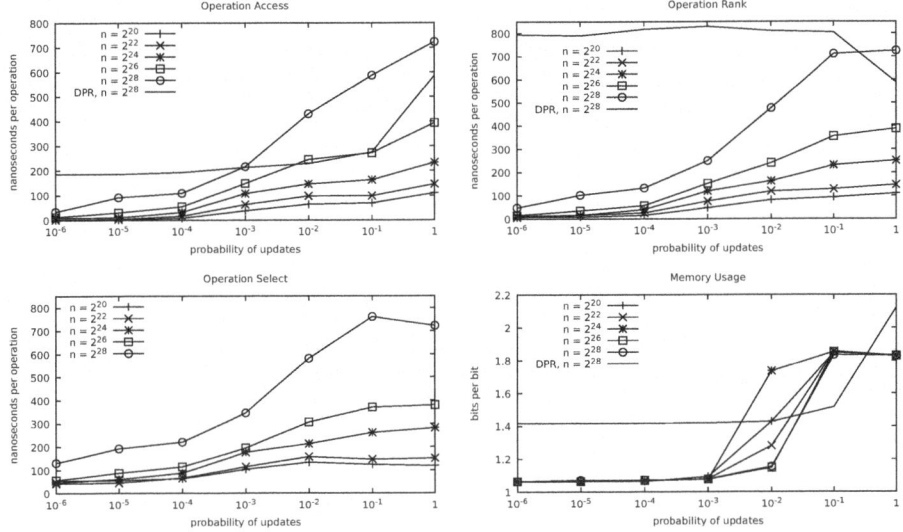

Fig. 2. Average time per operation when mixing queries access, rank, or select, with increasing proportions of updates (insert and delete), on various bitvector sizes. On the bottom right, memory usage of our data structure in bits per bit of the bitvector. DPR did not compile for select in our machine, but we expect times similar to those of rank.

random positions, $m/(2q)$ times, and queries in the other cases, running separate experiments with queries access, rank, and select (we did not mix different queries, as we do not expect insights from that). Each experiment was repeated 10 times (100 for $n = 2^{20}$ to gain more precision) and we show average user time.

In a static bitmap, access takes 1, rank 13, and select 42 nanoseconds for $n = 2^{20}$. For $n = 2^{28}$, the times are 14, 27, and 108 nanoseconds for access, rank, and select, respectively, which shows how caching affects even the constant-time algorithms. When all the operations we perform are updates, the times go from over 100 nanoseconds with $n = 2^{20}$ to over 700 nanoseconds for $n = 2^{28}$, one or two orders of magnitude slower than the basic operations. We obtained similar times when using previous results as input to the next operations to forbid parallel execution of consecutive queries.

Figure 2 shows the results for increasing values of $1/q$ (the update probability) between 10^{-6} and 1. Times are roughly linear in $\log(1/q)$ (note the logscale in $1/q$), as one would expect from our time complexity $O(\log(n/q)) = O(\log n + \log(1/q))$. DPR is about 15% faster than our implementation when $q = 1$ (all updates) and slower for rank with smaller $1/q$—by a large margin if $1/q \leq 10^{-2}$. For access, DPR is clearly faster if $1/q \geq 10^{-2}$, and clearly slower if $1/q \leq 10^{-4}$.

The figure also shows the use of memory of our data structure, in bits per bit of the bitvector. Note that in practice we always use less than $2n$ bits of space (plus, temporarily, at most n when flattening or splitting). It is clear that, until $1/q = 10^{-3}$, the space overhead is almost the same as for a single static bitvector, meaning that our structure is formed by just a few very long and shallow leaves. Since the leaves contain around 1,000–2,000 bits, this is the last value of $1/q$ for which it is still likely to reach static leaves. Note that a static leaf, even if short and deep, saves considerable time in the last part of the query, completing it in constant time where a dynamic leaf must be scanned. This explains a transition in the slope of times around $1/q = 10^{-3}$, and a saturation point at $1/q = 10^{-1}$, as by then most leaves are likely dynamic and their depth is near $\log n$.

A way to offer better times with higher values of $1/q$ would be to use smaller leaves, at the price of a higher space usage.

6 Conclusions

We have shown how to store a dynamic bitvector $B[1..n]$ within $4n + o(n)$ bits of space so that updates and queries can be solved in $O(\log(n/q))$ amortized time if queries are q times more frequent than updates. Our experiments are in line with our analysis and exhibit speedups of an order of magnitude—and more—compared to classic dynamic data structures, for $q \geq 10^2..10^4$.

Immediate applications of this result are implementations of dynamic sequences by means of wavelet trees or matrices, and tree topologies using parentheses or bits [3,9,12], among many others [11, Ch. 12]. Our results can also be extended to maintain dynamic arrays with cells of fixed width (supporting access, write, insert and delete), and sequences of variable-length elements. With the same mechanisms used in previous work [2,9,12], this yields a representation of bitvector $B[1..n]$, with entropy H, in $4nH + o(nH)$ bits while supporting the same operations in time $O(\log(n/q))$. In turn, such a representation implements searchable dynamic partial sums with indels [8].

References

1. Andersson, A.: Maintaining α-balanced trees by partial rebuilding. Int. J. Comput. Math. **38**(1–2), 37–48 (1991)
2. Blandford, D., Blelloch, G.: Compact representations of ordered sets. In: Proceedings of the 15th Annual ACM-SIAM Symposium on Discrete Algorithms (SODA), pp. 11–19 (2004)
3. Chan, H.L., Hon, W.K., Lam, T.W.: Compressed index for a dynamic collection of texts. In: Proceedings of the 15th Annual Symposium on Combinatorial Pattern Matching (CPM), pp. 445–456 (2004)
4. Clark, D.R.: Compact PAT trees. Ph. D. thesis, University of Waterloo, Canada (1996)
5. Dietz, P.: Optimal algorithms for list indexing and subset rank. In: Proceedings of the Workshop on Algorithms and Data Structures (WADS), pp. 39–46 (1989)
6. Dönges, S., Puglisi, S., Raman, R.: On dynamic bitvector implementations. In: Proceedings of the Data Compression Conference (DCC), pp. 252–261 (2022)
7. Fredman, M., Saks, M.: The cell probe complexity of dynamic data structures. In: Proceedings of the 21st Annual ACM Symposium on Theory of Computing (STOC), pp. 345–354 (1989)

8. Hon, W.K., Sadakane, K., Sung, W.K.: Succinct data structures for searchable partial sums. In: Proceedings of the 14th International Symposium on Algorithms and Computation (ISAAC), pp. 505–516 (2003)
9. Mäkinen, V., Navarro, G.: Dynamic entropy-compressed sequences and full-text indexes. ACM Trans. Algorithms **4**(3), 1–38 (2008)
10. Munro, J.I.: Tables. In: Proceedings of the 16th Conference on Foundations of Software Technology and Theoretical Computer Science (FSTTCS), pp. 37–42. LNCS 1180 (1996)
11. Navarro, G.: Compact Data Structures – A Practical Approach. Cambridge University Press (2016)
12. Navarro, G., Sadakane, K.: Fully-functional static and dynamic succinct trees. ACM Trans. Algorithms **10**(3), 1–39 (2014)
13. Nievergelt, J., Reingold, E.M.: Binary search trees of bounded balance. SIAM J. Comput. **2**(1), 33–43 (1973)
14. Prezza, N.: A framework of dynamic data structures for string processing. In: Proceedings of the 16th International Symposium on Experimental Algorithms (SEA), pp. 11:1–11:15 (2017)
15. Raman, R., Raman, V., Rao, S.S.: Succinct dynamic data structures. In: Proceedings of the 3rd International Symposium on Algorithms and Data Structures (WADS), pp. 426–437 (2001)

Compressed Graph Representations for Evaluating Regular Path Queries

Gonzalo Navarro and Josefa Robert(✉)

Millennium Institute for Foundational Research on Data (IMFD), Department of Computer Science, University of Chile, Santiago, Chile
josefarobert@proton.me

Abstract. Regular Path Queries (RPQs) are at the core of graph database query languages like SPARQL. They consist, essentially, of regular expressions that must match the sequence of edge labels of paths in the database graph. A way to answer them is to traverse the graph and the automaton of the RPQ in synchronization, reporting the graph nodes where the automaton reaches final states. We implement this approach on top of a compact graph representation that is particularly well suited for this task. The result is an index using considerably less space and/or query time than all existing approaches.

1 Introduction and Related Work

Graph databases offer a powerful way to model relationships within data, making them particularly useful in fields like social networking, bio-informatics, linguistics, and recommendation systems. This article focuses on directed labeled graph databases, where relationships have both direction and labels, and regular path queries (RPQs) over them, which search for paths of arbitrary length whose sequence of labels matches a given regular expression.

Consider, for example, the toy RDF [22] database from Fig. 1, which depicts a network of academics using relations of the form $x \xrightarrow{\text{expr}} y$, indicating that x is related to y by the relation expr. Then, a query like (Grace, coauthorOf+, x?) asks for direct and transitive collaborators of Grace ($x =$ Dan and $x =$ Eve).

RPQs are especially useful in scenarios where either the path's structure or length is not known in advance, or for studying the graph's topology, and represent a classic challenge in the field. RPQs can be extended to traverse edges in both directions, yielding what is known as two-way RPQs or 2RPQs. A central feature of SPARQL, the standard query language for RDF databases, are property path queries, which are a slight extension of 2RPQs. With the widespread adoption of SPARQL, (2)RPQs have become a popular feature [2]: out of 208 million SPARQL queries in the public logs from the Wikidata Query Service [21], about 24% use at least one RPQ feature [11]. Subsequence efforts,

---

Supported by ANID - Millennium Science Initiative Program - Code ICN17_002, and Fondecyt Grant 1-230755.

© The Author(s), under exclusive license to Springer Nature Switzerland AG 2025
Z. Lipták et al. (Eds.): SPIRE 2024, LNCS 14899, pp. 218–232, 2025.
https://doi.org/10.1007/978-3-031-72200-4_17

Subject	Predicate	Object
Alice	mentored	Bob
Alice	cited	Alice
Alice	cited	Dan
Bob	refereedFor	Dan
Dan	cited	Alice
Dan	coauthorOf	Grace
Dan	coauthorOf	Eve
Dan	cited	Bob
Eve	cited	Grace
Eve	mentored	Grace
Eve	mentored	Dan
Eve	cited	Bob
Eve	coauthorOf	Dan
Grace	refereedFor	Alice
Grace	coauthorOf	Dan

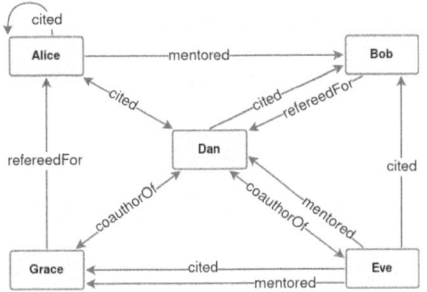

Fig. 1. The table on the left describes an RDF database of researchers and their relationships, while the graph on the right is its graphical representation.

like PGQL [28], Cypher [16], G-CORE [1], TigerGraph [13], and GQL [12], also support RPQ-like features.

An algorithmically sound approach to evaluate (2)RPQs [32] is to represent the regular expression of the RPQ as a finite automaton and use it to traverse the database graph in synchronization with the automaton. Formally, this is regarded as traversing a virtual "product graph" [23], whose nodes are the Cartesian product of the automaton and the graph nodes. Many systems that handle property path queries in SPARQL, instead, extend the relational algebra to support computing transitive closures, and then translate RPQs into (extended) relational algebra operations.

With big and growing available database graphs, indexing them within compact space is of interest in order to preserve memory space, increasing the chances of solving queries in main memory, which is much faster than the disk. While classic systems that solve graph queries, like Virtuoso [14] and Blazegraph [30], use 60 to 90 bytes per graph edge (bpe), recent research handles the most important queries within much less space. The Ring [3,4], for example, handles the core SPARQL queries, including RPQs, within 16 bpe; a faster version called Ring$_{AB}$ uses 28 bpe. An even more recent development based on Boolean matrices [5,6] uses as little as 4.3 bpe, though it is markedly slower than the Ring.

In this paper we build on the same product-graph approach as the Ring, but halve its space while sharply improving its speed on the most popular RPQs, where one end is fixed. Instead of using the Ring graph representation [8], which was developed with the aim of solving basic graph patterns (the other main kind of graph database queries), we resort to a compact representation of labeled graphs that had not been implemented so far [25, Sec. 9.1.4]. Ours is also, arguably, a more natural representation of the database graph, which may simplify the implementation of other graph algorithms (e.g., traversals) on it.

2 Basic Concepts

2.1 Labeled Graphs and Regular Path Queries

Let V and $\mathcal{L}$ be finite sets of nodes and labels, respectively. We assume these sets have been already integer-encoded as $V = \{1,\ldots,n\}$ and $\mathcal{L} = \{1,\ldots,\lambda\}$. A *directed edge-labeled graph* G is a finite set containing e triples $(s,p,o) \in V \times \mathcal{L} \times V$ representing the graph edges $s \xrightarrow{p} o$ from vertex s to vertex o with label p. In the RDF model [22], s is called a *subject*, p a *predicate*, and o an *object*.

A *path* ρ from a node x_0 to node x_k in G is a string $x_0 p_1 x_1 \cdots x_{k-1} p_k x_k$ such that $(x_{i-1}, p_i, x_i) \in G$ for $1 \leq i \leq k$. Given a path ρ, we denote word$(\rho) = p_1 \cdots p_k$ the string labeling path ρ. Two-way RPQs (2RPQs) also allow traversing reversed edges. Hence, we define the set of inverse labels as $\hat{\mathcal{L}} = \{\hat{p} : p \in \mathcal{L}\}$, and let $\mathcal{L}^{\leftrightarrow} = \mathcal{L} \cup \hat{\mathcal{L}}$ be the set of predicates and their inverses. We define the *inverse graph* as $\hat{G} = \{(y, \hat{p}, x) : (x, p, y) \in G\}$, and its *completion* as $G^{\leftrightarrow} = G \cup \hat{G}$. A *two-way regular expression* (2RE) is then formed from the following rules: ε is a 2RE; if $c \in \mathcal{L}^{\leftrightarrow}$, then c is a 2RE; if E, E_1 and E_2 are 2REs, then so are E^* (Kleene star), E_1/E_2 (concatenation), and $E_1 \mid E_2$ (disjunction). If E is a 2RE, we abbreviate E^*/E as E^+ and $\varepsilon|E$ as $E^?$.

The *language* $L(E)$ of a 2RE E over the alphabet of terminals $\mathcal{L}$ is the language of the regular expression E over $\mathcal{L}^{\leftrightarrow}$. We say that a path ρ *matches* a 2RE E if word$(\rho) \in L(E)$. A *two-way regular path query*, or 2RPQ for short, is a query of the form (x, E, y), which has as solution all the pairs of nodes (s, o) such that there is a path $\rho = s p_1 \ldots p_k o$ in $G^{\leftrightarrow}$ where word$(\rho) \in L(E)$; x and/or y can be constants (thus fixing the value of s and/or o, respectively), or variables.

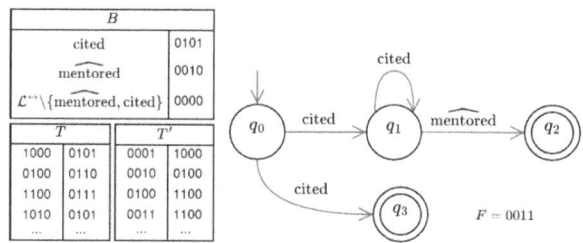

Fig. 2. Glushkov automaton for the regular expression (cited+/$\widehat{\text{mentored}}$)|cited.

2.2 Glushkov Automata

Let R be a regular expression on m symbols. The *Glushkov automaton* [10,17] of R is a particular non-deterministic finite automaton (NFA) $A_R = (Q, \mathcal{L}_R, \Delta, q_0, F)$ recognizing the language of R. It satisfies, in particular, that

(i) it has no ε-transitions; (ii) it has exactly $m+1$ states, that is, $|Q| = m+1$; (iii) all the transitions arriving at a state have the same label. The automaton can be constructed in $O(m^2)$ time and uses $\Theta(m^2)$ space. As an example, consider the set of terminals formed by the predicates in Fig. 1. The regular expression (cited+/$\widehat{\text{mentored}}$)|cited, where $\widehat{\text{mentored}}$ is the inverse of mentored, is represented by the Glushkov automaton of Fig. 2.

The properties of the Glushkov automaton allow for an efficient, bit-parallel processing [4,26], which uses the following data structures:

- D, a word containing $m+1$ bits, which represents a non-deterministic state of A_R, i.e., a subset of Q,
- $B[1..\sigma]$, a table containing bitvectors of length $m+1$, in which $B[c]$ indicates with 1 s the states targeted by transitions labeled c,
- $T[0..2^{m+1} - 1]$, a table storing in $T[X]$, where X is a $(m+1)$-bit argument representing a subset of Q, the set of states reachable from X in one step using any symbol.

We use the symbol & to denote the bit-wise and operator, and represent the final states F as a binary word. Then, it can be shown [26] that $T[D]$ & $B[c]$ is the bitvector of length $m+1$ indicating the states of Q that are reachable from state D using transitions labeled c. The *forward* traversal of A_R, which recognizes strings in $L(R)$ prefixing a sequence, starts with $D \leftarrow 2^{m+1}$ (which activates the initial state q_0), and then repeatedly (i) recognizes the word seen so far if D & $F \neq 0$ (i.e., a final state was reached), (ii) finishes if $D = 0$ (i.e., there are no active states in the NFA), and (iii) updates $D \leftarrow T[D]$ & $B[c]$ with the next symbol c.

It is also possible to recognize R by starting from the final states of the automaton and reading the input from right to left. For this, we first build a table $T'[0..2^{m+1} - 1]$ such that $T'[X]$ marks with 1 s the states that reach some state in X in one step. Then, we activate the final states with $D \leftarrow F$ and, repeatedly, (i) recognize the word seen so far if D & $2^{m+1} \neq 0$, (ii) finish if $D = 0$, and (iii) update $D \leftarrow T'[D$ & $B[c]]$ with the next symbol c.

The data structures use $O(2^m + |\mathcal{L}_R|)$ bits [27]. Precomputing B and T (or T') takes $O(2^m)$ time [27]. Figure 2 shows the tables for our example.

2.3 The Product-Graph Approach and Its Ring Implementation

Data Structures. The Ring [3] represents the completion $G^{\leftrightarrow}$ (not G) using the sequences L_p, C_o, L_s, and C_p defined next. Let us write $G^{\leftrightarrow} = \{(s_i, p_i, o_i) : 1 \leq i \leq 2e\}$ in such a way that the edges are lexicographically sorted, that is, first by o, in case of ties by s, and still in case of ties by p (we call this order OSP), and define the array $L_p = p_1, \ldots, p_{2e}$. Observe that L_p can be written as a concatenation of segments $L_p^1 \cdot L_p^2 \cdots L_p^n$, where L_p^o contains those p_i such that $o_i = o$. We define the bitvector $C_o = 10^{|L_p^1|}10^{|L_p^2|} \cdots 10^{|L_p^n|}$. Suppose now that $G^{\leftrightarrow}$ is sorted in order POS, and let $L_s = s_1, \ldots, s_{2e}$. We can similarly write $L_s = L_s^1 \cdots L_s^{2\lambda}$, with L_s^p containing those s_i such that $p_i = p$, and define $C_p = 10^{|L_s^1|} \cdots 10^{|L_s^{2\lambda}|}$.

Solving RPQs. To solve RQPs on the Ring [4,7], double-variable queries (x, R, y) are reduced to solving the single-variable queries (x, R, o) for every node o. Additionally, single-variable queries of the form (s, R, x) are reduced to solving the case $(x, \hat{R}, s)$, where $\hat{R}$ is the reversed regular expression.

To solve an RPQ (y, R, o) with variable y, the bit-parallel Glushkov automaton A_R for the regular expression R is constructed, requiring $O(2^m)$ time and $O(2^m + |\mathcal{L}_R|)$ space [27]. Then, the conceptual traversal of the product graph $A_R \times G^{\leftrightarrow}$ starts from (F, o). This is equivalent to simultaneously traversing A_R and $G^{\leftrightarrow}$, starting from the final states of A_R and from the node $o \in G^{\leftrightarrow}$. Each step consists of three parts:

1. Identify the predicates associated with the current object o (*i.e.*, labeling arrows that arrive at or leave from o), which are the elements of L_p^o.
2. For each identified predicate p, feed A_R with p in reverse traversal mode, and abandon this branch if A_R becomes out of active states.
3. Determine the subjects s of edges of the form $(s, p, o) \in G^{\leftrightarrow}$, which are the elements of L_s^p. Report s if the initial state of A_R is active.
4. Interpret s as an object o, and create a new branch that starts from step 1.

In addition, graph nodes are marked with the active states of A_R they have been already visited with, to avoid loops in the traversal of the product graph.

The Ring represents L_p and L_s using wavelet trees [20]. This allows obtaining the predicates and subjects in steps 1 and 3 in batch using the so-called backward search [15]. Further, it can efficiently process ranges of symbols in steps 1 and 3. Finally, they can integrate steps 1 and 2 so as to directly produce the predicates p that are associated with o *and* lead to active states in A_R, in a time they prove to be optimal in terms of computing the intersection of both sets.

A variant called Ring$_{AB}$ [4] uses more space but improves the time, by starting the traversal from a "split" point of R, not necessarily from its end, and traversing $G^{\leftrightarrow}$ in both direction from the split point. If the split point is an uncommon predicate, much fewer nodes need be visited.

3 Our Graph Representation

Data Structures. Our approach represents G (not $G^{\leftrightarrow}$) using sequences L, B_L, N, and B_N, which we define as follows. We sort $G = \{(s_i, p_i, o_i) : 1 \leq i \leq e\}$ with order SOP, and define the array $L = p_1, \ldots, p_e$. Array L is the concatenation of segments $L_1 \cdot L_2 \cdots L_n$, where L_s contains those p_i such that $s_i = s$. We encode the lengths of the subarrays L_s in the bitvector $B_L = 10^{|L_1|} \ldots 10^{|L_n|}$. The sequences N and B_N are defined similarly: we sort G with order PSO and define the array $N = o_1, \ldots, o_e$, which can also be decomposed as $N = N_1 \cdots N_\lambda$, with N_p containing those o_i such that $p_i = p$. Finally, we set $B_N = 10^{|N_1|} \ldots 10^{|N_\lambda|}$.

Figure 3 depicts the sequences N, L, B_L, and B_N for the database obtained by integer-encoding the example from Fig. 1. While our structures already exist [25, Sec. 9.1.4], our presentation here has the added value of exposing their relation with the Ring representation [4]: we store predicates with order SOP in L and

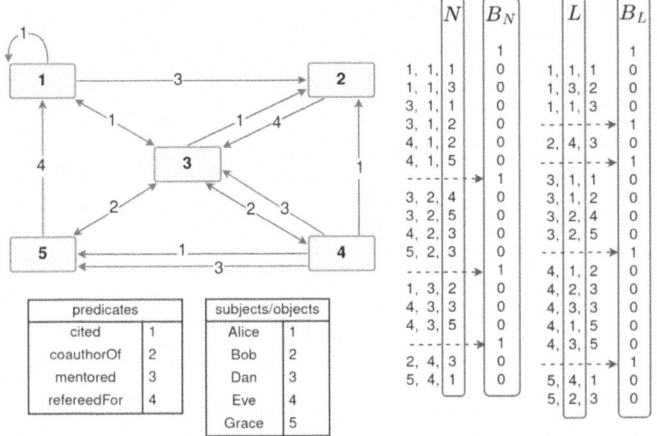

Fig. 3. On the left, the integer-encoded version of the graph from Fig. 1, along with its translation to the original terms. On the right we show how N is derived from the triples in order PSO. The arrows highlight the limits of the segments N_p and point to the 1 s in B_N. The equivalent diagram for L and B_L uses the triples in order SOP.

objects in order PSO in N; the Ring stores predicates in order OSP in L_p and subjects in order POS in L_s. The ring doubles our space because it stores the reverse edges (with reverse labels) explicitly.

We utilize plain bitvectors [24] to store B_L and B_N, and GMR-arrays [19] to store L and N. These choices allow storing bitvectors B using $(1 + o(1))|B|$ bits, and general arrays A with values in the range $[1..k]$ using $(1 + o(1))|A|\log|k|$ bits. Therefore, our representation of G uses $(1 + o(1))(e\log(\lambda n) + n + \lambda)$ bits, which, unlike the space of the Ring, is asymptotically optimal.

Note that we do not explicitly store the triples of G. To recover the original triples and other useful information about the graph, we can use RSA operations on the sequences L, B_L, N, and B_N. These are defined for any array A as follows:

- access(A, i) returns the letter $A[i]$, for any $1 \leq i \leq n$.
- rank$_a(A, i)$ returns the number of occurrences of a in $A[1..i]$, for any $1 \leq i \leq n$. We assume rank$_a(A, 0) = 0$.
- select$_a(A, j)$ returns the position of the j-th occurrence of a in A, for any $j \geq 0$. We assume select$_a(A, 0) = 0$ and select$_c(A, j) = n+1$ if $j >$ rank$_a(A, n)$.

Our data structure choices yield constant time RSA operations for B_L and B_N [24], constant time select for L and N, $O(\log \log n)$ time rank and access for N, and $O(\log \log \lambda)$ time rank and access for L [19].

Retrieving Graph Edges. We now detail how to recover the triple associated to a position in L or in N. If $(s, o, p) \in G$ is the triple corresponding to position i of L, then $p =$ access(L, i), whereas s and o can be recovered as follows:

$$s = \mathsf{select}_0(B_L, i) - i; \quad q = \mathsf{select}_1(B_N, p) - p; \quad j = q + \mathsf{rank}_p(L, i); \quad o = N[j]. \quad (1)$$

Similarly, if $(s, o, p) \in G$ is the triple corresponding to position j of N, then $o = N[j]$, whereas p and s can be recovered as follows:

$$p = \mathsf{select}_0(B_N, j) - j; \quad r = \mathsf{select}_1(B_N, p) - p; \quad i = \mathsf{select}_p(L, j - r); \quad (2)$$
$$s = \mathsf{select}_0(B_L, i) - i.$$

We note that Formulas (1) and (2) resemble the way the Ring recovers its triples using forward and backward direction, respectively [3].

Operations on the Structure The representation supports several useful queries on the graph. We describe next those we use to solve 2RPQs. We call $G_p = \{(s, p, o) \in G\}$ the triples of G labeled p.

- $\mathsf{neigh}_p(G, s)$ returns all neighbors of s in G_p, i.e., $\{o : (s, p, o) \in G\}$.
- $\mathsf{rneigh}_p(G, o)$ returns all reverse neighbors of o in G_p, i.e., $\{s : (s, p, o) \in G\}$.
- $\mathsf{sources}_p(G)$ returns the nodes s that are the origin of an edge with label p, i.e., $\{s : \exists o, (s, p, o) \in G\}$.
- $\mathsf{targets}_p(G)$ returns the nodes o that are the target of an edge with label p, i.e., $\{o : \exists s, (s, p, o) \in G\}$.

We now describe the algorithms for neigh_p, rneigh_p, $\mathsf{sources}_p$, and $\mathsf{targets}_p$.

The neighbors of s in G_p are exactly the elements o in the segment $N_{p,s}$ of N_p that corresponds to the edges $(s, p, o) \in G$. To identify $N_{p,s}$, we first note that the beginning of the segment N_p in N is $r = \mathsf{select}_1(B_N, p) - p$. Then, we get the offset at which the objects associated with s begin by counting the number of edges whose predicate is p and whose subject is some $s' < s$. This is done with

$$q_1 = \mathsf{rank}_p(L, l_1), \text{ where } l_1 = \mathsf{select}_1(B_L, s) - s.$$

Similarly, the start of $N_{p,s+1}$ in N_p is at

$$q_2 = \mathsf{rank}_p(L, l_2), \text{ where } l_2 = \mathsf{select}_1(B_L, s+1) - (s+1).$$

Since, in N_p, the subsegment $N_{p,s+1}$ immediately follows $N_{p,s}$, it follows that

$$\mathsf{neigh}_p(G, s) = N_p[q_1 + 1 .. q_2] = N[r + q_1 + 1 .. r + q_2].$$

This also yields that $\mathsf{outdegree}_p(s)$, the outdegree of s in G_p, is equal to $q_2 - q_1$. Overall, we compute any element of $\mathsf{neigh}_p(G, s)$ in time $O(\log \log n)$.

To obtain the reverse neighbors of o in G_p, we recall that $N_p = N[r_1 + 1 .. r_2]$, where $r_1 = \mathsf{select}_1(B_N, p) - p$ and $r_2 = \mathsf{select}_1(B_N, p+1) - (p+1)$. The number of reverse neighbors of o in G_p is then $\mathsf{indegree}_p(G, o) = \mathsf{rank}_o(N, r_2) - \mathsf{rank}_o(N, r_1)$. Moreover, the position j in N of the k-th reverse neighbor of o in G_p is

$$j = \mathsf{select}_o(N, l), \text{ where } l = \mathsf{rank}_o(N, r_1) + k.$$

Formula (2) then yields the corresponding subject s, which is then the kth element of $\mathsf{rneigh}_p(G, v)$, computed in total time $O(\log \log n)$.

To implement $\mathsf{sources}_p(G)$, we start by computing the position i of the first occurrence of p in L, with $i = \mathsf{select}_p(L, 1)$. We then report the subject s associated to position i of L using formula (1). To avoid adding duplicates (which arise if the same subject is connected to two different objects by the same label), we move forward until the end of L_s, which is at position $q = \mathsf{select}_1(B_L, s+1) - (s+1)$. Note that the last occurrence of p in L_s is the $\mathsf{rank}_p(L, q)$-th one in L. We iterate by looking for the next occurrence of p (at position $l = \mathsf{select}_p(L, \mathsf{rank}_p(L, q)+1)$) until we reach the end of L. The cost is $O(\log \log \lambda)$ per element reported.

It rests to implement $\mathsf{targets}_p(G)$. Note that the objects of the edges with label p are the elements of N_p. So, it is enough to collect the elements in $N_p = N[r_1 + 1 .. r_2]$ and then remove the duplicates with, for example, integer sorting. The total time is $O(|G_p| \log \log n)$.

4 Solving RPQs on Our Representation

4.1 Single-Variable 2RPQs

We start by discussing a symmetry in the queries. A pair (s, o) is a solution for the 2RPQ (s, R, x) iff (o, s) is a solution for $(y, \hat{R}, s)$, where $\hat{R}$ is the reverse of R. Hence, when solving single-variable queries, we can choose the position of the variable. Queries of the form $(s, R, x)/(x, R, o)$ involve traversing A_R forwards/backwards. This gives us an important degree of freedom: we can choose the direction to traverse A_R, with yields slightly different algorithms.

We first present the algorithm for a query (s, R, x) and then show how it is modified to solve the other class of queries. The basic strategy of the algorithm is analogous to that of the Ring [4], but the implementation of the ideas differs. The solutions to (s, R, x) are the ends of paths ρ in G starting at s such that word(ρ) is accepted by A_R. Thus, the problem boils down to traversing the product graph $A_R \times G$. Along our traversal we maintain a pair (D, v), meaning that we are visiting, at the same time, all the nodes $(q, v) \in A_R \times G$ such that q is represented in D. Our initial pair is $(2^{m+1}, s)$. From the current pair (D, v), we find all the labels p that leave some state in D. For each such label, we compute the new set of states D' using the formula seen in Sect. 2.2. In G, we jump from v to every node $v' \in \mathsf{neigh}_p(G, v)$ or $v' \in \mathsf{rneigh}_p(G, v)$, depending on whether p is an inverse label or not. For each resulting node v', we recursively continue our traversal by pair (D', v').

There are two critical checks that we make when considering a new pair (D, v). First, we have to determine whether v belongs to the solution. This is done by verifying that D contains an accepting state. Second, we need to avoid loops in the traversal of the product graph. To that end, we maintain a structure that keeps track of the states of A_R with which we have already visited every node of G. We now enter into details.

Avoiding Redundancies and Loops. If we are processing (D, v) and have previously visited another pair (D', v), we should remove from D the states that are also in D', as those have already been processed and may lead to loops. We maintain a table $\mathsf{seen}[1 .. n]$ containing in $\mathsf{seen}[v]$ the bitvector of length $m + 1$

that represents all the states that have been active in previously seen pairs of the form (D', v). This array is initialized as $\mathsf{seen}[i] = 0$ before the query starts. When arriving at (D, v), we remove from D the states in $\mathsf{seen}[v]$, and add those to $\mathsf{seen}[v]$, as follows (where $\sim$ and $|$ are the bitwise logical not and or, respectively):

$$D \leftarrow D \mathbin{\&} \sim\!\mathsf{seen}[v] \text{ and } \mathsf{seen}[v] \leftarrow \mathsf{seen}[v] \mid D.$$

Reporting Solutions. If $D \mathbin{\&} F \neq 0$, the automaton A_R accepts the path towards node v, so v is a solution to the query. Note, however, that v may be reachable from multiple accepting paths. To avoid reporting duplicated answers, we must verify that there is no previously seen pair (D', v), where D' contains a final state. This is done by verifying that $\mathsf{seen}[v] \mathbin{\&} F = 0$ before updating seen.

Computing the New Pairs. We now explain in detail how we produce the new pairs (D', v') from the current one, (D, v).

First, we find the set P_D of all the elements $p \in \mathcal{L}_R$ that label an edge in A_R leaving D. As seen in Sect. 2.2, a label p is in P_D iff $T[D] \mathbin{\&} B[p] \neq 0$. We can then obtain P_D by iterating through all the labels $p \in \mathcal{L}_R$ and keeping those for which the formula holds, which takes only $O(m)$ time. Since we may encounter D multiple times throughout the process, however, we define a lazy-initialized table $P[0\,..\,2^{m+1} - 1]$ and store P_D in $P[D]$ as a bitvector of length $m+1$.

For each $p \in P_D$, we generate the new set of active states $D' = T[D] \mathbin{\&} B[p]$. The set of nodes v' we reach by following p is, as explained, $\mathsf{neigh}_p(v)$ if $p \in \mathcal{L}$, and $\mathsf{rneigh}_{\hat{p}}(v)$ if $p \in \hat{\mathcal{L}}$. Those sets are computed as described in Sect. 3.

Queries of the Form (y, R, o). The solution to these queries is analogous, but we traverse the edges of A_R and of G backwards. Concretely, the initial NFA state is F and the final NFA state is 2^{m+1}, and their roles are exchanged everywhere in the description above. We start from pair (F, o) and report v when D contains the initial state. We compute the NFA edges backwards with the formula $D' = T'[D \mathbin{\&} B[p]]$, recall Sect. 2.2.

Finally, $P[D]$ now corresponds to the *ingoing* predicates, which are determined as those $p \in \mathcal{L}_R$ such that $D \mathbin{\&} B[p] \neq 0$. In G we move to the reverse neighbors of v with label p if $p \in \mathcal{L}$ and to its neighbors with label $\hat{p}$ if $p \in \hat{\mathcal{L}}$.

4.2 Double-Variable 2RPQs

It remains to treat the case of double-variable queries (x, R, y). The naive approach is to solve the single-variable queries (v, R, y) for all nodes $v \in V$ (or, symmetrically, the queries (x, R, v)). This is highly inefficient as many objects v may not lead to any solution. We improve this basic method by first obtaining a *feasible set* V', which is a (typically small) subset of V that contains all the nodes v such that (v, R, y) leads to at least one solution to (x, R, y). Let $\mathcal{L}'$ be the set of labels of edges leaving the initial state in A_R. We then define

$$V' = \bigcup_{p \in \mathcal{L}' \cap \mathcal{L}} \text{sources}_p(G) \ \cup \ \bigcup_{p \in \mathcal{L}' \cap \hat{\mathcal{L}}} \text{targets}_{\hat{p}}(G).$$

It is easy to see that every subject that is a solution to the query (x, R, o), $o \in V$, is in V'. Therefore, the solution set of (x, R, y) is equal to the collection of all solutions of the queries (v, R, y), $v \in V'$.

Let us now describe in detail how to get the elements of V'. First, the labels p connected to q_0 are those that lead to at least one active state when jumping from q_0, that is, those such that $T[2^{m+1}] \ \& \ B[p] \neq 0$. Then, for each such p, we need all subjects of edges in $G^{\leftrightarrow}$ with label p. These are given by $\text{sources}_p(G)$ if p is not inverted and by $\text{targets}_{\hat{p}}(G)$ otherwise.

5 Experimental Results

5.1 Benchmark

Benchmark Database. We use the Wikidata Graph Pattern Benchmark (WGPB) as our benchmark [31]. This graph has $e = 958,844,164$ edges, $n = 348,945,080$ nodes, and $\lambda = 5,419$ labels (*i.e.*, predicates). From the nodes, $|S| = 106,736,662$ act as subjects and $|O| = 295,611,216$ as objects. This amounts to a total of 10.7 GB in plain form (with 32-bit integers for each triple component, and thus 12 bytes per edge, or bpe) and 7.9 GB in packed form (*i.e.*, using $\lceil \log |S| \rceil + \lceil \log |P| \rceil + \lceil \log |O| \rceil$ bits, or 8.63 bpe).

Queries. In order to get challenging, real-world RPQs, the authors of the Ring [4] extracted all the RPQs posed to the Wikidata Query Service that threw timeout error, that is, that needed more than 60 s to complete, from the Wikidata Query Logs [21]. After filtering RPQs using Wikidata-specific features, mentioning constants not used in the dataset, having one label, normalizing variable names, and removing duplicates, this process yielded 1,952 unique queries. Furthermore, they only keep the 1,567 queries with less than 1 million unique results for comparability reasons (as Virtuoso has a hard-coded limit of $2^{20} \approx 1$ million results). All queries are run with a timeout of 60 s under set semantics (using DISTINCT in the case of SPARQL).

Systems Compared. We compare with the Ring and with its larger and faster version, Ring$_{AB}$. A completely different approach, which translates RPQs to operations on sparse Boolean matrices, has recently appeared [5,6]. We compare with both their uncompressed baseline (BM) and their compressed version that uses k^2-trees (k^2-BM). Additionally, we compare our algorithm to the following well-known platforms for managing and querying RDF databases.

1. *Blazegraph* is the official SPARQL endpoint used by Wikidata and by other large customers.
2. *Apache Jena* is a widely used graph database and the reference implementation of the SPARQL standard.

3. *Virtuoso* is a multi-model database that accommodates RDF data, which hosts the public DBpedia endpoint, among others [14].

Machine and Implementation. Our experiments were conducted on an isolated Intel(R) Xeon(R) CPU E5-2630 running at 2.30GHz, with 15 MB of cache and 384 GB of RAM. The operating system is GNU/Linux Devuan 2.1, with kernel 4.9.0-18-amd64. We used the SDSL library [18], which implements our bitvectors and GMR arrays, and the Glushkov automata implementation of the Ring authors [9]. Our implementation is written in C++11, using the compiler g++ version 6.3.0 and the flags -std=c++11, -O3 and -msse4.2. All experiments are single-threaded. Recall that our algorithm allows both traversal directions of the NFA. Given that we observed slightly better performance with forward traversal, we have used this direction in our experiments. The complete source code and the instructions for compiling it can be found in the repository [29].

5.2 Results

Index Construction. Constructing the integer-encoded database takes 5.2 h using the code by Arroyuelo et al. [4]. Constructing our index then takes 0.3 additional hours. The resulting index uses 6.87 GB, or 7.17 bpe. The total space used at query time is higher, 10.28 bpe. The excess is dominated by the seen table (3.11 bpe); the $O(2^m)$ space related to NFA preprocessing is negligible.

Table 1 shows the construction time and resulting space usage of our index compared to the other systems. Our index is the second most space-efficient, only surpassed by the k^2-BM. Note that the Ring uses slightly more than twice the space of our index, which is consistent with the fact that we do not duplicate the edges for dealing with the inverted predicates. BM has a similar space requirement as the Ring, while the rest of the systems use much more.

Our index exhibits a competitive construction time of 5.5 h, matching k^2-BM and ranking second only to Virtuoso's leading time of 3 h.

Table 1. Index space (in bpe) and construction time (in hours) for the different systems. The integer-encoding times are included in the construction.

	Ours	Ring	Ring$_{AB}$	BM	k^2-BM	Jena	Virtuoso	Blazegraph
Index space	7.17	16.41	27.93	16.45	4.33	95.83	60.07	90.79
Indexing time	5.5	7.5	8.3	10.7	5.5	37.4	3.0	39.4

Querying. Table 2 details the average, median, and the number of timeouts for the different types of queries across the competing algorithms. Figure 4 shows the performance distribution across all queries and specifically for double-variable queries, with the algorithms ordered on the x-axis according to their space usage.

Our index shows average performance, 0.6 s per query, is outperformed only by the 0.4 s of Ring$_{AB}$; we remind that this index uses 4 times more space than

Table 2. Performance comparison of the different algorithms, detailing the number of timeouts (execution time over 60 s) and the average and median times, not considering timeouts. The notation $1v/2v$ denotes single-/double-variable queries.

	Ours	Ring	Ring$_{AB}$	BM	k^2-BM	Jena	Virtuoso	Blazegraph
Average	0.61	0.95	0.41	1.39	3.25	4.51	2.08	3.23
Median	0.01	0.07	0.03	0.005	0.35	0.21	0.13	0.13
Timeout	2	4	1	14	39	84	1	41
Average $1v$	0.28	0.48	0.25	1.19	2.84	3.62	1.79	3.24
Median $1v$	0.009	0.06	0.03	0.005	0.35	0.19	0.11	0.13
Timeout $1v$	0	0	0	12	30	58	1	39
Average $2v$	7.41	10.91	3.66	5.45	11.92	22.83	8.17	2.98
Median $2v$	3.70	1.45	0.93	0.01	0.87	1.57	3.89	0.14
Timeout $2v$	2	4	1	2	9	26	0	6

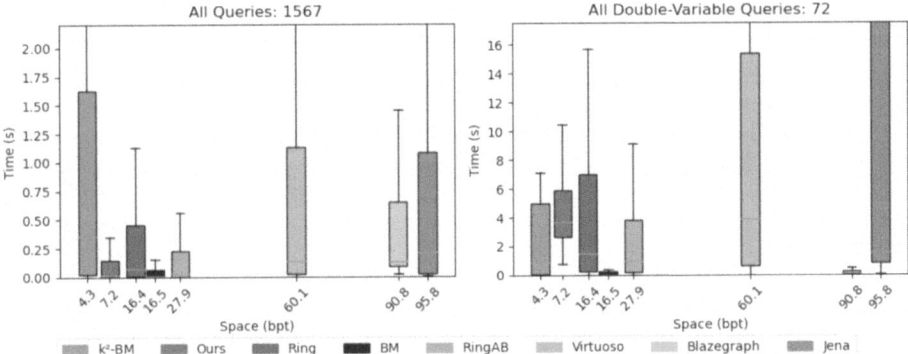

Fig. 4. Time taken by the different algorithms across all queries and for those with two variables (2v). The figure does not include the timeouts.

ours. The only index using less space than ours (60% of our space), k^2-BM, is 5 times slower on average. The other systems use more space and more average time than ours. The median times of our index are also far better than the others, except for BM, which in exchange uses twice as much space.

The advantage of our index is most evident in queries with a single variable, where it achieves nearly the best performance in both median and average times, while using less space. For double-variable queries, our index remains competitive but does not stand among the fastest, and its particular its median time is close to the highest. Despite its high median times, our index produces fewer timeouts, which demonstrates higher reliability for complex queries.

Figure 4 compares space and time distributions, showing on the left that our index is a key point in the Pareto-optimal curve, formed also by the k^2-BM (which uses 60% of the space but an order of magnitude more time) and the BM (which uses half the time but twice the space); Ring$_{AB}$ has a good average but not such a good distribution. On the right, for 2 variables, the Pareto-optimal curve is formed by the Boolean-Matrix-based indices. Our index does outperform k^2-BM on the average, but the latter distributes better.

6 Conclusions

We have demonstrated that a compact representation designed for labeled graphs outperforms in space, and competes in time, with the Ring index [4] at solving Regular Path Queries. The main advantages of our representation are (1) it helps navigating edges bidirectionally without duplicating data, whereas the Ring navigates only backwards; (2) it builds on sequence representations [19] that are faster but with less functionality than the wavelet trees [20] used by the Ring. Both decisions in the Ring, unidirectionality and use of wavelet trees, aim at matching intersection-time lower bounds, while our index only ensures $O(m)$ time (which in practice is low anyway). On one-variable queries (which are most popular) our index provides better time distribution, losing only on the average to a Ring variant that uses 4 times more space (Ring$_{AB}$). Our representation fills an important place in the Pareto-optimal curve for one-variable queries.

Our work also uncovers subtle relations of independent interest between the Ring structure and the labeled graph representation we build on [25, Sec. 9.1.4].

References

1. Angles, R., et al.: G-CORE: a core for future graph query languages. In: Proceedings of the 44th ACM International Conference on Management of Data (SIGMOD, pp. 1421–1432 (2018)
2. Angles, R., Arenas, M., Barceló, P., Hogan, A., Reutter, J.L., Vrgoc, D.: Foundations of Modern Query Languages for Graph Databases. ACM Comput. Surv. **50**(5), 68:1–68:40 (2017)
3. Arroyuelo, D., et al.: The ring: worst-case optimal joins in graph databases using (almost) no extra space. ACM Trans. Datab. Syst. **29**(2), 5 (2024)
4. Arroyuelo, D., Gómez-Brandón, A., Hogan, A., Navarro, G., Rojas-Ledesma, J.: Optimizing RPQs over a compact graph representation. Very Large Datab. J. **33**, 349–374 (2024)
5. Arroyuelo, D., Gómez-Brandón, A., Navarro, G.: Evaluating regular path queries on compressed adjacency matrices. In: Proceedings of the 30th International Symposium on String Processing and Information Retrieval (SPIRE), pp. 35–48 (2023)
6. Arroyuelo, D., Gómez-Brandón, A., Navarro, G.: Evaluating regular path queries on compressed adjacency matrices. arXiv preprint arXiv:2307.14930 (2024)
7. Arroyuelo, D., Hogan, A., Navarro, G., Rojas-Ledesma, J.: Time- and space-efficient regular path queries. In: Proceedings of the 38th IEEE International Conference on Data Engineering (ICDE), pp. 3091–3105 (2022)
8. Arroyuelo, D., Navarro, G., Reutter, J.L., Rojas-Ledesma, J.: Optimal joins using compressed quadtrees. ACM Trans. Database Syst. **47**(2), 8 (2022)
9. Arroyuelo, D., Gómez-Brandón, A., Hogan, A., Navarro, G., Rojas-Ledesma, J.: Ring-RPQ (2022). https://github.com/darroyue/Ring-RPQ

10. Berry, G., Sethi, R.: From regular expressions to deterministic automata. Theor. Comput. Sci. **48**(C), 117–126 (1986)
11. Bonifati, A., Martens, W., Timm, T.: Navigating the Maze of Wikidata Query Logs. In: Proceedings of the World Wide Web Conference (WWW), pp. 127–138 (2019)
12. Deutsch, A., et al.: Graph pattern matching in GQL and SQL/PGQ. In: Proceedings of the 48th ACM International Conference on Management of Data (SIGMOD), pp. 2246–2258 (2022)
13. Deutsch, A., Xu, Y., Wu, M., Lee, V.E.: Aggregation support for modern graph analytics in TigerGraph. In: Proceedings of the 46th ACM International Conference on Management of Data (SIGMOD), pp. 377–392 (2020)
14. Erling, O., Mikhailov, I.: RDF support in the Virtuoso DBMS. In: Pellegrini, T., Auer, S., Tochtermann, K., Schaffert, S. (eds.) Networked Knowledge - Networked Media, pp. 7–24. Springer, Heidelberg (2009). https://doi.org/10.1007/978-3-642-02184-8_2
15. Ferragina, P., Manzini, G.: Indexing compressed text. J. ACM **52**(4), 552–581 (2005). https://doi.org/10.1145/1082036.1082039
16. Francis, N., et al.: Cypher: an evolving query language for property graphs. In: Proceedings of the 44th ACM International Conference on Management of Data (SIGMOD), pp. 1433–1445 (2018)
17. Glushkov, V.M.: The abstract theory of automata. Russ. Math. Surv. **16**(5), 1–53 (1961)
18. Gog, S., Beller, T., Moffat, A., Petri, M.: SDSL - succinct data structure library (2016). https://github.com/simongog/sdsl-lite
19. Golynski, A., Munro, J., Rao Satti, S.: Rank/select operations on large alphabets: a tool for text indexing. In: Proceedings of the 17th Annual ACM-SIAM Symposium on Discrete Algorithms (SODA), pp. 368–373 (2006)
20. Grossi, R., Gupta, A., Vitter, J.S.: High-order entropy-compressed text indexes. In: Proceedings of the 14th Annual ACM-SIAM Symposium on Discrete Algorithms (SODA), pp. 841–850 (2003)
21. Malyshev, S., Krötzsch, M., González, L., Gonsior, J., Bielefeldt, A.: Getting the most out of Wikidata: semantic technology usage in Wikipedia's knowledge graph. In: Proceedings of the International Semantic Web Conference (ISWC), pp. 376–394 (2018)
22. Manola, F., Miller, E.: RDF Primer. W3C Recommendation (2004). http://www.w3.org/TR/rdf-primer/
23. Mendelzon, A.O., Wood, P.T.: Finding regular simple paths in graph databases. SIAM J. Comput. **24**(6), 1235–1258 (1995)
24. Munro, J.I.: Tables. In: Chandru, V., Vinay, V. (eds.) FSTTCS 1996. LNCS, vol. 1180, pp. 37–42. Springer, Heidelberg (1996). https://doi.org/10.1007/3-540-62034-6_35
25. Navarro, G.: Compact Data Structures—A Practical Approach. Cambridge University Press (2016)
26. Navarro, G., Raffinot, M.: New techniques for regular expression searching. Algorithmica **41**(2), 89–116 (2005)
27. Navarro, G., Raffinot, M.: New techniques for regular expression searching. Algorithmica **41**(2), 89–116 (2005). https://doi.org/10.1007/S00453-004-1120-3
28. van Rest, O., Hong, S., Kim, J., Meng, X., Chafi, H.: PGQL: a property graph query language. In: Proceedings of the 4th International Workshop on Graph Data Management: Experiences and Systems (GRADES), p. 7 (2016)

29. Robert, J.: A compact graph structure for efficiently solving RPQs (2023). https://github.com/j-rparra/navGraph
30. Thompson, B.B., Personick, M., Cutcher, M.: The bigdata®RDF graph database. In: Linked Data Management, pp. 193–237. Chapman and Hall/CRC (2014)
31. Vrandecic, D., Krötzsch, M.: Wikidata: a free collaborative knowledgebase. Commun. ACM **57**(10), 78–85 (2014)
32. Yakovets, N., Godfrey, P., Gryz, J.: Query planning for evaluating SPARQL property paths. In: Proceedings of the 42th ACM International Conference on Management of Data (SIGMOD), pp. 1875–1889 (2016)

Greedy Conjecture for the Shortest Common Superstring Problem and Its Strengthenings

Maksim S. Nikolaev[✉][iD]

Steklov Institute of Mathematics at St. Petersburg, Russian Academy of Sciences, St. Petersburg, Russia
makc-nicko@yandex.ru

Abstract. In the Shortest Common Superstring problem, one needs to find the shortest superstring for a set of strings. This problem is APX-hard, and many approximation algorithms were proposed, with the current best approximation factor of 2.466. Whereas these algorithms are technically involved, for more than thirty years the Greedy Conjecture remains unsolved, that states that the Greedy Algorithm "take two strings with the maximum overlap; merge them; repeat" is a 2-approximation. This conjecture is still open, and one way to approach it is to consider its stronger version, which may make the proof easier due to the stronger premise or provide insights from its refutation. In this paper, we propose two directions to strengthen the conjecture. First, we introduce the Locally Greedy Algorithm (LGA), that selects a pair of strings not with the largest overlap but with the *locally largest* overlap, that is, the largest among all pairs of strings with the same first or second string. Second, we change the quality metric: instead of length, we evaluate the solution by the number of occurrences of an arbitrary symbol.

Despite the double strengthening, we prove that LGA is a *uniform* 4-approximation, that is, it always constructs a superstring with no more than four times as many occurrences of an arbitrary symbol as any other superstring. At the same time, we discover the limitations of the greedy heuristic: we show that LGA is at least 3-approximation, and the Greedy Algorithm is at least uniform 2.5-approximation. These result show that if the Greedy Conjecture is true, it is not because the Greedy Algorithm is locally greedy or is uniformly 2-approximation.

Keywords: superstring · shortest common superstring · approximation · greedy algorithms · greedy conjecture

1 Introduction

In the Shortest Common Superstring problem (SCS) one is given a set of n input strings $\mathcal{S} = \{s_1, \ldots, s_n\}$ and needs to find the shortest string that contains all of them as substrings. SCS has applications in genome assembly [19,24]: modern technologies can extract only relatively short sequences of nucleotides, which leads to the problem of reconstructing an entire DNA from many such pieces. Without loss of generality, we can assume that no string of $\mathcal{S}$ is contained in any other.

SCS is NP-hard [10] and even APX-hard [4]. The classic way to get a practical solution for a difficult optimization problem is to use some kind of heuristic. Perhaps the simplest heuristic for SCS is *the greedy* one: to find a short superstring, merge strings with the longest *overlap*, that is, the longest suffix of one string that is also a prefix of another. An algorithm that uses this heuristic is called *the Greedy Algorithm* (GA). It operates as follows: while there are more than two strings, choose a pair with the longest overlap and merge it. The resulting string is clearly a superstring for the input strings.

GA is not deterministic, as it may merge any pair with the maximum overlap, and this can greatly affect the length of the result. To see this, consider a dataset $\mathcal{S} = \{\mathtt{ab}^n, \mathtt{b}^{n+1}, \mathtt{b}^n\mathtt{c}\}$, for which GA can construct both an optimal solution $\mathtt{ab}^{n+1}\mathtt{c}$ and an asymptotically twice as long solution $\mathtt{ab}^n\mathtt{cb}^{n+1}$, depending on the tie-breaking rule. This example shows, that the approximation factor of GA is at least two, and Tarhio and Ukkonen [21] conjectured that it equals two for any tie-braking rule. This conjecture is known as *the Greedy Conjecture* (GC).

Over the past thirty-odd years, this conjecture has not been resolved. This is strange, because usually the analysis of such simple greedy algorithms is also quite simple: either we understand that a given algorithm can construct an arbitrarily bad solution, or we can obtain an upper bound on its approximation factor and construct an example for which this bound turns out to be tight. The first constant upper bound of 4 on the GA approximation factor was obtained by Blum, Jiang, Li, Tromp, and Yannakakis [4]. This bound was improved to 3.5 by Kaplan and Shafrir [12], and the next two improvements: to 3.425 and to 3.396, respectively, were obtained recently by Englert, Matsakis, and Veselý [8,9].

Apart from that, the following results related to the Greedy Algorithm are known:

- GA is a factor $\frac{1}{2}$-approximation for *the maximum compression* [21,23], where compression is the difference between the total length of all input strings and the length of a given superstring;
- GC holds for strings of length no more than 3 [6];
- GC holds for strings of length exactly 4 [14];
- GC for the binary alphabet is equivalent to GC for the larger ones (see proof of Theorem 3 in [10]);
- All instantiations of GA (that is, GA with specified tie-braking rule) achieve the same factor of approximation [17].

The Greedy Conjecture is challenging even for rather simple special cases: for instance, the case of strings of length no more than 4 is uncharted territory.

In addition to the Greedy Algorithm, numerous approximation algorithms have been proposed (see Table 1). The current best upper bound of 2.466 on the approximation factor was recently obtained by Englert, Matsakis, and Veselý [9].

Table 1. Known upper bounds on approximation factors for SCS.

3.000	Blum, Jiang, Li, Tromp, Yannakakis [4]	1991
2.889	Teng, Yao [22]	1993
2.834	Czumaj, Gasieniec, Piotrow, Rytter [7]	1994
2.794	Kosaraju, Park, Stein [13]	1994
2.750	Armen, Stein [1]	1994
2.725	Armen, Stein [2]	1995
2.667	Armen, Stein [3]	1996
2.596	Breslauer, Jiang, Jiang [5]	1997
2.500	Sweedyk [20]	1999
2.500	Kaplan, Lewenstein, Shafrir, Sviridenko [11]	2005
2.500	Paluch, Elbassioni, van Zuylen [18]	2012
2.479	Mucha [16]	2013
2.475	Englert, Matsakis, Veselý [8]	2022
2.466	Englert, Matsakis, Veselý [9]	2023

1.1 Our Contribution

We present *the Locally Greedy Algorithm* (LGA), whose pseudocode is given in Algorithm 1.

Algorithm 1. Locally Greedy Algorithm

    Input: set of strings $\mathcal{S}$.
    Output: a superstring for $\mathcal{S}$.
1: **while** $\mathcal{S}$ contains at least two strings **do**
2:     extract from $\mathcal{S}$ any ordered pair (s,t), whose overlap is the longest among pairs of the form (s',t'), where $s' = s$ or $t' = t$
3:     add to $\mathcal{S}$ the shortest string with a prefix s and a suffix t
4: return the only string from $\mathcal{S}$

Clearly, GA is a special case of LGA, and, as GA, this algorithm makes kind of "greedy" moves, but in general they have little to do with the length minimization. As an example, consider a set $\{\mathtt{ab}^n, \mathtt{b}^n\mathtt{a}\}$. For this set, GA always constructs an optimal superstring $ab^n a$, while LGA may well construct a superstring $b^n ab^n$, which is asymptotically 2 times longer than the optimal one. Although it seems implausible that LGA is an approximation algorithm at all (and for the maximum compression it is indeed not: the example above shows that the total overlap of LGA can be arbitrarily less than the maximum), we modify the proof of Blum et al. [4] to obtain

$$|\mathrm{LGA}(\mathcal{S})| \leq 4 \cdot |\mathrm{OPT}(\mathcal{S})|,$$

where $|s|$ is the length of a string s.

This result shows that to obtain a constant factor approximation for the length of the result, it is not necessary to focus on the reducing the length itself. This means that LGA (and hence GA) can be a constant factor approximation not only for the shortest length, but also for other similar metrics that it does not directly minimize. We show that it is the case for *the number of occurrences of a given symbol*. Let $|s|_p$ be the number of occurrences of a symbol p.

Theorem 1. *For any symbol p and a superstring s for $\mathcal{S}$,*

$$|\mathrm{LGA}(\mathcal{S})|_p \leq 4 \cdot |s|_p.$$

We call this property a *uniform* factor 4 approximation. Since $|s| = \sum_p |s|_p$, where the summation is taken over all the letters of the alphabet, the uniform λ-approximation implies λ-approximation.

Theorem 1 may give the impression that the key to solving the Greedy Conjecture lies in the uniform approximation of GA, and we can try to prove it by showing that GA is a uniform 2-approximation. However, this is not the case.

Theorem 2. LGA *is at least 3-approximation.* GA *is at least uniform 2.5-approximation.*

As with λ-approximation, bounds on the uniform approximation factor for one instantiation of GA imply the same bounds for all instantiations:

Theorem 3. *If some instantiation of* GA *is a uniform λ-approximation, then all of them are uniform λ-approximations.*[1]

The second result of Theorem 2 seems particularly interesting: we do not know if GA is a 2-approximation, but we do know for sure that it is not a uniform 2-approximation. This means that if GA is indeed a 2-approximation, then it cannot be inefficient in terms of the number of occurrences for too many different symbols at once. Why does this happen, and how can it be controlled? On the other hand, this same result may mean that GA is not a 2-approximation, and the counterexample from Theorem 2 can serve as a starting point for constructing a counterexample to GC. Finally, although a uniform approximation cannot be used to prove the 2-approximation of GA, it can potentially be used to prove the 2.5- or 3-approximation (if GA is indeed a 2.5 or 3-approximation, of course).

---

[1] The proof of this result is based on the technique from [17] and is omitted due to the page limit. See the full version for the proof.

1.2 Structure of the Paper

In Sect. 2 we introduce the necessary notation and definitions. In Sect. 3 we prove Theorem 1. In Sect. 4 we prove Theorem 2. Finally, in Sect. 5 we prove the general result, which implies Theorem 1 as a special case.

2 Preliminaries

For non-empty strings s and t, by $\mathrm{ov}(s,t)$ we denote their *overlap*, that is, the longest string y, such that $s = xy$ and $t = yz$ for some *non-empty* strings x and z, which we denote by $\mathrm{pref}(s,t)$ and $\mathrm{suff}(s,t)$, respectively. A string $xyz = \mathrm{pref}(s,t)\,\mathrm{ov}(s,t)\,\mathrm{suff}(s,t)$ is *a merge* of s and t (see Fig. 1). By $d(s,t) = |\mathrm{pref}(s,t)|$ we denote *the distance* between s and t. By ε we denote the empty string.

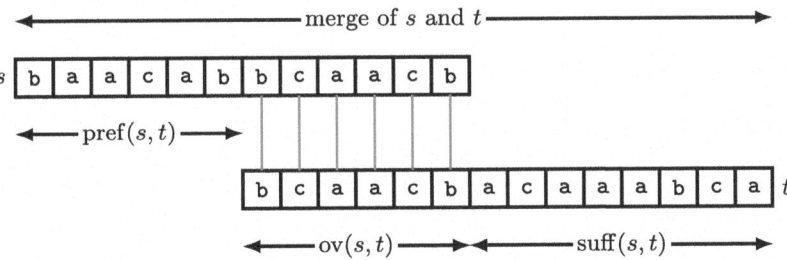

Fig. 1. Pictorial explanations of pref, suff, and ov functions.

Throughout the paper by $\mathcal{S} = \{s_1, \ldots, s_n\}$ we denote the set of n input strings. We assume that no input string is a substring of another, since such strings may be found and removed efficiently. In this case, SCS becomes *a permutation problem*: to find the shortest superstring, it is sufficient to find a permutation $(s_{\pi(1)}, \ldots, s_{\pi(n)})$ that gives the shortest string after merging the adjacent strings. For a given permutation of indices π, the length of the corresponding superstring $s(\pi)$ is equal to

$$|s_{\pi(1)}| + |\mathrm{suff}(s_{\pi(1)}, s_{\pi(2)})| + \cdots + |\mathrm{suff}(s_{\pi(n-1)}, s_{\pi(n)})|$$
$$= |\mathrm{pref}(s_{\pi(1)}, s_{\pi(2)})| + \cdots + |\mathrm{pref}(s_{\pi(n-1)}, s_{\pi(n)})| + |s_{\pi(n)}|$$
$$= \sum_i |s_i| - \sum_{i=1}^{n-1} |\mathrm{ov}(s_{\pi(i)}, s_{\pi(i+1)})|.$$

The last expression shows that finding the shortest superstring is equivalent to maximizing *the compression*, which is the sum of the overlaps of adjacent strings in a permutation. This naturally reduces SCS to the Longest Hamiltonian Path

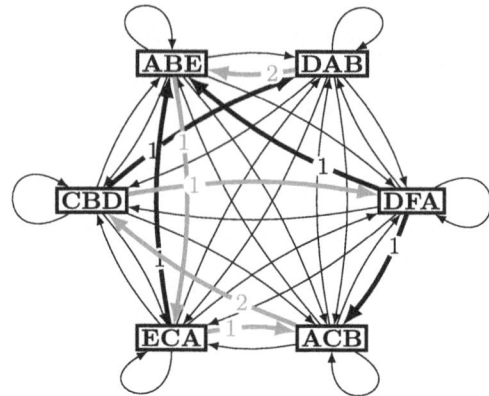

Fig. 2. An overlap graph for $\mathcal{S} = \{\texttt{ABE}, \texttt{DAB}, \texttt{DFA}, \texttt{ACB}, \texttt{ECA}, \texttt{CBD}\}$. Thin edges have zero weight. Orange path is a Hamiltonian path of the maximum weight 7, corresponding to the shortest superstring $\texttt{DABECACBDFA}$ of length $6 \cdot 3 - 7 = 11$.

problem in *an overlap graph* $OG(\mathcal{S})$, a complete directed graph (V, E) with self-loops, where $V = \mathcal{S}$, $E = \mathcal{S} \times \mathcal{S}$ and the length of a directed edge $(s, t) \in E$ is $|\operatorname{ov}(s,t)|$ (see Fig. 2).

The overlap graph has some crucial properties, which are widely used. One of them is *the triangle inequality*:

$$d(u, w) \leq d(u, v) + d(v, w). \tag{1}$$

Another one is *Monge inequality*:

Lemma 1 (Lemma 3.1 from [21]). *If* $|\operatorname{ov}(s,t)| \geq \max\{|\operatorname{ov}(s,t')|, |\operatorname{ov}(s',t)|\}$, *then* $|\operatorname{ov}(s,t)| + |\operatorname{ov}(s',t')| \geq |\operatorname{ov}(s,t')| + |\operatorname{ov}(s',t)|$.

For a cycle $C = v_1 \to \cdots \to v_k \to v_1$ in the overlap graph, by $|C|$ we denote its *length*:

$$|C| := \sum_i (|v_i| - |\operatorname{ov}(v_i, v_{(i+1) \bmod k})|) = \sum_i |\operatorname{pref}(v_i, v_{(i+1) \bmod k})|. \tag{2}$$

A set of edges is *a cycle cover* if both the in-degree and the out-degree of any node is equal to 1. This set is a collection of nonintersecting simple loops $C_1, \ldots, C_k$. The *maximum weight cycle cover* in the overlap graph is a cycle cover CC with the maximum total weight of its edges. Note, that the maximum weight cycle cover is also *the shortest* one, since $\sum_i |C_i| = \sum_i |s_i| - \sum_{(s,t) \in CC} |\operatorname{ov}(s,t)|$ and $\sum_i |s_i|$ is the same for all cycle covers.

Another property of the overlap graph is that nodes from the different cycles in the shortest cycle cover have bounded overlap.

Lemma 2 (Lemma 9 from [4]). *Let C_1 and C_2 be two cycles in the shortest cycle cover with $c_1 \in C_1$ and $c_2 \in C_2$. Then*

$$|\operatorname{ov}(c_1, c_2)| < |C_1| + |C_2|.$$

3 Uniform 4-Approximation of the Locally Greedy Algorithm

3.1 Pseudo-overlap Graph

In this section, we introduce a general result that implies Theorem 1 as a special case. Our approach is reminiscent to the one of Weinard and Schnitger [25] and Laube and Weinard [15]: they considered some properties of the overlap graph (namely, the triangle inequality, Monge inequality and the triple inequality) and proved that they are insufficient to prove the Greedy Conjecture, by providing an example of a graph with all these properties, but on which the Greedy Algorithm constructs a superstring that is too long. We also consider some properties of the overlap graph, but instead of proving insufficiency to achieve a factor 2 approximation, we prove sufficiency of these properties to achieve a factor 4 approximation.

Let $\mathcal{G} = (\mathcal{V}, \mathcal{V} \times \mathcal{V}, |\cdot|, \mathrm{w})$, where $\mathcal{V} = \{v_1, \ldots, v_n\}$, be a complete directed graph with self-loops and non-negative weights on nodes and edges: weight of a node v is $|v|$, weight of an edge (u, v) is $\mathrm{w}(u, v)$. For a subgraph G of $\mathcal{G}$ (that is, G is a pair (V, E), where $V \subset \mathcal{V}$ and $E \subset V \times V$) define *a weight* $\mathrm{w}(G) = \sum_{(u,v) \in E} \mathrm{w}(u,v)$ and *a length* $\|G\| := \sum_{v \in V} |v| - \mathrm{w}(G)$. For example, the weight of the path consisting of one edge $u \to v$ is $\mathrm{w}(u,v)$ and its length is $|u| - \mathrm{w}(u,v) + |v|$. Note that $\|u \to v \to w\| \neq \|u \to v\| + \|v \to w\|$, since $|v|$ is added once on the left side and twice on the right.

As in the previous section, we may define *the maximum weight cycle cover*, which is a cycle cover C with maximum $\mathrm{w}(C)$. Note, that C is also *the shortest cycle cover* with respect to the length $\|\cdot\|$, since $\|C\| = \sum_{v \in \mathcal{V}} |v| - \mathrm{w}(C)$ and $\sum_{v \in \mathcal{V}} |v|$ is the same for all cycle covers. Since $\mathrm{w} \geq 0$, C is also no longer than any Hamiltonian path in $\mathcal{G}$.

Let us call $\mathcal{G}$ *a pseudo-overlap graph* if it satisfies the following four properties:

P1. $|u| \geq \mathrm{w}(u,v)$ and $|u| \geq \mathrm{w}(v,u)$ for all $u, v \in \mathcal{V}$.
P2. Triangle inequality: $|u| - \mathrm{w}(u,v) + |v| - \mathrm{w}(v,w) \geq |u| - \mathrm{w}(u,w)$ for all $u, v, w \in \mathcal{V}$.
P3. Monge inequality: if $\mathrm{w}(u,v) \geq \max\{\mathrm{w}(u,v'), \mathrm{w}(u',v)\}$, then

$$\mathrm{w}(u,v) + \mathrm{w}(u',v') \geq \mathrm{w}(u,v') + \mathrm{w}(u',w).$$

P4. If C_1 and C_2 are two cycles in the maximum weight cycle cover on $\mathcal{G}$ with $c_1 \in C_1$ and $c_2 \in C_2$, then $\mathrm{w}(c_1, c_2) \leq \|C_1\| + \|C_2\|$.

Note that **P1** ensures that the length of any path or cycle is nonnegative, and **P2** ensures that a path $u \to v \to w$ is at least as long as $u \to w$. This means, in particular, that the shortest cycle cover on a complete subgraph $G = (V, V \times V)$ of $\mathcal{G}$ is no longer than the shortest cycle cover on $\mathcal{G}$.

Let us say that an edge (u,v) *dominates* (*strictly dominates*) another edge (u',v'), if $u' = u$ or $v' = v$ and $\mathrm{w}(u,v) \geq \mathrm{w}(u',v')$ ($\mathrm{w}(u,v) > \mathrm{w}(u',v')$). Call (u,v) *dominant*, if it dominates every edge, that begins in u or ends in v. Recall that $\mathcal{G}$ includes self-loops, so u may well be v, u', v', and so on.

Consider an algorithm PATH that goes through a list of all edges in G in a *dominance respecting order*, that is, if (u,v) strictly dominates (u',v'), then (u,v) will be processed earlier, and includes some of them in the final solution. Specifically, PATH *does not* include another edge (u,v) if and only if

R1. it is dominated by an already chosen edge,
R2. it is not dominated but it would form a cycle.

The resulting set of chosen edges forms a Hamiltonian path PATH($\mathcal{G}$).

Theorem 4. *Let SHP be the shortest Hamiltonian path in $\mathcal{G}$. Then $\|\text{PATH}(\mathcal{G})\| \leq 4 \cdot \|SHP\|$.*

The proof is in Sect. 5.

Corollary 1. *The Locally Greedy Algorithm is a factor 4 approximation.*

Proof. Note that LGA is a special case of PATH and the overlap graph is indeed a pseudo-overlap graph, where $|s|$ is the length of the corresponding string, $w(s,t) = |\text{ov}(s,t)|$ and the length of the path is the length of the corresponding superstring: **P1** and **P2** are obvious, **P3** is valid by Lemma 1 and **P4** is valid by Lemma 2.

3.2 Number of Occurrences Instead of Lengths

In this section, we prove that if we replace the length of a string by the number of occurrences of an arbitrary symbol p in it, the corresponding overlap graph remains a pseudo-overlap graph, which implies Theorem 1. More specifically, for any non-empty string p, let $\mathcal{G}_p$ be a graph such that $\mathcal{V} = \mathcal{S}$, $|s| = |s|_p$ and $w(s,t) = |\text{ov}(s,t)|_p$.

Theorem 5. *$\mathcal{G}_p$ is a pseudo-overlap graph.*

Proof. **P1.** This property clearly holds.
  P2. Since $\text{pref}(u,v)\,\text{pref}(v,w)$ contains $\text{pref}(u,w)$, the number of occurrences in it is at least the number of occurrences in $\text{pref}(u,w)$.
  P3. If $|\text{ov}(u,v)|_p \geq |\text{ov}(u,v')|_p + |\text{ov}(u',v)|_p$ then this property holds. If $|\text{ov}(u,v)|_p < |\text{ov}(u,v')|_p + |\text{ov}(u',v)|_p$ then there are strings x, y, z such that $\text{ov}(u,v) = xyz$, $\text{ov}(u,v') = yz$, $\text{ov}(u',v) = xy$. Since y is a suffix of u' and also a prefix of v', then $\text{ov}(u',v')$ contains y and hence $|\text{ov}(u',v')|_p \geq |y|_p$. Finally,

$$|\text{ov}(u,v)|_p = |xyz|_p = |xy|_p + |yz|_p - |y|_p \geq |\text{ov}(u,v')|_p + |\text{ov}(u',v)|_p - |\text{ov}(u',v')|_p.$$

  P4. For a string s by $s[i]$, we denote the i-th symbol of s. If $s[i] = p$, then i is *an occurrence* of p in s. Let $s(\pi)$ be a superstring, corresponding to the permutation $(s_{\pi(1)}, \ldots, s_{\pi(n)})$. Then

$$|s(\pi)|_p = \sum_{i=1}^{n-1} |\text{pref}(s_{\pi(i)}, s_{\pi(i+1)})|_p + |s_{\pi(n)}|_p \qquad (3)$$

Similarly, if $C = v_1 \to \cdots \to v_k \to v_1$ is a cycle in the overlap graph, by $|C|_p$ denote the following:

$$|C|_p := \sum_{i=1}^{k} |\operatorname{pref}(v_i, v_{(i+1) \bmod k})|_p. \qquad (4)$$

Lemma 3. *Let $C = v_1 \to \cdots \to v_k \to v_1$ be a cycle in the overlap graph and $v \in C$. If $l \leq r \leq |v|$ and $r - l < |C|$, then the number of occurrences of p in the substring $v[l, r]$ of v from l-th to r-th symbol is not greater than $|C|_p$.*

Proof. Without loss of generality, assume that $v = v_1$. To begin with, note, that if $s = w^k u$, $k > 0$, is a prefix of an infinite cyclic string w^∞ with the period w, then $|s[i, i + |w| - 1]|_p$ is the same for all i and is equal to $|w|_p$. Also, if $[l', r'] \subset [l, r]$, then $|s[l', r']|_p \leq |s[l, r]|_p$. Therefore, if $r - l < |w|$, then

$$|w|_p = |s[1, |w|]|_p \geq |s[l, r]|_p. \qquad (5)$$

Let $w = \operatorname{pref}(v_1, v_2) \operatorname{pref}(v_2, v_3) \ldots \operatorname{pref}(v_k, v_1)$. Then $v = v_1$ is a prefix of w^∞ and $|w| = |C|$, $|w|_p = |C|_p$. Therefore, if $r - l < |C|$, then by (5) $|v[l, r]|_p \leq |C|_p$.

Now we are ready to prove that property **P4** holds.

Lemma 4. *Let C_1 and C_2 be two cycles in the shortest cycle cover with $c_1 \in C_1$ and $c_2 \in C_2$. Then*

$$|\operatorname{ov}(c_1, c_2)|_p < |C_1|_p + |C_2|_p.$$

Proof. By Lemma 2, $|\operatorname{ov}(c_1, c_2)| \leq |C_1| + |C_2|$. If $|\operatorname{ov}(c_1, c_2)| \leq |C_i|$ for some $i \in \{1, 2\}$, then $\operatorname{ov}(c_1, c_2) = c_i[l, r]$ for some $l \leq r$, $r - l < |C_i|$ and Lemma 3 finishes the proof.

Otherwise, let $\operatorname{ov}(v, v') = xy$, where $|x| = |C_1|$ and $|y| \leq |C_2|$. Then $x = c_1[l_x, r_x]$, where $r_x - l_x = |C_1| - 1$, and $y = c_2[l_y, r_y]$, where $r_y - l_y < |C_2|$. Again, by Lemma 3, $|\operatorname{ov}(c_1, c_2)|_p = |x|_p + |y|_p \leq \|C_1\|_p + \|C_2\|_p$.

3.3 Proof of Theorem 1

Note that if $|\operatorname{ov}(s, t)| \geq |\operatorname{ov}(s, t')|$ then $|\operatorname{ov}(s, t)|_p \geq |\operatorname{ov}(s, t')|_p$ for any p. Similarly, if $|\operatorname{ov}(s, t)| \geq |\operatorname{ov}(s', t)|$ then $|\operatorname{ov}(s, t)|_p \geq |\operatorname{ov}(s', t)|_p$. This means that LGA is an instantiation of PATH for all $\mathcal{G}_p$ simultaneously.

Theorem 5 and the fact, that $|\operatorname{PATH}(\mathcal{G}_p)|_p$ is precisely $\|\operatorname{PATH}(\mathcal{G}_p)\|$, finish the proof.

4 Lower Bounds

4.1 LGA Is at Least 3-Approximation

The main source of suboptimality of LGA is that if there are two strings s and t such that s has only empty overlaps to the right and t has only empty overlaps to the left, it can merge them, even if there are much better options. In particular, if there are only two remaining strings, LGA can merge them in any order.

Consider a set $\mathcal{S} = \{ab^n, b^{n+1}, b^nc, b^{n-1}c^2\}$. The shortest solution $ab^{n+1}c^2$ has length $n+4$. Let us consider the following solution:

$$b^{n-1}c^2 \xrightarrow{2} ab^n \xrightarrow{1} b^nc \xrightarrow{3} b^{n+1},$$

where superscripts indicate the order of merges. Here the first merge is valid due to the largest overlap, the second is valid because $b^{n-1}c^2$ has empty overlap to the right with both remaining strings, and ab^n has empty overlap to the left with both remaining strings, and the last merge is always valid thanks to the observation from the beginning of the section. The corresponding superstring $b^{n-1}c^2ab^ncb^{n+1}$ has the length $3n+4$, which is asymptotically 3 times longer than $n+4$.

This example is similar to the dataset $\{ab^n, b^{n+1}, b^nc\}$ from the introduction, where GA could get twice as long a solution, but now there are two strings that we may place one to the left and one to the right of ab^nc, so the resulting solution is three times longer.

It is interesting that if we continue and consider, say, a dataset

$$\{ab^n, b^{n+1}, b^nc, b^{n-1}c^2, b^{n-2}c^3\},$$

then we will not be able to get a 4 times longer solution, because after we get the string $b^{n-2}c^3ab^nc$, it and $b^{n-1}c^2$ will have long overlaps on both sides.

4.2 GA Is at Least Uniform 2.5-Approximation

Consider a set $\mathcal{S} = \{\text{aaaab}, \text{aaabaa}, \text{aabaaba}, \text{baabaa}, \text{abaaaa}\}$. A solution

$$\text{aaaabaabaaaa}$$

has 2 occurrences of $p = b$.

Let us consider the following solution:

$$\text{baabaa} \xrightarrow{1} \text{aabaaba} \xrightarrow{4} \text{aaabaa} \xrightarrow{2} \text{abaaaa} \xrightarrow{3} \text{aaaab}.$$

The first merge is valid due to the longest possible overlap of length 5, the second and third are valid due to the longest possible overlaps of length 4, and the last is valid due to the longest possible overlap of length 1. The corresponding superstring baabaabaaabaaaab has 5 occurrences of b.

This example was found through computer search and creates more questions than answers. First, it is not clear whether this example can be tweaked to achieve uniform 3-approximation of GA.

Second, it seems like this example cannot be easily modified into a set for which GA would construct a solution more than twice as long as the shortest one. A natural approach in this direction would be to replace the symbol b with b^k, but in this case this would also require increasing the number of a's to keep the chosen overlaps maximal, and this does not result in more than two times longer solution.

5 Proof of Theorem 4

As was mentioned before, this proof is a modification of the proof of the 4-approximation of the Greedy Algorithm, presented in [4]. The overall structure is similar, and the main goal is to get rid of all arguments that appeal to the fact that the Greedy Algorithm merges pairs of strings with the largest overlap. For many statements below, their counterparts from [4] are indicated in brackets after the title.

Consider an algorithm CYC that goes through a list of all edges in $\mathcal{G}$ in the dominance respecting order and does not include an edge (u,v) in the final solution if and only if it is dominated by an already chosen edge. The resulting set of edges is a cycle cover on $\mathcal{G}$, and Monge inequality ensures that it is the shortest cycle cover.

Lemma 5 (Theorem 10). *Let C be a cycle cover constructed by* CYC. *Then C is a shortest cycle cover.*

Proof. Among the shortest cycle covers consider a cycle cover C' that has the maximum number of edges in common with C. We prove that C' is, in fact, equal to C.

Let (u, v') be an edge in $C' \setminus C$ with the maximum weight. Since (u, v') was not chosen by CYC, it is dominated by some edge in C. Without loss of generality, assume that this edge is of form (u, v). As $(u, v) \notin C'$, there must be an edge of form (u', v). Since $w(u, v) \geq w(u, v') \geq w(u', v)$, where second inequality holds by the definition of (u, v'), the edge (u, v) is also dominating (u', v). By Monge inequality, replacing (u, v') and (u', v) with (u, v) and (u', v') results in a cycle cover C'' with at least as large weight (and hence at most as large length) as C', that has at least one more edge in common with C. This contradicts the definition of C'.

Now let us return to the algorithm PATH. The resulting set of chosen edges PATH($\mathcal{G}$) forms a Hamiltonian path $v_{\pi(1)} \to \cdots \to v_{\pi(n)}$ for some permutation π. For convenience, let us renumber V so that $v_{\pi(i)}$ becomes v_i.

Call an edge (v_j, v_i), rejected because of **R2**, *a bad back edge*, which is "back" because necessarily $i \leq j$.

Observation 1 *If (v_j, v_i) is a bad back edge, then all edges (v_i, v_{i+1}), ..., (v_{j-1}, v_j) were processed earlier and edges (v_{i-1}, v_i) and (v_j, v_{j+1}) will be processed later.*

Let us say that a bad back edge (v_j, v_i) *spans an interval* $[i, j]$. Observation 1 implies that

Observation 2 (Lemma 13). *The intervals $[i, j]$ and $[i', j']$ spanned by two bad back edges are disjoint, or one contains the other.*

Thus, the bad back edges do not cross each other. Call a path $v_i \to \cdots \to v_j$, covered by some innermost bad back edge (v_j, v_i), a *culprit*. Call the edge included last between two successive culprits a *weak link*[2]. If the weak links are removed, the Hamiltonian path $s_1 \to \cdots \to s_n$ falls into *blocks*, each of which consists of the culprit as the middle segment and, possibly empty, *left* and *right* *extensions*. Thus $\mathcal{V}$ is divided into sets $\mathcal{V}_l$, $\mathcal{V}_m$ and $\mathcal{V}_r$ of the left, middle, and right nodes, respectively.

Observation 3 (Lemma 14). *Let (v_j, v_i) be a bad back edge. Node v_i is the left node or the first node of a culprit. Node v_j is either a right node or the last node of a culprit.*

Proof. If v_i is a right node, then an edge (v_{i-1}, v_i) lies on the left of the corresponding weak link. By observation 1 this weak link was included before (v_j, v_i) was processed, and hence before (v_{i-1}, v_i) was included. This contradicts the definition of a weak link. A similar argument holds for v_j.

Consider the graph $G_l = (V_l, E_l)$, where $V_l = \mathcal{V}_l \cup \mathcal{V}_m$ and E_l is a set of all chosen non-weak edges between left and middle nodes. Let C_l be the shortest cycle cover on $(V_l, V_l \times V_l)$. Let $V_r = \mathcal{V}_m \cup \mathcal{V}_r$ and define similarly the graph G_r and the shortest cycle cover C_r.

Let V be the disjoint union $V_l \sqcup V_r$, $E := E_l \sqcup E_m$ and $G := G_l \sqcup G_m$. Thus, each left or right node occurs in V once, while each middle node occurs twice. Add each weak link to E as an edge from the last node of the corresponding culprit/right extension in G_r to the first node of the corresponding culprit/left extension in G_l. Denote the resulting set of edges by E'. Note, that

$$\mathrm{w}(E') = \mathrm{w}(E_l) + \mathrm{w}(E_r) + \mathrm{w}(WL), \tag{6}$$

where WL is a set of weak links.

Consider $C = C_l \sqcup C_r$, a cycle cover on $(V, V \times V)$. Its weight equals

$$\mathrm{w}(C) = \mathrm{w}(C_l) + \mathrm{w}(C_r). \tag{7}$$

Each edge of C connects two V_l nodes or two V_r nodes, so all edges of C satisfy the assumptions of the following lemma.

Lemma 6 (Lemma 16). *Let C' be any cycle cover on V. Let (u,v) be an edge of $C' \setminus E'$ not in $V_r \times V_l$. Then (u,v) is dominated by either*

1. *an adjacent E' edge,*
2. *a bad back edge of the culprit with which it shares the head v and $v \in V_r$, or*
3. *a bad back edge of the culprit with which it shares the tail u and $u \in V_l$.*

---

[2] In [4], a weak link is *the shortest* edge between two successive culprits. Of course, the shortest edge is the last included in the case of GA. Authors remarked that weak links are also shorter than all the edges in the corresponding culprits, but they did not use this property.

Proof. Suppose first that (u,v) corresponds to a bad back edge. By Observation 3, v corresponds to a left node or to the first node of a culprit. In the latter case, (u,v) is dominated by the back edge of the culprit: since it is inside (u,v), it was processed before in the dominance respecting order. Therefore, either v is the first node of a culprit (and case (2) holds), or else $v \in V_l$. Similarly, either u is the last node of a culprit (and case (3) holds), or else $u \in V_r$. Since (u,v) is not in $V_r \times V_l$, it follows that case (2) or case (3) holds.

Suppose that (u,v) does not correspond to a bad back edge. Then it must be dominated by some edge chosen by PATH. If this edge is in E' then case (1) holds. If it is not in E', then (u,v) shares the head or tail with the bad back edge of some culprit, so (2) or (3) holds.

Although Lemma 6 ensures that each edge of C is dominated either by edge of E' or by bad back edge of culprit, it may be that some edges of E' dominate both of their adjacent edges of C. The following lemma shows that we can modify C into a new cycle cover C' with at least the same weight, so that each edge of E' dominates no more than one of its adjacent edges of C'.

Lemma 7 (Lemma 17). *Let C be any cycle cover on V such that $C \setminus E'$ does not contain edges of $V_r \times V_l$. Then there is a cycle cover C' such that*

1. *$C' \setminus E'$ has also no edges from $V_r \times V_l$,*
2. *$\mathrm{w}(C') \geq \mathrm{w}(C)$,*
3. *each edge in $E' \setminus C'$ dominates no more than one of its two adjacent C' edges.*

Proof. Since C already has the first two properties, it is sufficient to argue that if C violates property (3), then we can construct another cycle cover C' that satisfies properties (1) and (2), and has more edges in common with E'.

Let (u,v) be an edge from $E' \setminus C$ that dominates both adjacent C edges (u',v) and (u,v'). By Monge inequality, replacing edges (u',v) and (u,v') with (u,v) and (u',v') produces a cover C' with at least as much weight. To see that the new edge (u',v') is not in $V_r \times V_l$, observe that if $u' \in V_r$ then $v \in V_r$ ($C \setminus E'$ has no edges in $V_r \times V_l$), which implies that $u \in V_r$ (E' has no edges in $V_l \times V_r$ by construction), which implies that also $v' \in V_r$ because of $(u,v') \in C \setminus E'$.

By Lemmas 6 and 7, we can construct from the cycle cover C another cycle cover C' with at least as large weight, and such that each edge of C' is dominated by the edge of E' or by the bad back edge of the culprit. The edges of E' do not dominate more than one edge of C' and the bad back edges of the culprits do not dominate more than two. Thus, by (7),

$$\mathrm{w}(C_l) + \mathrm{w}(C_r) = \mathrm{w}(C) \leq \mathrm{w}(C') \leq \mathrm{w}(E') + 2\,\mathrm{w}(BC), \tag{8}$$

where BC is a set of bad back edges of culprits.

Let C_m be a cycle cover on $\mathcal{V}_m$, where each cycle is a culprit path P_i closed by the corresponding bad back edge.

Lemma 8 (Lemma 15). *C_m is the shortest cycle cover on $\mathcal{V}_m$.*

Proof. Recall the algorithm CYC that goes through a list of all edges in $\mathcal{V}_m \times \mathcal{V}_m$ in a dominance respecting order, induced by the order of PATH in $\mathcal{V}$, and does not include another edge if and only if it is dominated by an already chosen edge. Clearly, it constructs C_m. Therefore, by Lemma 5, C_m is the shortest cycle cover.

Note that $\|\text{PATH}(\mathcal{G})\| = \|E'\| - \sum_i \|P_i\|$ and $\sum_i \|P_i\| = \|C_m\| + \text{w}(BC)$. Then

$$\|\text{PATH}(\mathcal{G})\| = \|E'\| - \sum_i \|P_i\|$$

$$= \sum_{v \in V_r} |v| + \sum_{v \in V_l} |v| - \text{w}(E') - \sum_i \|P_i\|$$

$$\stackrel{(8)}{\leq} \sum_{v \in V_r} |v| + \sum_{v \in V_l} |v| - \text{w}(C_l) - \text{w}(C_r) + 2\,\text{w}(BC) - \sum_i \|P_i\|$$

$$= \|C_l\| + \|C_r\| + 2\,\text{w}(BC) - \sum_i \|P_i\|$$

$$\leq 2 \cdot \|SHP\| + \text{w}(BC) - \|C_m\|. \tag{9}$$

Lemma 9 (Theorem 8).

$$\text{w}(BC) - 2\|C_m\| \leq \|SHP\|. \tag{10}$$

Proof. Suppose $C_m = C_1 \cup \cdots \cup C_d$ and let $c_i \in C_i$ be a node with the maximum weight $|c_i|$. By **P1**, the weight of the bad back edge of C_i is not greater than $|c_i|$. By **P4**, the shortest cycle SC that goes through $\{c_1, \ldots, c_d\}$ has weight at most $2 \sum_i \|C_i\| = 2\|C_m\|$. Moreover, SC is not longer than the shortest Hamiltonian cycle on $\mathcal{V}$, which in turn is not longer than SHP. Therefore,

$$\text{w}(BC) - 2\|C_m\| \leq \sum_i |c_i| - 2\|C_m\| \leq \|SC\| \leq \|SHP\|.$$

Combining (9) and (10), we finally obtain

$$\|\text{PATH}(\mathcal{G})\| \leq 2 \cdot \|SHP\| + \text{w}(BC) - \|C_m\| \leq 3 \cdot \|SHP\| + \|C_m\| \leq 4 \cdot \|SHP\|.$$

Acknowledgments. The author would like to thank the anonymous reviewers for their valuable advice.

References

1. Armen, C., Stein, C.: A $2\frac{3}{4}$-Approximation Algorithm for the Shortest Superstring Problem. Tech. rep, Dartmouth College, Hanover (1994)
2. Armen, C., Stein, C.: Improved length bounds for the shortest superstring problem. In: Akl, S.G., Dehne, F., Sack, J.-R., Santoro, N. (eds.) WADS 1995. LNCS, vol. 955, pp. 494–505. Springer, Heidelberg (1995). https://doi.org/10.1007/3-540-60220-8_88

3. Armen, C., Stein, C.: A $2\frac{2}{3}$-approximation for the shortest superstring problem. In: Hirschberg, D., Myers, G. (eds.) Combinatorial Pattern Matching, LNCS, vol. 1075, pp. 87–101. Springer, Heidelberg (1996). https://doi.org/10.1007/3-540-61258-0_8
4. Blum, A., Jiang, T., Li, M., Tromp, J., Yannakakis, M.: Linear approximation of shortest superstrings. In: STOC 1991, pp. 328–336. ACM (1991)
5. Breslauer, D., Jiang, T., Jiang, Z.: Rotations of periodic strings and short superstrings. J. Algorithms **24**(2), 340–353 (1997)
6. Cazaux, B., Rivals, E.: Relationship between superstring and compression measures: new insights on the greedy conjecture. Discrete Appl. Math. **245**, 59–64 (2018)
7. Czumaj, A., Gasieniec, L., Piotrów, M., Rytter, W.: Parallel and sequential approximation of shortest superstrings. In: Schmidt, E.M., Skyum, S. (eds.) SWAT 1994. LNCS, vol. 824, pp. 95–106. Springer, Heidelberg (1994). https://doi.org/10.1007/3-540-58218-5_9
8. Englert, M., Matsakis, N., Veselý, P.: Improved approximation guarantees for shortest superstrings using cycle classification by overlap to length ratios. In: Proceedings of the 54th Annual ACM SIGACT Symposium on Theory of Computing, pp. 317–330 (2022)
9. Englert, M., Matsakis, N., Veselý, P.: Approximation guarantees for shortest superstrings: simpler and better. In: Iwata, S., Kakimura, N. (eds.) 34th International Symposium on Algorithms and Computation (ISAAC 2023). Leibniz International Proceedings in Informatics (LIPIcs), vol. 283, pp. 29:1–29:17. Schloss Dagstuhl – Leibniz-Zentrum für Informatik, Dagstuhl (2023)
10. Gallant, J., Maier, D., Storer, J.A.: On finding minimal length superstrings. J. Comput. Syst. Sci. **20**(1), 50–58 (1980)
11. Kaplan, H., Lewenstein, M., Shafrir, N., Sviridenko, M.: Approximation algorithms for asymmetric TSP by decomposing directed regular multigraphs. J. ACM **52**, 602–626 (2005)
12. Kaplan, H., Shafrir, N.: The greedy algorithm for shortest superstrings. Inf. Process. Lett. **93**(1), 13–17 (2005)
13. Kosaraju, S.R., Park, J.K., Stein, C.: Long tours and short superstrings. In: Proceedings of the 35th Annual Symposium on Foundations of Computer Science (SFCS 1994), pp. 166–177. IEEE Computer Society, Washington (1994)
14. Kulikov, A.S., Savinov, S., Sluzhaev, E.: Greedy conjecture for strings of length 4. In: Cicalese, F., Porat, E., Vaccaro, U. (eds.) CPM 2015. LNCS, vol. 9133, pp. 307–315. Springer, Cham (2015). https://doi.org/10.1007/978-3-319-19929-0_26
15. Laube, U., Weinard, M.: Conditional inequalities and the shortest common superstring problem. Int. J. Found. Comput. Sci. **16**(06), 1219–1230 (2005)
16. Mucha, M.: Lyndon words and short superstrings. In: SODA 2013, pp. 958–972. SIAM (2013)
17. Nikolaev, M.S.: All instantiations of the greedy algorithm for the shortest common superstring problem are equivalent. In: String Processing and Information Retrieval: 28th International Symposium, SPIRE 2021, Lille, 4–6 October 2021, Proceedings 28, pp. 61–67. Springer (2021)
18. Paluch, K., Elbassioni, K., van Zuylen, A.: Simpler approximation of the maximum asymmetric traveling salesman problem. In: STACS 2012. LIPIcs, vol. 14, pp. 501–506 (2012)
19. Pevzner, P.A., Tang, H., Waterman, M.S.: An eulerian path approach to dna fragment assembly. Proc. Natl. Acad. Sci. U.S.A. **98**(17), 9748–9753 (2001)
20. Sweedyk, Z.: $2\frac{1}{2}$-approximation algorithm for shortest superstring. SIAM J. Comput. **29**(3), 954–986 (1999)

21. Tarhio, J., Ukkonen, E.: A greedy approximation algorithm for constructing shortest common superstrings. Theor. Comput. Sci. **57**(1), 131–145 (1988)
22. Teng, S.H., Yao, F.: Approximating shortest superstrings. In: Proceedings of the 1993 IEEE 34th Annual Foundations of Computer Science (SFCS 1993), pp. 158–165. IEEE Computer Society, Washington (1993)
23. Turner, J.S.: Approximation algorithms for the shortest common superstring problem. Inf. Comput. **83**(1), 1–20 (1989)
24. Waterman, M.S.: Introduction to Computational Biology: Maps, Sequences and Genomes. CRC Press (1995)
25. Weinard, M., Schnitger, G.: On the greedy superstring conjecture. SIAM J. Discrete Math. **20**(2), 502–522 (2006)

Faster Computation of Chinese Frequent Strings and Their Net Frequencies

Enno Ohlebusch(✉), Thomas Büchler, and Jannik Olbrich

Institute of Theoretical Computer Science, Ulm University, 89069 Ulm, Germany
{enno.ohlebusch,thomas.buechler,jannik.olbrich}@uni-ulm.de

Abstract. A Chinese frequent string (CFS) is a repeated string in a Chinese text that has at least one net occurrence (i.e., an occurrence with the property that both its left-extension and its right-extension by one character are unique in the text). The net frequency of a CFS is the number of its net occurrences in the text. In this short paper, we improve recent work of Guo et al. [4] in such a way that the computation of Chinese frequent strings and their net frequencies becomes much faster.

1 Introduction

The notion of net frequency has been introduced by Lin and Yu [7] in the context of extracting "Chinese frequent strings" (CFSs) from a text. To learn more about CFSs, we recommend the reader to read the beginning of Sect. 2 in [7]. We here confine ourselves to the following citation: "A CFS is a Chinese string that is used frequently by people. There are about 13,000 Chinese characters. The number of acceptable combinations of several characters is very large, but only a few combinations are frequently used." In short, a CFS is a repeated string in a Chinese text that has at least one *net occurrence* [4], i.e. an occurrence which has the property that its extension by one character to the left (and to the right, respectively) is not a repeated string. The number of net occurrences of a CFS is called *net frequency* (NF) of the CFS. Detecting CFSs and their net occurrences/frequencies is "very useful in Chinese natural language processing and its related applications" [7], e.g. because it enables word segmentation without a dictionary (there are no explicit word boundaries in Chinese sentences) [5,7,13]. CFSs have also been used to enhance Web search [3], and for Chinese spelling error correction, among other natural language processing tasks [8].

Recently, Guo et al. [4] re-examined the original definition of NF and showed that it is equivalent to the definition used above. They presented two algorithms that efficiently compute all strings with positive NF (ALL_NF_REPORT) and, if desired, all their net occurrences (ALL_NF_EXTRACT). In this short paper, we improve their Algorithm 3 in such a way that it becomes much faster and uses less space. As pointed out by Guo et al. [4], strings with positive NF and maximal repeats are closely related but different from each other.

2 Preliminaries

Let S be a text of length n on an ordered alphabet Σ of size σ. We assume that S has the sentinel $\$ \in \Sigma$ at the end (and nowhere else), which is smaller than any other character. For $1 \leq i \leq n$, $S[i]$ denotes the *character at position i* in S. For $1 \leq i \leq j \leq n$, $S[i..j]$ denotes the *substring* of S starting at position i and ending at position j. Furthermore, S_i denotes the i-th suffix $S[i..n]$ of S.

The *suffix array* SA of the text S is an array of integers in the range 1 to n specifying the lexicographic ordering of the n suffixes of S, that is, it satisfies $S_{\mathsf{SA}[1]} < S_{\mathsf{SA}[2]} < \cdots < S_{\mathsf{SA}[n]}$. The suffix array can be built in linear time; we refer to the overview article of [12] for suffix array construction algorithms and to [11] for recent developments.

The Burrows and Wheeler transform [2] converts S into the string $\mathsf{BWT}[1..n]$ defined by $\mathsf{BWT}[i] = S[\mathsf{SA}[i] - 1]$ for all i with $\mathsf{SA}[i] \neq 1$ and $\mathsf{BWT}[i] = \$$ otherwise. To avoid tedious case distinctions, we assume $S[0] = \$$ from now on (but $S[0]$ is not part of the text).

The suffix array SA is often enhanced with the so-called LCP-array containing the lengths of longest common prefixes between consecutive suffixes in SA. Formally, the LCP-array is an array so that $\mathsf{LCP}[1] = -1 = \mathsf{LCP}[n+1]$ and $\mathsf{LCP}[i] = |\mathsf{lcp}(S_{\mathsf{SA}[i-1]}, S_{\mathsf{SA}[i]})|$ for $2 \leq i \leq n$, where $\mathsf{lcp}(u,v)$ denotes the longest common prefix between two strings u and v. As the suffix array, the LCP-array can be computed in linear time [6]. Abouelhoda et al. [1] introduced the concept of lcp-intervals. An interval $[i..j]$, where $1 \leq i < j \leq n$, in the LCP-array is called an *lcp-interval of lcp-value* ℓ (denoted by $\ell\text{-}[i..j]$) if

1. $\mathsf{LCP}[i] < \ell$,
2. $\mathsf{LCP}[k] \geq \ell$ for all k with $i+1 \leq k \leq j$,
3. $\mathsf{LCP}[k] = \ell$ for at least one k with $i+1 \leq k \leq j$,
4. $\mathsf{LCP}[j+1] < \ell$.

Every index k, $i < k \leq j$, with $\mathsf{LCP}[k] = \ell$ is called ℓ-*index* or *lcp-index*. Figure 1 shows an example. Abouelhoda et al. [1] showed that there is a one-to-one correspondence between the set of all lcp-intervals and the set of all internal nodes of the suffix tree of S (we assume a basic knowledge of suffix trees). A leaf in the suffix tree corresponds to a *singleton interval* $[k..k]$. For every index k, $1 \leq k \leq n$, $[k..k]$ is a singleton interval. The parent of a singleton interval is the smallest lcp-interval $[i..j]$ that contains k. In this case, we call $[k..k]$ the child of $[i..j]$. In Fig. 1, for example, the singleton interval $[2..2]$ is child of 1-$[2..5]$, $[5..5]$ is child of 4-$[4..5]$, and $[6..6]$ is child of 0-$[1..12]$. There is no singleton interval which has the lcp-interval 1-$[9..12]$ as parent.

3 Fast Computation of Net Frequencies

Let ω be a substring of S with $|\omega| = \ell$. The frequency of ω (often also called term frequency) is $f(\omega) = |\{ k : 1 \leq k \leq n - \ell + 1 \text{ and } S[k..k+\ell-1] = \omega \}|$, i.e. the number of occurrences of ω in S. The string ω is unique in S if $f(\omega) = 1$; it is a repeat if $f(\omega) \geq 2$.

k	SA	LCP	BWT	$S_{\mathsf{SA}[k]}$
1	12	-1	i	$
2	11	0	p	i$
3	8	1	s	ippi$
4	5	1	s	issippi$
5	2	4	m	ississippi$
6	1	0	$	mississippi$
7	10	0	p	pi$
8	9	1	i	ppi$
9	7	0	s	sippi$
10	4	2	s	sissippi$
11	6	1	i	ssippi$
12	3	3	i	ssissippi$

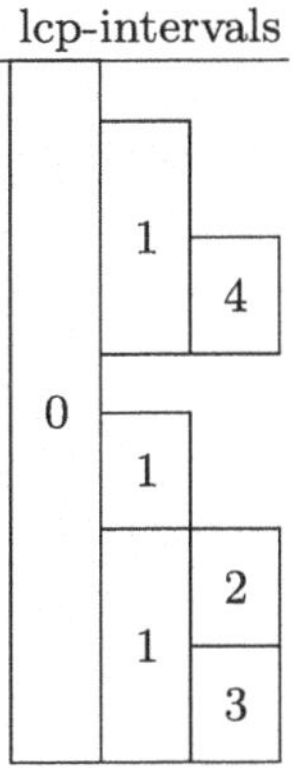

Fig. 1. Suffix array, LCP-array, BWT and lcp-intervals (each containing its lcp-value) of the text $S = mississippi\$$

Definition 1. *The* net frequency $\Phi(\omega)$ *of a repeat ω in S is defined by*

$$\Phi(\omega) = |\{(a,b) \in \Sigma \times \Sigma : f(a\omega) = 1 \text{ and } f(\omega b) = 1 \text{ and } f(a\omega b) = 1\}|$$

An occurrence $S[k..k+\ell-1]$ at position k in S of the length-ℓ repeat ω is called net occurrence *of ω in S if both its left-extension $S[k-1..k+\ell-1]$ and its right-extension $S[k..k+\ell]$ are unique.*[1]

In this terminology, the net frequency of a repeat ω is the number of its net occurrences in S. The net frequency of the (very short) string i in $S=mississippi\$$ is $\Phi(i) = 1$.[2] The occurrence at position 11 is a net occurrence, but those at positions 2, 5, and 8 are not. For instance, the right-extension $S[2..3] = is$ of the occurrence at position 2 has frequency 2 and the left-extension $S[7..8] = si$ of the occurrence at position 8 has frequency 2 as well.

Our next goal is to characterize net occurrences (Lemma 1). Let ω be a substring of S. The ω-interval is the interval $[i..j]$ so that ω is a common prefix of $S_{\mathsf{SA}[i]}, S_{\mathsf{SA}[i+1]}, \ldots, S_{\mathsf{SA}[j]}$, but neither of $S_{\mathsf{SA}[i-1]}$ nor of $S_{\mathsf{SA}[j+1]}$. Clearly, the ω-interval is a singleton interval (i.e. $i = j$) if and only if ω is unique in S.

Lemma 1. *An occurrence $S[k..k+\ell-1]$ of the length-ℓ repeat ω is a net occurrence if and only if (a) the ω-interval $[i..j]$ is an lcp-interval of lcp-value $\ell = |\omega|$, (b) the $S[k..k+\ell]$-interval is a singleton interval, and (c) $S[k-1]$ occurs exactly once in the sequence $\mathsf{BWT}[i], \mathsf{BWT}[i+1], \ldots, \mathsf{BWT}[j]$.*

Proof. "$\Rightarrow$": Suppose that $S[k..k+\ell-1]$ is a net occurrence of ω in S. Since ω is a repeat, the ω-interval $[i..j]$ is not a singleton interval. Since the right-extension $S[k..k+\ell]$ is unique, statement (b) holds. It is a direct consequence

---
[1] Strictly speaking, $f(S[k-1..k+\ell-1])$ is not defined for $k = 1$. In this case, we define that $f(S[k-1..k+\ell-1])$ is unique. The same applies to the definition of Φ.
[2] We do not use Chinese texts in examples because we do not have command of Chinese.

of (b) and the fact that ω is a repeat that there must be another occurrence $S[k'..k' + \ell - 1]$ of ω with $S[k..k + \ell] \neq S[k'..k' + \ell]$. That is, the characters immediately following the two occurrences are different. This implies that the ω-interval is an lcp-interval of lcp-value $\ell = |\omega|$, i.e. (a) holds true. Since the left-extension $S[k - 1..k + \ell - 1]$ is unique in S, the character $S[k - 1]$ occurs exactly once in $\mathsf{BWT}[i], \mathsf{BWT}[i + 1], \ldots, \mathsf{BWT}[j]$. This proves (c).

"$\Leftarrow$": According to (a), ω is a common prefix of the suffixes $S_{\mathsf{SA}[i]}, \ldots, S_{\mathsf{SA}[j]}$ and $i < j$, so ω is a repeat of length ℓ. It is a direct consequence of (b) that the right-extension $S[k..k+\ell]$ of the occurrence $S[k..k+\ell-1]$ is unique. Furthermore, if $S[k - 1]$ occurs exactly once in $\mathsf{BWT}[i], \mathsf{BWT}[i + 1], \ldots, \mathsf{BWT}[j]$, then the left-extension $S[k - 1..k + \ell - 1]$ is also unique.

Let us illustrate Lemma 1 with a small example. The string i of length 1 is a repeat in $S = mississippi\$$. We consider the occurrence $S[11]$ of i. (a) The i-interval is the lcp-interval $[2..5]$ of lcp-value 1; see Fig. 1. (b) $S[11..12] = i\$$ and the $i\$$-interval is the singleton interval $[2..2]$. (c) $S[10] = p$ occurs exactly once in $\mathsf{BWT}[2], \ldots, \mathsf{BWT}[5]$. Consequently, $S[11]$ is a net occurrence of i in S according to Lemma 1.

Algorithm 1 is based on Lemma 1. It is a variation of the well-known stack-based algorithm that enumerates all lcp-intervals in $O(n)$ time [1,6] (which was also used by Guo et al. [4]). In Algorithm 1, push (pushes an element onto the stack) and pop (pops an element from the stack and returns that element) are the usual stack operations, while top provides a pointer to the topmost element of the stack. The elements on the stack are triples $\langle lcp, lb, list \rangle$. When such a triple is popped from the stack, then $[lb..rb]$ is an lcp-interval of lcp-value lcp. Its right boundary is $rb = k - 1$, where k is the current value of the for-loop variable. Moreover, $list$ contains all indices j with $lb \leq j \leq rb$ so that the singleton interval $[j..j]$ is a child of $[lb..rb]$. This is because $k - 1$ (i.e., the singleton interval $[k-1..k-1]$) is inserted into the list of its parent interval when the body of the for-loop is executed for a value k. To verify this claim, we use a case distinction. Let $m := \mathsf{LCP}[k]$.

1. If $\mathsf{LCP}[k] = top().lcp$, then k is an lcp-index of the lcp-interval $top()$. This is only possible if $\mathsf{LCP}[k - 1] = m$ (note that $\mathsf{LCP}[k - 1] > m$ would imply $top().lcp > m$ and $\mathsf{LCP}[k - 1] < m$ would imply $top().lcp < m$). As a consequence of $\mathsf{LCP}[k - 1] = m = \mathsf{LCP}[k]$, $top()$ is the parent of $[k - 1..k - 1]$; cf. [9, Lemma 4.3.5]. Thus, $k - 1$ is correctly added to $top().list$.
2. If $\mathsf{LCP}[k] > top().lcp$, then neither the preceding if-then-statements nor the while-loop is executed. Hence $dealt\_with = false$ (at line 23). Furthermore, a new lcp-interval is found and pushed onto the stack. This lcp-interval starts at index $k - 1$ and has lcp-value m. Thus it contains the singleton interval $[k-1..k-1]$. If there was a smaller lcp-interval containing $[k-1..k-1]$, then it would have $k-1$ as its left boundary and an lcp-value $> m$. This, however, is impossible. Consequently, $k - 1$ is correctly added to the $top().list$.
3. If $\mathsf{LCP}[k] < top().lcp$, then $k - 1$ is the right boundary of the lcp-interval $top()$. Therefore, $[k - 1..k - 1]$ is a child of $top()$. So $k - 1$ is correctly added to the $top().list$ and $dealt\_with$ is set to $true$. Subsequently, lcp-intervals are

Algorithm 1: This procedure computes the net frequency of all substrings of S with positive net frequency.

Input: LCP-array, BWT
1. **for** $c \leftarrow 1$ **to** σ **do**
2. $\quad penultimate[c] \leftarrow 0$
3. $\quad last[c] \leftarrow 0$
4. $push(\langle 0, 1, [\ ]\rangle)$
5. **for** $k \leftarrow 2$ **to** $n+1$ **do**
6. $\quad c \leftarrow \mathsf{BWT}[k-1]$
7. $\quad penultimate[c] \leftarrow last[c]$
8. $\quad last[c] \leftarrow k - 1$
9. $\quad$ **if** $\mathsf{LCP}[k] = top().lcp$ **then**
10. $\quad\quad top().list \leftarrow add(top().list, k-1)$
11. $\quad\quad$ **continue** // with the next iteration of the for-loop
12. $\quad dealt\_with \leftarrow false$
13. $\quad$ **if** $\mathsf{LCP}[k] < top().lcp$ **then**
14. $\quad\quad top().list \leftarrow add(top().list, k-1)$
15. $\quad\quad dealt\_with \leftarrow true$
16. $\quad lb \leftarrow k - 1$
17. $\quad$ **while** $\mathsf{LCP}[k] < top().lcp$ **do**
18. $\quad\quad \langle \ell, i, list \rangle \leftarrow pop()$
19. $\quad\quad \texttt{computeNetFrequency}(\ell, i, list, penultimate, last)$
20. $\quad\quad lb \leftarrow i$
21. $\quad$ **if** $\mathsf{LCP}[k] > top().lcp$ **then**
22. $\quad\quad push(\langle \mathsf{LCP}[k], lb, [\ ]\rangle)$
23. $\quad\quad$ **if** $dealt\_with = false$ **then**
24. $\quad\quad\quad top().list \leftarrow add(top().list, k-1)$

popped from the stack until $\mathsf{LCP}[k] > top().lcp$. Observe that $k-1$ is not added to the *list* of the new topmost element of the stack (at line 23) because $dealt\_with = true$.

When a triple $\langle \ell, i, list \rangle$ is popped from the stack in line 18, then the interval $[i..k-1]$ (where k is the current value of the for-loop variable) is an lcp-interval of lcp-value ℓ. That is, the string $\omega = S[\mathsf{SA}[i]..\mathsf{SA}[i] + \ell - 1]$ is a common prefix of all the suffixes in the interval $[i..k-1]$. Moreover, *list* contains all singleton intervals for which $[i..k-1]$ is the parent. That is, for each $j \in list$ the occurrence $S[\mathsf{SA}[j]..\mathsf{SA}[j] + \ell - 1]$ satisfies conditions (a) and (b) of Lemma 1. According to Lemma 1 it is a net occurrence of ω if it also satisfies condition (c). Lines 1-3 and 6-8 of Algorithm 1 provide a means to test condition (c) in constant time. The algorithm maintains two arrays, *last* and *penultimate*, of size $\sigma = |\Sigma|$. It is an invariant of the for-loop that an entry $last[c]$ stores the position of the last occurrence of the character c in $\mathsf{BWT}[1..k-1]$. Similarly, an entry $penultimate[c]$ stores the position of the second to last occurrence of

Algorithm 2: This procedure checks whether an occurrence of a candidate string has a unique left-extension. If this is the case, it reports the net frequency and the net occurrences of the string.

```
1  Function computeNetFrequency(ℓ, i, list, penultimate, last)
2      Φ ← 0; posNetFreq ← ∅
3      foreach j ∈ list do
4          c ← BWT[j]
5          if j = last[c] and penultimate[c] < i then
6              Φ ← Φ + 1; posNetFreq ← posNetFreq ∪ {j}
       // ALL_NF_REPORT
7      if Φ > 0 then
8          output: the length-ℓ string starting at position SA[i] has NF Φ
       // ALL_NF_EXTRACT
9      foreach j ∈ posNetFreq do
10         output: the length-ℓ string at position SA[j] is a net occurrence
```

the character c in $\mathsf{BWT}[1..k-1]$. Suppose a triple $\langle \ell, i, list \rangle$ is popped from the stack. As explained above, it must be checked for each $j \in list$ whether the character $c = S[\mathsf{SA}[j]-1] = \mathsf{BWT}[j]$ appears exactly once in $\mathsf{BWT}[i..k-1]$. The procedure *computeNetFrequency* in Algorithm 2 makes use of the fact that this is the case if and only if $j = last[c]$ and $penultimate[c] < i$. To see why this is correct, observe that $j = last[c]$ if and only if there is no other occurrence of c in $\mathsf{BWT}[j+1..k-1]$. Now if $j = last[c]$, then there is no other occurrence of c in $\mathsf{BWT}[i..k-1]$ if and only if $penultimate[c] < i$ (if $penultimate[c] \geq i$, then the inequalities $i \leq penultimate[c] < last[c] \leq k-1$ imply that there are at least two occurrences of c in $\mathsf{BWT}[i..k-1]$).

The worst-case time complexity of our algorithm is $O(n)$. This is because the enumeration of all lcp-intervals takes $O(n)$ time, exactly n elements are added to the lists, and the test (Algorithm 2) takes constant time for every element. Apart from the initialization of the arrays *penultimate* and *last*, our algorithm has an alphabet-independent linear runtime; cf. [10]. It should be pointed out that the runtime can be superlinear if the algorithm would output all strings explicitly. Note that ALL_NF_REPORT in Algorithm 2 outputs an implicit representation of the string $S[\mathsf{SA}[i]..\mathsf{SA}[i]+\ell-1]$ in form of the triple $(\mathsf{SA}[i], \ell, \Phi)$. As shown in [9, Lemma 5.3.6] for longest repeats in de Bruijn sequences (which is also applicable here because each longest repeat in a de Bruijn sequence over a binary alphabet has NF 2), the overall output size can be proportional to $n \log n$ if the strings are output explicitly as sequences of characters.

4 Experiments

We implemented our algorithm in C++ using the divsufsort library[3] for constructing the suffix array (Guo et al. [4] also used divsufsort). The implementation is publicly available on GitHub.[4] In our experiments, we used the files dna, proteins, and english.1024MB from the Pizza&Chili Corpus[5] as input. Our algorithm was compiled with GCC 11.1.0 and the optimization flags -O3 -DNDEBUG -march=native. All experiments were conducted on a Linux-5.4.0 machine with an AMD EPYC 7742 (64 cores, 128 threads) processor and 256GB of RAM. The reported wall clock time and the memory consumption were measured externally by GNU Time. We compared our implementation with Algorithms 2 and 3 from Guo et al. [4]. In order to do that, we had to make minor changes to their implementation,[6] which only affected the output of the net frequencies. Furthermore, we created an alternative main file that calls the algorithms. (The original main file only calls Algorithm 1 from Guo et al. [4].)

In our experiments, we measured memory consumption, the run-time of the algorithm that computes the net frequencies, as well as the total program time. The latter includes the construction of the suffix array, LCP-array, BWT, and other related processes. The results are shown in Table 1. In comparison, our implementation is more than 10 times faster and requires less than a quarter of the memory in all test cases. For instance, for the 1 GB file of English text, our program computes all net frequencies in approximately 4 minutes, requiring 14 GB of memory.

Table 1. Evaluation of our algorithm as well as Algorithms 2 and 3 from Guo et al. [4]. The input text files dna, protein, and english were of size 386 MB, 1.2 GB, and 1 GB, respectively

	time for NF in seconds			total time in min:sec			memory usage in GB		
	dna	prot.	engl.	dna	prot.	engl.	dna	prot.	engl.
Our algorithm	32	53	43	1:36	4:49	4:03	5	15	14
Guo et al. Alg. 2	169	238	180	18:24	53:32	46:03	22	67	60
Guo et al. Alg. 3	193	323	283	18:59	55:21	47:56	22	67	60

---

[3] https://github.com/y-256/libdivsufsort.
[4] https://github.com/thomas-buechler-ulm/net-frequencies.
[5] https://pizzachili.dcc.uchile.cl/texts.html.
[6] https://github.com/Peakergzf/string-net-frequency/.

References

1. Abouelhoda, M.I., Kurtz, S., Ohlebusch, E.: Replacing suffix trees with enhanced suffix arrays. J. Discrete Algorithms **2**(1), 53–86 (2004)
2. Burrows, M., Wheeler, D.J.: A block-sorting lossless data compression algorithm. Research Report 124, Digital Systems Research Center (1994)
3. Chen, Y.-R., Hung, M.-C., Yang, D.-L.: Using data mining to construct an intelligent web search system. Int. J. Comput. Process. Orient. Lang. **16**(02), 143–170 (2003)
4. Guo, P., Eades, P., Wirth, A., Zobel, J.: Exploiting new properties of string net frequency for efficient computation. In: Proceedings of the 35th Annual Symposium on Combinatorial Pattern Matching, vol. 296 of Leibniz International Proceedings in Informatics, pp. 16:1–16:16 (2024)
5. Jiang, T.-J., Liu, S.-H., Sung, C.-L., Hsu, W.-L.: Term contributed boundary feature using conditional random fields for Chinese word segmentation task. In: Proceedings of the 22nd Conference on Computational Linguistics and Speech Processing, pp. 143–156 (2010)
6. Kasai, T., Lee, G., Arimura, H., Arikawa, S., Park, K.: Linear-time longest-common-prefix computation in suffix arrays and its applications. In: Amir, A. (ed.) CPM 2001. LNCS, vol. 2089, pp. 181–192. Springer, Heidelberg (2001). https://doi.org/10.1007/3-540-48194-X_17
7. Lin, Y.-J., Yu, M.-S.: Extracting Chinese frequent strings without a dictionary from a Chinese corpus and its applications. J. Inf. Sci. Eng. **17**, 805–824 (2001)
8. Lin, Y.-J., Yu, M.-S.: The properties and further applications of Chinese frequent strings. Int. J. Comput. Linguist. Chin. Lang. Process. **9**(1), 113–128 (2004)
9. Ohlebusch, E.: Bioinformatics Algorithms: Sequence Analysis, Genome Rearrangements, and Phylogenetic Reconstruction. Oldenbusch Verlag (2013)
10. Ohlebusch, E., Beller, T.: Alphabet-independent algorithms for finding context-sensitive repeats in linear time. J. Discrete Algorithms **34**, 23–36 (2015)
11. Olbrich, J., Ohlebusch, E., Büchler, T.: Generic non-recursive suffix array construction. ACM Trans. Algorithms **20**(2), 1–42 (2024)
12. Puglisi, S.J., Smyth, W.F., Turpin, A.: A taxonomy of suffix array construction algorithms. ACM Comput. Surv. **39**(2), 4 (2007)
13. Shen, M., Kawahara, D., Kurohashi, S.: Chinese word segmentation and unknown word extraction by mining maximized substring. Inf. Media Technol. **11**, 181–212 (2016)

Faster Algorithms for Ranking/Unranking Bordered and Unbordered Words

Jakub Radoszewski(✉) , Wojciech Rytter , and Tomasz Waleń

University of Warsaw, Warsaw, Poland
{jrad,rytter,walen}@mimuw.edu.pl

Abstract. We show how the arithmetic structure of the set of borders (periods) of a word can be used to substantially reduce complexity of an interesting problem in combinatorics on words. A word w is a *bordered word* if it has a non-empty proper border (a prefix which is a suffix); equivalently, it has a period smaller than $|w|$. Words which are not bordered are called *unbordered*. The problem of ranking/unranking such words of a given length n over an alphabet of size k was considered by Gabric (*Inf. Process. Lett.*, 2024). We improve his results as follows: complexity of ranking is reduced by a factor $nk/\log n$ and complexity of unranking by $n^2k/\log n$ (for large alphabets these improvement factors are $n^2/\log n$ and $n^3/\log n$, respectively). We use the unit-cost RAM model (the same model was used by Gabric).

Keywords: string border · unbordered string · bordered string · ranking algorithm · unranking algorithm

1 Introduction

A prefix of a word w is called a *border* of w if and only if it is also a suffix of w. A border of a word w is called *nontrivial* if it is non-empty and proper, that is, shorter than w. A word w is called *bordered* if it has a nontrivial border; otherwise w is called *unbordered*. Unbordered words were considered already in the 1970s, by Nielsen [16] under the name of bifix-free words and by Ehrenfeucht and Silberger [4]. Unbordered words are also known as primary words [14]. The relationship between the length of a maximal unbordered factor of a word and its period was studied in [1,3,10]. Combinatorial bounds on the length of a maximal unbordered factor of a word as well as algorithms computing such factor were shown in [2,8,12,13].

By Σ_k we denote an alphabet of size k; for simplicity, we fix Σ_k as $\{0,\ldots,k-1\}$. By $\text{Bordered}_k(n)$ and $\text{Unbordered}_k(n)$ we denote the subsets of Σ_k^n consisting of bordered and unbordered words, respectively. These sets were considered combinatorially in [9,18]; they are also interesting due to their relation to palindromes.

Supported by the Polish National Science Center, grant no. 2022/46/E/ST6/00463.

© The Author(s), under exclusive license to Springer Nature Switzerland AG 2025
Z. Lipták et al. (Eds.): SPIRE 2024, LNCS 14899, pp. 257–271, 2025.
https://doi.org/10.1007/978-3-031-72200-4_20

Remark 1. ([7,17]) The number of length-n unbordered words over a k-letter alphabet equals the number of length-n words over a k-letter alphabet having no even palindromic prefix. If n is odd, then it also equals the number of length-n k-ary words with no nontrivial palindromic prefix of odd length. The number of length-n unbordered words equals the number of length-$2n$ prime palstars (a *prime palstar* is an even length palindrome which is not a nontrivial composition of nonempty even length palindromes).

The ranking and unranking can be considered in general with respect to an arbitrarily defined ordering of a set of objects. In many applications, the underlying order does not matter, for example in [15] such ordering is defined by the ranking function. In this paper we consider ranking and unranking with respect to the *lexicographic* order of words. For a word $w \in \text{Bordered}_k(n)$, we denote
$$\text{rankB}_k(w) = |\{u \in \text{Bordered}_k(n) : u \leq w\}|.$$
We consider the following problems.

RANKING BORDERED WORDS

Input: positive integers k and n and a word $w \in \text{Bordered}_k(n)$

Output: $\text{rankB}_k(w)$.

UNRANKING BORDERED WORDS

Input: positive integers r and k, n

Output: $\text{unrankB}_{k,n}(r) = w$, where $w \in \text{Bordered}_k(n)$, $\text{rankB}_k(w) = r$, if there is any such w.

Similarly we define $\text{rankU}_k(w)$, $\text{unrankU}_{k,n}(r)$, for $w \in \text{Unbordered}_k(n)$.

Example 1. The sorted list of binary unbordered words of length 6 is:

000001 000011 000101 000111 001011 001101 001111 010011 010111 011111

100000 101000 101100 110000 110010 110100 111000 111010 111100 111110.

We have

$\text{rankU}_2(010011) = 8$, $\text{unrankU}_{2,6}(5) = 001011$ and $\text{unrankU}_{2,6}(15) = 110010$.

Our main result is as follows.

Theorem 1. *The problems of ranking/unranking bordered and unbordered words can be solved in $\mathcal{O}(n^2 \log n)$ time and $\mathcal{O}(n)$ space in the unit-cost RAM model.*

In [6] the ranking problems were solved in $\mathcal{O}(n^3 k)$ time and the unranking problems in $\mathcal{O}(n^4 k \log k)$ time. It was already noticed in [6] that "It seems believable that the ranking and unranking algorithms in this paper can be improved by a factor of n". Our Theorem 1 shows a much greater improvement, by a factor

$nk/\log n$ for ranking and by a factor $n^2 k \log k / \log n$ for unranking. For k that is of order n, the improvements are $n^2/\log n$ and n^3, respectively.

Python implementations of the algorithms behind Theorem 1 are available at https://github.com/twalen/RankBorderedWords. Performance tests against the implementations provided by Gabric [6] translated into Python confirm that our algorithms are significantly faster; the graphs are provided in Appendix A.

Structure of the Paper. The preliminary Sect. 2 presents some useful properties and facts related to the arithmetic structure of borders and the numbers $\mathbf{B}_k(u, n)$, where $\mathbf{B}_k(u, n)$ is the cardinality of the set of all bordered length-n words with a prefix u.

Then, in Sect. 3, we present efficient algorithms computing the numbers $\mathbf{B}_k(u, n)$. We present two rather unrelated algorithms: in Sect. 3.2 for the case $|u| \geq n/2$ (long u) and in Sect. 3.3 for the opposite short case, reducing it to the case of long u.

The ranking and unranking algorithms are presented in Sect. 4, where we further reduce complexity by using additional properties of the compacted structure of the sets of borders. In the unranking algorithm, Gabric used as a main subroutine the ranking algorithm. Here we use the numbers $\mathbf{B}_k(u, n)$ directly, omitting "expensive" use of ranking.

2 Preliminaries

We assume that letters of words are numbered starting from 1. By ε we denote the empty word. For a word $w = w_1 \cdots w_{|w|}$, by $w[i \mathinner{.\,.} j]$ we denote the word $w_i \cdots w_j$. We also denote $[i \mathinner{.\,.} j] = \{i, \ldots, j\}$.

2.1 Borders and Arithmetic Representations of Sets of Borders

Let w be a word of length n. Gabric [6] introduced two binary arrays constructed for w, the *unbordered prefix indicator* array $\mathbf{a}_w[1 \mathinner{.\,.} n]$ and the *border indicator* array $\mathbf{b}_w[1 \mathinner{.\,.} n]$, such that for each i:

- $\mathbf{a}_w[i] = 1$ if and only if $w[1 \mathinner{.\,.} i]$ is unbordered,
- $\mathbf{b}_w[i] = 1$ if and only if w has a border of length i.

Lemma 1. *([6, Lemmas 1 and 2]) Given a word w of length n, the unbordered prefix indicator $\mathbf{a}_w[1 \mathinner{.\,.} n]$ and the border indicator $\mathbf{b}_w[1 \mathinner{.\,.} n]$ of w can be computed in $\mathcal{O}(n)$ time.*

The border array of a word w of length n, $B_w[1 \mathinner{.\,.} n]$, contains for every i as $B_w[i]$ the length of the longest proper border of $w[1 \mathinner{.\,.} i]$. The border array can be computed in $\mathcal{O}(n)$ time as the failure function of the Morris-Pratt pattern matching algorithm [11]. An integer p is called a period of a word w if $0 < p < |w|$ and $w[i] = w[i+p]$ for all $i \in [1 \mathinner{.\,.} |w|-p]$. Word w is called *periodic* if its smallest period p satisfies $2p \leq |w|$; otherwise w is called *aperiodic*. The length of the longest proper border equals $|w| - p$, where p is the smallest period of w.

For a word w, by $\mathbf{R}(w)$ we denote a representation of the set of lengths of nontrivial borders of w as a union of disjoint arithmetic progressions such that each aperiodic border forms a singleton sequence and for every two periodic borders b, b', they belong to the same arithmetic progression if and only if their smallest periods are the same. Moreover, in the second case, the difference of the common progression is exactly the smallest period of b and b'. By $|\mathbf{R}(w)|$ we denote the number of arithmetic progressions in the representation.

Example 2. If $w = (a^4b)^5a^4$, then $\mathbf{R}(w) = \{\,\{1\}, \{2,3,4\}, \{9\}, \{14,19,24\}\,\}$.

Lemma 2. *We are given a word w of length n. Then the following holds.*

(a) $|\mathbf{R}(w)| = \mathcal{O}(\log n)$.
(b) After $\mathcal{O}(n)$ time and space preprocessing, the representation $\mathbf{R}(w[1..i])$ for any $i \in [1..n]$ can be reported in $\mathcal{O}(\log i)$ time.

Proof. (a): Let $b_1, b_2, \ldots, b_m$ be all nontrivial borders of w in the order of increasing lengths. For every $i \in [1..m]$, b_i is a border of b_{i+1}.

Hence, if $2|b_i| \geq |b_{i+1}|$, then b_{i+1} is periodic. Thus there are $\mathcal{O}(\log n)$ aperiodic borders of w.

If b_{i+1} is periodic, its smallest period is $p = |b_{i+1}| - |b_i|$ (since b_i is the longest border of b_{i+1}). By Fine and Wilf's Periodicity Lemma [5], all borders $b_{i+1}, b_i, b_{i-1}, \ldots$ of length at least $2p$ have smallest period p, so the lengths of these borders form an arithmetic progression. Let b_j be the longest border that does not belong to this progression. Then $|b_j| \leq \frac{2}{3}|b_i|$. This concludes that the number of arithmetic progressions representing periodic borders is also $\mathcal{O}(\log n)$.

Algorithm ComputeR(i)

$R := \emptyset$; $j := B_w[i]$;

while $j > 0$

$\quad k := B_w[j]$;

$\quad$ // j, k are lengths of two longest borders of $w[1..i]$

$\quad$ **if** $2k < j$ **then**   // $w[1..j]$ is aperiodic

$\quad\quad$ insert $\{k\}$ to R; $j := k$;

$\quad$ **else**

$\quad\quad p := j - k$;   // p is the smallest period of $w[1..j]$

$\quad\quad$ insert $\{j, j-p, \ldots, 2p + (j \bmod p)\}$ to R;

$\quad\quad$ // inserted set consists of lengths of periodic borders of $w[1..j]$ with period p

$\quad\quad j := p + (j \bmod p)$;

return R;

Point (b) is an implementation of the above combinatorial argument, as shown in the algorithm ComputeR. The preprocessing of w consists in computing the border array B_w [11]. It is known that the lengths of all nontrivial borders of $w[1..i]$ can be computed by iterating the border array B_w starting from $B_w[i]$.

After every step of the while-loop, j decreases by at least $j/3$. Hence, the algorithm works in $\mathcal{O}(\log i)$ time. □

Remark 2. Representations $\mathbf{R}(w[1..i])$ for all $i \in [1..n]$ can be reported on-line in $\mathcal{O}(\log n)$ time per representation without any preprocessing. This would be slightly more complicated and would not improve the final time complexity of our ranking/unranking algorithms.

2.2 Bordered Words with Fixed Prefixes

For a word u of length at most n we denote $\text{Ext}_n(u) = u \cdot \Sigma_k^{n-|u|}$, and denote

$$\mathbf{B}_k(u,n) = |\text{Ext}_n(u) \cap \text{Bordered}_k(n)|.$$

In other words, $\mathbf{B}_k(u,n)$ is the cardinality of the set of all bordered length-n words with prefix u.

Example 3. $\mathbf{B}_2(01,4) = 3$ since $\text{Ext}_4(01) \cap \text{Bordered}_2(4) = \{0100, 0101, 0110\}$.

In particular, $\mathbf{B}_k(\varepsilon,n) = |\text{Bordered}_k(n)|$ and $k^n - \mathbf{B}_k(\varepsilon,n) = |\text{Unbordered}_k(n)|$.

The following theorem from [6] gives a combinatorial characterization of $\mathbf{B}_k(u,n)$ in terms of indicator sequences.

Theorem 2. *([6, Theorem 5]) Let u be a nonempty word over the alphabet Σ_k, for $k \geq 2$. Then, if $n/2 \leq |u| \leq n$,*

$$\mathbf{B}_k(u,n) = \sum_{i=1}^{d} \mathbf{a}_u[i] k^{d-i} + \sum_{i=d+1}^{\lfloor n/2 \rfloor} \mathbf{a}_u[i] \cdot \mathbf{b}_u[i-d] \text{ where } d = n - |u|. \quad (1)$$

Computation of $\text{rankB}_k(w)$ directly relies on computation of $\mathbf{B}_k(u,j)$ values for u related to prefixes of w and $j \leq n$ (see Eq. (4) in Sect. 4).

3 Computing $\mathbf{B}_k(u,n)$

Gabric [6] showed how to compute $\mathbf{B}_k(u,n)$ in $\mathcal{O}(n^2)$ time. We improve the time complexity to $\mathcal{O}(n \log n)$ in the worst case.

We compute separately $\mathbf{B}_k(u,n)$ for long prefixes ($|u| > n/2$) and short prefixes ($|u| \leq n/2$). We start with an arithmetic problem.

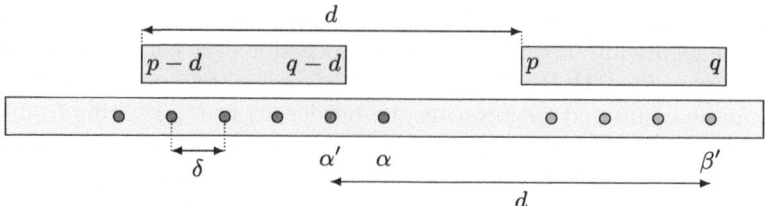

Fig. 1. Illustration of computing $\mathsf{SUM}_{a,b}(p,q,d)$ in case $\{i : b[i] \neq 0\}$ is a single progression with difference δ (the red dots). We find the last red position, denoted by α', in the interval $[p-d\mathinner{.\,.}q-d]$. Let $\beta' = \alpha' + d$. Then, in this example, $\mathsf{SUM}(p,q,d) = a[\beta'] + a[\beta' - \delta] + a[\beta' - 2\delta] + a[\beta' - 3\delta]$. It is enough to keep for each t the "prefix" sum $\sum_{0 < t - h\delta \leq t} a[t - h\delta]$

3.1 SUM Queries

For sequences a, b of integers, we define a query

$$\mathsf{SUM}_{a,b}(p,q,d) = \sum_{i=p}^{q} a[i] \cdot b[i-d].$$

Such queries are called here *SUM queries*. In other words, each such query is a scalar product of a fragment of a and a fragment of b of the same length.

Lemma 3. *Assume that we are given binary sequences a, b of length n, such that the set $\{i : b[i] = 1\}$ is represented as a union of k disjoint arithmetic progressions. We are to answer m SUM queries. All the queries can be answered in $\mathcal{O}((n+m)k)$ time and $\mathcal{O}(n+m)$ space.*

Proof. Let us first consider the case $k = 1$ (one arithmetic progression). Assume the set $\{i : b[i] = 1\}$ is represented as an arithmetic progression

$$\{\alpha - \lambda\delta, \alpha - (\lambda - 1)\delta, \ldots, \alpha\}.$$

We use an array $\mathsf{Sum}_{a,\delta}[1\mathinner{.\,.}n]$ such that

$$\mathsf{Sum}_{a,\delta}[i] = a[i] + a[i-\delta] + a[i-2\delta] + \cdots + a[i \bmod \delta].$$

Given δ, such an array can be computed iteratively in $\mathcal{O}(n)$ time as follows:

$$\mathsf{Sum}_{a,\delta}[i] = \begin{cases} 0 & \text{if } i \leq 0 \\ a[i] & \text{if } i \leq \delta \\ a[i] + \mathsf{Sum}_{a,\delta}[i-\delta] & \text{otherwise.} \end{cases}$$

With the auxiliary array, a query $\mathsf{SUM}_{a,b}(p,q,d)$ can be answered in constant time as follows.

We compute the arithmetic progression

$$\{\alpha - \lambda\delta, \alpha - (\lambda - 1)\delta, \ldots, \alpha\} \cap [p-d\mathinner{.\,.}q-d].$$

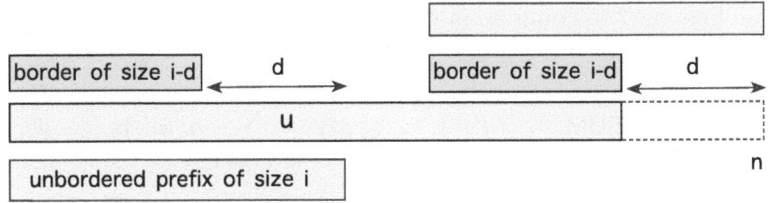

Fig. 2. Illustration of the situation when $\mathbf{a}_u[i]\cdot\mathbf{b}_u[i-d] = 1$, where $n = |u|+d$. The value of $\mathbf{B}_k''(u,n)$ equals the number of unbordered prefixes of u of length $i \in [d+1 \mathinner{.\,.} \lfloor n/2 \rfloor]$, such that $i - d$ is a length of a border of u

If it is non-empty, let us express it as

$$\{\alpha' - \lambda'\delta, \alpha' - (\lambda' - 1)\delta, \ldots, \alpha'\}.$$

Then (see also Fig. 1):

$$\mathsf{SUM}_{a,b}(p,q,d) := \mathsf{Sum}_{a,\delta}[\alpha' + d] - \mathsf{Sum}_{a,\delta}[\alpha' - (\lambda' + 1)\delta + d]. \tag{2}$$

If $k > 1$, we can consider every arithmetic sequence in $\{i : b[i] = 1\}$ separately. For each such sequence $\{\alpha - \lambda\delta, \alpha - (\lambda - 1)\delta, \ldots, \alpha\}$, we compute the array $\mathsf{Sum}_{a,\delta}$ and for each of the SUM queries, we use the operation (2) but with `+=` instead of assignment.

This can be done thanks to the assumption that all the arithmetic sequences are disjoint. Thus we obtain $\mathcal{O}(nk + mk)$ time. We can reuse the space of the $\mathsf{Sum}_{a,\delta}$ arrays, achieving $\mathcal{O}(n+m)$ total space. □

3.2 Computing $\mathbf{B}_k(u,n)$ for $|u| \geq n/2$

Denote $d = n - |u|$. We define two partial sums (see Fig. 2).

$$\mathbf{B}_k'(u,n) = \sum_{i=1}^{d} \mathbf{a}_u[i] k^{d-i}, \quad \mathbf{B}_k''(u,n) = \sum_{i=d+1}^{\lfloor n/2 \rfloor} \mathbf{a}_u[i] \cdot \mathbf{b}_u[i-d].$$

By Eq. (1), we have $\mathbf{B}_k(u,n) = \mathbf{B}_k'(u,n) + \mathbf{B}_k''(u,n)$.

Lemma 4. *For a nonempty word u, we can compute in $\mathcal{O}(|u| \log |\mathbf{R}(u)|)$ time and $\mathcal{O}(|u|)$ space the values $\mathbf{B}_k(u,n)$ for all $n \in [|u| \mathinner{.\,.} 2|u|]$.*

Proof. We compute the unbordered prefix indicator $\mathbf{a}_u[1 \mathinner{.\,.} |u|]$ and the border indicator $\mathbf{b}_u[1 \mathinner{.\,.} |u|]$ of u in $\mathcal{O}(|u|)$ time (Lemma 1). Then we compute the values $\mathbf{B}'$ iteratively as follows:

$$\mathbf{B}_k'(u,|u|) = 0 \text{ and } \mathbf{B}_k'(u,n) = \mathbf{B}_k'(u,n-1) \cdot k + \mathbf{a}_u[n - |u|] \text{ for } n > |u|.$$

Over all $n \in [|u| \mathinner{.\,.} 2|u|]$ this takes $\mathcal{O}(|u|)$ time under unit-cost RAM.

We use Lemma 2 to compute in $\mathcal{O}(|u|)$ time a representation $\mathbf{R}(u)$ of the set $\{i : \mathbf{b}_u[i] = 1\}$ as a union of disjoint arithmetic progressions. All the values

$$\mathbf{B}_k''(u,n) = \mathsf{SUM}_{\mathbf{a}_u, \mathbf{b}_u}(d+1, \lfloor n/2 \rfloor, d) = \sum_{i=d+1}^{\lfloor n/2 \rfloor} \mathbf{a}_u[i] \cdot \mathbf{b}_u[i-d]$$

for $n \in [|u| \mathinner{\ldotp\ldotp} 2|u|]$, $d = n - |u|$, can be computed in $\mathcal{O}(|u| \log |\mathbf{R}(u)|)$ time and $\mathcal{O}(|u|)$ space using Lemma 3. □

3.3 Computing $\mathbf{B}_k(u,n)$ for $|u| < n/2$

Let $\mathbf{U}_k(u,n)$ be the number of unbordered words of length n with prefix u.[1]

Gabric [6, Theorem 5] proposed a recursive formula for $\mathbf{B}_k(u,n)$ in the case that $|u| < n/2$. We obtain a simpler formula using values $\mathbf{U}_k(u,n)$; it is a generalization of a formula from [16].

Lemma 5. *If $|u| < n/2$, then*

$$\mathbf{U}_k(u,n) = \begin{cases} k \cdot \mathbf{U}_k(u, n-1) & \text{if } n \text{ is odd} \\ k \cdot \mathbf{U}_k(u, n-1) - \mathbf{U}_k(u, n/2) & \text{otherwise.} \end{cases} \quad (3)$$

Proof. We use the following claim.

Claim 1. Let $a, b \in \Sigma_k$ and $x, y \in \Sigma_k^m$. Then:

(1) if xay is unbordered, then xy is unbordered;
(2) $xaby$ is bordered and xay is unbordered if and only if $xa = by$ and xa is unbordered.

Proof. The shortest nontrivial border of a word w, if it exists, has length at most $|w|/2$. (Otherwise, w would have a period smaller than $|w|/2$ which implies a border of length at most $|w|/2$.)

Part (1): If xy is bordered, then the shortest nontrivial border u of xy satisfies $|u| \le |x|$. Consequently, u is a border of xay.

Part (2): If $xaby$ is bordered and its shortest nontrivial border is u, then $|u| \le |xa|$. If $|u| \le |x|$, then u would be a border of xay. Hence, $u = xa = by$. Moreover, u is itself unbordered, as its nontrivial border would be a border of $xaby$. □

By the claim, the formula (3) is true for odd n. Due to the claim, if n is even and $|u| < n/2$, then $\mathbf{U}_k(u,n)$ equals $\mathbf{U}_k(u, n-1)$ minus the number of unbordered words with prefix u of the form $xaby$ such that $xaby$ is bordered and xay is unbordered. The latter number is the number of words $xaby$ such that $xa = by$ and xa is unbordered. The number of unbordered words xa equals $\mathbf{U}_k(u, n/2)$ since $|xa| = n/2$. This completes the proof. □

Due to $\mathbf{B}_k(u,n) = k^{n-|u|} - \mathbf{U}_k(u,n)$ and Lemma 4 we have the following fact.

---
[1] The sequence of numbers $\mathbf{U}_2(\varepsilon, n)$ is listed in the On-Line Encyclopedia of Integer Sequences as https://oeis.org/A003000.

Corollary 1. *If $|u| < n/2$ then $\mathbf{B}_k(u, n)$ can be computed in $\mathcal{O}(n \log |\mathbf{R}(u)|)$ time and $\mathcal{O}(n)$ space.*

Now, Lemma 4 and Corollary 1 imply directly the main result of this section.

Lemma 6. *If u is a non-empty word over the alphabet Σ_k and $n \geq |u|$, then we can compute $\mathbf{B}_k(u, n)$ in $\mathcal{O}(n \log |\mathbf{R}(u)|)$ time and $\mathcal{O}(n)$ space.*

In particular, by Lemma 2(a) the time complexity in Lemma 6 is $\mathcal{O}(n \log n)$. We need the refined bound with $\log |\mathbf{R}(u)|$ instead of $\log n$ to shave off a $\log n$ factor in the final time complexity.

As a remark, Lemma 6 allows to compute all values $\mathbf{B}_k(u, j)$, for $j \in [|u| \mathinner{.\,.} n]$, in $\mathcal{O}(n \log |\mathbf{R}(u)|)$ time.

4 Ranking and Unranking

It was shown in [6] that $\mathrm{rankB}_k(w)$ for $w \in \mathrm{Bordered}_k(n)$ can be expressed in terms of $\mathbf{B}_k$ values as follows

$$\mathrm{rankB}_k(w) = 1 + \sum_{i=1}^{n} \sum_{c < w[i]} \mathbf{B}_k(w[1 \mathinner{.\,.} i-1] \cdot c, n). \quad (4)$$

Example 4. For $w = 10110$ we have three words of the form $w[1 \mathinner{.\,.} i-1]c$, where $i \in [1 \mathinner{.\,.} n]$ and $c \in \Sigma_k$, $c < w[i]$: 0, 100, 1010.
The set $0 \cdot \Sigma_2^4 \cap \mathrm{Bordered}_2(5)$ of size 10 consists of

$$00000, 00010, 00100, 00110, 01000, 01001, 01010, 01100, 01101, 01110.$$

We have also

$$100 \cdot \Sigma_2^2 \cap \mathrm{Bordered}_2(5) = \{10001, 10010, 10011\},$$

$$1010 \cdot \Sigma_2^1 \cap \mathrm{Bordered}_2(5) = \{10101\}.$$

Hence, according to Eq. (4), we have

$\mathrm{rankB}_2(10110) = 1 + \mathbf{B}_2(0, 5) + \mathbf{B}_2(100, 5) + \mathbf{B}_2(1010, 5) = 1 + 10 + 3 + 1 = 15.$

Equation (4) reduces computation of rank to computations of $\mathbf{B}_k(u, n)$ for multiple words u. However, it requires, if applied directly, $\mathcal{O}(kn)$ instances of computing $\mathbf{B}_k$. This gave in [6] the total time $\mathcal{O}(n^3 k)$. Using Lemma 6 we can immediately reduce it by a factor $n/\log n$ and get time complexity $\mathcal{O}(n^2 k \log n)$.

In this section we show how to reduce the number of components of the sum in Eq. (4) from $\mathcal{O}(nk)$ to $\mathcal{O}(n \log n)$. Consequently, the total complexity is reduced by another factor $n/\log n$ for large k. Furthermore, with some care we are able to avoid having two $\log n$ factors in the complexity. We need to use additional properties of periods and borders expressed in the next two crucial lemmas.

Lemma 7. *If u is a word over alphabet Σ_k and $ua, ub \in \text{Unbordered}_k(|u|+1)$, then for all $n > |u|$ and $a, b \in \Sigma_k$ we have $\mathbf{B}_k(ua, n) = \mathbf{B}_k(ub, n)$.*

Proof. It suffices to check that Equations (1) (long u) and (3) (short u) give the same result when applied to ua and to ub, for every value of $n > |u|$.

We have $\mathbf{a}_{ua}[1 \mathinner{.\,.} |u|] = \mathbf{a}_{ub}[1 \mathinner{.\,.} |u|] = \mathbf{a}_u$ and $\mathbf{a}_{ua}[|u|+1] = \mathbf{a}_{ub}[|u|+1] = 1$. Hence, $\mathbf{a}_{ua} = \mathbf{a}_{ub}$. Moreover, $\mathbf{b}_{ua}$ and $\mathbf{b}_{ub}$ consist of only zeros as ua, ub are unbordered. Equation (1) only depends on the arrays $\mathbf{a}, \mathbf{b}$, so indeed it gives the same values for every n such that $|u|+1 \leq n \leq 2|u|+2$.

Hence, $\mathbf{U}_k(ua, n) = \mathbf{U}_k(ub, n)$ holds for all $n \in [|u|+1 \mathinner{.\,.} 2|u|+2]$. Equation (3) computes $\mathbf{U}_k(u, n)$ based on $\mathbf{U}_k(u, n')$ for $n' < n$ and the formula does not depend on the particular word u. Therefore, it gives the same values for ua and ub. Finally, we have $\mathbf{B}_k(ua, n) = \mathbf{B}_k(ub, n)$ for all $n > 2|u|+2$. □

For a word w of length n we define

$$\mathbf{X}(w, i) = \{\, a \in \Sigma_k : w[1 \mathinner{.\,.} i] \cdot a \text{ is a bordered word}\,\}.$$

We assume that the letters in these sets are sorted. We exploit the following crucial fact.

Lemma 8. *For a given word w of length n and any $i \in [1 \mathinner{.\,.} n-1]$ we have:*

(a) $|\mathbf{X}(w, i)| = \mathcal{O}(\log i)$,
(b) After $\mathcal{O}(n)$-time preprocessing, for every $i \in [1 \mathinner{.\,.} n]$ a set $\mathbf{X}(w, i)$ can be computed in $\mathcal{O}(\log i \log \log i)$ time.
(c) $\sum_{a \in \mathbf{X}(w, i)} |\mathbf{R}(w[1 \mathinner{.\,.} i] \cdot a)| = \mathcal{O}(\log i)$.

Proof. We prove each point separately.

(a) If $a \in \mathbf{X}(w, i)$, then there is a border u of $w[1 \mathinner{.\,.} i]$ such that ua is a border of $w[1 \mathinner{.\,.} i] \cdot a$. If $u = \varepsilon$, then $a = w[1]$. Otherwise, $|u| \in G$ for some arithmetic progression $G \in \mathbf{R}(w[1 \mathinner{.\,.} i])$. Let $G = \{\alpha, \alpha+\delta, \ldots, \alpha+\lambda\delta\}$. For $j \in [0 \mathinner{.\,.} \delta-1]$, we have $w[\alpha+j\delta+1] = w[\alpha+1]$ by the common periodicity of all the borders represented in this arithmetic progression. Hence, $a \in \{b, c\}$, where $b = w[\alpha+\lambda\delta+1]$ and $c = w[\alpha+1]$; see Fig. 3. Hence, there are at most $2|\mathbf{R}(w[1 \mathinner{.\,.} i])|+1$ letters in $\mathbf{X}(w, i)$ (we are adding 1 due to $a = w[1]$).

(b) We define an *i-special border* as the longest or second longest in its progression in $\mathbf{R}(w[1 \mathinner{.\,.} i])$. By point (a), $\mathbf{X}(w, i)$ is a union of $\{w[z+1] : z \text{ is i-special}\}$ and possibly $\{w[1]\}$. By Lemma 2, after $\mathcal{O}(n)$ preprocessing, the set $\mathbf{R}(w[1 \mathinner{.\,.} i])$ can be computed in $\mathcal{O}(\log i)$ time. Finally, we need $\mathcal{O}(|\mathbf{X}(w, i)| \log |\mathbf{X}(w, i)|)$ time to sort the set $\mathbf{X}(w, i)$ and remove duplicates.

(c) There are at most $2 \cdot |\mathbf{R}(w[1 \mathinner{.\,.} i])|$ i-special borders. The sets $\mathbf{R}(w[1 \mathinner{.\,.} i] \cdot a)$ are disjoint for distinct a's, so each arithmetic progression in $\mathbf{R}(w[1 \mathinner{.\,.} i])$ is partitioned into at most two arithmetic progressions, each of which (with all its elements incremented) belongs to one set $\mathbf{R}(w[1 \mathinner{.\,.} i] \cdot a)$. □

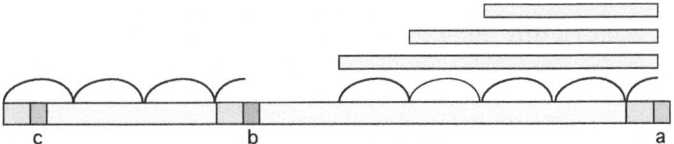

Fig. 3. Assume u is a nonempty border of $w[1..i]$. The yellow boxes correspond to the unique arithmetic progression G of lengths of borders of $w[1..i]$ in $\mathbf{R}(w[1..i])$ containing $|u|$. If ua is a border of $w[1..i] \cdot a$, then $a = b$ or $a = c$, where $b = w[q+1]$, $c = w[(q+1) \bmod p]$, q is the length of longest border in G and p is its period (Color figure online)

4.1 Ranking

Lemma 8 is used to reduce the number of components of the sum in Eq. (4). This sum is split into the sum for $c \in \mathbf{X}(w, i-1)$ and for $c \notin \mathbf{X}(w, i-1)$.

The first sum has only $\mathcal{O}(\log n)$ components whereas by Lemma 7 the second sum can be computed using only one call to Lemma 8(b).

For a word w and $i \in [0..|w|-1]$ we define

$$\mathbf{Y}(w, i) = \mathbf{X}(w, i) \cap \{a \in \Sigma_k : a < a_{i+1}\}$$
$$\widetilde{\mathbf{Y}}(w, i) = \{a \in \Sigma_k : a < a_{i+1}\} \setminus \mathbf{X}(w, i)$$

If $\widetilde{\mathbf{Y}}(w, i) \neq \emptyset$, then we define $\mathsf{Repr}(w, i)$ as any fixed single element of $\widetilde{\mathbf{Y}}(w, i)$. The symbol $\mathsf{Repr}(w, i)$ in our computation represents all letters smaller than w_{i+1} outside $\mathbf{Y}(w, i)$. There are $|\widetilde{\mathbf{Y}}(w, i)|$ such letters. If $c_i = \mathsf{Repr}(w, i)$, then by Lemma 7,

$$\forall\, (a \in \widetilde{\mathbf{Y}}(w, i))\ \mathbf{B}_k(w[1..i]\, a, n) = \mathbf{B}_k(w[1..i]\, c_i, n).$$

We are now ready to show the ranking part of our main result.

Theorem 3. *The problems of ranking bordered and unbordered words can be solved in $\mathcal{O}(n^2 \log n)$ time and $\mathcal{O}(n)$ space in the unit-cost RAM model.*

Proof. For ranking bordered words we apply the following algorithm $\mathsf{RANK\_B}(w)$. Its correctness follows by Eq. (4) which gives a summation formula for $\mathrm{rank}\mathbf{B}_k(w)$ and Lemma 7, which shows that all components of the sum that correspond to $a \in \widetilde{\mathbf{Y}}(w, i)$ are equal. Let us analyze the complexity of the algorithm.

By Lemma 8(b), the set $\mathbf{X}(w, i)$ can be computed in $\mathcal{O}(\log n \log \log n)$ time after $\mathcal{O}(n)$ preprocessing. This set, given in a sorted order, allows to compute the set $\mathbf{Y}(w, i)$, the size $|\widetilde{\mathbf{Y}}(w, i)|$, and the representative $\mathsf{Repr}(w, i)$ if $|\widetilde{\mathbf{Y}}(w, i)| > 0$, in $\mathcal{O}(\log n)$ time. The cost of these computations will be much smaller than the cost of computing $\mathbf{B}_k$ below.

By Lemma 6, for a given $i \in [0..n-1]$ the total cost of computing $\mathbf{B}_k(w[1..i] \cdot a, n)$, for all $a \in \mathbf{Y}(w, i)$, is $\mathcal{O}(n \sum_{a \in \mathbf{X}(w,i)} |\mathbf{R}(w[1..i] \cdot a)|)$. By Lemma 8(c), this sum is just $\mathcal{O}(n \log n)$. By Lemmas 6 and 2, the cost of computing $\mathbf{B}_k(w[1..i] \cdot c_i, n)$, for $c_i = \mathsf{Repr}(w, i)$, is also $\mathcal{O}(n \log n)$.

> **Algorithm** RANK_B(w)
>
> $rank := 1$;
>
> **for** $i := 0$ **to** $n - 1$ **do**
>
>   compute $\mathbf{Y}(w,i)$ and $|\widetilde{\mathbf{Y}}(w,i)|$;
>
> $rank := rank + \sum_{a \in \mathbf{Y}(w,i)} \mathbf{B}_k(w[1\mathinner{.\,.}i] \cdot a, n)$;
>
>   **if** $|\widetilde{\mathbf{Y}}(w,i)| > 0$ **then**
>
> $c_i := \mathsf{Repr}(w,i)$;
>
> $rank := rank + |\widetilde{\mathbf{Y}}(w,i)| \cdot \mathbf{B}_k(w[1\mathinner{.\,.}i] \cdot c_i, n)$;
>
> **return** $rank$;

For a fixed $i \in [0\mathinner{.\,.}n-1]$ we spend $\mathcal{O}(n \log n)$ total time. Over all i, we get $\mathcal{O}(n^2 \log n)$ time. The space is $\mathcal{O}(n)$ by Lemmas 6 and 8.

For ranking unbordered words, the same formula as Eq. (4) can be used with rankB and $\mathbf{B}$ replaced with rankU and $\mathbf{U}$, respectively. We can implement the formula in $\mathcal{O}(n^2 \log n)$ time and $\mathcal{O}(n)$ space by replacing each $\mathbf{B}$ with $\mathbf{U}$ in algorithm RANK_B, thus obtaining an algorithm RANK_U. □

4.2 Unranking

In [7], the unranking algorithm used multiple ranking. Here we avoid it. Altogether we reduce the complexity obtained in [7] by a factor $n^3/\log n$ for large alphabets of size $k = \Omega(n)$. We start with a straightforward version of the algorithm. The general structure of a slow version of the algorithm is as follows.

> **Algorithm** SLOW_UNRANK_B(n, k, r)
>
> $w := $ empty word;
>
> **for** $i := 1$ **to** n **do**
>
> $c := \min\{d \in \Sigma_k : \sum_{a \leq d} \mathbf{B}_k(w \cdot a, n) \geq r\}$;
>
> $s := \sum_{a < c} \mathbf{B}_k(w \cdot a, n)$;
>
> $w := w \cdot c$; $r := r - s$;
>
> **return** w;

We define $\mathbf{X}(w, m)$ for a string w of length m, since the letters on positions $j > m$ are not important in definition of $\mathbf{X}(w, m)$.

We consider the following problem; see Fig. 4. We apply the problem to an array $W[1\mathinner{.\,.}k]$ such that $W[a] = \mathbf{B}_k(w \cdot a, n)$.

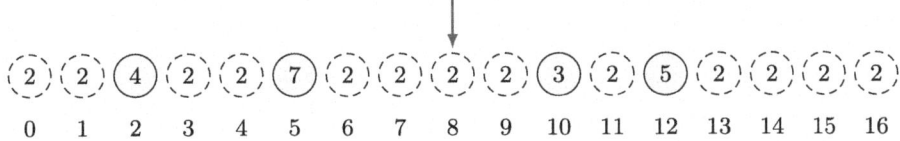

Fig. 4. Example of the `ImplicitMinPrefix` query. The input consists of the set $X = \{2, 5, 10, 12\}$ (colored nodes), the values of W are inside nodes, $y = 2$, and $r = 24$. The output is the position $c = 9$ and the sum $s = 23$ up to that position

---

`ImplicitMinPrefix`(W, X, y, r)

Input: A number r and an *implicitly given* table $W[a]$, for $a \in \Sigma_k$, of nonnegative integers such that there are known values of $W[a]$ for $a \in X$, where $X \subseteq \Sigma_k$ is a given set, and we know a value y such that $W[a] = y$ if $a \notin X$.

Output: A pair of integers (c, s),
where $c = \min\{d \in \Sigma_k : \sum_{a \leq d} W[a] \geq r\}$, $s = \sum_{a < c} W[a]$

We only need to scan positions of X in increasing order to answer each `ImplicitMinPrefix` query. This implies the following fact

Fact 1. *We can compute each `ImplicitMinPrefix` query in $\mathcal{O}(|X|+1)$ time.*

Theorem 4. *The problems of unranking bordered and unbordered words can be solved in $\mathcal{O}(n^2 \log n)$ time and $\mathcal{O}(n)$ space in the unit-cost RAM model.*

Proof. A pseudocode of the fast version of the algorithm is given below.

Algorithm UNRANK_B(n, k, r)

$w :=$ empty word;

for $i := 1$ **to** n **do**

    Compute $\mathbf{X}(w, i - 1)$; // $i - 1 = |w|$

    for each $b \in \mathbf{X}(w, i - 1)$ **do** Compute $\mathbf{B}_k(w \cdot b, n)$;

    if $\mathbf{X}(w, i - 1) \neq \Sigma_k$ **then**

        $y := \mathbf{B}_k(w \cdot z, n)$, where z is any letter not in $\mathbf{X}(w, i - 1)$;

    Let $W[a] = \mathbf{B}_k(w \cdot a, n)$ for each $a \in \Sigma_k$;

    $(c, s) :=$ `ImplicitMinPrefix`$(W, \mathbf{X}(w, i - 1), y, r)$;

    $w := w \cdot c$; $r := r - s$;

return w;

There are $\mathcal{O}(n)$ main iterations in the algorithm. In each of them we compute the value of $\mathbf{X}(w, i-1)$ and the values of $\mathbf{B}_k(w \cdot b, n)$ for all $b \in \mathbf{X}(w, i-1)$. It takes $\mathcal{O}(n \log n)$ time per iteration, due to Lemmas 6 and 8.

Consequently, the whole algorithm works in $\mathcal{O}(n^2 \log n)$ time. The space complexity is just $\mathcal{O}(n)$.

The algorithm UNRANK_B can be easily transformed into the one unranking unbordered words within asymptotically the same complexity. □

A Performance Tests

See Fig. 5.

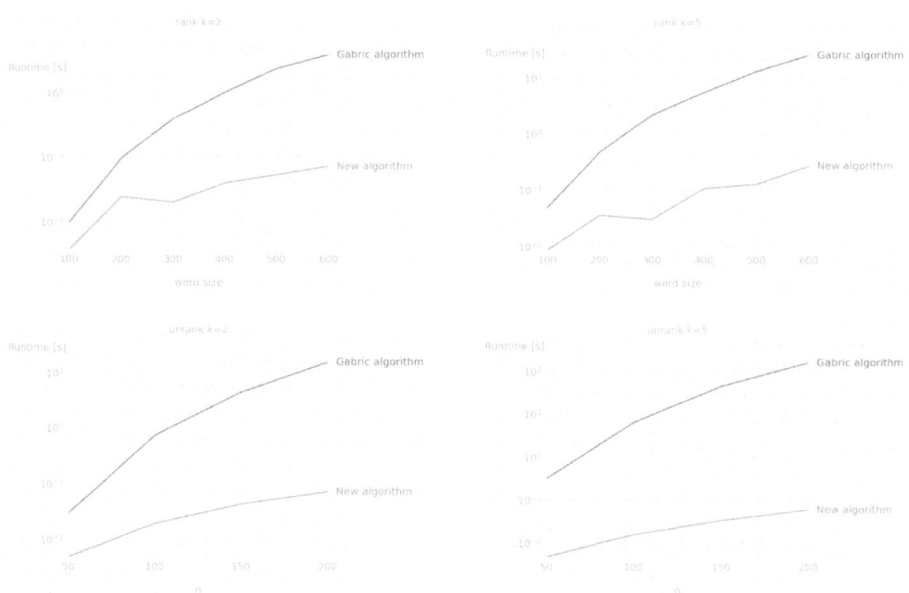

Fig. 5. Top: comparison of running times of ranking algorithms for bordered words, ours vs Gabric's [6], for alphabet size $k = 2$ (left) and $k = 5$ (right). Bottom: comparison of unranking algorithms for bordered words. Experiments were performed on random words using Python 3.12.0 on Apple Macbook Pro with CPU M1 Pro

References

1. Assous, R., Pouzet, M.: Une caracterisation des mots periodiques. Discret. Math. **25**(1), 1–5 (1979). https://doi.org/10.1016/0012-365X(79)90146-8
2. Cording, P.H., Gagie, T., Knudsen, M.B.T., Kociumaka, T.: Maximal unbordered factors of random strings. Theor. Comput. Sci. **852**, 78–83 (2021). https://doi.org/10.1016/J.TCS.2020.11.019

3. Duval, J.-P.: Relationship between the period of a finite word and the length of its unbordered segments. Discret. Math. **40**(1), 31–44 (1982). https://doi.org/10.1016/0012-365X(82)90186-8
4. Ehrenfeucht, A., Silberger, D.M.: Periodicity and unbordered segments of words. Discret. Math. **26**(2), 101–109 (1979). https://doi.org/10.1016/0012-365X(79)90116-X
5. Fine, N.J., Wilf, H.S.: Uniqueness theorems for periodic functions. Proc. Am. Math. Soc. **16**(1), 109–114 (1965). https://doi.org/10.2307/2034009
6. Gabric, D.: Ranking and unranking bordered and unbordered words. Inf. Process. Lett. **184**, 106452 (2024). https://doi.org/10.1016/j.ipl.2023.106452
7. Gabric, D., Shallit, J.O.: Borders, palindrome prefixes, and square prefixes. Inf. Process. Lett. **165**, 106027 (2021). https://doi.org/10.1016/J.IPL.2020.106027
8. Gawrychowski, P., Kucherov, G., Sach, B., Starikovskaya, T.: Computing the longest unbordered substring. In: Iliopoulos, C., Puglisi, S., Yilmaz, E. (eds.) SPIRE 2015. LNCS, vol. 9309, pp. 246–257. Springer, Cham (2015). https://doi.org/10.1007/978-3-319-23826-5_24
9. Harju, T., Nowotka, D.: Counting bordered and primitive words with a fixed weight. Theor. Comput. Sci. **340**(1), 273–279 (2005). https://doi.org/10.1016/J.TCS.2005.03.040
10. Holub, S., Nowotka, D.: The Ehrenfeucht-Silberger problem. J. Comb. Theory, Ser. A **119**(3), 668–682 (2012). https://doi.org/10.1016/J.JCTA.2011.11.004
11. Knuth, D.E., Morris, Jr., J.H., Pratt, V.R.: Fast pattern matching in strings. SIAM J. Comput. **6**(2), 323–350 (1977). https://doi.org/10.1137/0206024
12. Kociumaka, T., Kundu, R., Mohamed, M., Pissis, S.P.: Longest unbordered factor in quasilinear time. In: Hsu, W.-L., Lee, D.-T., Liao, C.-S. (eds.) 29th International Symposium on Algorithms and Computation, ISAAC 2018, 16–19 December 2018, Jiaoxi, Yilan, volume 123 of LIPIcs, pp. 70:1–70:13. Schloss Dagstuhl - Leibniz-Zentrum für Informatik (2018). https://doi.org/10.4230/LIPICS.ISAAC.2018.70
13. Loptev, A., Kucherov, G., Starikovskaya, T.: On maximal unbordered factors. In: Cicalese, F., Porat, E., Vaccaro, U. (eds.) CPM 2015. LNCS, vol. 9133, pp. 343–354. Springer, Heidelberg (2015). https://doi.org/10.1007/978-3-319-19929-0_29
14. Lothaire, M.: Combinatorics on Words. Cambridge Mathematical Library, 2nd edn. Cambridge University Press (1997)
15. Myrvold, W.J., Ruskey, F.: Ranking and unranking permutations in linear time. Inf. Process. Lett. **79**(6), 281–284 (2001). https://doi.org/10.1016/S0020-0190(01)00141-7
16. Nielsen, P.T.: A note on bifix-free sequences (corresp.). IEEE Trans. Inf. Theory **19**(5), 704–706 (1973). https://doi.org/10.1109/TIT.1973.1055065
17. Rampersad, N., Shallit, J.O., Wang, M.: Inverse star, borders, and palstars. Inf. Process. Lett. **111**(9), 420–422 (2011). https://doi.org/10.1016/J.IPL.2011.01.018
18. Régnier, M.: Enumeration of bordered words. Le langage de la vache-qui-rit. RAIRO Theor. Inf. Appl. **26**, 303–317 (1992). https://doi.org/10.1051/ITA/1992260403031

Computing String Covers in Sublinear Time

Jakub Radoszewski[1] and Wiktor Zuba[2]

[1] University of Warsaw, Warsaw, Poland
jrad@mimuw.edu.pl
[2] CWI, Amsterdam, The Netherlands
wiktor.zuba@cwi.nl

Abstract. In the word RAM model, a string T of length n over an integer alphabet of size σ can be represented in $\mathcal{O}(n/\log_\sigma n)$ space. We show that a representation of all covers of T can be computed in the optimal $\mathcal{O}(n/\log_\sigma n)$ time; in particular, the shortest cover can be computed within this time. We also design an $\mathcal{O}(n(\log \sigma + \log \log n)/\log n)$-sized data structure that computes in $\mathcal{O}(1)$ time any element of the so-called (shortest) cover array of T, that is, the length of the shortest cover of any given prefix of T. As a by-product, we describe the structure of the cover array of Fibonacci strings. On the negative side, we show that the shortest cover of a length-n string cannot be computed using $o(n/\log n)$ operations in the PILLAR model of Charalampopoulos, Kociumaka, and Wellnitz (FOCS 2020).

Keywords: Cover · Quasiperiod · Cover array · Packed string matching · PILLAR model

1 Introduction

A string C is called a *cover* (or a *quasiperiod*) of a string T if each position in T lies within an occurrence of C in T. A cover is called *proper* if it is shorter than the covered string. A string that does not have a proper cover is called *superprimitive* (see [7]). The shortest cover of a string of length n can be computed in $\mathcal{O}(n)$ time [1]. Furthermore, all covers of a length-n string can be computed in $\mathcal{O}(n)$ time [36]. A cover of a string is a prefix of the string, so a string of length n indeed has at most n covers.

The lengths of all covers of a string of length n can be represented using $\mathcal{O}(\log n)$ disjoint arithmetic progressions [17]. For a string T, we denote such a representation as $\mathsf{Covers}(T)$. A similar representation is well known to exist for the set of all borders of a string (see, e.g., [15]).

J. Radoszewski—Supported by the Polish National Science Center, grant no. 2022/46/E/ST6/00463
W. Zuba—Supported by the European Union's Horizon 2020 research and innovation programme under the Marie Skłodowska-Curie Grant Agreement No. 101034253.

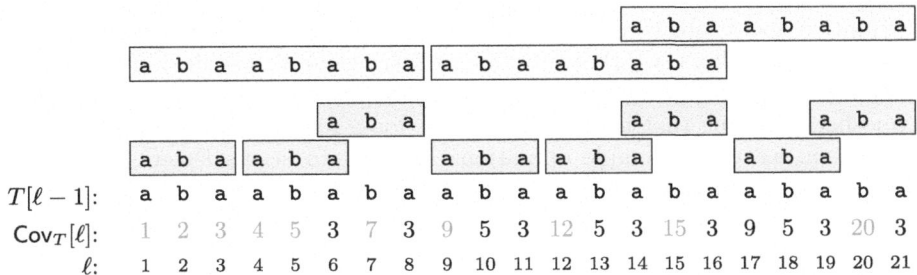

Fig. 1. Both proper covers (aba, abaababa) and the cover array of a Fibonacci string T. Values $\mathsf{Cov}_T[\ell] = \ell$ corresponding to superprimitive prefixes $T[0\mathinner{.\,.}\ell)$ are shown in gray

We consider the standard word RAM model with machine word composed of $\omega \geq \log_2 n$ bits. In this model, a string of length n over an alphabet of size σ can be represented using $\mathcal{O}(n/\log_\sigma n)$ machine words, that is, $\mathcal{O}(n \log \sigma)$ bits, in a so-called packed representation; see [5]. In Sect. 3 we show the following result that improves upon [1,7,36] in the case that the string is over a small alphabet.

Theorem 1. *A representation* $\mathsf{Covers}(T)$ *of all the covers of a string T of length n over an alphabet of size σ given in a packed form, consisting of $\mathcal{O}(\log n)$ arithmetic progressions, can be computed in $\mathcal{O}(n/\log_\sigma n)$ time.*

The representation $\mathsf{Covers}(T)$ can be transformed in $\mathcal{O}(n/\log n)$ time to a Boolean array of size n, represented in a packed form, that stores for every $\ell \in [1 \mathinner{.\,.} n]$ a Boolean value that determines if a length-ℓ prefix of T is a cover of T.

The (shortest) cover array of a string T, $\mathsf{Cov}_T[1 \mathinner{.\,.} |T|]$, stores for every position ℓ of T the length of the shortest cover of a length-ℓ prefix of T as $\mathsf{Cov}_T[\ell]$; see Fig. 1. The cover array is the output of Breslauer's on-line algorithm computing shortest covers [7]. The cover array was also considered in [16]. We give a sublinear-sized representation of this array.

Theorem 2. *Let T be a string of length n over an integer alphabet of size σ. There exists a data structure using space $\mathcal{O}(n(\log \sigma + \log \log n)/\log n)$ that, given $\ell \in [1 \mathinner{.\,.} n]$, returns $\mathsf{Cov}_T[\ell]$ in $\mathcal{O}(1)$ time.*

Our results extend the list of basic stringology problems for which representing the input in a packed form allows to obtain an $o(n)$-time solution; see [2,5,10,13,29,37].

As a by-product, we give a characterization of the cover arrays of Fibonacci strings (Theorem 5).

We also consider covers in the PILLAR model. This model was introduced in [11] with the aim of unifying approximate pattern matching algorithms across different settings. In this model, we consider a collection $\mathcal{X}$ of strings and assume that certain primitive PILLAR operations can be performed efficiently. The set of primitive operations consists of computing the length of the longest common prefix (LCP) or suffix ($\mathsf{LCP_R}$) of substrings of strings in $\mathcal{X}$, so-called internal pattern matching (IPM) queries that ask for the set of occurrences of one substring in another substring that is at most twice as long, represented as an arithmetic progression, as well as simple operations allowing to access letters of strings. (For a formal definition, see Sect. 5.)

The strength of the PILLAR model lies in the fact that efficient implementations of its primitives are known in many different settings:

- In the *standard setting*, in which all strings in the collection $\mathcal{X}$ are substrings of a given string of length n over an integer alphabet of size σ, each PILLAR operation on its substrings can be performed in $\mathcal{O}(1)$ time after $\mathcal{O}(n)$ preprocessing [6,21,34] and even after just $\mathcal{O}(n/\log_\sigma n)$ preprocessing [29,33].
- In the *dynamic setting*, the collection $\mathcal{X}$ can be updated dynamically under edit operations (insertions, deletions, substitutions) with each edit operation and each PILLAR operation performed in $\mathcal{O}(\log^{\mathcal{O}(1)} N)$ time, where N is the total size of $\mathcal{X}$ [11,20,25,30].
- In the *fully compressed setting*, given a collection $\mathcal{X}$ of straight-line programs (SLPs) of total size n generating strings of total length N, each PILLAR operation can be performed in $\mathcal{O}(\log^2 N \log \log N)$ time after $\mathcal{O}(n \log N)$-time preprocessing [11].
- An efficient implementation of the PILLAR operations is also known in the *quantum setting* [26,28].

Thus if a problem can be solved fast in the PILLAR model, it immediately implies its efficient solutions in all the above mentioned settings. For example, the fact that an $\mathcal{O}(\log n)$-sized representation of all the periods (equivalently, borders) of a length-n string can be computed in $\mathcal{O}(\log n)$ time in the PILLAR model [33,34, period query] implies that a representation of the periods of a dynamic string can be updated in $\mathcal{O}(\log^{\mathcal{O}(1)} N)$ time per operation and that a representation of all periods of a fully compressed string of length N generated by an SLP of size n can be computed in $\mathcal{O}(n \log^{\mathcal{O}(1)} N)$ time. In the case of covers, some efficient algorithms were designed for each of the above mentioned non-standard settings separately:

- In the *internal setting*, which is a special case of the standard setting, after $\mathcal{O}(n \log n)$ preprocessing of a length-n string T, one can compute a representation of all covers of any substring of T in $\mathcal{O}(\log n \log \log n)$ time and the shortest cover of any substring in $\mathcal{O}(\log n)$ time [4,17].
- In a restricted dynamic setting in which each edit operation is reverted immediately after it is performed, the shortest cover can be updated in $\mathcal{O}(\log n)$ time [35]. No algorithm is known for computing covers in the fully dynamic setting.
- In the fully compressed setting, a representation of all covers of a length-N string specified by an SLP of size n with derivation tree of height h can be computed in $\mathcal{O}(nh(n + \log^2 N))$ time [27]; with the technique of balancing SLPs [24], the time complexity becomes $\mathcal{O}(n \log N(n + \log^2 N))$.

The $\mathcal{O}(n)$-time algorithms for computing covers of a length-n string [1,7,36] perform only single-letter comparisons and thus work also in the PILLAR model. If there was a (much) more efficient algorithm computing the shortest cover of a string in the PILLAR model, one would immediately improve or generalize all the above results, including our Theorem 1. We show that, contrary to the case of periods, no such efficient algorithm for covers exists. A proof of Theorem 3 is given in Sect. 5.

Theorem 3. *There is no algorithm in the PILLAR model that solves any of the following problems for a length-n binary string T in $o(n/\log n)$ time:*

- *check if T is superprimitive;*
- *check if a given prefix of T is a cover of T.*

Consequently, computing the shortest cover or a representation of all covers of a string requires $\Omega(n/\log n)$ time in the PILLAR model.

2 Preliminaries

By $[i\mathinner{.\,.} j] = [i\mathinner{.\,.} j+1)$ we denote an integer interval $\{i, i+1, \ldots, j\}$. We assume that letters of a string T are numbered from 0 to $|T|-1$, i.e., $T = T[0]\cdots T[|T|-1]$. By $T[i\mathinner{.\,.} j] = T[i\mathinner{.\,.} j+1)$ we denote a substring $T[i]\cdots T[j]$. If T is given in a packed form, then a packed representation of its substring $T[i\mathinner{.\,.} j]$ can be computed in $\mathcal{O}((j-i+1)/\log_\sigma n)$ time using standard word RAM operations. A substring $T[i\mathinner{.\,.} j]$ is called a prefix if $i = 0$ and a suffix if $j = |T| - 1$. A string B that occurs in T as a prefix and as a suffix is called a border of T.

A positive integer p is called a period of string U if $U[i] = U[i+p]$ holds for all $i \in [0\mathinner{.\,.}|U|-p)$. A string U has a period $p \in [1\mathinner{.\,.}|U|]$ if and only if it has a border of length $|U| - p$. A string U is called *periodic* if the smallest period p of U satisfies $2p \leq |U|$. Otherwise, U is called *aperiodic*. We use the following so-called periodicity lemma.

Lemma 1. *(Fine and Wilf, [22]) If a string U has periods p and q and $p + q \leq |U|$, then $\gcd(p,q)$ is a period of U.*

We also use the following known corollary of the periodicity lemma.

Lemma 2. *([8,38]) If $|X| < |Y| < 2|X|$ are two strings and X has at least three occurrences in Y as a substring, then X is periodic.*

For a string X and non-negative integer k, by X^k we denote a concatenation of k copies of X. A non-empty string U is *primitive* if $U = X^k$ implies that $k = 1$. A string of the form X^2 is called a square. If X is primitive, the square X^2 is said to be *primitively rooted*.

Lemma 3. *(Three Squares Lemma, [3,19]) If a string U has primitively rooted square prefixes X^2, Y^2, Z^2 such that $|X| < |Y| < |Z|$, then $|Z| > |X| + |Y|$.*

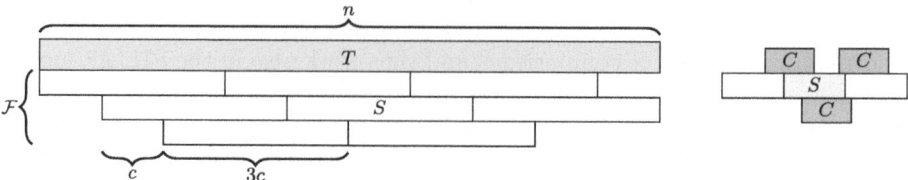

Fig. 2. Algorithm for checking a single candidate for a short cover. For each string S in $\mathcal{F}$ we check if the occurrences of C therein cover its middle part

3 Sublinear-Time Covers

Let $c = \lfloor \frac{1}{6} \log_\sigma n \rfloor$. To show Theorem 1 we divide the set Covers(T) into two subsets:

- SCovers(T) = Covers(T) $\cap$ $[1\mathinner{\ldotp\ldotp} c]$ of short cover lengths, and
- LCovers(T) = Covers(T) $\cap$ $[c+1\mathinner{\ldotp\ldotp} n]$ = Covers(T) \ SCovers(T) of long cover lengths

and compute their representations separately. If $c = 0$, there are only long covers.

Lemma 4. *For a string T of length n over integer alphabet of size σ, a representation of* SCovers(T) *can be computed in $\mathcal{O}(n/\log_\sigma n)$ time.*

Proof. Let $\mathcal{F}$ be the set of all the factors of T of length $3c$ that start at positions that are multiples of c. If the length of T is not a multiple of c, when computing $\mathcal{F}$ we extend T with at most $c - 1$ arbitrary letters. Let us notice that every string in $\mathcal{F}$ fits in a machine word, so those strings can be treated as integers.

To compute $\mathcal{F}$, we first construct a Boolean array over all the possible length-$3c$ strings, and then iterate through the length-$3c$ substrings of T starting at positions that are multiples of c, addressing the array directly through those integer representations. The size of the array as well as $|\mathcal{F}|$ is bounded by $\sigma^{3c} \leq \sigma^{\frac{1}{2}\log_\sigma n} = \sqrt{n}$. Iterating through all the considered substrings takes $\mathcal{O}(n/c)$ time. Hence, this computation takes $\mathcal{O}(n/\log_\sigma n)$ time in total. □

The next claim resembles to some extent a property of seeds; cf. [32, Lemma 2.2]. See Fig. 2 for an illustration.

Claim. For $i \in [1\mathinner{\ldotp\ldotp} c]$, $C = T[0\mathinner{\ldotp\ldotp} i)$ is a cover of T if and only if C is a border of T and the occurrences of C in each string S in $\mathcal{F}$ cover the middle length-c part of S.

Proof. ($\Rightarrow$) Assume that C is a cover of T. As C has to cover the first and the last position of T, C is a border of T. For every integer multiple ic of c such that $(i+1)c < |T|$, occurrences of C in T have to cover $U = T[ic\mathinner{\ldotp\ldotp}(i+1)c)$. These occurrences need to be contained in $S = T[(i-1)c\mathinner{\ldotp\ldotp}\min((i+2)c, n-1)]$. We have $S \in \mathcal{F}$ (possibly after appending $3c - |S|$ letters) and C covers the middle length-c part of S.

($\Leftarrow$) Assume that C is a border of T and the occurrences of C in each string S in $\mathcal{F}$ cover the middle length-c part of S. Since C is a border of T, its occurrences cover substrings $T[0 \mathinner{.\,.} |C|)$ and $T[n - |C| \mathinner{.\,.} n)$. Moreover, for every integer multiple ic of c such that $(i+1)c < |T|$, occurrences of C in T cover $T[ic \mathinner{.\,.} (i+1)c)$. Hence, occurrences of C in T cover all positions of T. □

Now to compute $\mathsf{SCovers}(T)$ we iterate through all the c lengths of candidates for a short cover independently.

For a single candidate $C = T[0 \mathinner{.\,.} i)$, it is enough to check if C is a suffix of T, which can be done in $\mathcal{O}(1)$ time, and then for each substring $S \in \mathcal{F}$ check if the occurrences of C in S cover its middle part. The latter can be done naively in $\mathcal{O}(|C| + |S|)$ time, which sums up to $\mathcal{O}(c \cdot |\mathcal{F}|) = \mathcal{O}(\sqrt{n} \cdot \log_\sigma n)$ time for each i, and $\mathcal{O}(\sqrt{n} \cdot \log_\sigma^2 n) = o(n/\log_\sigma n)$ time in total for all the candidates.

The result is reported in the form of the $\mathcal{O}(c) = \mathcal{O}(\log n)$ lengths of short covers (arithmetic progressions of length 1).

In the computation of long covers, we use internal pattern matching. In particular, we use IPM queries which, given two substrings X, Y of T such that $|X| \leq |Y| \leq 2|X|$, return the set of occurrences of X in Y represented as an arithmetic progression. Moreover, we use period queries that return the set of all periods (equivalently, borders) of any substring of T. We need to apply a period query only to T itself. Such a query returns, for every d being an integer power of two, the set of lengths of all borders of T of length between d and $2d$ represented as an arithmetic progression. If an arithmetic progression has length greater than two, then the difference p of the progression is the common shortest period of all the borders represented by this progression.

Theorem 4. *[33] Assume that a string T of length n over integer alphabet of size σ is given in a packed form. After $\mathcal{O}(n/\log_\sigma n)$-time preprocessing, one can answer an IPM query for substrings X, Y of T in $\mathcal{O}(|Y|/|X|)$ time and a period query for T in $\mathcal{O}(\log n)$ time.*

Lemma 5. *For a string T of length n over integer alphabet of size σ, a representation of $\mathsf{LCovers}(T)$ can be computed in $\mathcal{O}(n/\log_\sigma n)$ time.*

Proof. As we already noticed, a cover of T is in particular its border. We ask a period query of Theorem 4 to compute a representation of the set of all borders of T. We disregard arithmetic progressions such that all their elements are smaller than c, as they can only correspond to the case of short covers that was already considered. Moreover, if an arithmetic progression contains elements smaller than c (there can be at most one such progression remaining), we trim it so that it contains only border lengths greater than c.

For each arithmetic progression there exists a cut-off value t such that all borders of length at most t represented by the progression are covers of T, while the longer ones are not covers of T. (This is because a shorter border from the progression is a cover of a longer border from the progression.) It is sufficient to compute this cut-off value for each progression.

We consider the progressions separately. Let us consider a progression Γ of border lengths in $[d\mathbin{..}2d]$. For the two shortest borders B_1, B_2 represented by the progression Γ (or fewer if progression Γ contains at most one element), we use IPM queries for B_i and substrings of T of length $2|B_i| - 1$ starting at positions $\equiv 0 \pmod{|B_i|}$ to find a representation of the set of occurrences of B_i in T as $\mathcal{O}(n/d)$ arithmetic progressions. This representation allows us to easily check in $\mathcal{O}(n/d)$ time if B_i is a cover of T.

If any of B_1 and B_2 exists and is not a cover of T, we can safely ignore all the remaining borders in progression Γ. Otherwise, if at least three borders are represented by Γ, we know that the difference p of the progression is the common smallest period of all borders represented by Γ. In $\mathcal{O}(n/d)$ time we partition the already computed occurrences of B_1 in T into maximal arithmetic progressions of consecutive occurrences with difference p. Let Δ be the minimum length of such an arithmetic progression of occurrences of B_1. Then exactly the Δ shortest borders of the progression Γ are covers of T, or all borders if the progression contains less than Δ elements.

Let us argue for the correctness of the algorithm. Let $B_1, B_2, \ldots, B_r$ be all borders represented by progression Γ ordered by increasing lengths. We have $|B_{i+1}| = |B_i| + p$ for all $i \in [1\mathbin{..}r]$. It suffices to note that (1) $B_{\Delta'}$ for $\Delta' = \min(\Delta, r)$ is a cover of T and (2) $B_{\Delta+1}$, if it exists, is not a cover of T.

As for (1), as all arithmetic progressions with difference p of occurrences of B_1 in T have length at least Δ', they imply occurrences of $B_{\Delta'}$ that cover the same set of positions of T as the occurrences of B_1, i.e., all positions.

As for (2), assume that $B_{\Delta+1}$ exists (with $\Delta \geq 2$) and let $i, i+p, \ldots, i+(\Delta-1)p$ be a maximal arithmetic progression of occurrences of B_1 in T. The previous arithmetic progression has its last element smaller than $i - (|B_1| - p)$, as otherwise, by Lemma 1, B_1 would have a period smaller than p. Similarly, the next arithmetic progression starts at a position greater than $i + (\Delta-1)p + (|B_1| - p)$. An occurrence of $B_{\Delta+1}$ in T implies an arithmetic progression of occurrences of B_1 with difference p and $\Delta + 1$ elements starting at the same position. Thus none of the occurrences of B_1 at positions $i, i+p, \ldots, i+(\Delta-1)p$ extend to an occurrence of $B_{\Delta+1}$, the last position of T covered by an occurrence of $B_{\Delta+1}$ from a previous arithmetic progression is smaller than $i+p$, and the first position of T covered by an occurrence of $B_{\Delta+1}$ from a next arithmetic progression is greater than $i+(\Delta-1)p$. Hence, position $i+p$ of T is not covered by occurrences of $B_{\Delta+1}$. This proves (2) and concludes correctness of the algorithm.

Overall, an arithmetic progression of border lengths in $[d\mathbin{..}2d]$ is processed in $\mathcal{O}(n/d)$ time. It suffices to consider d such that $2d \geq c$, i.e., $d = 2^i$ for $i \geq (\log_2 c) - 1$. The time complexity is thus proportional to:

$$\sum_{i=\lfloor \log_2 c \rfloor - 1}^{\lfloor \log_2 n \rfloor} \frac{n}{2^i} \leq \frac{n}{2^{\lfloor \log_2 c \rfloor - 1}} \sum_{i=0}^{\infty} \frac{1}{2^i} = \mathcal{O}(n/c) = \mathcal{O}(n/\log_\sigma n),$$

as desired. □

Theorem 1 follows directly from Lemmas 4 and 5.

Remark 1. The algorithm for computing long covers works in $\mathcal{O}(n/\log_\sigma n)$ time in the PILLAR model. For a constant σ, the complexity matches our lower bound of Theorem 3. However, the computation of short covers in Lemma 4 works in $\Theta(n)$ time in the PILLAR model.

4 Sublinear Data Structure for Cover Array

4.1 Why Representing the Cover Array in Sublinear Space Can Be a Challenge

The cover array may require $\Theta(n \log n)$ bits to be represented in a straightforward manner. In particular, the array may contain $\Theta(n)$ different values; this is true even if we disregard trivial positions i such that $\mathsf{Cov}_T[i] = i$ and positions i such that $T[0 \mathrel{..} i)$ is periodic, as shown in the following Example 1.

Example 1. Let $T = \mathtt{a}^{2m}\mathtt{ba}^{3m}\mathtt{ba}^{2m}$ for positive integer m be a string of length $\Theta(m)$. Then all prefixes of T of length at least $2m+1$ are aperiodic and the last $m+1$ positions of the array Cov_T contain the following lengths of proper covers: $3m+1, 3m+2, \ldots, 4m+1$.

Example 1 might still not be fully convincing that a sublinear-sized representation of the cover array is not obvious. Indeed, the cover array of the string family from Example 1 has an especially simple structure (a prefix consisting only of ones, a substring with an arithmetic sequence with difference 1 corresponding to superprimitive prefixes, and a suffix with an arithmetic sequence with difference 1). Below in Corollary 1 we give a different example, that in a Fibonacci string all but a logarithmic number of prefixes have a proper cover and (except for a short prefix) no *two* consecutive positions of the cover array form an arithmetic sequence of difference 1. The cover array of a Fibonacci string contains a logarithmic number of different values.

Let us recall that the Fibonacci strings are defined as follows: $\mathsf{Fib}_0 = \mathtt{b}$, $\mathsf{Fib}_1 = \mathtt{a}$, and $\mathsf{Fib}_m = \mathsf{Fib}_{m-1}\mathsf{Fib}_{m-2}$ for $m > 1$. All covers of whole Fibonacci strings (as well as other types of quasiperiodicity) were characterized in [14] (see also [40] for similar results on Tribonacci strings). Moreover, a complete characterization of the lengths of shortest covers of cyclic shifts of Fibonacci strings was shown [18]. However, apparently, the structure of the cover array of Fibonacci strings was not studied before. The theorem below shows the recursive structure of the array; see Fig. 1 for a concrete example.

Let Fib be the infinite Fibonacci string. For any $m > 0$, Fib_m is a prefix of Fib, and hence also $\mathsf{Cov}_{\mathsf{Fib}_m}$ is a prefix of $\mathsf{Cov}_{\mathsf{Fib}}$. Thus it is enough to characterize the values of $\mathsf{Cov}_{\mathsf{Fib}}$. Let $F_k = |\mathsf{Fib}_k|$.

Theorem 5. *In the corner cases* $\mathsf{Cov}_{\mathsf{Fib}}[\ell]$ *is equal to*

- ℓ *if* $\ell \leq 2$,
- 3 *if* $\ell = F_k$ *for odd* $k \geq 3$,
- 5 *if* $\ell = F_k$ *for even* $k \geq 4$,

– ℓ if $\ell = F_k - 1$ or $\ell = 2F_k - 1$ for $k \geq 4$.

Otherwise, $\mathsf{Cov}_{\mathsf{Fib}}[\ell] = \mathsf{Cov}_{\mathsf{Fib}}[\ell - F_{k-1}]$, where $F_k < \ell < F_{k+1}$.

Proof. It is well-known that for any $m \geq 1$, $\mathsf{LCP}(\mathsf{Fib}_{m+1}, \mathsf{Fib}_{m-1}\mathsf{Fib}_m) = F_{m+1} - 2$; see e.g. [31]. In particular, F_{m-1} is a period of $\mathsf{Fib}_{m+1}[0 \mathinner{.\,.} F_{m+1} - 2) = (\mathsf{Fib}_{m-1}\mathsf{Fib}_m)[0 \mathinner{.\,.} F_{m+1} - 2) = (\mathsf{Fib}_{m-1}\mathsf{Fib}_{m-1}\mathsf{Fib}_{m-2})[0 \mathinner{.\,.} F_{m+1} - 2)$, but not a period of $\mathsf{Fib}_{m+1}[0 \mathinner{.\,.} F_{m+1} - 1)$. Equivalently, $\mathsf{Fib}[0 \mathinner{.\,.} \ell - F_{m-1})$ is a border of $\mathsf{Fib}[0 \mathinner{.\,.} \ell)$ if and only if $F_{m-1} < \ell \leq F_{m+1} - 2$.

The values of $\mathsf{Cov}_{\mathsf{Fib}}[F_k]$ as well as of $\mathsf{Cov}_{\mathsf{Fib}}[\ell]$ for $\ell \leq 2$ follow from [14]. From the same paper we know that Fib_{k-2} is the longest proper border of Fib_k. Moreover, $\mathsf{Fib}[0 \mathinner{.\,.} F_{k-2} - 1)$ is the longest border of $\mathsf{Fib}[0 \mathinner{.\,.} F_k - 1)$. Indeed, an existence of a longer border (of length different than $F_{k-1} - 1$) would result in $\mathsf{Fib}[0 \mathinner{.\,.} F_k - 2)$ having period 1 by the periodicity lemma (as it already has periods F_{k-1} and F_{k-2}; see [31]).

We will prove by induction that $\mathsf{Cov}_{\mathsf{Fib}}[F_k - 1] = F_k - 1$ for $k \geq 4$. The base case holds. By the above, the only candidate for the length of a proper cover of $\mathsf{Cov}_{\mathsf{Fib}}[F_k - 1]$ is $\mathsf{Cov}_{\mathsf{Fib}}[F_{k-2} - 1]$, which equals $F_{k-2} - 1$ by induction. Prefix $\mathsf{Fib}[0 \mathinner{.\,.} F_{k-2} - 1)$ has occurrences at positions 0 and F_{k-2} in Fib, but the position $F_{k-2} - 1$ remains uncovered; existence of yet another occurrence that contains this position would result in a long overlap of occurrences which, in turn, would result in the string $\mathsf{Fib}[0 \mathinner{.\,.} F_{k-2} - 1)$ being periodic, which is not the case. Hence, $\mathsf{Cov}_{\mathsf{Fib}}[F_k - 1] = F_k - 1$.

Next we prove by induction that $\mathsf{Cov}_{\mathsf{Fib}}[2F_k - 1] = 2F_k - 1$ for $k \geq 4$. Similarly, $\mathsf{Fib}[0 \mathinner{.\,.} F_k - 1)$ is the longest border of $\mathsf{Fib}[0 \mathinner{.\,.} 2F_k - 1)$. By the inductive hypothesis, $\mathsf{Cov}_{\mathsf{Fib}}[F_k - 1] = F_k - 1$ is the only candidate for the length of a proper cover of $\mathsf{Fib}[0 \mathinner{.\,.} 2F_k - 1)$. By exactly the same argument as in the previous case, the position $F_k - 1$ in Fib is not covered by this candidate. Thus $\mathsf{Cov}_{\mathsf{Fib}}[2F_k - 1] = 2F_k - 1$.

We have $\mathsf{Cov}_{\mathsf{Fib}}[6] = 3$. Now, for $k \geq 5$, let $\ell \in [F_k + 1 \mathinner{.\,.} 2F_{k-1} - 2] \cup [2F_{k-1} \mathinner{.\,.} F_{k+1} - 2]$. As noted, $\mathsf{Fib}[0 \mathinner{.\,.} \ell - F_{k-1})$ is a border of $\mathsf{Fib}[0 \mathinner{.\,.} \ell)$ (since $\ell \leq F_{k+1} - 2$). Additionally, if $\ell \leq 2F_{k-1} - 2$, string $\mathsf{Fib}[0 \mathinner{.\,.} \ell - F_{k-1})$ also appears in Fib at position F_{k-2} (by the LCP equality from the beginning of the proof). Those two or three occurrences cover all the positions of $\mathsf{Fib}[0 \mathinner{.\,.} \ell)$, hence a cover of $\mathsf{Fib}[0 \mathinner{.\,.} \ell - F_{k-1})$ is also a cover of $\mathsf{Fib}[0 \mathinner{.\,.} \ell)$. At the same time a shortest cover of $\mathsf{Fib}[0 \mathinner{.\,.} \ell)$ has to be a cover of a border that is a cover, hence $\mathsf{Cov}_{\mathsf{Fib}}[\ell] = \mathsf{Cov}_{\mathsf{Fib}}[\ell - F_{k-1}]$. □

Corollary 1. *For any $m \geq 1$, the array $\mathsf{Cov}_{\mathsf{Fib}_m}$ contains $\Theta(m)$ different values. Only $\Theta(m)$ prefixes of Fib_m are superprimitive. Moreover, for all $\ell \in [5 \mathinner{.\,.} F_m)$, we have $\mathsf{Cov}_{\mathsf{Fib}_m}[\ell] + 1 \neq \mathsf{Cov}_{\mathsf{Fib}_m}[\ell + 1]$.*

Proof. The first two statements follow readily from Theorem 5. As for the third statement, among the distinct values in $\mathsf{Cov}_{\mathsf{Fib}}$ from position 5 onwards, the only pairs of consecutive numbers are $(3, 4)$ and $(4, 5)$. (This is because for large enough k, values $F_{k+1} - 1$ and $2F_k - 1$ differ by more than 1.) Therefore, if $\ell \geq 5$ would be the smallest position such that $\mathsf{Cov}_{\mathsf{Fib}}[\ell + 1] = \mathsf{Cov}_{\mathsf{Fib}}[\ell] + 1$, then ℓ or

$\ell + 1$ would be equal to F_k for some $k \geq 5$. By the recursion in Theorem 5, if $\ell + 1 = F_k$, then $\mathsf{Cov}_{\mathsf{Fib}}[\ell] = \ell > 5$ and $\mathsf{Cov}_{\mathsf{Fib}}[\ell + 1] \in \{3,5\}$, so two consecutive values are not possible. If $\ell = F_k$ and $\mathsf{Cov}_{\mathsf{Fib}}[\ell] = 3$, then $\mathsf{Cov}_{\mathsf{Fib}}[\ell + 1] = 9$ by easy induction, so again two consecutive values on consecutive positions are not possible. □

The recursive characterization of Theorem 5 allows to compute any element of the cover array of Fib_m in $\mathcal{O}(\log n)$ time, where $n = F_m$, without additional space. By Corollary 1, the cover array of Fib_m has only $\mathcal{O}(\log n)$ different values, which allows one to store the cover array of Fib_m in a packed form in $\mathcal{O}(n \log \log n / \log n)$ space so that its elements can be retrieved in $\mathcal{O}(1)$ time. In the next subsection we show that an equally space-efficient representation exists for every string over a constant-sized alphabet.

4.2 Proof of Theorem 2

Let us recall the statement of the theorem:

Theorem 2. *Let T be a string of length n over an integer alphabet of size σ. There exists a data structure using space $\mathcal{O}(n(\log \sigma + \log \log n)/\log n)$ that, given $\ell \in [1 \mathinner{.\,.} n]$, returns $\mathsf{Cov}_T[\ell]$ in $\mathcal{O}(1)$ time.*

Proof. Before we describe the data structure, let us give some intuition. A schematic illustration is provided in Fig. 3.

Assume that $\mathsf{Cov}_T[\ell] = c$ with $c < \ell$. That is, string $C = T[0 \mathinner{.\,.} c)$ is a proper shortest cover of a prefix $T[0 \mathinner{.\,.} \ell)$. If the second occurrence of C in T is at position $j > 0$, then $U^2 = T[0 \mathinner{.\,.} 2j)$ is a square. Further, $j > c/2$, as otherwise C would be periodic. Hence, C is a prefix of $T[0 \mathinner{.\,.} 2j)$. This concludes that the square $T[0 \mathinner{.\,.} 2j)$ is primitively rooted, as otherwise C would be periodic. By Lemma 3, there are only $\mathcal{O}(\log n)$ primitively rooted square prefixes of T. Thus, if C is a proper shortest cover of a prefix of T, we can assign to C one of $\mathcal{O}(\log n)$ primitively rooted square prefixes of T.

Let $P = T[0 \mathinner{.\,.} p)$ be the shortest aperiodic prefix of T such that $p \geq j$. As $T[0 \mathinner{.\,.} c) = C$ is aperiodic, p is well-defined and C has a prefix P. Thus, if C is a proper shortest cover of a prefix of T, this allows to assign to C one of $\mathcal{O}(\log n)$ aperiodic prefixes of T.

For $k = \ell - c$, we have $T[k \mathinner{.\,.} k+p) = P$. We observe that there can be no further occurrence of P in T at a position in $(k \mathinner{.\,.} \ell - p]$ (that is, no further occurrence of P in $T[0 \mathinner{.\,.} c)$). Indeed, such an occurrence would be a substring of C, so it would imply an occurrence of P in T at a position in $[1 \mathinner{.\,.} j)$. By Lemma 2, this would contradict the fact that P is aperiodic. In summary, if C is a proper shortest cover of a prefix $T[0 \mathinner{.\,.} \ell)$, then C can be uniquely identified by the rightmost occurrence in $T[0 \mathinner{.\,.} \ell)$ of the aperiodic prefix P of T that is assigned to C. Moreover, the occurrence is at one of the positions in $(\ell - 2p \mathinner{.\,.} \ell)$, as $2p \geq 2j > c$.

Data structure: Let $j_1, \ldots, j_t$ be the half lengths of all primitively rooted square prefixes of T. By Lemma 3, we have $t = \mathcal{O}(\log n)$. The data structure

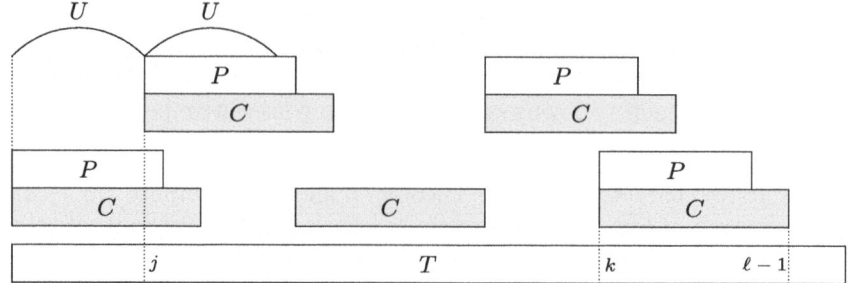

Fig. 3. The intuition behind the proof of Theorem 2. A proper shortest cover C of $T[0\mathinner{..}\ell)$ implies a primitively rooted square prefix U^2 of T. The rightmost occurrence in $T[0\mathinner{..}\ell)$ of the shortest aperiodic prefix P of T of length at least $|U|$ implies the rightmost occurrence of C in $T[0\mathinner{..}\ell)$

stores t lengths of aperiodic prefixes of T, $p_1, \ldots, p_t$. For every $i \in [1\mathinner{..}t]$, p_i is the length of the shortest aperiodic prefix of $T[0\mathinner{..}2j_i)$ of length at least j_i. (It is known that such a prefix exists, as $T[0\mathinner{..}2j_i - 1)$ is aperiodic by Lemma 1.)

For each $\ell \in [1\mathinner{..}n]$, we store a bit $sp[\ell]$ that equals 1 if and only if $T[0\mathinner{..}\ell)$ is superprimitive. If $sp[\ell] = 0$, a number $pref[\ell] \in [1\mathinner{..}t]$ is stored that determines the aperiodic prefix $P = T[0\mathinner{..}p_{pref[\ell]})$ of T that corresponds to the shortest cover C of $T[0\mathinner{..}\ell)$, as discussed above. Precisely, if j_i is the position of the second occurrence of C in T, then $pref[\ell] = i$. Finally, a data structure for IPM queries in T is stored.

Overall, provided that the arrays sp and $pref$ are stored in a packed form, the space complexity is $\mathcal{O}(\log n + n \log \log n / \log n + n / \log n + n / \log_\sigma n) = \mathcal{O}(n \log \log n / \log n + n / \log_\sigma n)$, as required.

Queries: To compute $\mathsf{Cov}_T[\ell]$, we first check if $sp[\ell] = 1$ and, if that is the case, return ℓ. Otherwise, we ask an IPM query to compute the righmost occurrence of $P = T[0\mathinner{..}p_{pref[\ell]})$ in $T[\ell - 2 \cdot p_{pref[\ell]} + 1\mathinner{..}\ell)$. As P is aperiodic, there are at most two such occurrences. We select as k the starting position of the rightmost occurrence. The shortest cover of $T[0\mathinner{..}\ell)$ is $T[k\mathinner{..}\ell)$ (i.e., $\mathsf{Cov}_T[\ell] = \ell - k$).

By [33], the query time complexity is $\mathcal{O}(1)$. This concludes the proof. □

An example of the data structure and queries can be found in Example 2.

Example 2. The table in Fig. 4 contains the arrays sp and $pref$ from the data structure of Theorem 2 for $T = \mathsf{Fib}_7$ from Fig. 1. T has three primitively rooted square prefixes, Fib_3^2, Fib_4^2 and Fib_5^2. Hence, $j_1 = 3$, $j_2 = 5$, $j_3 = 8$, and since Fibonacci strings are aperiodic, $p_1 = 3$, $p_2 = 5$, $p_3 = 8$.

Let us consider a query for $\ell = 17$, $sp[\ell] = 0$. $P = T[0\mathinner{..}p_3) = \mathsf{Fib}_5$ appears only once in $T[2\mathinner{..}17)$, at position $k = 8$. We have $\mathsf{Cov}_T[\ell] = 17 - 8 = 9$.

ℓ:	1	2	3	4	5	6	7	8	9	10	11	12	13	14	15	16	17	18	19	20	21
$T[\ell-1]$:	a	b	a	a	b	a	b	a	a	b	a	a	b	a	b	a	a	b	a	b	a
$\mathrm{Cov}_T[\ell]$:	1	2	3	4	5	3	7	3	9	5	3	12	5	3	15	3	9	5	3	20	3
$sp[\ell]$:	1	1	1	1	1	0	1	0	1	0	0	1	0	0	1	0	0	0	0	1	0
$pref[\ell]$:						1		1		2	1		2	1		1	3	2	1		1

Fig. 4. A concrete example for data structure of Theorem 2

5 Lower Bound on the Complexity in the PILLAR Model

5.1 The PILLAR Model

Let us start by formally introducing the primitives of the PILLAR model [12]. The argument strings are fragments of strings in a given collection $\mathcal{X}$:

- Extract(S, ℓ, r): Retrieve string $S[\ell..r]$.
- LCP(X, Y), LCP$_\mathsf{R}(X, Y)$: Compute the length of the longest common prefix/suffix of X and Y.
- IPM(X, Y): Assuming that $|Y| \leq 2|X|$, compute the starting positions of all exact occurrences of X in Y, expressed as an arithmetic progression.
- Access(S, i): Retrieve the letter $S[i]$;
- Length(S): Compute the length $|S|$ of the string S.

The runtime of algorithms in this model can be expressed in terms of the number of primitive PILLAR operations (and additional operations not performed on the strings themselves).

5.2 Lower Bound

We now proceed with proving the following theorem.

Theorem 3. *There is no algorithm in the PILLAR model that solves any of the following problems for a length-n binary string T in $o(n/\log n)$ time:*

- *check if T is superprimitive;*
- *check if a given prefix of T is a cover of T.*

We focus on checking if a string over an alphabet $\{\mathsf{a}, \mathsf{b}\}$ is covered by its border aba. Strings covered by aba are formed of concatenations of strings of a form $(\mathsf{ab})^k\mathsf{a}$ for $k \geq 1$; equivalently, strings that have aba as a border and do not contain a substring bb or aaa.

For infinitely many positive integers n, we show a strategy for an adversary to answer $Cn/\log n$ PILLAR queries on a length-n binary string, for a certain constant $C > 0$, after which the adversary still has the choice of fixing the string in two ways: in one T has a cover aba, and in the other T is superprimitive (i.e., it has no proper cover).

We define a morphism $\phi : \{0,1\} \mapsto \{\mathtt{a},\mathtt{b}\}$:

- $\phi(0) = \mathtt{abababa\ aba\ ababa} = (\mathtt{ab})^3\mathtt{a}(\mathtt{ab})\mathtt{a}(\mathtt{ab})^2\mathtt{a}$
- $\phi(1) = \mathtt{abababa\ ababa\ aba} = (\mathtt{ab})^3\mathtt{a}(\mathtt{ab})^2\mathtt{a}(\mathtt{ab})\mathtt{a}$

Both $\phi(0)$ and $\phi(1)$ have length 15 and have a cover $\mathtt{aba}$. Thus $\phi(S)$, for any string S over alphabet $\{0,1\}$, has a cover $\mathtt{aba}$.

Let us recall that a de Bruijn sequence of order k over an alphabet Σ is a string of length $|\Sigma|^k + k - 1$ such that its every substring of length k is distinct. It is well-known that such sequences exist for every finite alphabet Σ and integer $k \geq 1$ [9].

Let B_k be a de Bruijn sequence of order k over the binary alphabet $\{0,1\}$. We apply the morphism ϕ to B_k to obtain a string T_k over alphabet $\{\mathtt{a},\mathtt{b}\}$ of length $15 \cdot (2^k + k - 1)$. Due to the property of de Bruijn sequences, each substring of T of length at least $15(k+1) - 1$ is distinct. Indeed, every "aligned" substring of length $15k$ starting at a position divisible by 15 in T_k is distinct, and every substring of T_k of length $15(k+1) - 1$ contains an "aligned" substring of length $15k$.

Due to this property, an answer to an LCP or LCP$_\mathsf{R}$ query on T_k for two different positions is always bounded from above by $15(k+1) - 2$. Similarly for the IPM queries; if we query for a substring of length at least $15(k+1) - 1$, then we do not gain any interesting information (the only occurrence of the substring is the one used to ask the query). On the other hand, by asking an IPM query for a shorter substring we only gain information about a part of T_k of length at most $30(k+1) - 2$.

Formally, the strategy of the adversary for a text T of length $n = |T_k|$ is as follows. Queries Extract, Access, Length, LCP, LCP$_\mathsf{R}$ are answered as in T_k. An IPM(X,Y) query for $|X| < 15(k+1) - 1$ is also answered as in T_k. Finally, to answer an IPM(X,Y) query for $|X| \geq 15(k+1) - 1$, we refer to the fragments $T[i_x \mathinner{.\,.} j_x] = X$ and $T[i_y \mathinner{.\,.} j_y] = Y$ and return an occurrence of X in Y at position $i_y - i_x$ if $[i_x \mathinner{.\,.} j_x] \subseteq [i_y \mathinner{.\,.} j_y]$ and no occurrence otherwise.

We say that a position $i \in [0 \mathinner{.\,.} n)$ of T has been touched if the algorithm has performed (1) an Access query on $T[i]$, or (2) an LCP$(T[i_x \mathinner{.\,.} j_x], T[i_y \mathinner{.\,.} j_y])$ query such that $i \in [i_x \mathinner{.\,.} i_x + \ell) \cup [i_y \mathinner{.\,.} i_y + \ell)$ where ℓ is the result of the LCP query, or (3) similarly an LCP$_\mathsf{R}$ query such that i belongs to the computed LCP$_\mathsf{R}$ of one of the two queried substrings of T, or (4) an IPM$(T[i_x \mathinner{.\,.} j_x], T[i_y \mathinner{.\,.} j_y])$ query for $j_x - i_x + 1 < 15(k+1) - 1$ such that $i \in [i_x \mathinner{.\,.} j_x] \cup [i_y \mathinner{.\,.} j_y]$. In total, after q PILLAR operations, fewer than $45q(k+1) \leq 90kq$ positions of T have been touched. Thus after $q = \lfloor 2^k/(6k) \rfloor$ operations, there still exists a position in T that has not been touched. Assume i is such a position. Then the adversary can make the choice to set $T[i]$ as $T_k[i]$ or as the letter different from $T_k[i]$; all the remaining untouched positions are set as in T_k. If $T[i] = T_k[i]$, $T = T_k$ has a cover $\mathtt{aba}$. If $T[i] \neq T_k[i]$, T contains exactly one substring $\mathtt{a}^s$ for some $s \in [3 \mathinner{.\,.} 5]$, or exactly one substring b^t, for some $t \in [2..3]$. It is easy to see that in this case T is superprimitive.

There exists a constant $C > 0$ (for example, $C = 1/180$) such that the selected value of q satisfies $q \geq Cn/\log n$. Theorem 3 is proved.

6 Open Problems

It remains open if the data structure of Theorem 2 can be constructed in sublinear time or if its space complexity can be decreased to $\mathcal{O}(n/\log n)$ for $\sigma = \mathcal{O}(1)$.

Future work also includes designing sublinear-time algorithms for other notions of quasiperiodicity for which $\mathcal{O}(n)$-time algorithms are already known, for a length-n string over an integer alphabet; this includes, for example, seeds [32,39], enhanced covers [23], and partial covers [39].

References

1. Apostolico, A., Farach, M., Iliopoulos, C.S.: Optimal superprimitivity testing for strings. Inf. Process. Lett. **39**(1), 17–20 (1991). https://doi.org/10.1016/0020-0190(91)90056-N
2. Bannai, H., Ellert, J.: Lyndon arrays in sublinear time. In: Gørtz, I.L., Farach-Colton, M., Puglisi, S.J., Herman, G. (eds.) 31st Annual European Symposium on Algorithms, ESA 2023, 4–6 September 2023, Amsterdam. LIPIcs, vol. 274, pp. 14:1–14:16. Schloss Dagstuhl - Leibniz-Zentrum für Informatik (2023). https://doi.org/10.4230/LIPICS.ESA.2023.14
3. Bannai, H., Mieno, T., Nakashima, Y.: Lyndon words, the three squares lemma, and primitive squares. In: Boucher, C., Thankachan, S.V. (eds.) SPIRE 2020. LNCS, vol. 12303, pp. 265–273. Springer, Heidelberg (2020). https://doi.org/10.1007/978-3-030-59212-7_19
4. Belazzougui, D., Kosolobov, D., Puglisi, S.J., Raman, R.: Weighted ancestors in suffix trees revisited. In: Gawrychowski, P., Starikovskaya, T. (eds.) 32nd Annual Symposium on Combinatorial Pattern Matching, CPM 2021, 5–7 July 2021, Wrocław. LIPIcs, vol. 191, pp. 8:1–8:15. Schloss Dagstuhl - Leibniz-Zentrum für Informatik (2021). https://doi.org/10.4230/LIPICS.CPM.2021.8
5. Ben-Kiki, O., Bille, P., Breslauer, D., Gasieniec, L., Grossi, R., Weimann, O.: Towards optimal packed string matching. Theor. Comput. Sci. **525**, 111–129 (2014). https://doi.org/10.1016/J.TCS.2013.06.013
6. Bender, M.A., Farach-Colton, M.: The LCA problem revisited. In: Gonnet, G.H., Panario, D., Viola, A. (eds.) LATIN 2000. LNCS, vol. 1776, pp. 88–94. Springer, Heidelberg (2000). https://doi.org/10.1007/10719839_9
7. Breslauer, D.: An on-line string superprimitivity test. Inf. Process. Lett. **44**(6), 345–347 (1992). https://doi.org/10.1016/0020-0190(92)90111-8
8. Breslauer, D., Galil, Z.: Finding all periods and initial palindromes of a string in parallel. Algorithmica **14**(4), 355–366 (1995). https://doi.org/10.1007/BF01294132
9. de Bruijn, N.G.: A combinatorial problem. Indagationes Math. **8**, 461–467 (1946). http://www.dwc.knaw.nl/DL/publications/PU00018235.pdf

10. Charalampopoulos, P., Kociumaka, T., Pissis, S.P., Radoszewski, J.: Faster algorithms for longest common substring. In: Mutzel, P., Pagh, R., Herman, G. (eds.) 29th Annual European Symposium on Algorithms, ESA 2021, 6–8 September 2021, Lisbon (Virtual Conference). LIPIcs, vol. 204, pp. 30:1–30:17. Schloss Dagstuhl - Leibniz-Zentrum für Informatik (2021). https://doi.org/10.4230/LIPICS.ESA.2021.30
11. Charalampopoulos, P., Kociumaka, T., Wellnitz, P.: Faster approximate pattern matching: a unified approach. In: Irani, S. (ed.) 61st IEEE Annual Symposium on Foundations of Computer Science, FOCS 2020, Durham, 16–19 November 2020, pp. 978–989. IEEE (2020). https://doi.org/10.1109/FOCS46700.2020.00095
12. Charalampopoulos, P., Kociumaka, T., Wellnitz, P.: Faster approximate pattern matching: a unified approach. In: 61st IEEE Annual Symposium on Foundations of Computer Science, FOCS 2020, pp. 978–989. IEEE (2020). https://doi.org/10.1109/FOCS46700.2020.00095. Full version: arXiv:2004.08350v2
13. Charalampopoulos, P., Pissis, S.P., Radoszewski, J.: Longest palindromic substring in sublinear time. In: Bannai, H., Holub, J. (eds.) 33rd Annual Symposium on Combinatorial Pattern Matching, CPM 2022, 27–29 June 2022, Prague. LIPIcs, vol. 223, pp. 20:1–20:9. Schloss Dagstuhl - Leibniz-Zentrum für Informatik (2022). https://doi.org/10.4230/LIPICS.CPM.2022.20
14. Christou, M., Crochemore, M., Iliopoulos, C.S.: Quasiperiodicities in Fibonacci strings. Ars Comb. **129**, 211–225 (2016). https://arxiv.org/abs/1201.6162
15. Crochemore, M., et al.: The maximum number of squares in a tree. In: Kärkkäinen, J., Stoye, J. (eds.) CPM 2012. LNCS, vol. 7354, pp. 27–40. Springer, Heidelberg (2012). https://doi.org/10.1007/978-3-642-31265-6_3
16. Crochemore, M., Iliopoulos, C.S., Pissis, S.P., Tischler, G.: Cover array string reconstruction. In: Amir, A., Parida, L. (eds.) CPM 2010. LNCS, vol. 6129, pp. 251–259. Springer, Heidelberg (2010). https://doi.org/10.1007/978-3-642-13509-5_23
17. Crochemore, M., et al.: Internal quasiperiod queries. In: Boucher, C., Thankachan, S.V. (eds.) SPIRE 2020. LNCS, vol. 12303, pp. 60–75. Springer, Heidelberg (2020). https://doi.org/10.1007/978-3-030-59212-7_5
18. Crochemore, M., et al.: Shortest covers of all cyclic shifts of a string. Theor. Comput. Sci. **866**, 70–81 (2021). https://doi.org/10.1016/J.TCS.2021.03.011
19. Crochemore, M., Rytter, W.: Squares, cubes, and time-space efficient string searching. Algorithmica **13**(5), 405–425 (1995). https://doi.org/10.1007/BF01190846
20. Duyster, A., Kociumaka, T.: Logarithmic-time internal pattern matching queries in compressed and dynamic texts. In: Lipták, Z., et al. (eds.) SPIRE 2024. LNCS, vol. 14899, pp. 102–117, Springer, Cham (2024). https://doi.org/10.1007/978-3-031-72200-4_8
21. Farach, M.: Optimal suffix tree construction with large alphabets. In: 38th Annual Symposium on Foundations of Computer Science, FOCS 1997, Miami Beach, 19–22 October 1997, pp. 137–143. IEEE Computer Society (1997). https://doi.org/10.1109/SFCS.1997.646102
22. Fine, N.J., Wilf, H.S.: Uniqueness theorems for periodic functions. Proc. Am. Math. Soc. **16**(1), 109–114 (1965). https://doi.org/10.1090/S0002-9939-1965-0174934-9
23. Flouri, T., et al.: Enhanced string covering. Theor. Comput. Sci. **506**, 102–114 (2013). https://doi.org/10.1016/J.TCS.2013.08.013
24. Ganardi, M., Jeż, A., Lohrey, M.: Balancing straight-line programs. J. ACM **68**(4), 27:1–27:40 (2021). https://doi.org/10.1145/3457389

25. Gawrychowski, P., Karczmarz, A., Kociumaka, T., Lacki, J., Sankowski, P.: Optimal dynamic strings. In: Czumaj, A. (ed.) Proceedings of the Twenty-Ninth Annual ACM-SIAM Symposium on Discrete Algorithms, SODA 2018, New Orleans, 7–10 January 2018, pp. 1509–1528. SIAM (2018). https://doi.org/10.1137/1.9781611975031.99
26. Hariharan, R., Vinay, V.: String matching in Õ(sqrt(n)+sqrt(m)) quantum time. J. Discrete Algorithms **1**(1), 103–110 (2003). https://doi.org/10.1016/S1570-8667(03)00010-8
27. I, T., et al.: Detecting regularities on grammar-compressed strings. Inf. Comput. **240**, 74–89 (2015). https://doi.org/10.1016/J.IC.2014.09.009
28. Jin, C., Nogler, J.: Quantum speed-ups for string synchronizing sets, longest common substring, and k-mismatch matching. In: Bansal, N., Nagarajan, V. (eds.) Proceedings of the 2023 ACM-SIAM Symposium on Discrete Algorithms, SODA 2023, Florence, 22–25 January 2023, pp. 5090–5121. SIAM (2023). https://doi.org/10.1137/1.9781611977554.CH186
29. Kempa, D., Kociumaka, T.: String synchronizing sets: sublinear-time BWT construction and optimal LCE data structure. In: Charikar, M., Cohen, E. (eds.) Proceedings of the 51st Annual ACM SIGACT Symposium on Theory of Computing, STOC 2019, Phoenix, 23–26 June 2019, pp. 756–767. ACM (2019). https://doi.org/10.1145/3313276.3316368
30. Kempa, D., Kociumaka, T.: Dynamic suffix array with polylogarithmic queries and updates. In: Leonardi, S., Gupta, A. (eds.) STOC 2022: 54th Annual ACM SIGACT Symposium on Theory of Computing, Rome, 20–24 June 2022, pp. 1657–1670. ACM (2022). https://doi.org/10.1145/3519935.3520061
31. Knuth, D.E., Jr., J.H.M., Pratt, V.R.: Fast pattern matching in strings. SIAM J. Comput. **6**(2), 323–350 (1977). https://doi.org/10.1137/0206024
32. Kociumaka, T., Kubica, M., Radoszewski, J., Rytter, W., Waleń, T.: A linear-time algorithm for seeds computation. ACM Trans. Algorithms **16**(2), 27:1–27:23 (2020). https://doi.org/10.1145/3386369
33. Kociumaka, T., Radoszewski, J., Rytter, W., Waleń, T.: Internal pattern matching queries in a text and applications. arXiv preprint arXiv:1311.6235 (2013)
34. Kociumaka, T., Radoszewski, J., Rytter, W., Waleń, T.: Internal pattern matching queries in a text and applications. In: Indyk, P. (ed.) Proceedings of the Twenty-Sixth Annual ACM-SIAM Symposium on Discrete Algorithms, SODA 2015, San Diego, 4–6 January 2015, pp. 532–551. SIAM (2015). https://doi.org/10.1137/1.9781611973730.36
35. Mitani, K., Mieno, T., Seto, K., Horiyama, T.: Shortest cover after edit. In: Inenaga, S., Puglisi, S.J. (eds.) 35th Annual Symposium on Combinatorial Pattern Matching, CPM 2024, 25–27 June 2024, Fukuoka. LIPIcs, vol. 296, pp. 24:1–24:15. Schloss Dagstuhl - Leibniz-Zentrum für Informatik (2024). https://doi.org/10.4230/LIPICS.CPM.2024.24
36. Moore, D.W.G., Smyth, W.F.: A correction to "An optimal algorithm to compute all the covers of a string". Inf. Process. Lett. **54**(2), 101–103 (1995). https://doi.org/10.1016/0020-0190(94)00235-Q
37. Munro, J.I., Navarro, G., Nekrich, Y.: Text indexing and searching in sublinear time. In: Gørtz, I.L., Weimann, O. (eds.) 31st Annual Symposium on Combinatorial Pattern Matching, CPM 2020, 17–19 June 2020, Copenhagen. LIPIcs, vol. 161, pp. 24:1–24:15. Schloss Dagstuhl - Leibniz-Zentrum für Informatik (2020). https://doi.org/10.4230/LIPICS.CPM.2020.24

38. Plandowski, W., Rytter, W.: Application of Lempel-Ziv encodings to the solution of words equations. In: Larsen, K.G., Skyum, S., Winskel, G. (eds.) ICALP 1998. LNCS, vol. 1443, pp. 731–742. Springer, Heidelberg (1998). https://doi.org/10.1007/BFB0055097
39. Radoszewski, J.: Linear time construction of cover suffix tree and applications. In: Gørtz, I.L., Farach-Colton, M., Puglisi, S.J., Herman, G. (eds.) 31st Annual European Symposium on Algorithms, ESA 2023, 4–6 September 2023, Amsterdam. LIPIcs, vol. 274, pp. 89:1–89:17. Schloss Dagstuhl - Leibniz-Zentrum für Informatik (2023). https://doi.org/10.4230/LIPICS.ESA.2023.89
40. Singh, M.: Quasiperiodicity in Tribonacci Word (2020). https://hal.science/hal-02141636. Working paper or preprint

LZ78 Substring Compression with CDAWGs

Hiroki Shibata[1] and Dominik Köppl[2]

[1] Department of Informatics, Kyushu University, Fukuoka, Japan
shibata.hiroki.753@s.kyushu-u.ac.jp
[2] University of Yamanashi, Kofu, Japan
dkppl@yamanashi.ac.jp

Abstract. The Lempel–Ziv 78 (LZ78) factorization is a well-studied technique for data compression. It and its derivates are used in compression formats such as `compress` or `gif`. While most research focuses on the factorization of plain data, not much research has been conducted on indexing the data for fast LZ78 factorization. Here, we study the LZ78 factorization in the substring compression model, where we are allowed to index the data and have to return the factorization of a substring specified at query time. In that model, we propose an algorithm that works in CDAWG-compressed space, computing the factorization with a logarithmic slowdown compared to the optimal time complexity.

Keywords: Lossless data compression · LZ78 Factorization · Substring compression · CDAWG

1 Introduction

The substring compression problem [10] is to preprocess a given input text T such that computing a compressed version of a substring of $T[i..j]$ can be done efficiently. This problem has been originally stated for the Lempel–Ziv-77 (LZ77) factorization [31], but extensions to the generalized LZ77 factorization [18], the Lempel–Ziv 78 (LZ78) factorization [21] as well as two of its derivates [22], the run-length encoded Burrows–Wheeler transform (RLBWT) [2], and the relative LZ factorization [20, Sect. 7.3] have been studied. Given n is the length of T, a trivial solution would be to precompute the compressed output of $T[\mathcal{I}]$ for all intervals $\mathcal{I} \subset [1..n]$. This however gives us already $\Omega(n^2)$ solutions to store.

For an appealing solution, we want to be able to index a large amount of data efficiently within a fraction of space; two criteria (speed and space) that are likely to be anti-correlated. However, as far as we are aware of, the substring compression problem has not yet been studied with compressed space bounds that can be sub-linear for compressible input data. Our main target is therefore a solution that works in compressed space and can answer a query in time linear in the output size with a polylogarithmic term on the text length. In this paper, we build upon the line of research on LZ78 factorization algorithms that superimpose the LZ trie on the suffix tree [15,21,22,25], which all use $\Omega(n)$ space for storing the suffix tree. Here, we make the algorithmic idea of the superimposition compatible with the compact directed acyclic word graph (CDAWG) [9], trading a tiny time-penalty with a large space improvement for compressible texts. Table 1 gives an overview of known solutions for the problem we tackle.

Table 1. Solutions for computing the LZ78 substring compression for a substring with z LZ78 facors of a string of length n with e CDAWG edges. Extra space means the additional working space required for processing queries in addition to the index.

method	space		time
	index	extra space	query
naive	$\mathcal{O}(n^2)$	$\mathcal{O}(z)$	$\mathcal{O}(z)$
[21]	$\mathcal{O}(n)$	$\mathcal{O}(z)$	$\mathcal{O}(z)$
Theorem 2	$\mathcal{O}(e)$	$\mathcal{O}(z)$	$\mathcal{O}(z \lg n)$

Our contribution fits into the line of research focussed on data compression with the CDAWG. Given e and z are the number of CDAWG edges and the number of LZ78 factors, respectively, in that line, [5] proposed a straight-line program (SLP), which can be computed in $\mathcal{O}(e)$ time taking $\mathcal{O}(e)$ space. Given an SLP of size $\mathcal{O}(g)$, [3] showed how to compute LZ78 from that SLP in $\mathcal{O}(g + z \lg z)$ time and space. Combining both solutions, we can compute LZ78 from the CDAWG in $\mathcal{O}(e + z \lg z)$ time and space. Recently, [1] showed how to compute, among others, the RLBWT and LZ77 in $\mathcal{O}(e)$ time and space.

2 Preliminaries

With lg we denote the logarithm $\log_2$ to base two. Our computational model is the word RAM model with machine word size $\Omega(\lg n)$ bits for a given input size n. Accessing a word costs $\mathcal{O}(1)$ time.

Let T be a text of length n whose characters are drawn from an integer alphabet $\Sigma = [1..\sigma]$ with $\sigma \leq n^{\mathcal{O}(1)}$. Given $X, Y, Z \in \Sigma^*$ with $T = XYZ$, then X, Y and Z are called a *prefix*, *substring* and *suffix* of T, respectively. We call $T[i..]$ the i-th suffix of T, and denote a substring $T[i] \cdots T[j]$ with $T[i..j]$. A *parsing dictionary* is a set of strings. A parsing dictionary $\mathcal{D}$ is called *prefix-closed* if it contains, for each string $S \in \mathcal{D}$, all prefixes of S as well. A *factorization* of T of size z partitions T into z substrings $F_1 \cdots F_z = T$. Each such substring F_x is called a *factor* and x its *index*.

LZ78 Factorization. Stipulating that F_0 is the empty string, a factorization $F_1 \cdots F_z = T$ is called the *LZ78 factorization* [33] of T iff, for all $x \in [1..z]$, the factor F_x is the longest prefix of $T[|F_1 \cdots F_{x-1}| + 1..]$ such that $F_x = F_y \cdot c$ for some $y \in [0..x-1]$ and $c \in \Sigma$, that is, F_x is the longest possible previous factor F_y appended by the following character in the text. The dictionary for computing F_x is $\mathcal{D}_x := \{F_y \cdot c : y \in [0..x-1], c \in \Sigma\}$, which is prefix-closed. Formally, F_x starts at $\mathsf{dst}_x := |F_1..F_{x-1}| + 1$ and $y = \mathrm{argmax}\{|F_{y'}| : F_{y'} = T[\mathsf{dst}_x..\mathsf{dst}_x + |F_{y'}| - 1]\}$. We say that y and F_y are the *referred index* and the *referred factor* of the factor F_x, respectively. The LZ78 factorization of $T = \mathsf{babac}$ is $F_0, F_1, \ldots, F_4 = \epsilon, \mathsf{b}, \mathsf{a}, \mathsf{ba}, \mathsf{c}$. The referred factor of $F_3 = F_1 \mathsf{a}$ is F_1; F_3's referred index is 1.

$\mathcal{D}_x$ is often implemented by the *LZ trie*, which represents each LZ factor as a node; the root represents the factor F_0. The node representing the factor F_y has a child representing the factor F_x connected with an edge labeled by a character $c \in \Sigma$ if and only if $F_x = F_y c$. To see the connection of the LZ trie and $\mathcal{D}_x$, we observe that adding any new leaf to the LZ trie storing $\{F_1, \ldots, F_{x-1}\}$ gives an element of $\mathcal{D}_x$, and vice versa we can obtain any element of $\mathcal{D}_x$ by doing so. A crucial observation is that every path from the LZ trie root downwards visits nodes in increasing LZ factor index order.

Suffix Tree. Given a tree, with an *s-t path* we denote the path from a node s to a node t. All trees in this paper are considered non-empty with a root node, which we denote by root. The *suffix trie* of T is the trie of all suffixes of T. There is a one-to-one relationship between the suffix trie leaves and the suffixes of T. The *suffix tree* [32] ST of T is the tree obtained by compacting the suffix trie of T. The string stored in an ST edge g is called the *label* of g. The *string label* of a node v is defined as the concatenation of all edge labels on the root-v path; its *string depth* is the length of its string label. The leaf corresponding to the i-th suffix $T[i..]$ is labeled with the *suffix number* $i \in [1..n]$. The *locus* of a substring S of T is the place we end up when reading S from ST starting at root. The locus of S is either an ST node, or on an ST edge (called an *implicit node* because it is represented by a suffix trie node). The left of Fig. 1 gives an example of ST.

Reading the suffix numbers stored in the leaves of ST in leaf-rank order gives the suffix array [24]. We denote the suffix array of T by SA. Since the ST leaves are sorted in SA order, an ST node v can be uniquely represented by an SA range $[i..j]$ such that the k-th leaf is in the subtree of v for all $k \in [i..j]$.

Centroid Path Decomposition. The centroid path decomposition [14] of a tree is defined as follows. For each internal node, we call its child whose subtree is the largest among all its siblings (ties are broken arbitrarily if there are multiple such children) a *heavy* node, while we call all other children *light* nodes. Additionally, we make root a light node. A *heavy path* is a maximal-length path from a light node u to the parent of a leaf containing, except for u, only heavy nodes. Since heavy paths do not overlap, we can contract all heavy paths to single nodes and thus form the centroid-path decomposed tree whose nodes are the heavy paths that are connected by the light edges. The centroid-path decomposed tree is helpful because the number of light nodes on a path from root to a leaf is

$\mathcal{O}(\lg n)$, which means that a path from root to a leaf contains only $\mathcal{O}(\lg n)$ nodes. This can be seen from the fact that the subtree size of a light node is at most half of the subtree size of its parent; thus when visiting a light node during a top-down traversal in the tree, we at least half the number of nodes we can visit from then on. Consequently, a root-leaf path in a centroid-path decomposed tree has $\mathcal{O}(\lg n)$ nodes.

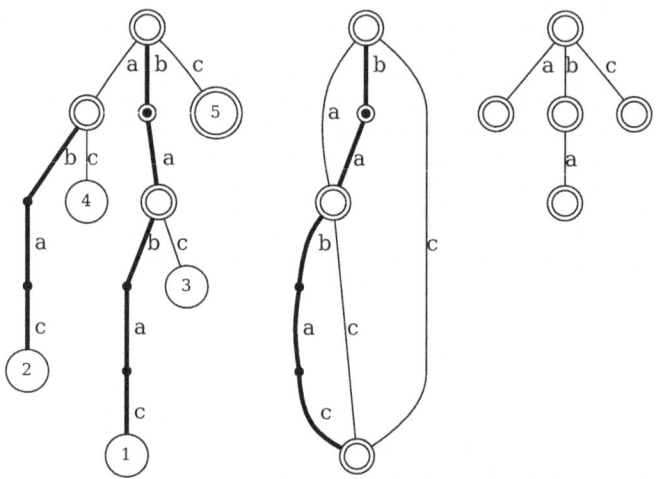

Fig. 1. ST (left), CDAWG (center), and the LZ trie (right) for $T := \mathtt{babac}$. The LZ78 factorization of T is $F_0, F_1, \ldots, F_4 = \epsilon, \mathtt{b}, \mathtt{a}, \mathtt{ba}, \mathtt{c}$. We superimpose the suffix trie and the DAWG on ST and CDAWG, respectively, by drawing implicit nodes with black dots on the edges. We additionally encircle vertices corresponding to LZ78 factors (thus showing explicit nodes as double circles). Bold and thin lines represent, respectively, the heavy and light edges of the centroid path decomposition. The CDAWG sink represents the set of strings $\{\mathtt{c}, \mathtt{ac}, \mathtt{bac}, \mathtt{abac}, \mathtt{babac}\}$, which can be read on the root-sink paths. Only a part of these strings are LZ78 factors.

CDAWG. In what follows, we adapt LZ78-substring-compression techniques to work with the CDAWG instead of ST. The CDAWG of T, denoted by CDAWG, is the minimal compact automaton that recognizes all suffixes of T [9,11]. The CDAWG of T is the minimization of ST, in which (a) all leaves are merged to a single node, called sink, and (b) all nodes, except sink, are in one-to-one correspondence with the maximal repeats of T [28], where a maximal repeat S is a substring of T having two occurrences $a_1 S b_1$ and $a_2 S b_2$ in T with $a_1 \neq a_2$ and $b_1 \neq b_2$. When transforming ST into CDAWG, multiple ST nodes can collapse into a single CDAWG node, and we say that such a CDAWG node *corresponds* to these collapsed ST nodes. We denote the number of CDAWG edges by e. With $\bar{e}$, we denote the number of edges of the CDAWG of the reverse of T. The number of CDAWG edges e can be regarded as a compression measure. For highly repetitive text, e can become asymptotically smaller than the text length n. In general, we can bound e with $e \in \mathcal{O}(n)$ and $e \in \Omega(\lg n)$. The upper bound is obtained from the fact that the number of ST edges is at most $2n - 1$; the lower bound is obtained from the fact that $g \in \mathcal{O}(e)$ and $g \in \Omega(\lg n)$, where g is the size of smallest grammar that produces T [5, Lemma 1]. Furthermore, there is a string family that achieves $e \in \Theta(\lg n)$ [29]. The right of Fig. 1 gives an example of CDAWG.

3 Factorization Algorithm

The aim of this paper is to compute, after indexing the input text T in a pre-processing step, upon request for a provided interval $[i..j] \subset [1..n]$, the LZ78 factorization of $T[i..j]$, in compressed space with time bounded linearly in the output size and logarithmic in the text length. For that, we propose two algorithms, where the first one simulates ST with CDAWG, and thus directly applies the techniques of our pre-cursors working on ST. For the last algorithm, we show how to drop the need for the ST functionality to improve the time bounds.

3.1 Superimposing CDAWG

In the following, we show how to adapt the ST superimposition by the LZ trie from [25] to CDAWG. The main observation of [25, Sect. 3] is that the LZ trie is a connected subgraph of the suffix trie containing its root because LZ78 is prefix-closed. The compacted suffix trie, i.e., ST, however contains not all suffix trie nodes. In fact, the locus of each factor F_x is either an ST node v or lies on an ST edge g. In the former case, v corresponds to the LZ node v' representing F_x in the sense that both have the same string depth. In the latter case, the locus of F_x can be witnessed by the lower node the edge g connects to (by storing information about the length of F_x at that node). Thus, we can represent the LZ trie with a marking of ST nodes. The marking is done dynamically while computing the factorization as we mark the locus of each factor after having it processed. By marking the ST root node, we identify the LZ trie root with the ST root. To find the factor lengths, we perform a traversal from the leaf λ to its lowest marked ancestor, where λ is the leaf whose suffix number corresponds to the starting position of the factor we want to compute. Thus, we process the leaves in the order of their suffix numbers while computing the factorization.

To translate this technique to CDAWG, we no longer move to different leaves since all leaves are contracted to sink. This is no problem if we keep track of the starting position of the factor we want to compute. However, an obstacle is that a CDAWG node can have multiple parents. Given we superimpose the LZ trie on CDAWG such that an explicit LZ (trie) node v is stored in its corresponding CDAWG node v'. Unlike the case for ST, we have in general no information about the actual string length of v because v' can have multiple paths leading to root. Figure 1 presents an example for which we cannot superimpose the LZ trie on CDAWG. In what follows, we propose two different solutions.

3.2 First Approach: Plug&Play Solution

Our solutions make the idea of the superimposition more implicit by modeling the LZ trie with a weighted segment tree data structure whose intervals correspond to ST nodes. In detail, we augment each LZ trie node with an SA range. For explicit LZ trie nodes having a corresponding ST node v, its SA range is the SA range of v. Otherwise, its range is the SA range of the ST node directly below. SA ranges of LZ trie nodes can be nested but are not overlapping due to the tree topology of ST. This makes it feasible to model the lowest marked ancestor data structure used in the precursor algorithms with a weighted segment tree data structure that represents each LZ trie by its SA range and its LZ index as weight. For the example in Fig. 1, we end with storing the weighted intervals $(0, [1..5]), (1, [3..4]), (2, [1..2]), (3, [3..4]), (4, [5..5])$, where the first components denote the weights (i.e., the factor indices). In particular, we can make use of the following data structure for a *stabbing-max query*, i.e., for a given query point q, to find the interval with the highest weight containing q in a set of weighted intervals.

Lemma 1 ([27]). *Given a set of z intervals in the range $[1..n]$ with weights in $[1..n]$, there exists a linear-space data structure that answers stabbing-max queries in $\mathcal{O}(\lg z/\lg\lg z)$ time. This data structure supports insertions and deletions of a weighted interval in $\mathcal{O}(\lg z)$ and $\mathcal{O}(\lg z/\lg\lg z)$ amortized time, respectively.*

With the data structure of Lemma 1, it is already possible to compute the LZ78 factorization without constructing the LZ trie in $\mathcal{O}(z \lg z)$ time with ST. For that we maintain the intervals of all computed LZ factors in an instance Stab of this data structure such that we can identify the factor index by the returned interval. We additionally index F_0 with interval $[1..n]$ and weight 0 for determining non-referencing factors. By doing so, given we want to compute a factor $F_x = T[\mathsf{dst}_x..\mathsf{dst}_x + |F_x| - 1]$, we can determine its reference y by querying Stab. If $y > 0$, then $F_x = F_y \cdot T[\mathsf{dst}_x + |F_y|]$. It is left to determine the interval of F_x, which we need to add to Stab. For that, we find the locus of F_x in the suffix tree, which can be done with a weighted level ancestor data structure in constant time [7,17].

This approach can be directly rewritten for CDAWG. To this end, we make use of the $\mathcal{O}(e + \bar{e})$-words representation of [6] and [5], which represents an ST node v with $\mathcal{O}(\lg n)$ bits of information, namely: (a) v's corresponding CDAWG node, (b) the string length of v, and (c) v's SA range. Their representation supports the following ST operations: (a) suffixlink(v, i) returns the ST node after taking i suffix links starting from v, in $\mathcal{O}(\lg n)$ time; (b) strAncestor(v, d) returns the highest ancestor of an ST node v with string depth of at least d, in $\mathcal{O}(\lg n)$ time.

Additionally, it is known that the number of runs r in the BWT is upper bounded by e [6]. Hence, in $\mathcal{O}(e)$ space, we can store the run-length compressed FM-index (RLFM)-index [23]. Given SA$[i]$, RLFM can recover $T[i-1]$ in $\mathcal{O}(\lg n)$ time. By storing RLFM-index in both directions, we can sequentially extract characters in $\mathcal{O}(\lg n)$ time, which we use for matching the next factor in

CDAWG—remembering that each ST node representation also stores the corresponding SA range.

Let us recall that for computing a factor $F_x = T[\mathsf{dst}_x..\mathsf{dst}_x + |F_x| - 1]$, the only thing left undone is to find its Stab interval. For that, we stipulate the invariant that when computing F_x, we have selected the SA leaf λ whose suffix number is dst_x. To ensure this invariant for F_{x+1}, we call suffixlink$(\lambda, |F_x|)$ to obtain the needed SA leaf. Finally, we find the locus of F_x by strAncestor$(\lambda, |F_x|)$. Since each ST node stores its SA range, we have all information for adding the interval of F_x to Stab, and are done. The time complexity is dominated by the ST simulation of CDAWG.

Theorem 1. *For a text T of length n, there exists a data structure of size $\mathcal{O}(e + \bar{e})$, which can, given an interval $\mathcal{I} \subset [1..n]$, compute the LZ78 factorization of $T[\mathcal{I}]$ in $\mathcal{O}(z \lg n)$ time with $\mathcal{O}(z)$ extra space, where z is the number of computed factors.*

3.3 Second Approach: Climbing Upwards

In what follows, we show how to get rid of the dependency on the ST simulation, which costs us $\mathcal{O}(\lg n)$ time per query and made it necessary to also store the CDAWG of the inverted text. Instead of simulating the ST leaf with suffix number dst_x for computing factor F_x, we select sink and search for a path to root of length $\ell := n - \mathsf{dst}_x + 1$. This also means that instead of the top-down traversals as in the previous subsection, we climb up CDAWG from sink. To this end, we use the centroid path decomposition and some definitions.

Centroid Path Decomposition. By applying the centroid path decomposition on ST, we obtain a centroid-path decomposed tree whose nodes are the heavy paths of ST and its edges the remaining light ST edges. Each root-leaf path in the centroid-path decomposed tree has a length of $\mathcal{O}(\lg n)$. [5] observed that the CDAWG edges corresponding to the ST heavy edges form a spanning tree of CDAWG. We apply the centroid path decomposition to the spanning tree of heavy edges again. We denote the heavy edges obtained by the second centroid path decomposition as the heavy edges of the CDAWG, and all other edges of the CDAWG as the light edges. After the second centroid path decomposition, the heavy edges form a set of disjoint paths, and each root-sink path in CDAWG visits at most $2 \lg n \in \mathcal{O}(\lg n)$ light edges. Figure 1 gives an example of the centroid path decomposition and the correspondence between ST and CDAWG.

To speed up the CDAWG traversal for the computation of the factorization, we want to skip heavy edges. For that, we accumulate the information about LZ nodes of all heavy nodes in a heavy path P and store this information directly in P such that we only require to query a heavy path instead of all its heavy nodes. A linear sink-root traversal in CDAWG thus visits $\mathcal{O}(\lg n)$ light nodes and heavy paths. We can perform this traversal efficiently with some preprocessing:

Node Lengths. Let len(u) for a CDAWG node u denote the set of the string lengths of all root-u paths in CDAWG. Actually, the set len(u) is an interval. This can be seen by the fact that if there are root-u paths with labels X and Y for $X \in \Sigma^*$ and Y being a suffix of X, then any suffix Z of X longer than Y has the same occurrences as X and Y in T, implying that these occurrences all follow the same characters, and therefore we can also reach u from root by reading Z. As a consequence, we can represent len(u) in $\mathcal{O}(1)$ words by using both interval ends, and augment each CDAWG node u with len(u) without violating our space budget.

Node Distances. For two CDAWG nodes u and v on the same heavy path, let dist(u,v) be their string depth distance, which is well-defined because either u is the parent of v or vice versa (otherwise they cannot belong to the same heavy path).

Upward Navigation. Recall that our aim is, after determining a factor $F_x = T[\mathsf{dst}_x..\mathsf{dst}_x+|F_x|-1]$ with Stab, to find its interval for indexing F_x with Stab. For that, we climb up CDAWG from sink and search a root-sink path P of length $\ell := n - \mathsf{dst}_x + 1$, which is the string depth of the ST leaf having suffix number dst_x. Such a path P is uniquely defined since the ST nodes collapsed to a CDAWG node have all distinct string depths. In particular, ST nodes with the same string depth cannot have isomorphic subtrees, and therefore no two root-v paths can share the same length (substituting v with non-root ST nodes).

For upward navigation, we augment each node v with a binary search tree B_v. For each parent u of v connected by a light edge (u,v), we store (u,v) with key $\min(\text{len}(u)) + c(u,v)$ in B_v, where $c(u,v)$ is the number of characters on the edge (u,v). With B_v, we can find the last edge (u,v) of the root-v path P of string length ℓ in $\mathcal{O}(\lg e)$ time. After climbing up to u, the remaining prefix of P is a root-u path P' of string length $\ell - c(u,v)$.

Now, a CDAWG ancestor u of v in the same heavy path can be a node in P if and only if $\ell - \text{dist}(u,v) \in \text{len}(u)$. Finding the highest possible such ancestor can be done with exponential search in $\mathcal{O}(\lg e)$ time. We end up with a CDAWG ancestor u of v in P that is connected to its parent node w in P via a light edge (or $u = $ root, and we terminate the traversal). We can find w with B_u, and recurse on w belonging to another heavy path closer to the root node. In total, we visit $\mathcal{O}(\min(\lg n, e)) = \mathcal{O}(\lg n)$ heavy paths and light nodes. On each heavy path or light node we process, we spend $\mathcal{O}(\lg e)$ time. Thus, the total time per factor is $\mathcal{O}(\lg n \lg e)$.

Finding the SA Range. Given we process factor F_x, we use the above procedure to find the ST locus of F_x represented by CDAWG. For that, we stop the climbing when we reach the shortest path P with a string length of at least $|F_x|$. However, unlike the previous approach, we do not have the SA ranges at hand. To compute them, we perform the following pre-computation step: We let each CDAWG node store (a) the number of ST leaves in the subtree rooted at one of its collapsed ST nodes (this is well-defined because all these collapsed ST nodes have the same tree topology) and (b) the number of ST leaves of its lexicographically preceding sibling nodes, which we call the *aggregated* CDAWG *value*. Additionally, each heavy path stores from bottom up the prefix-sums of the aggregated CDAWG values of the nodes such that we can get for the i-th node on a heavy path the number of all leaves of all lexicographically preceding siblings of the descendant nodes of the i-th node belonging to the same heavy path. This whole pre-preprocessing helps us find the SA range of F_x as follows: We use a counter c accumulating the leftmost border of the SA range we want to compute. For that, we increment c when climbing up to a light node by its aggregated CDAWG value. Additionally, when we leave a heavy node, we use the prefix-sum stored in its respective heavy path to perform the computation in constant time per light node or heavy path. When we reach the CDAWG node v representing the locus of F_x, c gives us the left border of the SA range we want to compute. However, the length of this SA range is given by the subtree size stored in v. This concludes our algorithm.

Speeding Up by Interval-Biased Search Trees. The above time can be improved from $\mathcal{O}(\lg n \lg e)$ to $\mathcal{O}(\lg n)$ by implementing (a) B_v and (b) the exponential search in each heavy path with *interval-biased search trees*.

Lemma 2 ([8, Lemma 3.1]). *Given a sequence of integers $\ell_1 \leq \cdots \leq \ell_m$ from a universe $[0..u]$, the interval-biased search tree is a data structure of $\mathcal{O}(m)$ space that can compute, for an integer p given a query time, the predecessor of p in $\mathcal{O}(\lg(u/x))$ time, where $x =$ successor$(p) -$ predecessor(p) is the difference between the predecessor* predecessor(p) *and successor* successor(p) *of p in $\{\ell_1, \ldots, \ell_m\}$.*

We note that there are faster predecessor data structures with time related to the distance of the query element to the predecessor such as [4,12], which however do not improve the total running time, which is dominated by the number of nodes $\mathcal{O}(\lg n)$ we visit.

For the former (a), denoting B_v as $B.$ for any node v, during a sink-root traversal, a query of $B.$ always leads us to a higher node v such that the next search in $B.$ is bounded by $\max(\text{len}(v))$, and therefore the query times in Lemma 2 lead to a telescoping sum of $\mathcal{O}(\lg n)$ total time.

For the latter (i.e., (b) the heavy paths), we let each heavy path maintain an interval-biased search tree storing its CDAWG nodes. A node u is stored with the key dist(u, v'), where v' is the deepest node in the heavy path. At query time we have the desired path-length ℓ and len$(u) = [\min(\text{len}(u))..\max(\text{len}(u))]$ available such that we can query for the highest node u_1 with dist$(u_1, v) = $ dist$(u_1, v') - $ dist$(v, v') \leq \ell - \min(\text{len}(u_1))$, i.e., dist$(u_1, v') \leq \ell - \min(\text{len}(u_1)) + $ dist(v, v') and the highest node u_2 with dist$(u_2, v') \geq \ell - \max(\text{len}(u_2)) + $ dist(v, v'). Then the deepest node among u_1 and u_2 is the highest ancestor of v that is still in P and is a member of the same heavy edge. The time complexity forms like for (a) a similar telescoping sum if we add to each key dist(u, v') the maximum depth of a heavy path such that each heavy path visit shrinks the search domain to be upper bounded by the last obtained key.

Theorem 2. *For a text T of length n, there exists a data structure of size $\mathcal{O}(e)$, which can, given an interval $\mathcal{I} \subset [1..n]$, compute the LZ78 factorization of $T[\mathcal{I}]$ in $\mathcal{O}(z \lg n + z \lg z) \subset \mathcal{O}(z \lg n)$ time and $\mathcal{O}(z)$ extra space, where z is the number of computed factors.*

4 Experiments

In what follows, we empirically evaluate CDAWGs on real-world text strings for computing LZ78 substring compression. To this end, we first highlight details of our CDAWG implementation in Sect. 4.1. Subsequently, we describe our experimental settings in Sect. 4.2. Finally, we report the memory consumption, running time, and the distribution of the number of edges on paths between the CDAWG root and its sink in Sect. 4.3.

4.1 Deviation from Theory

We implement a simplification of our CDAWG-based index proposed in Sect. 3.3. In particular, we omit the centroid path decomposition because we empirically observed that the average number of edges on the root-sink paths is small on our datasets. We will discuss this observed phenomenon in detail in Sect. 4.3. To reduce complexity, we implemented the branches of each internal node, instead of an interval-biased search tree, by a sorted list on which we do a binary search to find the edge with the right label. We also deviate from theory at the implementation of the stabbing-max data structure, for which we use splay trees [30]. With a splay tree built on z intervals, the times for answering a query or adding an interval are $\mathcal{O}(\lg z)$ amortized each, and the space is $\mathcal{O}(z)$ words. Therefore, replacing the original data structure with splay trees does not worsen the space of our index, and keeps the time within $\mathcal{O}(z \lg z)$. Splay trees provide fast access to frequent elements by rearrange their structure adaptively on each query, and thus can exploit skewed distributions unlike common balanced trees such as AVL trees. The reason for using splay trees is that vertices of the splay tree are sequentially inserted at positions adjacent to the vertex that becomes the root of the splay tree by the previous query. By doing so, chances are high that a splay tree query only involves the very upper part of the tree, making the implementation practically fast in most cases.

For comparison, we also implement a simplification of the ST-based index. Our implementation differs from the method proposed in [21] in the following two points: (i) we omit the weighted-ancestor data structure, and, (ii) we use a stabbing-max data structure instead of a lowest marked ancestor data structure. The first change is because the average number of edges on the root-leaf paths of ST is small on our datasets, similar to the root-sink paths of CDAWG. The second change is aimed at reducing memory consumption. The stabbing-max data structure requires only $\mathcal{O}(z)$ space, whereas the lowest marked ancestor data structure requires $\mathcal{O}(n)$ space in addition to ST.

Implementation Details. We maintain the nodes and the edges of CDAWG separately in two arrays A_V and A_E. We store nodes in A_V an arbitrary order, while we store edges in A_E in a sorted order based on two criteria. First, we partially sort the edges in groups sorted by the A_V index of the connecting child node. Second, for a fixed child node v, an edge (u, v) is sorted by the key $\min(\text{len}(u)) + c(u, v)$ within its groups of edges sharing the same child node v. This arrangement makes it possible to perform binary search on the edge array for simulating the binary search trees B_v, which we here no longer need. Given a node v, to jump into the range $[\ell..r]$ of the edge array of edges connecting to v for querying B_v, we let v store ℓ. We can do so by letting node $A_V[i]$ store (V1) the sum of the number of children over all preceding nodes (summing up the number of children of node $A_V[j]$ for each $j \in [1..i-1]$). We then also know the right end of the interval $[\ell..r]$ by querying the subsequent node in the A_V. Additionally, each node stores (V2) $\max(\text{len}(v))$ and (V3) the number of paths from v to the sink. An edge (u, v) from a node u to its child v is represented as a tuple of three integers: (E1) the index of u's entry in A_V, (E2) the string length of (u, v), and (E3) the prefix sum of u's aggregated CDAWG values (defined in Sect. 3.3). Therefore, both a node and an edge store three integers each. Following Sect. 3.3, we use these integers as follows: (E1) to select the parent node of v returned by B_v, (E2) to simulate a query on B_v via binary search with (V2), and to compute the string depth of the updated path when moving upwards to the returned parent, and (E3) with (V3) to determine the SA range of the factor we want to compute. In addition, we store the first character of each edge label for each edge incident to root to provide efficient random access to T. To restore $T[i]$ from CDAWG, we first compute the path $(e_1, e_2, \ldots, e_k)$ representing $T[i..n]$. Then, we obtain $T[i]$ by taking the first character of the edge label of e_1. Storing these labels takes $\sigma \lg \sigma$ bits in total. Therefore, the overall memory consumption is $3p(|V| + |E|) + \sigma \lg \sigma$ bits, where $p \in \Omega(\lg n)$ is the size of an integer in bits.

Our suffix tree consists of three parts: an array of nodes, an array of pointers to all leaves sorted by their suffix numbers, and the raw input text. Each node v stores its string depth, the index of v's parent node in the node array, and v's SA range. With the node array and the pointers to the leaves, we can determine the SA range for Stab. We do not need SA because for computing the LZ78 substring compression we only need to compute the SA range corresponding to an LZ78 factor, not the actual SA values. The overall memory consumption is $4p|V| + n \lg \sigma$ bits.

4.2 Experimental Settings

We have implemented our CDAWG- and ST-based LZ78 substring compression algorithms in C++. The source code is available at https://github.com/shibh308/CDAWG-LZ78. For simplification, we assume that the input is interpreted in byte alphabet ($\lg \sigma = 8$) and $n \leq 2^{32}$ (thus $p = 32$). Table 3 gives characteristics of the used input texts.

Table 2. Sizes and memory usage of CDAWG and ST of each dataset. Memory is measured in mebibytes (MiB). ST has approximately $2n$ vertices and edges regardless of the dataset

dataset	CDAWG size				memory usage	
	$|V|$	$|E|$	$|V|/n$	$|E|/n$	CDAWG	ST
SOURCES	19.70e6	66.33e6	0.147	0.494	984.5	3970.6
DNA	68.39e6	178.91e6	0.510	1.333	2830.2	4069.3
ENGLISH	30.49e6	102.21e6	0.227	0.761	1518.6	3920.0
FIB	38	74	2.831e−7	5.513e−7	1.28e−3	4736.0

E: set of edges, V: set of nodes, $n = 2^{27}$

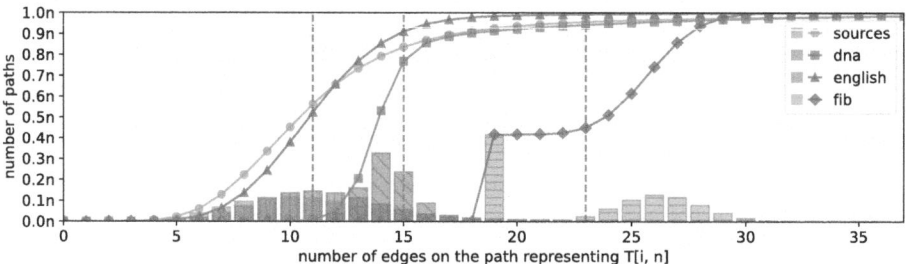

Fig. 2. The histogram of the number of edges on all paths between root and sink. The dashed vertical line for each dataset represents the average number of edges on all root-sink paths. The curve represents the cumulative sum of the histogram. Note that the number of all root-sink paths is $n = 134,217,728$ for all datasets

In one experiment instance, we construct the CDAWG or the ST of an input text and answer some LZ78 substring compression queries. As input texts, we used SOURCES, DNA, and ENGLISH from the Pizza&Chilli Corpus [13], and the length-n prefix of the (infinite) Fibonacci string FIB. It is known that the CDAWG of the length-n prefix of the Fibonacci string has only $O(\lg n)$ edges [29]. We fixed $n = 2^{27} = 134,217,728$, and generated our input texts by extracting the first 2^{27} bytes (=128MiB) from each dataset of the text collection.

We compiled our source code with GCC 12.2.0 using the -O3 option, and ran all experiments on a machine with Debian 12, Intel(R) Xeon(R) Platinum 8481C processor, and 64GiB of memory.

We first construct the CDAWG and the ST for each text string and compute its memory consumption. We also measure the distribution of the number of edges on the root-sink paths of CDAWG. Note that we did not measure the construction time because we did not focus on efficient construction. After construction, we let CDAWG and ST answer LZ78 substring compression queries. For each $\alpha \in \{2^3, 2^4, \ldots, 2^{27}\}$, we choose ten substrings of length α from the text uniformly at random, and compute the LZ78 compression of these substrings. We calculate the average memory consumption of the stabbing-max data structure and the elapsed time excluding the maximum and minimum values.

Table 3. The alphabet size and repetitive measures on the first 128MiB of each dataset ($n = 134,217,728$). Columns σ, e, r, z_{77}, and z_{78} represent the alphabet size, the number of edges in the CDAWG, the number of phrases of run-length encoded Burrows–Wheeler transform, and the number of factors of LZ77 and LZ78 factorization, respectively. Note that $e \in \Omega(\max\{r, z_{77}\})$ holds for any text [26]

dataset	σ	e	r	z_{77}	z_{78}
SOURCES	227	$0.494n$	$0.233n$	$0.058n$	$0.102n$
DNA	16	$1.333n$	$0.626n$	$0.069n$	$0.080n$
ENGLISH	218	$0.761n$	$0.360n$	$0.072n$	$0.105n$
FIB	2	74	20	41	267813

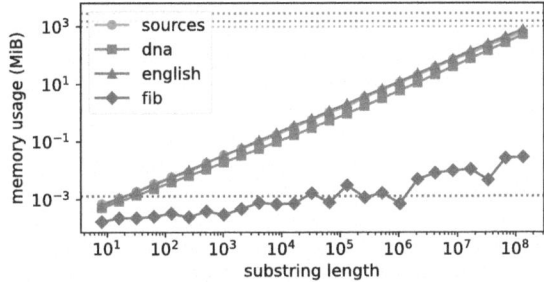

Fig. 3. Average memory usages of the stabbing-max data structure depicted by solid lines. The dotted line with the same respective color represents the memory usage of the CDAWG of the respective dataset.

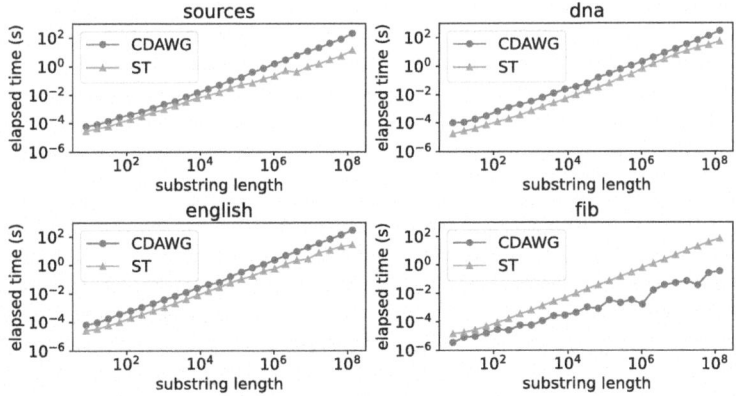

Fig. 4. Average elapsed time for LZ78 substring compression with CDAWG or ST.

4.3 Experimental Results

Table 2 indicates the size and memory consumption of the CDAWG and ST approaches. For all datasets, CDAWG consumes less memory than the ST. How-

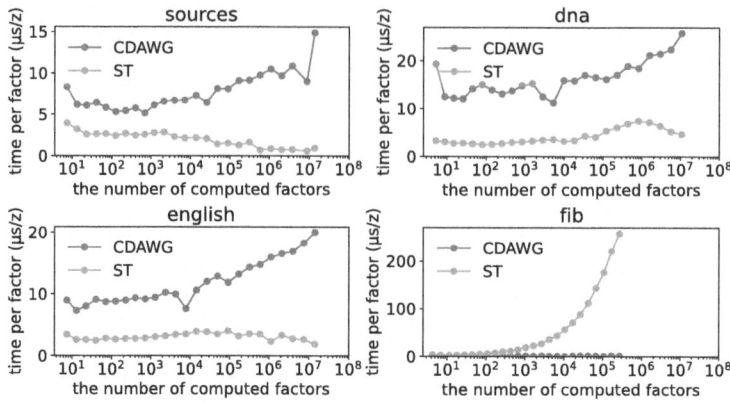

Fig. 5. Average elapsed time divided by the number of computed factors for LZ78 substring compression with CDAWG or ST.

ever, CDAWG takes more space than the input itself because each edge and vertex consists of multiple integers. Even more severe, the number of CDAWG edges alone is higher than n for DNA. For FIB, CDAWG compresses the input text exponentially. The memory usage is about 10^5 and 3.7×10^6 times less than the raw text and ST, respectively.

Figure 2 shows the distribution of the number of edges on all root-sink paths. We observe that the average number of edges on a root-sink path is about 10–15 and almost all paths have at most 20 edges in real-world datasets. Therefore, we can regard the number of paths as almost $O(\lg n)$. For plain CDAWGs without centroid path decomposition, path extraction can take $\mathcal{O}(n)$ time at worst. However, from this result, we empirically constitute that such cases are rare in practice, and both with and without centroid path decomposition, the running times are almost the same. Note that the average number of edges on the root-sink paths of CDAWG is the same as the number of root-leaf paths of ST, so this fact also applies to ST.

Figure 3 plots the memory usage of the used stabbing-max data structures. The memory consumption increases almost linearly with the increase in the length α of the queried substring. The memory usage of the stabbing-max data structure is about 50–80% lower than the memory consumption of CDAWG except for FIB even if $n = \alpha$ (i.e., the substring to compress in the whole text). Therefore, for real-world datasets, the major memory bottleneck is the CDAWG.

Figure 4 shows the elapsed time for LZ78 substring compression using STs and CDAWGs, and Fig. 5 shows the elapsed time divided by the number of computed factors. The elapsed time increases almost linearly with the increase in substring length. In the real-world datasets, the CDAWG is about 2–20 times slower than the ST. This is because the traversal of the ST is simpler than that of the CDAWG. In contrast, CDAWG computes the LZ78 substring compression about 5–300 times faster than ST in FIB. We speculate that an effective CDAWG-based compression has a positive impact on cache-friendly memory access.

5 Conclusion

We propose a method to compute the LZ78 substring compression in CDAWG-compressed space. Our method compresses substrings with z LZ78 factors in $\mathcal{O}(z \lg n)$ time and $\mathcal{O}(e)$ space. We conducted experiments on various types of data, and empirically evaluated that our method performs LZ78 substring compression efficiently with space improvements compared to ST-based methods. For that, we slightly deviated from theory by omitting the centroid path decomposition and sophisticated data structures. As future work, we plan to make our approach based on stabbing-max queries independent of the CDAWG, which is possible since we require only access to the input text, the suffix array, and its inverse. Elaborated compressed index data structures like [16,19] seem to support these access operations, which can take space asymptotically smaller than the CDAWG.

Acknowledgements. This work was supported by JSPS KAKENHI Grant Number JP23H04378.

References

1. Arimura, H., Inenaga, S., Kobayashi, Y., Nakashima, Y., Sue, M.: Optimally computing compressed indexing arrays based on the compact directed acyclic word graph. In: Nardini, F.M., Pisanti, N., Venturini, R. (eds.) SPIRE 2023. LNCS, vol. 14240, pp. 28–34. Springer, Heidelberg (2023). https://doi.org/10.1007/978-3-031-43980-3_3
2. Babenko, M.A., Gawrychowski, P., Kociumaka, T., Starikovskaya, T.: Wavelet trees meet suffix trees. In: Proceedings of SODA, pp. 572–591 (2015)
3. Bannai, H., Gawrychowski, P., Inenaga, S., Takeda, M.: Converting SLP to LZ78 in almost Linear Time. In: Fischer, J., Sanders, P. (eds.) CPM 2013. LNCS, vol. 7922, pp. 38–49. Springer, Heidelberg (2013). https://doi.org/10.1007/978-3-642-38905-4_6
4. Belazzougui, D., Boldi, P., Vigna, S.: Predecessor search with distance-sensitive query time. ArXiv CoRR arxiv:1209.5441 (2012)
5. Belazzougui, D., Cunial, F.: Representing the suffix tree with the CDAWG. In: Proceedings of CPM, vol. 78 of LIPIcs, pp. 7:1–7:13 (2017)
6. Belazzougui, D., Cunial, F., Gagie, T., Prezza, N., Raffinot, M.: Composite repetition-aware data structures. In: Cicalese, F., Porat, E., Vaccaro, U. (eds.) CPM 2015. LNCS, vol. 9133, pp. 26–39. Springer, Cham (2015). https://doi.org/10.1007/978-3-319-19929-0_3
7. Belazzougui, D., Kosolobov, D., Puglisi, S.J., Raman, R.: Weighted ancestors in suffix trees revisited. In: Proceedings of CPM, vol. 191 of LIPIcs, pp. 8:1–8:15 (2021)
8. Bille, P., Landau, G.M., Raman, R., Sadakane, K., Satti, S.R., Weimann, O.: Random access to grammar-compressed strings and trees. SIAM J. Comput. **44**(3), 513–539 (2015)

9. Blumer, A., Blumer, J., Haussler, D., Ehrenfeucht, A., Chen, M.T., Seiferas, J.I.: The smallest automaton recognizing the subwords of a text. Theor. Comput. Sci. **40**, 31–55 (1985)
10. Cormode, G., Muthukrishnan,: Substring compression problems. In: Proceedings of SODA, pp. 321–330 (2005)
11. Crochemore, M., Vérin, R.: Direct construction of compact directed acyclic word graphs. In: Apostolico, A., Hein, J. (eds.) CPM 1997. LNCS, vol. 1264, pp. 116–129. Springer, Heidelberg (1997). https://doi.org/10.1007/3-540-63220-4_55
12. Ehrhardt, M., Mulzer, W.: Delta-fast tries: local searches in bounded universes with linear space. In: WADS 2017. LNCS, vol. 10389, pp. 361–372. Springer, Cham (2017). https://doi.org/10.1007/978-3-319-62127-2_31
13. Ferragina, P., González, R., Navarro, G., Venturini, R.: Compressed text indexes: from theory to practice. ACM J. Exp. Algor. **13**, 1.12:1 – 1.12:31 (2008)
14. Ferragina, P., Grossi, R., Gupta, A., Shah, R., Vitter, J.S.: On searching compressed string collections cache-obliviously. In: Proceedings of PODS, pp. 181–190 (2008)
15. Fischer, J., Tomohiro, I., Köppl, D., Sadakane, K.: Lempel–Ziv factorization powered by space efficient suffix trees. Algorithmica **80**(7), 2048–2081 (2018)
16. Gagie, T., Navarro, G., Prezza, N.: Optimal-time text indexing in BWT-runs bounded space. In: Proceedings of SODA, pp. 1459–1477 (2018)
17. Gawrychowski, P., Lewenstein, M., Nicholson, P.K.: Weighted ancestors in suffix trees. In: Schulz, A.S., Wagner, D. (eds.) ESA 2014. LNCS, vol. 8737, pp. 455–466. Springer, Heidelberg (2014). https://doi.org/10.1007/978-3-662-44777-2_38
18. Keller, O., Kopelowitz, T., Feibish, S.L., Lewenstein, M.: Generalized substring compression. Theor. Comput. Sci. **525**, 42–54 (2014)
19. Kempa, D., Kociumaka, T.: Collapsing the hierarchy of compressed data structures: Suffix arrays in optimal compressed space. In: Proceedings of FOCS, pp. 1877–1886 (2023)
20. Kociumaka, T.: Efficient Data Structures for Internal Queries in Texts. PhD thesis, University of Warsaw (2018)
21. Köppl, D.: Non-overlapping LZ77 factorization and LZ78 substring compression queries with suffix trees. Algorithms **14**(2)(44), 1–21 (2021)
22. Köppl, D.: Computing LZ78-derivates with suffix trees. In: Proceedings of DCC, pp. 133–142 (2024)
23. Mäkinen, V., Navarro, G.: Succinct suffix arrays based on run-length encoding. Nord. J. Comput. **12**(1), 40–66 (2005)
24. Manber, U., Myers, E.W.: Suffix arrays: a new method for on-line string searches. SIAM J. Comput. **22**(5), 935–948 (1993)
25. Nakashima, Y., Tomohiro, I., Inenaga, S., Bannai, H., Takeda, M.: Constructing LZ78 tries and position heaps in linear time for large alphabets. Inf. Process. Lett. **115**(9), 655–659 (2015)
26. Navarro, G.: Indexing highly repetitive string collections, part I: repetitiveness measures. ACM Comput. Surv. **54**(2), 29:1–29:31 (2021)
27. Nekrich, Y.: A dynamic stabbing-max data structure with sub-logarithmic query time. In: Asano, T., Nakano, S., Okamoto, Y., Watanabe, O. (eds.) ISAAC 2011. LNCS, vol. 7074, pp. 170–179. Springer, Heidelberg (2011). https://doi.org/10.1007/978-3-642-25591-5_19
28. Raffinot, M.: On maximal repeats in strings. Inf. Process. Lett. **80**(3), 165–169 (2001)
29. Rytter, W.: The structure of subword graphs and suffix trees of Fibonacci words. Theor. Comput. Sci. **363**(2), 211–223 (2006)

30. Sleator, D.D., Tarjan, R.E.: Self-adjusting binary search trees. J. ACM **32**(3), 652–686 (1985)
31. Storer, J.A., Szymanski, T.G.: The macro model for data compression (extended abstract). In: Proceedings of STOC, pp. 30–39 (1978)
32. Weiner, P.: Linear pattern matching algorithms. In: Proceedings of SWAT, pp. 1–11 (1973)
33. Ziv, J., Lempel, A.: Compression of individual sequences via variable-rate coding. IEEE Trans. Inf. Theory **24**(5), 530–536 (1978)

2d Side-Sharing Tandems with Mismatches

Shoshana Marcus[1], Dina Sokol[2(✉)], and Sarah Zelikovitz[3]

[1] Department of Mathematics and Computer Science, Kingsborough Community College of the City University of New York, Brooklyn, NY, USA
shoshana.marcus@kbcc.cuny.edu

[2] Department of Computer and Information Science, Brooklyn College and The Graduate Center, City University of New York, Brooklyn, NY, USA
sokol@sci.brooklyn.cuny.edu

[3] Department of Computer Science, College of Staten Island and The Graduate Center, City University of New York, Staten Island, NY, USA
sarah.zelikovitz@csi.cuny.edu
http://www.sci.brooklyn.cuny.edu/ sokol,
http://www.cs.csi.cuny.edu/ zelikovi/

Abstract. One form of 2d periodicity is encapsulated by the definitions of 2d side-sharing tandems and runs. A 2d side-sharing tandem consists of two adjacent non-overlapping occurrences of the same rectangular block, and a side-sharing run is a maximally extended chain of side-sharing tandems. Furthering our understanding of 2d periodicity has long been an important goal, with motivation in the fields of image matching and multi-dimensional compression schemes. Much research has been accomplished on exact 2d periodicity, however, there have been few results on approximate 2d periodicity. In this paper we introduce several versions of approximate side-sharing tandems with k mismatches along with efficient algorithms for locating them in a rectangular array.

Keywords: 2-dimensional · k-mismatch tandem repeat · side-sharing

1 Introduction

A 2d side-sharing tandem, defined by Apostolico and Brimkov [5], is composed of two adjacent non-overlapping occurrences of the same rectangular block, called the *root*. This structure is depicted in Fig. 1. In a sense, this is the 2-dimensional generalization of a *square* in 1-dimension, which is a string that consists of two consecutive occurrences of a primitive substring (e.g. aa, abcabc). Squares are well-investigated objects in combinatorics of strings [1,7,9,10] and they form the building blocks of larger repetitions.

Apostolico and Brimkov [6] presented an algorithm that locates all side-sharing tandems in $O(n^3 \log n)$ time. Charalampopoulos et al. [8] demonstrated $\Theta(n^3)$ bounds for the number of *distinct* primitively-rooted side-sharing tandems

© The Author(s), under exclusive license to Springer Nature Switzerland AG 2025
Z. Lipták et al. (Eds.): SPIRE 2024, LNCS 14899, pp. 306–320, 2025.
https://doi.org/10.1007/978-3-031-72200-4_23

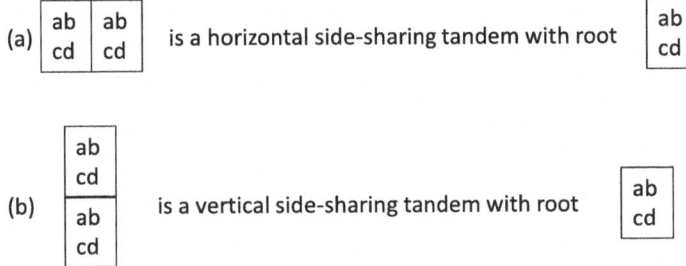

Fig. 1. Two possible configurations of a 2d side-sharing tandem

in an $n \times n$ array and also contributed an algorithm that runs in time proportional to these upper bounds. Gawrychowski et al. [12] construct an infinite family of $n \times n$ 2d arrays over the binary alphabet with $\Omega(n^3)$ distinct primitively-rooted side-sharing tandems.

A string r is *periodic* if its longest prefix that is also a suffix is at least half the length of r (e.g. abbabbab). A *run* (or maximal repetition) in a string s is a substring of s that is periodic and cannot be extended at all to the right or left, e.g., ababa is a maximal repetition in the string abaababac. Recently, the notion of a *run of side-sharing tandems* was introduced, as a maximally extended chain of 2d tandems [18]. Formally, a *2d horizontal run*, or *h-run*, in a 2d array M, is a horizontally periodic subarray with horizontal period (or *h-period*) h, in which extension by one subcolumn to the right or to the left yields an array with h-period h' such that $h' > h$. An algorithm is provided by Marcus et al. [18] to locate all runs of side-sharing tandems in an $n \times n$ input array in $O((n^2 + occ) \log n / \log \log n)$ time, where occ is the number of runs.

Heretofore, there has been a lot of work on finding exact repetitions in 2d input, however, the problem of inexact 2d repetitions has remained mostly open. An initial result that allows mismatches in 2d *corner-sharing* tandems is given in [2]. The algorithm searches an $n \times n$ input for corner-sharing tandems with up to k mismatches in time $O(n^3 \, k \log n)$. In this paper we tackle the problem of searching in an $n \times n$ input array for all subarrays that are inexact *side-sharing* tandems. Our goal is to search for *approximate* 2d side sharing tandems, allowing some degree of error between the copies, specifically, up to k mismatches in each output. The efficiency of the algorithms is contingent on the fact that the work done is mostly proportional to the amount of output obtained, subsequently denoted by occ.

There are many different ways of allowing errors in a 2d run, hence we consider several different definitions. The ensuing sections present the definitions and algorithms of the extensions of several known definitions of 1d tandems with mismatches. In Sect. 2.1, we extend the k-mismatch tandem repeat to 2d, where essentially we are looking for approximate side-sharing tandems with exactly two copies of the period (akin to squares). Then, in Sect. 2.2, we generalize the definition from squares to runs. In Sect. 3, we define the 2d version of k-mismatch *globally-defined* runs, and finally, in Sect. 4, we present a definition and algorithm for 2d k-maximal approximate runs (*2d k-MARs*). The second column in Table 1 summarizes the time complexities of the algorithms contributed in the current paper.

Table 1. The first column presents the state of the art in 1d, and the second column summarizes the contributions of this paper. For 1d algorithms, the input text is a string of length n, and for 2d algorithms, the input text is an $n \times n$ array. Both take the integer k as input, representing the upper limit for the number of mismatches in each output.

1d Problem and Time Complexity	2d Definition and Time Complexity	Sect.
k-mismatch tandem repeats $O(nk \log(n/k))$ [16]	k-mismatch side-sharing tandems $O(n^2 k \log k + occ)$	2.1
runs of k-mismatch tandem repeats $O(nk \log k + occ)$ [14]	k-mismatch side-sharing runs $O(n^4)$	2.2
k-mismatch *globally- defined* runs $O(nk \log k + occ)$ [14]	globally defined k-mismatch side-sharing runs $O(k(n^2 \log n + occ))$	3
k-maximal approximate runs $O(nk^2 \log^2 k + occ)$ [4]	k modified positions side-sharing runs $O(k^2 \log k(n^2 \log(n/k) + occ))$	4

2 k Mismatches Between Adjacent Periods

Each of the definitions presented in this paper extend a known definition of 1d approximate repeats over the Hamming distance. Thus, for clarity of exposition, we first describe each definition in 1-dimension, and then extend it to 2d. The definitions deal only with horizontal 2d side-sharing tandems; vertical 2d definitions are analogous and work exactly the same way. We use $H(u,v)$ to denote the Hamming distance between two strings u and v of equal length. In Sect. 2.1 the definition is limited to tandems that have two adjacent occurrences of the approximate root. We extend the discussion to *runs* of 2d k-mismatch tandems in Sect. 2.2.

2.1 k-Mismatch Side-Sharing Tandems

Definition 1. *(1d)* [16] *A k-mismatch tandem repeat t with period p is a string $t = \hat{u}\bar{u}$ such that $|\hat{u}| = |\bar{u}| = p$ and $H(\hat{u}, \bar{u}) \leq k$.*

Let U_i represent the string that is row i of a 2d array U. We define the Hamming distance between two 2d arrays U and V, each of size $r \times c$, as: $H(U,V) = \sum_{i=1}^{r} H(U_i, V_i)$, that is, the sum of the Hamming distances of their rows.

Definition 2. *(2d)* *A 2d k-mismatch side-sharing tandem with h-period p is a 2d array $T = \hat{U}\bar{U}$ such that the number of columns in both $\hat{U}$ and $\bar{U}$ equals p, and $H(\hat{U}, \bar{U}) \leq k$.*

In other words, a 1d k-mismatch tandem repeat t of length $2p$ has at most k locations t_j, $1 \leq j \leq p$, for which $t_j \neq t_{j+p}$, and a 2d k-mismatch side-sharing tandem T of size $r \times 2p$ has at most k locations for which $T_{i,j} \neq T_{i,j+p}$, $1 \leq i \leq r$, $1 \leq j \leq p$. Other than tallying, there is no relation between mismatches among rows.

Example 1. The following is a 2-mismatch side-sharing tandem with h-period 4.

```
aabc abbc
baba baaa
```

When reporting exact tandems, the concept of primitivity is very important since the primitive period encodes the information about all non-primitive periods. For example, the tandem *abababab* with period 4 is uninteresting as its period is non-primitive and is encoded in the period of size 2. However, once mismatches are introduced, the concept of primitivity becomes unclear. For example, in the tandem *abaabcabaabc* the 2-mismatch tandem of period 3 does not encode the tandem of period 6 with 0 mismatches. Thus, most of the algorithms that find tandems with mismatches ignore the concept of primitivity. This results in an upper bound on the number of tandems in 1d of $\Theta(n^2)$, and in 2d a bound of $\Theta(n^4)$. Nonetheless, most algorithms that search for approximate tandems report all output regardless of primitivity of the period.

The two stages in the following algorithm outline will be followed by all the algorithms in this paper, with the implementation of the stages differing for each varying definition.

Algorithm Outline:

Step 1: Run a 1d algorithm in each row to find all 1d tandems with k mismatches.

Step 2: Attempt to extend a known 2d output down one row by linking it to a 1d tandem.

Observation 1. *When linking approximate tandems on different rows, it is not relevant to link together different periods, because once we change periods, the positions of mismatches change (see Example 2).*

Example 2. The following is a 2d side-sharing tandem with h-period 6 and 0 mismatches. In row one there is a 2-mismatch 1d tandem with period 3 whose mismatches disappear when the period becomes 6. Note that the 1d algorithms do report the tandem with period 6.

```
big bag big bag
aaa bbb aaa bbb
```

Algorithm for k-mismatch Side Sharing Tandems (k-SST)

Step 1: Run a 1d algorithm for k-mismatch tandem repeats To locate the 1d tandem with period p that may occur at location (i, j), we can use k

LCP's [11] (otherwise known as "Kangaroo jumps") to compare the substring starting at location (i, j) with the substring starting at location $(i, j + p)$. If the two matching parts are p long, then we have a tandem.

Time: $O(n^3 k)$ since we are performing k LCP's for each of $O(n^2)$ locations, for each of $O(n)$ possible period sizes p.

A well-known speedup for finding repetitions that is attributed to Main and Lorentz [17] is to reduce the number of iterations by searching for all repetitions that cross certain *anchor* points, rather than searching for all repetitions starting at each point.

Main and Lorentz (ML) use the center of the string as the anchor point, finding all repeats that cross the center in the first iteration. In the following iteration, the string is divided into two halves, and each half is recursively searched. To find repeats that cross the center, the longest common extension is computed in both the forward and reverse directions, and combined to report the repeat. This idea is used in [15–17].

Subsequently, Kolpakov and Kucherov (KK) improved on this by using the Lempel-Ziv (LZ) factorization [22,23] to define the anchor points. Papers that search for repetitions that cross anchor points delineated by LZ include exact matching [13], mismatches [4,14], and edit distance [21], with a survey in [15].

For the current section, we apply the algorithm of [14] "Finding approximate repetitions under Hamming distance," Sect. 3, which we refer to as KK for the remainder of this paper. We apply KK to each row of the 2d input array. Using the output of the algorithm for each period p, we mark the start of each k-mismatch tandem in the appropriate OUTPUT$_p$ array.

We store the output from Step 1 in $n/2$ output arrays, one for each h-period $1 \leq p \leq n/2$, such that OUTPUT$_p[i,j]$ = the number of mismatches in the 1d tandem beginning at location (i,j) with period p. A value $> k$ indicates no output for the location.

Step 2: Linking Rows

The second step addresses the question: How do we combine information from row to row? The input to this stage is the set of OUTPUT arrays, and thus we process each column in each OUTPUT array conglomerating mismatches and reporting each 2d tandem that occurs with h-period p at location (i,j) in T. Each 2d tandem is clearly delineated in an OUTPUT array by its left and right borders, hence, we process each column in each OUTPUT array with the classical linear time subarray summing algorithm. For each column j of OUTPUT$_p$, beginning in row one, sum the values at each location in the column, reporting each height as long as the sum $\leq k$. Once the sum becomes $> k$, remove the first value, and continue moving down one location at a time.

Time for Step 2: $O(n^3)$ due to the constant time processing of each value in each OUTPUT array.

To save time initializing and processing the output arrays, we can replace the arrays with n^2 linked lists, one per column of each period. Each node in the linked list stores the row i that its value represents, and therefore, when processing a given column, we only continue the summing when nodes on the linked list are in consecutive rows.

Improved time for k-SST Algorithm: Step 1 has time complexity $O(nk \log k + occ)$ for each row, and overall $O(n^2 k \log k + occ)$. In Step 2, all numbers being processed constitute output of height 1. Since each value processed is already known to be valid output, and we process each value in constant time, we charge the summing of values to the output. This yields overall time of $O(n^2 + occ)$, and combining with Step 1, the time is $O(n^2 k \log k + occ)$ to find occ k-mismatch side-sharing tandems.

2.2 k-Mismatch Side-Sharing Runs

Definition 3. *(1d) [14] A string r of length n is called a* run *of k-mismatch tandem repeats of period $p \leq n/2$, iff for every $1 \leq i \leq n - 2p + 1$, the substring $r_{i...i+2p-1}$ is a k-mismatch tandem repeat of period p.*

Definition 4. *(2d) A 2d k-mismatch side-sharing run with h-period p is an $r \times c$ array R such that each $r \times 2p$ subarray T of R is a k-mismatch side-sharing tandem.*

In other words, both in 1d and 2d, we define a *run* as a continuous chain of adjacent tandems with k mismatches, i.e. there are no more than k mismatches between any two adjacent copies within the run. When searching for 1d (or 2d) runs occurring in a longer string (or 2d array), we are always only interested in runs that are *maximal*, i.e. that cannot be extended by one character (or column) either to the right or to the left maintaining the same period (or h-period). For the remainder of this paper we assume the concept of maximality is inherent in the term *run*. See Fig. 2 for an example of 1d and 2d k-mismatch runs.

All the information about k-mismatch runs can be gleaned from the OUTPUT arrays in Sect. 2.1. Starting at a given location, for a given period, we can link together neighboring tandems with a specific height. The number of mismatches in a given run is unbounded as each run is allowed to have xk mismatches, where x is the number of periods. The time complexity of this algorithm would be proportional to the number of 2d tandems rather than the number of 2d runs, and therefore cannot be expressed in terms of the output size. A loose upper bound is $O(n^4)$ since we process each row, for each height, and each period, horizontally linking together at most n adjacent k-mismatch tandems.

#mism	3	3	2	1	0	1	2	2													
index	1	2	3	4	5	6	7	8	9	10	11	12	13	14	15	16	17	18	19	20	21
row 1	a	b	b	b	b	b	b	a	x	x	x	b	b	b	a	x	x	x	y	y	b
							1	a	b	b	b	b	b	b	a	x	x	x	b	b	b
							2	b	b	b	b	b	b	a							
							3	b	b	b	b	b	a	x							
	starts of 1d k-mismatch tandems						4	b	b	b	b	a	x	x							
							5	b	b	b	a	x	x	x							
							6	b	b	a	x	x	x	b							
							7	b	a	x	x	x	b	b							
							8	a	x	x	x	b	b	b							
row 2	a	b	a	a	a	a	a	b	a	a	a	a	a	a	a	a	a	a	a	a	a
#mism for 2d	3	3	3	2	1	2	3	3													

Fig. 2. The figure shows a 1d k-mismatch run as row one of a 2d k-mismatch run, $k = 3$, period $p = 7$. In row one, every substring of length fourteen has ≤ 3 mismatches between its first half and second half, i.e. positions 1–8 are starts of 3-mismatch tandems. The number of mismatches in each of these tandems is written above the index. In 2d, every subarray of height two, width fourteen, is a 2d 3-mismatch tandem. The number of mismatches for each consecutive 2d tandem is written at the bottom, underneath its beginning location.

3 Globally Defined k-Mismatch Side-Sharing Runs

Definition 5. *[1d] [14] A string r of length n is called a k-mismatch globally-defined run of period $p \leq n/2$, iff $H(r_{1\ldots n-p}, r_{p+1\ldots n}) \leq k$. Equivalently, $r_{1\ldots n}$ is a k-mismatch globally-defined run of period p, if the number of locations i, $1 \leq i \leq n - p$, such that $r_i \neq r_{i+p}$ is at most k.*

Definition 6. *[2d] A 2d array R of size $r \times c$ is a globally-defined k-mismatch side-sharing run with h-period p if the Hamming distance between the prefix and the suffix with $n - p$ columns is no more than k. Equivalently, R has at most k locations for which $R_{i,j} \neq R_{i,j+p}$, $1 \leq i \leq r$, $1 \leq j \leq c - p$.*

Recall that there is an inherent assumption of maximality of these runs. The run of 2d 3-mismatch tandems shown in Fig. 2 does not satisfy the definition of a globally-defined 3-mismatch run, since in row one when you align the prefix from position $1\ldots 14$ against the suffix $8\ldots 21$ there are a total of five mismatches.

As pointed out in [16] for 1d, since the longest common extensions computed extend as far as possible, the algorithm that finds tandems can be used directly to find runs. In fact, the algorithm of KK that we use above for Step 1 to locate all 1d k-mismatch tandems, indeed locates all 1d k-mismatch globally-defined runs. Thus, one approach would be to run the algorithm of KK on each row of the input, and then link the runs in different rows as in [18] to construct the

2d output. However, we reject this approach since a lot of redundant work will result. There may be overlapping runs with the same period in a given string. Thus, the run-linking would be done multiple times for these overlapping runs which actually share many positions of mismatches.

A better approach would be to merge the original mismatch lists found in the algorithm, before the reporting began. In this way, we will process each mismatch once as we attempt to extend a run downwards. However, the KK algorithm will not be useful for this since the anchor points in each row are different as they depend on the specific LZ factorization of the row. Yet, Main and Lorentz's framework uses the same anchor points in each string, and thus the anchor points in each row align in the same column. Thus, instead of storing all start points of output to merge in the next step, we store the $k+1$ mismatch positions in each direction for each anchor and each period. Then, when merging, we process aligned anchor points in different rows.

Algorithm for 2d Globally Defined k-Mismatch Runs

Step 1: Run the algorithm of [16] on each row to find the 1d k-mismatch runs. The algorithm follows the framework of ML already described in Sect. 2.1. In each iteration with anchor anc, each period p is searched for separately. The algorithm fills in two arrays:

1. fwd_extension$[0 \ldots k]$ such that fwd_extension$[k']$ represents the length of the longest prefix match between the suffix beginning at anc and the suffix beginning at $anc + p$ with k' errors.
2. rev_extension$[0 \ldots k]$ such that rev_extension$[k']$ represents the length of the longest suffix match between the prefix ending at $anc - 1$ and the prefix ending at $anc + p - 1$ with k' errors.

Then, the algorithm considers each pair of numbers k_1, k_2 whose sum is k, allowing k_1 mismatches to the right and k_2 mismatches to the left, i.e.:
if fwd_extension$[k_1]$+rev_extension$[k_2] \geq p$
report a run extending from the start of the common suffix to the end of the common prefix.

Step 2: How do we merge the data from row to row? The key idea here is that instead of merging each output from row to row, if an anchor has output for a given period p, we merge its forward and reverse extension arrays with the arrays found on the next row for the same anchor and same period p. Then, for each height, we iterate through the (k_1, k_2) pairs, allowing k_1 mismatches in the forward direction and k_2 mismatches for the reverse direction.

The pseudocode for finding globally-defined k-mismatch runs appears in Algorithm 1. Starting with row one, we have our anchor points. These are the same underneath in row two. Hence, for each anchor point in row one, for each period p, we merge the forward and reverse extension mismatch lists of row one, respectively, with the lists of row two. Now we have 1d mismatch lists for height two, and we loop considering k_1 in the forward direction and $k - k_1$ in

Algorithm 1. (2d) Find globally-defined k-mismatch side-sharing runs

Input: 2d text T and integer k
Output: All globally-defined k-mismatch side-sharing runs in T
Run a 1d algorithm to find all globally defined k-mismatch runs in each row of T using ML anchors. Report all output found, and store the `fwd_extension` and `rev_extension` arrays for each iteration that had output.

for each row i **do**
   for each anchor point a **do**
      for each period p **do**
         hasOutput=false
         if there was 1d output for the triple (i,a,p) **then**
            $height = 1$
            hasOutput=true
         end
         while hasOutput **do**
            merge the `fwd_extension` array of the previous iteration (rows $i \ldots i + height - 1$) with the `fwd_extension` of the row underneath $(i + height)$
            merge the `rev_extension` array of the previous iteration with the `rev_extension` of the row underneath
            $height + +$
            hasOutput=false
            // Consider each pair (k_1, k_2) such that $k_1 + k_2 = k$
            for $k_1 = 0$ to k **do**
               $k_2 \leftarrow k - k_1$
               $\ell_1 \leftarrow$ `merged_fwd_extension[`k_1`]`
               $\ell_2 \leftarrow$ `merged_rev_extension[`k_2`]`
               if $\ell_1 + \ell_2 \geq p$ **then**
                  report the run beginning in row i of height $height$, extending from column $a - \ell_2 + 1 \ldots a + \ell_1$
                  hasOutput=true
               end
            end
         end
      end
   end
end

the reverse direction. If there is any output, then we continue to height three, etc. In the example in Fig. 3, for $k = 2$ and $p = 4$, two runs are found from the merged mismatch lists using the pairs (k_1, k_2) of $(0, 2)$ and $(1, 1)$ since in the merged lists: `merged_fwd_extension[0]`+`merged_rev_extension[2]`=$4 \geq p$ and `merged_fwd_extension[1]`+`merged_rev_extension[1]`=$4 \geq p$.

Time for k-mismatch globally defined runs: We consider each height a different output, especially since the left and right boundaries usually change as the height increases. Therefore, we will charge each downward extension (in the while loop of Algorithm 1) to the output. For each row, anchor and period, finding the k mismatches in each direction is all part of the cost of the 1d algorithm [16],

												mismatch lists for row 1:					
				anchor					anchor+p			fwd extension:	0	1	2	3	index=# of errors
index	1	2	3	4	5	*6*	7	8	9	*10*	11	12		2				length of extension
row 1	a	b	b	c	a	**b**	b	b	a	c	b	b						
	←				←				→								
												rev extension:	0	1	2	3	index=# of errors
													0	2	6		length of extension
												mismatch lists for row 2:					
				anchor					anchor+p			fwd extension:	0	1	2	3	index=# of errors
index	1	2	3	4	5	*6*	7	8	9	*10*	11	12		1	2			length of extension
row 2	b	a	b	a	b	a	c	a	b	a	c	c						
	←				←				→								
												rev extension:	0	1	2	3	index=# of errors
													3	6			length of extension
index	1	2	3	4	5	*6*	7	8	9	*10*	11	12	merged fwd extension:					
row 1	a	b	b	c	a	**b**	b	b	a	c	b	b		0	1	2	3	index=# of errors
row 2	b	a	b	a	b	a	c	a	b	a	c	c		1	2			length of extension
	←				←				→								
												merged rev extension:					
													0	1	2	3	index=# of errors
													0	2	3	6	length of extension

Fig. 3. Shown are examples of globally-defined k-mismatch side-sharing runs of height two, $k = 2$. The left character in each mismatch pair is bold, and the right character is red. The individual mismatch lists for the forward and reverse extension of each row are shown, as well as the merged lists. For $p = 4$, the run that extends from location 1 to 9 is found when considering $k_1 = 0, k_2 = 2$, as merged_fwd_extension[0]+merged_rev_extension[2]=4 $\geq p$, and the run that extends from location 4 to 12 is found when considering $k_1 = 1, k_2 = 1$, as merged_fwd_extension[1]+merged_rev_extension[1]=4$\geq p$. Note that starting at location 3, the run is *not valid output* since the number of errors is $3 > k$.

$O(kn \log n)$. The cost for each downward extension is $O(k)$ since we are merging lists of size $O(k)$ and we never let the lists get larger than $2k + 2$. This $O(k)$ time is charged to the output, since we only attempt a downward extension if there was output for a given anchor point. The resulting overall time complexity is then $O(k(n^2 \log n + occ))$ to find occ globally-defined k-mismatch runs.

4 k Modified Positions Side-Sharing Runs (2d k-MARs)

Amit et al. [4] introduced the concept of a k-maximal approximate run based upon counting *modified positions* rather than mismatches. This gives a more interesting definition of an approximate run, since the number of errors is bounded over the entire maximal run.

Definition 7. *[4] A set of (one or more) positions are called* modified position(s) *if replacement of the characters at each of these positions converts an approximate run into an exact run.*

To illustrate the difference between a mismatch in a run and a modified position, consider the string *abbabbabbabcabcabc* and period $p = 3$. This is a run with one mismatch; however, it is a run with three modified positions. The goal is to choose the smallest set of modified positions whose modification can yield an exact run. Sometimes, we are given a parameter k indicating that we may allow a set consisting of up to k modified positions.

Definition 8. *[1d] [4] A k-maximal approximate run (k-MAR) is a non-empty substring x of a string t that is a maximal run with at most k modified positions.*

In a k-MAR, a location within a period is called a "column" since each location yields a column in an alignment of all the periods. A column in the current paper refers to a column in a 2d array. In order to remain unambiguous, we call a "column" from Amit et al. a *pa-column* standing for a period alignment column.

Summary of 1d Find k-MARs Algorithm (Amit et al. [3])

The algorithm of Amit et al. [3] uses the framework of ML[1], described in Sect. 2.1. Each iteration processes a substring of the input string, a given period p, an anchor position *anc*, and reports k-MARs with period p that include position *anc*.

The Parikh matrix [19] is used to keep track of the tallies of the characters in each pa-column. A *Parikh matrix*, $P_{i,j}^p = P[1..|\Sigma|, 1..p]$, is a two-dimensional array defined over a substring $s_i \ldots s_j$ and a period length p. An entry $P_{i,j}^p[let, col]$ contains the number of occurrences of $let \in \Sigma$ in the pa-column *col* in the given substring with period p. For each pa-column, the majority character[2], i.e. the one that occurs more than any other (breaking ties arbitrarily) at that position within the period, is considered the correct character, and all other positions would be counted as modified positions.

The algorithm works by maintaining two pointers, ℓ and r, into the string, which represent possible leftmost and rightmost boundaries of a k-MAR, respectively. The leftmost k-MAR can be found easily in $O(k^2)$ time and it is used to initialize ℓ and r and the Parikh matrix for the iteration. The algorithm then attempts to move ℓ, possibly removing modified positions, updates the Parikh matrix, and decides how to increase r.

The efficiency of the algorithm relies on the fact that modified positions can only occur in a pa-column that contains a mismatch. Thus, the mismatches are precomputed using LCP queries, and placed in two lists *LeftMismatchList* and *RightMismatchList*, for those mismatches that occur to the left and right of the anchor, respectively. A *zone* is defined as the substring from one mismatch position to the next. Note that there are no mismatches in a zone. The 1d FIND-KMAR algorithm uses several additional tricks to efficiently process the $O(k)$

---

[1] The journal version [4] uses KK as an improvement, however, we use ML due to the necessity of aligning anchor points among rows, thus, we reference the conference version of the paper.

[2] also called the "winner".

| row 1 | x | b | c | b | b | c | b | b | c | b | b | c | b | b | c | b | b | c | b | b | c | a | b | c |
| row 2 | a | b | c | a | b | c | a | b | c | a | b | c | a | b | c | x | b | c | a | b | c | a | b | c |

Fig. 4. Shown is an example of text that is input to the FINDKMAR procedure for $p = 3$. The mismatch positions are red. Note that two of the problematic pa-columns of the Amit et al. algorithm are in row one and one problematic pa-column will be in row two. The 2d zones are marked with curly braces, each zone extending from one position of mismatch up until (not including) the next position of mismatch. This entire array is a 2d k-MAR with $k = 3$. For $k = 2$ there is a 2d k-MAR starting at position 2. (Color figure online)

pa-columns that may contain mismatches, called the *problematic pa-columns*, working both within zones and between zones.

Definition 9. *[2d] A 2d horizontal side-sharing k-maximal approximate run (2d k-MAR) is an h-run that contains at most k modified positions.*

It is tempting to outline an algorithm that first uses the 1d algorithm of [4] to find k-MARs, and then links together k-MARs in different rows. However, the run-linking often shrinks the width of the runs being linked. Notice that the majority character at a position can change when the width of the k-MAR shrinks. For example, consider the k-MAR *abcabcabcabbabb*. The two modified positions are clearly the last b in the two substrings *abb*. If we were to cut off the first 6 characters in order to link this to a run on a different row, then b becomes the majority character and the c is the modified letter. Thus, each time we move down one row, and attempt to link output from one row to the next, it is necessary to compute the Parikh matrix anew. Therefore, we cannot apply Amit et al. as a black box. However, we can use a similar idea that we used in Sect. 3; rather than extending output, store the mismatch lists that were used to compute the 1d output, and attempt to extend mismatch lists. In this way we can reduce the 2d problem to 1 dimension. Pseudocode for the 2d k-MARs algorithm is found in Algorithm 2.

The FINDKMAR algorithm can be called essentially as a black box. The only modification is that it makes sure to work within the specific row of the mismatch that is being added or removed. Thus, each mismatch is augmented with the row that it occurs in. The mismatches are ordered by column, and merging the lists among rows preserves that ordering. The Parikh matrix still stores $O(k)$ pa-columns, each one occurs within a particular row. The subroutine's left (ℓ) and right (r) pointers that delineate the current output essentially point to columns in the 2d input. In addition, a 2d *zone* represents the 2d array between consecutive mismatches. See Fig. 4 for an example of a 2d k-MAR found by our algorithm. The 2d zones as well as the mismatch positions are labelled.

Time for 2d k-MARs: The data structures of Amit et al. are first constructed for each row separately, and then additionally for each extension to a new height. Each extension reduces to the 1d problem, and over all rows in each extension

---
Algorithm 2. Find 2d k-MARs

Input: 2d text T and integer k
Output: All 2d k-MARs in T
Run a 1d algorithm that finds all k-MARs within each row of T using ML anchors [3]. Report the 1d output, and store the original positions of mismatches found.
for *each row i* **do**
    for *each anchor point a* **do**
        for *each period p* **do**
            hasOutput=false
            if *there was 1d output for the triple (i, a, p)* **then**
                $height = 1$
                hasOutput=true
            end
            while hasOutput **do**
                merge the mismatch lists from the previous iteration (which represent rows $i \ldots i + height - 1$) with the mismatch lists of the next row ($i + height$), limiting the number of mismatches to $2k + 1$ in each direction from the anchor.
                run the subroutine FINDKMAR of Amit et al.[3] on (i, a, p) and the merged mismatch lists
                if *output is found* **then**
                      $height + +$
                      report output found with height $height$
                end
                else
                      hasOutput=false
                end
            end
        end
    end
end

---

there are still $O(k)$ problematic pa-columns. Thus, the overall work remains the same, even though the mismatches being processed may occur in different rows. The conference version of Amit et al. [3] uses Main and Lorentz and has time complexity $O(nk^2 \log k \log(n/k))$, since each call to FINDKMAR has time $O(k^2 \log k)$ and there are $O(n \log(n/k))$ iterations. Thus the 2d time complexity is $O(k^2 \log k(n^2 \log(n/k) + occ))$ to find all occ 2d k-MARs.

5 Conclusion

In this paper we presented several new definitions and algorithms for 2d side sharing tandems with mismatches. The algorithms are efficient and in line with similar results in 1d. One open question that we have is: can we improve the time complexity for finding runs of k-mismatch tandem repeats? Other interesting problems include considering other definitions of k-mismatches, e.g. k mismatches against a *consensus* [20]. Furthermore, other metrics, such as edit distance, can be considered. How can we define a side-sharing tandem that allows insertions and deletions?

Acknowledgement. Shoshana Marcus was partially supported for work on this project by PSC-CUNY Awards 66369–00 54 and 67488-00 55, jointly funded by The Professional Staff Congress and The City University of New York. Dina Sokol was partially supported by the United States - Israel Binational Science Foundation grant 2018141.

Disclosure of Interests. The authors have no competing interests to declare that are relevant to the content of this article.

References

1. Allouche, J.-P.: Algebraic combinatorics on words: by Lothaire ISBN: 0-521-81220-8. Semigroup Forum **70**(1), 154–155 (2005)
2. Amir, A., Butman, A., Landau, G.M., Marcus, S., Sokol, D.: Double string tandem repeats. Algorithmica **85**(1), 170–187 (2023). https://doi.org/10.1007/s00453-022-01016-9
3. Amit, M., Crochemore, M., Landau, G.M.: Locating all maximal approximate runs in a string. In: Fischer, J., Sanders, P. (eds.) Combinatorial Pattern Matching, 24th Annual Symposium, CPM 2013, Bad Herrenalb, Germany, June 17-19, 2013. Proceedings. Lecture Notes in Computer Science, vol. 7922, pp. 13–27. Springer (2013https://doi.org/10.1007/978-3-642-38905-4_4, https://doi.org/10.1007/978-3-642-38905-4_4
4. Amit, M., Crochemore, M., Landau, G.M., Sokol, D.: Locating maximal approximate runs in a string. Theor. Comput. Sci. **700**, 45–62 (2017). https://doi.org/10.1016/j.tcs.2017.07.021
5. Apostolico, A., Brimkov, V.E.: Fibonacci arrays and their two-dimensional repetitions. Theoret. Comput. Sci. **237**(1–2), 263–273 (2000). https://doi.org/10.1016/S0304-3975(98)00182-0
6. Apostolico, A., Brimkov, V.E.: Optimal discovery of repetitions in 2d. Discret. Appl. Math. **151**(1–3), 5–20 (2005). https://doi.org/10.1016/j.dam.2005.02.019
7. Bland, W., Smyth, W.F.: Three overlapping squares: the general case characterized and applications. Theor. Comput. Sci. **596**, 23–40 (2015). https://doi.org/10.1016/J.TCS.2015.06.037
8. Charalampopoulos, P., Radoszewski, J., Rytter, W., Walen, T., Zuba, W.: The number of repetitions in 2d-strings. In: Grandoni, F., Herman, G., Sanders, P. (eds.) 28th Annual European Symposium on Algorithms, ESA 2020, 7–9 September 2020, Pisa (Virtual Conference). LIPIcs, vol. 173, pp. 32:1–32:18. Schloss Dagstuhl - Leibniz-Zentrum für Informatik (2020). https://doi.org/10.4230/LIPIcs.ESA.2020.32
9. Crochemore, M.: An optimal algorithm for computing the repetitions in a word. Inf. Process. Lett. **12**(5), 244–250 (1981)
10. Deza, A., Franek, F., Jiang, M.: A computational substantiation of the d-step approach to the number of distinct squares problem. Discret. Appl. Math. **212**, 81–87 (2016). https://doi.org/10.1016/J.DAM.2016.04.025
11. Galil, Z., Giancarlo, R.: Improved string matching with k mismatches. SIGACT News **17**(4), 52–54 (1986). https://doi.org/10.1145/8307.8309

12. Gawrychowski, P., Ghazawi, S., Landau, G.M.: Lower bounds for the number of repetitions in 2d strings. In: Lecroq, T., Touzet, H. (eds.) SPIRE 2021. LNCS, vol. 12944, pp. 179–192. Springer, Heidelberg (2021). https://doi.org/10.1007/978-3-030-86692-1_15
13. Kolpakov, R.M., Kucherov, G.: Finding maximal repetitions in a word in linear time. In: 40th Annual Symposium on Foundations of Computer Science, FOCS 1999, 17–18 October, 1999, New York, pp. 596–604. IEEE Computer Society (1999). https://doi.org/10.1109/SFFCS.1999.814634
14. Kolpakov, R.M., Kucherov, G.: Finding approximate repetitions under Hamming distance. Theor. Comput. Sci. **303**(1), 135–156 (2003). https://doi.org/10.1016/S0304-3975(02)00448-6
15. Kucherov, G., Sokol, D.: Approximate tandem repeats. In: Encyclopedia of Algorithms, pp. 106–109. Springer, New York (2016). https://doi.org/10.1007/978-1-4939-2864-4_24
16. Landau, G.M., Schmidt, J.P., Sokol, D.: An algorithm for approximate tandem repeats. J. Comput. Biol. **8**(1), 1–18 (2001). https://doi.org/10.1089/106652701300099038
17. Main, M.G., Lorentz, R.J.: An O(n log n) algorithm for finding all repetitions in a string. J. Algorithms **5**(3), 422–432 (1984). https://doi.org/10.1016/0196-6774(84)90021-X
18. Marcus, S., Sokol, D., Zelikovitz, S.: Runs of side-sharing tandems in rectangular arrays. In: Pedreira, O., Estivill-Castro, V. (eds.) SISAP 2023. LNCS, vol. 14289, pp. 88–102. Springer, Cham (2023). https://doi.org/10.1007/978-3-031-46994-7_8
19. Parikh, R.J.: On context-free languages. J. ACM **13**, 570–581 (1966)
20. Sagot, M., Myers, E.W.: Identifying satellites in nucleic acid sequences. In: Istrail, S., Pevzner, P.A., Waterman, M.S. (eds.) Proceedings of the Second Annual International Conference on Research in Computational Molecular Biology, RECOMB 1998, New York, 22–25 March 1998, pp. 234–242. ACM (1998). https://doi.org/10.1145/279069.279120
21. Sokol, D., Benson, G., Tojeira, J.: Tandem repeats over the edit distance. Bioinformatics **23**(2), 30–35 (2007). https://doi.org/10.1093/BIOINFORMATICS/BTL309
22. Ziv, J., Lempel, A.: A universal algorithm for sequential data compression. IEEE Trans. Inf. Theory **23**(3), 337–343 (1977). https://doi.org/10.1109/TIT.1977.1055714
23. Ziv, J., Lempel, A.: Compression of individual sequences via variable-rate coding. IEEE Trans. Inf. Theory **24**(5), 530–536 (1978). https://doi.org/10.1109/TIT.1978.1055934

Faster and Simpler Online/Sliding Rightmost Lempel-Ziv Factorizations

Wataru Sumiyoshi[1], Takuya Mieno[2], and Shunsuke Inenaga[3]([✉])

[1] Department of Information Science and Technology, Kyushu University, Fukuoka, Japan
sumiyoshi.wataru.342@s.kyushu-u.ac.jp
[2] Department of Computer and Network Engineering, University of Electro-Communications, Chofu, Japan
tmieno@uec.ac.jp
[3] Department of Informatics, Kyushu University, Fukuoka, Japan
inenaga.shunsuke.380@m.kyushu-u.ac.jp

Abstract. We tackle the problems of computing the *rightmost* variant of the Lempel-Ziv factorizations in the online/sliding model. Previous best bounds for this problem are $O(n \log n)$ time with $O(n)$ space, due to Amir et al. [IPL 2002] for the online model, and due to Larsson [CPM 2014] for the sliding model. In this paper, we present faster $O(n \log n / \log \log n)$-time solutions to both of the online/sliding models. Our algorithms are built on a simple data structure named *BP-linked trees*, and on a slightly improved version of the range minimum/maximum query (RmQ/RMQ) data structure on a dynamic list of integers. We also present other applications of our algorithms.

Keywords: suffix trees · Lempel-Ziv factorizations · closed factorizations

1 Introduction

1.1 Online Rightmost LZ-Factorizations and LPF Arrays

The *longest previous factor array*[1] LPF of a string S of length n is an array of length n such that, for each $1 \leq i \leq n$, LPF$[i]$ stores the length ℓ_i of the longest suffix of $S[1..i]$ that occurs at least twice in $S[1..i]$. The LPF array has a close relationship to the Lempel-Ziv (LZ) factorization [23], that is a basic and powerful tool for a variety of string processing tasks including data compression [34] and finding repetitions [19].

We consider a variant of LPF arrays with *rightmost reference*, denoted RLPF, where each RLPF$[i]$ also stores the distance $d = i - j$ to the rightmost previous ending position j ($j < i$) of the longest repeating length-ℓ_i suffix of

---

[1] Our definition of online LPF arrays follows from the literature [27,28].

$S[1..i]$. Computing the rightmost references is motivated by encoding each factor in the LZ-factorization with less bits [12], and has attracted much attention. The state-of-the-art *offline* algorithm for the rightmost LZ-factorization runs in $O(n(\log \log \sigma + \frac{\log \sigma}{\sqrt{\log n}}))$ time with $O(n \log \sigma)$ bits of space, where σ is the alphabet size [6]. Bille et al. [8] proposed an algorithm for computing a $(1+\epsilon)$-approximated version of the rightmost LZ-factorization for any $\epsilon > 0$. Ellert et al. [11] considered the rightmost version of the *LZ-End* factorization [20], a variant of the LZ-factorization designed for fast random access.

The other common method for limiting the distance from each factor to a previous occurrence is the *sliding* model, where only the previous occurrences of each factor within the preceding sliding window of fixed size $d \geq 1$ are considered [7,30]. The LZ-factorization in the sliding model is used in the real-world compression software's including zip and 7zip. Sliding suffix tree algorithms [21,24,29] are able to compute the LZ-factorization in the sliding model in $O(n \log \sigma)$ time with $O(d)$ words of working space. Bille et al. [8] presented another algorithm for sliding LZ-factorization that runs in $O(\frac{n}{d}\mathsf{sort}(d) + z \log \log \sigma)$ time with $O(d)$ words of working space, where z is the number of factors and $\mathsf{sort}(d)$ denotes the time for sorting the d characters in each of the $O(\frac{n}{d})$ blocks on the input string.

In this paper, we consider the three following problems:

Problem (1): The rightmost LPF array in the online model.
Problem (2): The rightmost LZ-factorization in the online model.
Problem (3): The rightmost LZ-factorization in the sliding model.

Amir et al. [4] proposed an algorithm for (1) that works in $O(n \log n)$ time with $O(n)$ words of space. Their key data structure is the *timestamped suffix tree*, which is based on Weiner's online suffix tree construction [32] and is augmented with an online range minimum query data structure. Larsson [22] presented an algorithm for (2) running in $O(n \log n)$ time with $O(n)$ words of space, that is based on Ukkonen's online suffix tree construction [31]. To the best of our knowledge, none of the existing algorithms provides an efficient solution to (3), where *both* of the rightmost and sliding properties are required.

1.2 Our New Online/Sliding Algorithms for Rightmost LZ and LPF

We consider a simple data structure named *BP-linked trees* capable of maintaining a representation of balanced parentheses (BP) of a dynamic rooted tree. Basically, our BP-linked trees are equivalent to an intermediate data structure used in the so-called *Euler tour trees* [18] that maintain the Euler tours of dynamic trees: Our BP-linked trees can be seen as a representation of the Euler tours of the input trees. In our BP-linked tree, the BP is maintained as a doubly-linked list, which can be updated in $O(1)$ worst-case time given the locus of the inserted/deleted node on the explicitly stored tree. By maintaining our BP-linked tree on top of the suffix tree, we achieve an online algorithm for computing rightmost LPF arrays in $O(n \log n / \log \log n)$ time with $O(n)$ words of space, thus achieving a faster online solution for (1). In addition, we show how our algorithm can be modified to solve (2) in the same complexity as (1), and in $O(n \log d / \log \log d)$ time with $O(d)$ words of working space for (3).

The $\log n / \log \log n$ (resp. $\log d / \log \log d$) term in our time complexities comes from *range minimum/maximum queries (RmQ/RMQ)* on a dynamic list of n integers (resp. d integers) - to compute the rightmost LZ-factorization and LPF array, we use RmQ/RMQ to retrieve the rightmost previous occurrence of a given locus in the online/sliding suffix tree. While those bounds for dynamic RmQ/RMQ can already be achieved by the use of Brodal et al.'s *path minimum/maximum queries* data structure on a dynamic tree [9] *in the amortized sense*, this paper shows how their data structure can be modified to perform updates and queries in the same *worst-case time bounds* in the case of dynamic lists, after sublinear-time preprocessing (Lemma 2).

The simple framework of our algorithms allows one to obtain very simple alternative solutions to the existing ones: By using folklore dynamic RmQ/RMQ data structures based on binary search trees in place of the aforementioned advanced RmQ/RMQ data structures, the same run times as the methods of Amir et al. [4] for (1) and Larsson [22] for (2) can readily be achieved. It appears that this version of our BP-linked trees with binary search trees is basically equivalent to the so-called Euler tour trees [18] that support updates and queries on dynamic input trees in $O(\log n)$ time each.

We also present other applications of our algorithms in Sect. 5.

1.3 Related Work for Dynamic BP Maintenance

In the problem of maintaining the BP $\mathcal{B}$ for a *dynamic* tree, one is required to efficiently support the following operations and queries:

- insert: add a new node to $\mathcal{B}$;
- delete: remove an existing non-root node from $\mathcal{B}$;
- leftmost leaf: return the left parenthesis "(" corresponding to a given node;
- rightmost leaf: return the right parenthesis ")" corresponding to a given node;
- parent: return the nearest enclosing parentheses for a given node;
- rank i: return the number of left/right parentheses in $\mathcal{B}[1..i]$;
- select i: return the ith left/right parenthesis in $\mathcal{B}$.

This problem was already studied at least in early 80's, in the context of maintaining a dynamic set of nesting intervals [17]. Since then, it has also appeared in various important problems including dynamic dictionary matching [3,10] and (compressed) suffix trees of dynamic collection of strings [3,10,26].

Navarro and Sadakane [26] proposed a data structure of $2n + o(n)$ bits of space that supports all the above queries and operations in worst-case $O(\log n / \log \log n)$ time. Chan et al. [10] showed an amortized $\Omega(\log n / \log \log n)$-time lower bound for the dynamic BP-maintenance via a reduction from the dynamic subset rank problem on a set $\mathcal{S}$ of integers [16]. Chan et al. reduce a subset rank query on $\mathcal{S}$ to finding the nearest enclosing parentheses in $\mathcal{B}$ (i.e.

finding the parent node), which can further be reduced to a constant number of rank/select queries in $\mathcal{B}$. Thus, any algorithm for dynamic BP-maintenance *which supports rank/select queries* must use (amortized) $\Omega(\log n / \log \log n)$ time.

Our BP-linked trees deal with a simpler version of the dynamic BP-maintenance problem where all the operations and queries, *excluding rank and select queries*, are supported. Our BP-linked trees are a simple pointer-based data structure, which occupies $O(n)$ words of space and performs insertions, deletions, accessing the leftmost/rightmost leaf, and the parent, in worst-case $O(1)$ time each.

2 Preliminaries

2.1 Strings

Let Σ denote an ordered *alphabet* of size σ. An element of Σ^* is called a *string*. The length of a string $S \in \Sigma^*$ is denoted by $|S|$. The *empty string* ε is the string of length 0. For string $S = xyz$, x, y, and z are called the *prefix*, *substring*, and *suffix* of S, respectively. Let $\mathsf{Prefix}(S)$, $\mathsf{Substr}(S)$, and $\mathsf{Suffix}(S)$ denote the sets of prefixes, substrings, and suffixes of S, respectively. For a string S of length n, $S[i]$ denotes the ith symbol of S and $S[i..j] = S[i]\cdots S[j]$ denotes the substring of S that begins at position i and ends at position j for $1 \leq i \leq j \leq n$. For convenience, let $S[i..j] = \varepsilon$ for $i > j$. The *reversed string* of a string S is denoted by S^R, that is, $S^R = S[|T|]\cdots S[1]$.

For a string S, the strings in $\mathsf{Prefix}(S) \cap \mathsf{Substr}(S[2..|S|])$ and the strings in $\mathsf{Suffix}(S) \cap \mathsf{Substr}(S[1..|S|-1])$ are called *repeating prefixes* and *repeating suffixes* of S, respectively. Let $\mathsf{lrp}(S)$ and $\mathsf{lrs}(S)$ denote the longest repeating prefix and the longest repeating suffix of S, respectively.

2.2 Model of Computation

This paper assumes the standard *word RAM model* with word size $\Theta(\log n)$, where n is the length of the input string.

2.3 Suffix Trees

The *suffix tree* [32] of a string S, denoted $\mathsf{STree}(S)$, is a path-compressed trie representing $\mathsf{Suffix}(S)$ such that

(1) Each internal node has at least two children;
(2) Each edge is labeled by a non-empty substring of S;
(3) The labels of out-going edges of the same node begin with distinct characters.

Each leaf of $\mathsf{STree}(S)$ is associated with the beginning position of its corresponding suffix of S. For a node v of $\mathsf{STree}(S)$, let $\mathsf{str}(v)$ denote the string label of the path from the root to v. Each node v stores its string depth $|\mathsf{str}(v)|$. The *locus* of a substring $w \in \mathsf{Substr}(S)$ in $\mathsf{STree}(S)$ is the position where w is spelled out from the root. The locus of w is said to be an *explicit node* if $w = \text{str}(v)$ for some node v in $\text{STree}(S)$. Otherwise, i.e. the locus of w is on an edge, then it is said to be an *implicit node*. The number of explicit nodes in $\text{STree}(S)$ is at most $n-1$, where $n = |S|$, while there are $O(n^2)$ implicit nodes in $\text{STree}(S)$. We can represent $\text{STree}(S)$ in $O(n)$ space by representing each edge label x with a pair (i,j) of positions in S such that $S[i..j] = x$.

2.4 Online/Sliding Rightmost LPF Arrays and LZ-Factorizations

The *online longest previous factors problem* is, given the ith character $S[i]$ of an online input string S, to compute the longest suffix $S[i-\ell_i+1..i]$ of $S[1..i]$ that occurs at least twice in $S[1..i]$. The *rightmost longest previous factor array* of a string S of length n, denoted RLPF, is an array of length n such that

$$\text{RLPF}[i] = \begin{cases} (0,1) & \text{if } i \text{ is the first occurrence of character } S[i] \text{ in } S \\ (\ell_i, i-j) & \text{otherwise,} \end{cases}$$

where $\ell_i = |\text{lrs}(S[1..i])|$ and $j = \max\{j' \mid S[i-\ell_i+1..i] = S[j'-\ell_i+1..j'], j' < i\}$.

A sequence $S = f_1, \ldots, f_z$ of z non-empty strings is called the *Lempel-Ziv (LZ) factorization* of string S of length n if (1) f_k is a fresh character not occurring to its left in S, or (2) f_k is the longest prefix of the suffix $f_k \cdots f_z = S[|f_1 \cdots f_{k-1}|+1..n]$ of S that has a previous occurrence beginning in $f_1 \cdots f_{k-1} = S[1..|f_1 \cdots f_{k-1}|]$. In the *rightmost* LZ-factorization of S, each factor f_k of type (2) is encoded by a pair $(|f_k|, x)$ such that $x = |f_1 \cdots f_k| - j$ is the distance to the ending position j of the rightmost previous occurrence of f_k in $S[1..|f_1 \cdots f_k|]$.

Example 1. The following table shows RLPF of string $S = \text{abaababaabba}$:

i	1	2	3	4	5	6	7	8	9	10	11	12
$S[i]$	a	b	a	a	b	a	b	a	a	b	b	a
$\text{RLPF}[i]$	(0,1)	(0,1)	(1,2)	(1,1)	(2,3)	(3,3)	(2,2)	(3,2)	(4,5)	(5,5)	(1,1)	(2,4)

The rightmost LZ-factorization of S is $(0, \text{a}), (0, \text{b}), (1, 2), (3, 3), (4, 5), (2, 4)$.

Let $d \geq 1$ denote the window size of fixed length. A sequence $S = g_1, \ldots, g_m$ of m non-empty strings is called the *sliding LZ-factorization* of a string S of length n w.r.t. window size d, if each factor g_k is the longest prefix of the suffix $|g_k \cdots g_m| = S[|g_1 \cdots g_{k-1}|+1..n]$ of S that has a previous occurrence beginning in the sliding window $W_k = S[\max\{1, |g_1 \cdots g_{k-1}| - d + 1\}..|g_1 \cdots g_{k-1}|]$.

3 Data Structures

This section introduces data structures for dynamic trees which are core components of our rightmost LZ algorithms.

3.1 BP-Linked Trees

Let T be a rooted ordered tree having N nodes. Let $\mathsf{BP}(\mathsf{T}) \in \{(,)\}^{2N}$ be the BP-representation of T. In this paper, we implement $\mathsf{BP}(\mathsf{T})$ using a doubly-linked list. For each node v in T, let $(_v$ and $)_v$ denote the (and) that correspond to v in $\mathsf{BP}(\mathsf{T})$. A *BP-linked tree* is a tree T augmented with its BP-representation $\mathsf{BP}(\mathsf{T})$ such that each node v of T has pointers to $(_v$ and $)_v$ in $\mathsf{BP}(\mathsf{T})$.

We consider the following edit operations on T: (1) inserting a leaf, or a new root as the parent of the old root, (2) inserting an internal node by splitting an edge, and (3) deleting a non-root node. We remark that our tree T is explicitly stored, and the input of each operation is given as a locus on the tree T (not on $\mathsf{BP}(\mathsf{T})$). The next lemma follows:

Lemma 1. *Given a tree-editing operation, we can update a BP-linked tree in worst-case $O(1)$ time.*

Proof. First we consider the case where a leaf v is inserted. Let u be the parent of v. If v is the leftmost child of u, then we take the pointer of u to access $(_u$ in $\mathsf{BP}(\mathsf{T})$, and then insert $(_v$ and $)_v$ immediately to the right of $(_u$. Otherwise, let x be v's neighbor to the left. Then, in a similar way as before, insert $(_v$ and $)_v$ immediately to the right of $)_x$. Also, when a new root r is inserted, we just prepend $(_r$ and append $)_r$ to $\mathsf{BP}(\mathsf{T})$.

Second we consider the case where an internal node v is inserted. Suppose that an edge $e = (u, w)$ is split into two edges $e_1 = (u, v)$ and $e_2 = (v, w)$. We take the pointer of w to access $(_w$ in $\mathsf{BP}(\mathsf{T})$, and insert $(_v$ immediately to the left of $(_w$. We also take the right pointer of w to access $)_w$ in $\mathsf{BP}(\mathsf{T})$, and then insert $)_v$ immediately to the right of $)_w$.

Third we consider the case where a non-root node v is deleted. Then we just delete $(_v$ and $)_v$ from $\mathsf{BP}(\mathsf{T})$. Note that if u is the parent of v and v has k children $w_1, \ldots, w_k$, then new parent of $w_1, \ldots, w_k$ becomes u after the deletion.

It is clear that each of these operations takes $O(1)$ worst-case time. □

3.2 Subtree Minimum Queries

In this subsection, we propose dynamic data structures with worst-case update/query time for *range minimum queries (RmQs)* on a linear list and for *subtree minimum queries (SmQs)* on a rooted and weighted tree.

Dynamic Range Minimum Queries. A *dynamic range minimum query* (RmQ) data structure on a linear-linked-list of integers supports the following:

- insert(u, v, x): insert a new node v with value x as the next node of u;
- delete(v): delete node v from the list;
- update(v, x): update the value of node v to x;
- RmQ(u, v): return a node with the smallest value in the path (u, v).

Brodal et al. [9] presented a dynamic RmQ data structure for a linear-linked-list[2] of n integers, which takes $O(n)$ space and supports the above queries and updates in *amortized* $O(\log n / \log \log n)$ time each in the RAM model. Below we make a few changes to their method in order to obtain *worst-case* time guarantees:

Lemma 2. *After $o(n)$-time preprocessing, we can maintain a dynamic RmQ data structure on a linear-linked-list of n integers which takes $O(n)$ space and supports each query/operation in worst-case $O(\log n / \log \log n)$ time.*

Proof. Let L be the dynamic list of integers. Let $B = \lfloor \log^\varepsilon n \rfloor \geq 1$ for some small constant $0 < \varepsilon < 1$. We build a *q*-heap* (Corollary 3.4 of [33]) on top of the dynamic list L, which is a variant of B-trees of order B and supports predecessor queries, insertions, and deletions over L in *worst-case* $O(\log n / \log \log n)$ time each, after $o(n)$-time preprocessing. Note that updating a value of an element in L can be simulated by combining an insertion and a deletion. Also, as in Theorem 2 of [9], we precompute lookup-tables of total size $o(n)$ in order to support RmQ, insert, delete and update inside any list of size $O(B)$, which represents a node of the q*-heap, in worst-case $O(1)$ time in the RAM model. Then we maintain, for each node of the q*-heap, the list consisting of the minima of its children by using the lookup-tables. Given a range minimum query, we can answer the query by visiting at most $O(\log n / \log \log n)$ nodes of the q*-heap, similar to the standard method for 1D-range trees (see [25] for example). □

Dynamic Subtree Minimum Queries. We introduce subtree minimum queries (SmQs) on a rooted and weighted tree.

Definition 1. *A subtree minimum query (SmQ) on a rooted and weighted tree* T *is, given a node v in* T*, to compute a node having the minimum weight in the subtree rooted at v.*

For the static case, we can easily answer any query in constant time after storing the answer to each node by traversing the tree.

We focus on a dynamic case, where tree-editing operation mentioned in Sect. 3.1 will be applied to the tree. Furthermore, we consider update operations, i.e., updating the weight of a node to a new weight. We show the next lemma.

Lemma 3. *After $o(n)$-time preprocessing, we can maintain a dynamic SmQ data structure on a rooted and weighted tree with n nodes which takes $O(n)$ space and supports each query/operation in worst-case $O(\log n / \log \log n)$ time. Also, the time complexity per each query/operation is optimal.*

---

[2] They actually presented a data structure for *Path Minimum Queries* for an edge-weighted dynamic tree, which is a generalization of RmQs for a dynamic linear list. Since such a general setting is not needed for our purpose, we cite their result as a dynamic RmQ data structure and make some changes to it for simplicity.

Proof. Let T be the input tree. Further let weight(v) be the weight of v for each node v in T. The SmQs on T can be reduced to the RmQs on BP(T) as follows: For each node v of T, the weight of "$(_v$" is assigned weight(v) and the weight of "$)_v$" is assigned ∞. By doing this reduction, it follows that for any node v in T, if RmQ for pair "$(_v$", "$)_v$" returns "$(_u$", then node u is an answer of SmQ for v. Since we can maintain T as a BP-linked tree for any given tree-editing operation in $O(1)$ time (Lemma 1), we can maintain the BP(T) with weights in $O(1)$ time as well. Also, by Lemma 2, the RmQ data structure on BP(T) can be maintained in worst-case $O(\log n / \log \log n)$ time for each query/editing operation. Therefore, we obtain the desired upper bound.

To prove the lower bound, we reduce the *priority searching problem* [1] to the dynamic SmQ problem. Let $S \subseteq \{1, \ldots, n\}$ be a set of integers with priorities. A priority $p(x)$ of an integer x is a positive integer at most n. The priority searching problem on S supports (1) insertion of an integer x with priority $p(x)$ to S, (2) deletion of an integer x from S, and (3) searching for the integer $y \leq x$ in P for given x such that $p(y)$ is maximized. For any instance S of the priority searching problem, we can consider the path graph G_S of size $|S|$ obtained by connecting the elements in S linearly. The weight of each element is the priority of the element. Clearly, any query/update of the priority searching on S can be simulated by a query/update of the dynamic SmQ on G_S. □

4 Online/Sliding Rightmost LZ Factorizations

In this section, we present our algorithms for Problems (1)–(3). We begin with our key data structure.

4.1 BP-Linked Suffix Trees

We call the suffix tree of string S augmented with its BP-representation a *BP-linked suffix tree* and denote it by BPSTree(S). See Fig. 1 for a concrete example of BPSTree(S). Note that the BP-linked suffix tree is similar to the *timestamped suffix tree* proposed by Amir et al. [4]. However, the BP-linked suffix tree is superior to the timestamped suffix tree in the following sense: Our BP-linked suffix trees support a node deletion in worst-case $O(1)$ time, while the timestamped suffix trees can require $\Omega(n)$ time for a node deletion in the worst case to maintain their *rightmost/leftmost leaves pointers for all nodes*.

By combining Lemma 1 with the known online suffix tree construction algorithms, we immediately obtain the following results:

Corollary 1. *For a string S of length n, using $O(n)$ working space, one can update* BPSTree(S) *to* BPSTree(cS) *and find the locus of* lrp(cS) *in* BPSTree(cS) *for a given character $c \in \Sigma$*

(a) in worst-case $O(\log \log n + (\log \log \sigma)^2 / \log \log \log \sigma)$ time for an integer alphabet of size $\sigma = n^{O(1)}$ with Fischer and Gawrychowski's algorithm [14, 15];

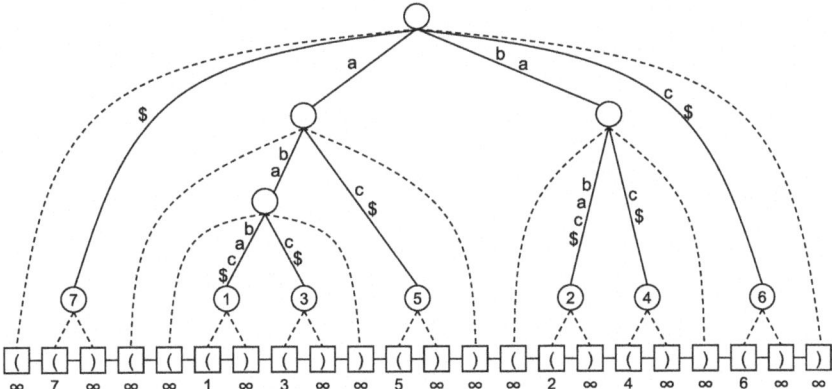

Fig. 1. The BP-linked suffix tree of string $S = \mathtt{ababac\$}$.

(b) in amortized $O(\log \sigma)$ time for a general ordered alphabet of size σ with Weiner's algorithm [32].

Corollary 2. *For a string S of length n over a general ordered alphabet of size σ, using $O(n)$ working space, one can update $\mathsf{BPSTree}(S)$ to $\mathsf{BPSTree}(Sc)$ and find the locus of $\mathsf{lrs}(Sc)$ in $\mathsf{BPSTree}(Sc)$ for a given character $c \in \Sigma$ in amortized $O(\log \sigma)$ time with Ukkonen's algorithm [31].*

Also, we employ our dynamic SmQ data structure (Lemma 3) to the BP-linked suffix trees. This gives us the following:

Lemma 4. *For an online string of length n, there exists a data structure of size $O(n)$ which supports,*

(a) in worst-case $O(\log n / \log \log n)$ time for an integer alphabet of size $\sigma = n^{O(1)}$ after $o(n)$-time preprocessing;

(b) in amortized $O(\log \sigma + \log n / \log \log n)$ time for a general ordered alphabet of size σ,

the following queries and updates:

- *Given an implicit or explicit node v on the current suffix tree, find the leftmost occurrence of $\mathsf{str}(v)$ in the current string;*
- *Update the data structure when a new character is prepended.*

Proof. Let S be the input string. Since we use a Weiner-type of construction where a new character c is prepended to S, we can assume that the right-end of S terminates with a end-maker $\$$, with which all the suffixes of S are represented by the leaves of $\mathsf{STree}(S)$.

We consider Case (a). Let $\mathsf{weight}(v)$ be the weight of v for each node v in $\mathsf{STree}(S)$. For each leaf ℓ, we set $\mathsf{weight}(\ell)$ to the beginning position of the suffix corresponding to ℓ. For each non-leaf node v, we set $\mathsf{weight}(v) = \infty$. By applying Lemma 3 to this weighted suffix tree, we can answer the query in $O(\log n/\log\log n)$ time. Also, the auxiliary data structures can be updated in worst-case $O(\log n/\log\log n)$ time by Corollary 1-(a) and Lemma 3.

Case (b) can be proven similarly with Corollary 1-(b). □

4.2 Online Rightmost LPF

Here we present our algorithm for Problem (1).

Theorem 1 (Online Rightmost LPF). *For a string S of length n, there exist online algorithms which use $O(n)$ space and compute $\mathsf{RLPF}[i]$ for each $1 \leq i \leq n$*

(a) in worst-case $O(\log n/\log\log n)$ time after $o(n)$-time preprocessing for an integer alphabet of size $\sigma = n^{O(1)}$;

(b) in amortized $O(\log \sigma + \log n/\log\log n)$ time for a general order alphabet of size σ.

Proof. Let us consider Case (a). Since $\mathsf{lrs}(S[1..i]) = \mathsf{lrp}((S[1..i])^R) = \mathsf{lrp}(S^R[n-i+1..n])$, the problem is reducible to computing the locus p_j of $\mathsf{lrp}(S^R[j..n])$ on $\mathsf{STree}(S^R[j..n])$ for decreasing $j = n, \ldots, 1$, and finding the leaf in the subtree under p_j that has the second smallest value. For this sake we can use (1) of Corollary 1 and Lemma 4. Since $\sigma = n^{O(1)}$, we have $\log\log n + (\log\log\sigma)^2/\log\log\log\sigma \in O(\log n/\log\log n)$. Thus $\mathsf{RLPF}[i]$ can be computed in worst-case $O(\log n/\log\log n)$ time each, after $o(n)$-time preprocessing. Case (b) can be shown similarly. □

4.3 Online Rightmost LZ-Factorization

In this subsection, we present our algorithm for Problem (2).

Theorem 2 (Online Rightmost LZ). *For a string S of length n over a general order alphabet of size σ, there exists an online algorithm which uses $O(n)$ space and computes the rightmost LZ-factorization of S in amortized $O(\log \sigma + \log n/\log\log n)$ time per character.*

Proof. We use a standard technique with Ukkonen's online suffix tree construction with Corollary 2. Suppose we have computed the first $k-1$ factors $f_1, \ldots, f_{k-1}$, and that we have built $\mathsf{BPSTree}(S[1..i])$ where $i = |f_1 \cdots f_{k-1}| + 1$ is the beginning position of the next factor f_k. If $S[i]$ is a fresh character, then clearly $f_k = S[i]$. Otherwise, we perform the following. We grow the BP-liked suffix tree while reading subsequent characters $S[i+\ell-1]$ for increasing $\ell = 2, 3, \ldots$ until we find the smallest $\ell^* \geq 2$ such that $\||\mathsf{lrs}(S[1..i+\ell^*-1])| < \ell^*$ (see Fig. 2). When we find such ℓ^*, it turns out that $f_k = S[i..i+\ell^*-2]$ since $S[i..i+\ell^*-2]$ has a previous occurrence beginning at some position in $S[1..i-1]$ and $S[i..i+\ell^*-1]$ does not. Now, we search for the rightmost previous occurrence of f_k by using $\mathsf{BPSTree}(S[1..i+\ell^*-1])$. Since $\||\mathsf{lrs}(S[1..i+\ell^*-1])| \leq \ell^*-1 = |f_k|$, all the occurrences of f_k are represented by leaves or the *active point* that is the locus corresponding to the longest repeating suffix. Thus the rightmost previous occurrence

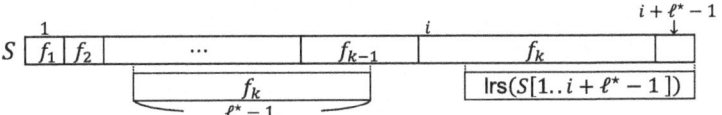

Fig. 2. Illustration for Theorem 2.

of f_k can be obtained by querying RMQs $O(1)$ times for the leaves under the locus of f_k. The above procedures for f_k can be done in $O(|f_k|\log\sigma + \log n/\log\log n)$ time except for the time for maintaining the BP-linked suffix trees that takes $O(1)$ amortized time per character. □

4.4 Sliding Rightmost LZ

In this subsection, we present our algorithm for Problem (3).

Theorem 3 (Sliding Rightmost LZ). *For an online string of length n over a general ordered alphabet of size σ and a fixed window size d, one can compute the sliding window rightmost LZ-factorization in amortized $O(\log\sigma + \log d/\log\log d)$ time per character, using $O(d)$ total space.*

Proof. We use a similar strategy to the case of online rightmost LZ-factorization from Theorem 2, with a variant of Corollary 2 using a sliding suffix tree algorithm (cf. [21,24,29]). Suppose that we have computed the first $k-1$ factors $g_1, \ldots, g_{k-1}$, and that we have maintained BPSTree(W_i) where $W_i = S[i-d..i-1]$ is the current window of width d. If $S[i]$ does not occur in W_i, then clearly $g_k = S[i]$. Otherwise, as in Theorem 2, we grow the BP-liked suffix tree while reading subsequent characters $S[i+\ell-1]$ for increasing $\ell = 2, 3, \ldots, 2d$ until the value ℓ reaches $2d$ or we find the smallest $\ell^* \geq 2$ such that $|\mathsf{lrs}(S[i-d..i+\ell^*-1])| < \ell^*$. If such ℓ^* is found, then $g_k = S[i..i+\ell^*-2]$ and we can retrieve the rightmost previous occurrence of g_k as in Theorem 2. Otherwise, $\ell = 2d$ and $|\mathsf{lrs}(S[i-d..i+2d-1])| \geq 2d$ hold, and we then stop growing the suffix tree. Let $g' = S[i..i+2d-1]$ be the length-$2d$ suffix of the extended window $S[i-d..i+2d-1]$. Let p be the difference between the beginning positions of the occurrence of $\mathsf{lrs}(S[i-d..i+2d-1])$ as suffix and its (arbitrary) previous occurrence. Now $p \leq d$ holds since $\mathsf{lrs}(S[i-d..i+2d-1]) \geq 2d$. Then, g' also appears p positions to the left, i.e., at position $i-p$, and thus, p is a period of g' and $p \leq |g'|/2$. The longest right-extension of g' with period p is g_k (see Fig. 3). Such extension can be computed in $O(|g_k|)$ time with $O(d)$ space by naive character comparisons in S as follows: for incremental $j = 0, 1, 2, \ldots$, we compare character $S[i+2d+j]$ to $S[i+(j \bmod p)]$ instead of $S[i+2d+j-p]$ until a mismatch is found. By doing this, no matter how large j becomes, every character comparison is possible by retaining only the extended window $S[i-d..i+2d-1]$ of size $3d$ and a single character $S[i+2d+j]$.

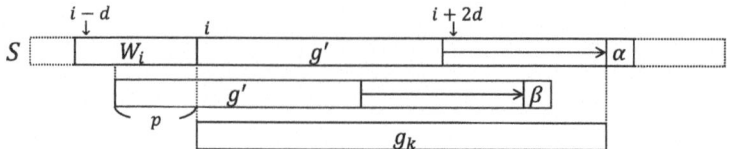

Fig. 3. Illustration for Theorem 3. String g' of length $2d$ has period p. When $\alpha \neq \beta$ where β is p characters before α, the next factor g_k is determined since $g_k\alpha$ cannot occur before it due to the periodicity of $g_k\beta$.

At each kth step, we use only $O(d)$ space for the BP-linked suffix tree of an extended window of length at most $3d$ and some auxiliary $O(1)$ working space. While we may need to compare $\omega(d)$ characters in S when g_k is much longer than $2d$, we do not need to store the characters outside of the extended window. Thus, such character-comparisons can be done within $O(d)$ space. Then, to proceed to the $(k+1)$th step, we move to the next window of size d, namely, the length-d suffix of $S[1..|g_1 g_2 \cdots g_k|]$. □

5 Other Applications of BP-Linked Suffix Trees

In this section, we present other applications of our BP-linked (suffix) trees, which are online computation of *closed factorizations* of a given string.

5.1 Online Longest Closed Factorizations

A string w is *closed* if w is a character, or the longest border b of w occurs exactly twice in w as prefix and suffix [13]. The *longest closed factorization* $\mathsf{LCF}(S) = g_1, \ldots, g_k$ of a string S is a factorization of S such that each g_i is the longest closed suffix of $S[1..|g_1 \cdots g_i|]$. The *longest closed factor array* LCFA of a string S of length n is an array of length n such that $\mathsf{LCFA}[i]$ stores the length of the last factor of $\mathsf{LCF}(S[1..i])$ and the size of $\mathsf{LCF}(S[1..i])$ for $1 \leq i \leq n$. $\mathsf{LCF}(S)$ can readily be obtained from LCFA for S.

Alzamel et al. [2] showed the following property:

Lemma 5. ([2]) *For a string S, if $g_1, \ldots, g_k = \mathsf{LCF}(S)$, then $g_k = S[i..|S|]$, where i is the second rightmost occurrence of $\mathsf{lrs}(S)$ in S. Also, $\mathsf{lrs}(S)$ is the longest border of g_k.*

Alzamel et al. [2] employ Ukkonen's online suffix tree and rely on RMQ on a dynamic list of leaves, for computing LCFA online. The inputs of their RMQ is given as a pair l, r of two integers representing an interval $[l, r]$ in the sorted list of leaves in the online suffix tree, where l and r are the lexicographical ranks of the leftmost and rightmost leaves in the subtree rooted at the active point. However, in [2] the authors do not describe how to explicitly maintain the ranks of leaves on a growing suffix tree as integers. We remark that even a single leaf insertion to the suffix tree can change the ranks of $\Omega(n)$ existing leaves.

However, as we have observed previously, by the use of our online BP-linked suffix tree, maintaining the ranks of the leaves in a growing suffix tree is no more necessary for performing RMQs under the active point. Due to Lemma 5, we can use a similar strategy as in Theorem 1 by noting that the second rightmost occurrence, which is the second leftmost occurrence in the reversed string, can be found with a constant number of RmQs. Thus we have:

Theorem 4. *For a string S of length n, there exist online algorithms which use $O(n)$ space and compute* LCFA$[i]$ *for each $1 \leq i \leq n$*

(a) *in worst-case $O(\log n/\log \log n)$ time after $o(n)$-time preprocessing for an integer alphabet of size $\sigma = n^{O(1)}$;*

(b) *in amortized $O(\log \sigma + \log n/\log \log n)$ time for a general order alphabet of size σ.*

Our result in Theorem 4 can be seen as an online alternative to the offline solution in the literature [5], with the same complexity.

5.2 Online Minimum Closed Factorizations

The closed factorization $g_1, \ldots, g_k$ of a string S is called the *minimum closed factorization* of S if the number k of factors is smallest [5]. Let mcf(S) denote the size of the minimum closed factorization of S.

Theorem 5. *For a string S of length n, there exist online algorithms which use $O(n)$ space and compute the minimum closed factor array* MCFA$[i] = $ mcf$(S[1..i])$ *for each $1 \leq i \leq n$, with $\ell_i = |\mathsf{lrs}(S[1..i])|$,*

(a) *in worst-case $O(\ell_i \log n/\log \log n)$ time after $o(n)$-time preprocessing for an integer alphabet of size $\sigma = n^{O(1)}$;*

(b) *in amortized $O(\log \sigma + \ell_i \log n/\log \log n)$ time for a general order alphabet of size σ.*

Proof. Consider Case (a). We find the locus for $\mathsf{lrp}(S[1..i]^R)$ in BPSTree$(S[1..i]^R)$ in worst-case $O(\log n/\log \log n)$ time with Corollary 1-(a). Let $v_1, \ldots, v_{\ell_i}$ be the explicit/implicit nodes on the path from the root to the locus for $\mathsf{lrp}(S[1..i]^R)$. For each v_j, we perform a constant number of RmQs to find the second leftmost occurrence of str(v_j) with Lemma 4 in worst-case $O(\log n/\log \log n)$ time. Then, we can compute mcf$(S[1..i]^R)[i]$ by dynamic programming in $O(\ell_i)$ time.

Case (b) can be obtained with Corollary 1-(b). □

Alzamel et al. [2] claimed a solution with $O(\ell_i(\log \sigma + \log n))$ worst-case running time for each i, which is based on Ukkonen's algorithm. Although amortized, our algorithm is faster than theirs also in the case of general ordered alphabets.

Acknowledgements. This work was supported by JSPS KAKENHI Grant Numbers JP23H04381, JP24K20734 (TM) and JP20H05964, JP23K24808, JP23K18466 (SI). The authors thank the anonymous referee for a pointer to reference [18] that introduced the Euler tour trees.

References

1. Alstrup, S., Husfeldt, T., Rauhe, T.: Marked ancestor problems. In: 39th Annual Symposium on Foundations of Computer Science, FOCS 1998, 8-11 November 1998, Palo Alto, pp. 534–544. IEEE Computer Society (1998). https://doi.org/10.1109/SFCS.1998.743504
2. Alzamel, M., Iliopoulos, C.S., Smyth, W.F., Sung, W.: Off-line and on-line algorithms for closed string factorization. Theor. Comput. Sci. **792**, 12–19 (2019)
3. Amir, A., Farach, M., Idury, R.M., Poutré, J.A.L., Schäffer, A.A.: Improved dynamic dictionary matching. Inf. Comput. **119**(2), 258–282 (1995)
4. Amir, A., Landau, G.M., Ukkonen, E.: Online timestamped text indexing. Inf. Process. Lett. **82**(5), 253–259 (2002). https://doi.org/10.1016/S0020-0190(01)00275-7
5. Badkobeh, G., et al.: Closed factorization. Discret. Appl. Math. **212**, 23–29 (2016)
6. Belazzougui, D., Puglisi, S.J.: Range predecessor and Lempel-Ziv parsing. In: SODA 2016, pp. 2053–2071 (2016)
7. Bell, T.C.: Better OPM/L text compression. IEEE Trans. Commun. **34**(12), 1176–1182 (1986)
8. Bille, P., Cording, P.H., Fischer, J., Gørtz, I.L.: Lempel-Ziv compression in a sliding window. In: CPM 2017. LIPIcs, vol. 78, pp. 15:1–15:11 (2017)
9. Brodal, G.S., Davoodi, P., Srinivasa Rao, S.: Path minima queries in dynamic weighted trees. In: Dehne, F., Iacono, J., Sack, J.-R. (eds.) WADS 2011. LNCS, vol. 6844, pp. 290–301. Springer, Heidelberg (2011). https://doi.org/10.1007/978-3-642-22300-6_25
10. Chan, H., Hon, W., Lam, T.W., Sadakane, K.: Compressed indexes for dynamic text collections. ACM Trans. Algorithms **3**(2), 21 (2007)
11. Ellert, J., Fischer, J., Pedersen, M.R.: New advances in rightmost Lempel-Ziv. In: Nardini, F.M., Pisanti, N., Venturini, R. (eds.) SPIRE 2023. LNCS, vol. 14240, pp. 188–202. Springer, Cham (2023). https://doi.org/10.1007/978-3-031-43980-3_15
12. Ferragina, P., Nitto, I., Venturini, R.: On the bit-complexity of Lempel-Ziv compression. SIAM J. Comput. **42**(4), 1521–1541 (2013)
13. Fici, G.: A classification of Trapezoidal words. In: WORDS 2011. EPTCS, vol. 63, pp. 129–137 (2011)
14. Fischer, J., Gawrychowski, P.: Alphabet-dependent string searching with wexponential search trees. arXiv preprint arXiv:1302.3347 (2013)
15. Fischer, J., Gawrychowski, P.: Alphabet-dependent string searching with wexponential search trees. In: CPM 2015, pp. 160–171 (2015)
16. Fredman, M.L., Saks, M.E.: The cell probe complexity of dynamic data structures. In: STOC 1989, pp. 345–354. ACM (1989)
17. Güting, R.H., Wood, D.: The parenthesis tree. Inf. Sci. **27**(2), 151–162 (1982)
18. Henzinger, M.R., King, V.: Randomized fully dynamic graph algorithms with polylogarithmic time per operation. J. ACM **46**(4), 502–516 (1999)
19. Kolpakov, R.M., Kucherov, G.: Finding maximal repetitions in a word in linear time. In: FOCS 1999, pp. 596–604 (1999)
20. Kreft, S., Navarro, G.: LZ77-like compression with fast random access. In: DCC 2010, pp. 239–248 (2010)
21. Larsson, N.J.: Extended application of suffix trees to data compression. In: DCC 1996, pp. 190–199 (1996)
22. Larsson, N.J.: Most recent match queries in on-line suffix trees. In: Kulikov, A.S., Kuznetsov, S.O., Pevzner, P. (eds.) CPM 2014. LNCS, vol. 8486, pp. 252–261. Springer, Cham (2014). https://doi.org/10.1007/978-3-319-07566-2_26

23. Lempel, A., Ziv, J.: On the complexity of finite sequences. IEEE Trans. Inf. Theory **22**(1), 75–81 (1976)
24. Leonard, L., Inenaga, S., Bannai, H., Mieno, T.: Constant-time edge label and leaf pointer maintenance on sliding suffix trees (2024)
25. Mäkinen, V., Belazzougui, D., Cunial, F., Tomescu, A.I.: Genome-Scale Algorithm Design: Bioinformatics in the Era of High-Throughput Sequencing, 2nd edn. Cambridge University Press (2023). http://www.genome-scale.info/
26. Navarro, G., Sadakane, K.: Fully functional static and dynamic succinct trees. ACM Trans. Algorithms **10**(3), 16:1–16:39 (2014)
27. Okanohara, D., Sadakane, K.: An online algorithm for finding the longest previous factors. In: Halperin, D., Mehlhorn, K. (eds.) ESA 2008. LNCS, vol. 5193, pp. 696–707. Springer, Heidelberg (2008). https://doi.org/10.1007/978-3-540-87744-8_58
28. Prezza, N., Rosone, G.: Faster online computation of the succinct longest previous factor array. In: Anselmo, M., Della Vedova, G., Manea, F., Pauly, A. (eds.) CiE 2020. LNCS, vol. 12098, pp. 339–352. Springer, Cham (2020). https://doi.org/10.1007/978-3-030-51466-2_31
29. Senft, M.: Suffix tree for a sliding window: an overview. In: WDS 2005, vol. 5, pp. 41–46 (2005)
30. Storer, J.A., Szymanski, T.G.: Data compression via textual substitution. J. ACM **29**(4), 928–951 (1982)
31. Ukkonen, E.: On-line construction of suffix trees. Algorithmica **14**(3), 249–260 (1995)
32. Weiner, P.: Linear pattern matching algorithms. In: 14th Annual Symposium on Switching and Automata Theory, pp. 1–11 (1973)
33. Willard, D.E.: Examining computational geometry, van Emde Boas trees, and hashing from the perspective of the fusion tree. SIAM J. Comput. **29**(3), 1030–1049 (2000). https://doi.org/10.1137/S0097539797322425
34. Ziv, J., Lempel, A.: A universal algorithm for sequential data compression. IEEE Trans. Inf. Theory **23**(3), 337–343 (1977)

Space-Efficient SLP Encoding for $O(\log N)$-Time Random Access

Akito Takasaka and Tomohiro I$^{(\boxtimes)}$

Kyushu Institute of Technology, 680–4 Kawazu, Iizuka, Fukuoka 820-8502, Japan
takasaka.akito977@mail.kyutech.jp tomohiro@ai.kyutech.ac.jp

Abstract. A Straight-Line Program (SLP) $\mathcal{G}$ for a string $\mathcal{T}$ is a context-free grammar (CFG) that derives $\mathcal{T}$ only, which can be considered as a compressed representation of $\mathcal{T}$. In this paper, we show how to encode $\mathcal{G}$ in $n\lceil \lg N \rceil + (n+n')\lceil \lg(n+\sigma) \rceil + 4n - 2n' + o(n)$ bits to support random access queries of extracting $\mathcal{T}[p..q]$ in worst-case $O(\log N + q - p)$ time, where N is the length of $\mathcal{T}$, σ is the alphabet size, n is the number of variables in $\mathcal{G}$ and $n' \leq n$ is the number of symmetric centroid paths in the DAG representation for $\mathcal{G}$.

Keywords: Data compression · Grammar compression · Random access data structures

1 Introduction

A Straight-Line Program (SLP) $\mathcal{G}$ for a string $\mathcal{T}$ is a context-free grammar (CFG) that derives $\mathcal{T}$ only. The idea of grammar compression is to take $\mathcal{G}$ as a compressed representation of $\mathcal{T}$, which is a useful scheme to capture repetitive substrings in $\mathcal{T}$. In fact the output of many practical dictionary-based compressors like RePair [6] and LZ77 [7,17] can be considered as or efficiently converted to an SLP. Moreover SLPs have gained popularity for designing algorithms and data structures to work directly on compressed data. For more details see survey [8] and references therein.

One of the most fundamental tasks on compressed string is to support random access without explicitly decompressing the whole string. Let $\mathcal{T}$ be a string of length N over an alphabet Σ of size σ and $\mathcal{G}$ be an SLP that derives $\mathcal{T}$ with V being the set of variables. For simplicity we assume that SLPs are in the normal form such that every production rule is of the form $X \to YZ \in (V \cup \Sigma)^2$. If we store the length of the string derived from every variable in $n \lg N$ bits in addition to the information of production rules, it is not difficult to see that we can access $\mathcal{T}[p]$ for any position $1 \leq p \leq N$ in $O(h)$ time, where h is the height of the derivation tree of $\mathcal{G}$: We can simulate the traversal from the root to the p-th leaf of the derivation tree of $\mathcal{G}$ while deciding if the current node contains the target leaf in its left child or not. This simple random access algorithm is good enough if $\mathcal{G}$ is well balanced, i.e., $h = O(\log N)$, but h could be as large as n in the worst case.

To solve this problem Bille et al. [2] showed that there is a data structure of $O(n \log N)$ bits of space that can retrieve any substring of length ℓ of $\mathcal{T}$ in $O(\log N + \ell)$ time. Belazzougui et al. [1] showed that the query time can be improved to $O(\log N + \ell/\log_\sigma N)$ by adding some other data structures of $O(n \log N)$ bits to accelerate accessing $O(\log_\sigma N)$ consecutive characters. Verbin and Yu [15] studied lower bounds of random access data structures on grammar compressed strings and showed that the $O(\log N)$ term in the time complexities cannot be significantly improved in general with $\text{poly}(n)$-space data structures.

Another approach is to transform a given grammar into a balanced grammar of height $O(\log N)$ and apply the above mentioned simple random access algorithm. Ganardi et al. [5] showed that any SLP can be transformed in linear time into a balanced grammar without increasing its order in size. The result was refined in [4] for contracting SLPs, which have a stronger balancing argument. These results are helpful not only for random access but also for other operations that can be done depending on the height of the derivation tree. However the constant-factor blow-up in grammar size can be a problem in space-sensitive applications.

In order to keep space usage small in practice, it is important to devise a space-economic way to encode $\mathcal{G}$ and auxiliary data structures as such importance has been highlighted by increasing interest of succinct data structures. Since the righthand side of each production rule can be stored in $2\lceil \lg(n+\sigma) \rceil$ bits, a naive encoding for $\mathcal{G}$ would use $2n\lceil \lg(n+\sigma) \rceil + n\lceil \lg N \rceil$ bits of space, which is far from an information-theoretic lower bound $\lg(n!) + 2n + o(n)$ bits for representing $\mathcal{G}$ [13]. Maruyama et al. proposed an asymptotically optimal encoding of $n\lceil \lg(n+\sigma) \rceil + 2n + o(n)$ bits [9] for $\mathcal{G}$, which can be augmented with additional $n\lceil \lg \frac{N}{n} \rceil + 2n + o(n)$ bits to support $O(h+\ell)$-time random access. Another practical encoding for random access was studied in [3] but its worst-case query time is still $O(h+\ell)$.

In this study, we propose a novel space-efficient SLP encoding that supports random access queries in worst-case $O(\log N + \ell)$ time. In so doing we simplify some ideas of [2] and adjust them to work with succinct data structures. For example, we replace the heavy-path decomposition of a Directed Acyclic Graph (DAG) with the symmetric centroid decomposition proposed in [5]. We decompose the DAG representation of the derivation tree of $\mathcal{G}$ into disjoint Symmetric Centroid paths (SC-paths) so that every path from the root to a leaf passes through $O(\log N)$ distinct SC-paths. We augment each SC-path with compacted binary tries to support interval-biased search, which leads to $O(\log N)$-time random access. Under a standard Word-RAM model with word size $\Omega(\lg N)$ we get the following result:

Theorem 1. *Let $\mathcal{T}$ be a string of length N over an alphabet of σ. An SLP $\mathcal{G}$ for $\mathcal{T}$ can be encoded in $n\lceil \lg N \rceil + (n+n')\lceil \lg(n+\sigma) \rceil + 4n - 2n' + o(n)$ bits of space while allowing to retrieve, given $1 \leq p \leq q \leq N$, the substring $\mathcal{T}[p..q]$ in $O(\log N + q - p)$ time, where n is the number of variables of $\mathcal{G}$ and $n' \leq n$ is the number of SC-paths in the DAG representation for $\mathcal{G}$.*

2 Preliminaries

2.1 Basic Notations

For two integers i and j with $i \le j$, let $[i..j]$ represents the integer interval from i to j, i.e. $[i..j] := \{i, i+1, \ldots, j-1, j\}$. If $i > j$, then $[i..j]$ denotes the empty interval. Also let $[i..j) := [i..j-1]$ and $(i..j] := [i+1..j]$. We use lg to denote the binary logarithm, i.e., the logarithm to the base 2.

Let Σ be a finite *alphabet*. An element of Σ^* is called a *string* over Σ. The length of a string w is denoted by $|w|$. The empty string ε is the string of length 0, that is, $|\varepsilon| = 0$. Let $\Sigma^+ = \Sigma^* - \{\varepsilon\}$. The concatenation of two strings x and y is denoted by $x \cdot y$ or simply xy. When a string w is represented by the concatenation of strings x, y and z (i.e. $w = xyz$), then x, y and z are called a *prefix*, *substring*, and *suffix* of w, respectively. A substring x of w is called *proper* if $|x| < |w|$.

The i-th character of a string w is denoted by $w[i]$ for $1 \le i \le |w|$, and the substring of a string w that begins at position i and ends at position j is denoted by $w[i..j]$ for $1 \le i \le j \le |w|$, i.e., $w[i..j] = w[i]w[i+1]\cdots w[j]$. For convenience, let $w[i..j] = \varepsilon$ if $j < i$.

2.2 Straight-Line Programs (SLPs)

Let $\mathcal{T}$ be a string of length N over Σ. A *Straight-Line Program (SLP)* $\mathcal{G}$ for $\mathcal{T}$ is a context-free grammar that derives $\mathcal{T}$ only. Let V be the *variables* (non-terminals) of $\mathcal{G}$. We use a term *symbol* to refer to an element in $(V \cup \Sigma)$. We obtain $\mathcal{T}$ by recursively replacing the starting variable of $\mathcal{G}$ according to the production rules of $\mathcal{G}$ until every variable is turned into a sequence of *characters* (terminals). To derive $\mathcal{T}$ uniquely, the derivation process of $\mathcal{G}$ should be deterministic and end without loop. In particular, for each variable X there is exactly one production rule that has X in its lefthand side, which we call the production rule of X. We denote by $\mathsf{R}(X)$ the righthand side of the production rule of X. For simplicity we assume that SLPs are in the normal form such that $\mathsf{R}(X) \in (V \cup \Sigma)^2$. For any symbol x, let $\langle x \rangle$ denote the string derived from x, i.e., $\langle x \rangle = x$ if $x \in \Sigma$, and otherwise $\langle x \rangle = \langle \mathsf{R}(x)[1] \rangle \langle \mathsf{R}(x)[2] \rangle$. We extend this notation so that $\langle w \rangle := \langle w[1] \rangle \langle w[2] \rangle \cdots \langle w[|w|] \rangle$ for a string w over $(V \cup \Sigma)^*$. For a symbol x, the *derivation tree* of x is the rooted tree that represents the derivation process from x to $\langle x \rangle$. The derivation tree of the starting symbol is called the derivation tree of $\mathcal{G}$. Note that the derivation tree of $\mathcal{G}$ can be represented by a Directed Acyclic Graph (DAG) with $n + \sigma$ nodes and $2n$ edges, which we call the DAG of $\mathcal{G}$ and denote by $\mathcal{D}_\mathcal{G}$. Since there is a natural one-to-one correspondence from symbols to nodes of $\mathcal{D}_\mathcal{G}$, we will sometimes use them interchangeably.

We set some assumptions under which we study the space needed to store $\mathcal{G}$. We assume that symbols are identified by integers with $V = [1..n]$ and $\Sigma = [n+1..n+\sigma]$ so that each of them is represented in $\lceil \lg(n+\sigma) \rceil$ bits, where $n := |V|$ and $\sigma := |\Sigma|$. For a terminal symbol associated with an integer $i \in [n+1..n+\sigma]$, its original code on computer is assumed to be obtained easily, e.g., by computing

$i - n$ or storing a mapping table whose space usage is excluded from our space complexity.

2.3 Succinct Data Structures

For a bit string $B \in \{0,1\}^*$, we consider the following queries:

- For any $b \in \{0,1\}$ and $i \in [1..|B|]$, $\mathsf{rank}_b(B,i)$ returns the number of occurrences of b in $B[1..i]$. For convenience, we let $\mathsf{rank}_b(B,i)$ return 0 if $i < 1$, and $|B|$ if $i > |B|$.
- For any $b \in \{0,1\}$ and $j \in [1..\mathsf{rank}_b(B,|B|)]$, $\mathsf{select}_b(B,j)$ returns the position i such that $\mathsf{rank}_b(B,i) = j$ and $B[i] = b$. For convenience, we let $\mathsf{select}_b(B,j)$ return 0 if $j \leq 0$, and $|B|+1$ if $j > \mathsf{rank}_b(B,|B|)$.

We use the following succinct data structure on bit strings:

Lemma 1 ([12]). *For a bit string $B \in \{0,1\}^n$, there is a data structure of $n + o(n)$ bits that supports* rank *and* select *queries in $O(1)$ time.*

We also consider succinct data structures for rooted ordered full binary trees in which all internal nodes have exactly two children of left and right. Notice that a full binary tree with m nodes can be represented in m bits instead of $2m$ bits needed in the case of arbitrary rooted ordered trees, and there is a data structure to support various operations in $m + o(m)$ bits. Although there could be several ways to encode the topology of the tree, we employ the post-order encoding B in which every node is identified by its post-order rank. In this paper we use the queries listed in the following lemma, which is not new as it has been used in the literature (e.g., [9, 14]). We give its proof for the sake of completeness.

Lemma 2. *For a full binary tree with m nodes, there is a data structure of $m + o(m)$ bits to support the following queries in $O(1)$ time for a node v, where every node involved in the queries is identified by its post-order rank.*

- isleaf(v) *returns if v is a leaf node.*
- lchild(v) *returns the left child of v.*
- rchild(v) *returns the right child of v.*
- rmleaf(v) *returns the rightmost leaf in the subtree rooted at v.*
- leafrank(v) *returns the number of leaves up to and including a leaf v.*

Proof. We store a bit string $B[1..m]$ of length m such that the i-th bit is 0 if and only if the node with post-order rank i is a leaf node. We store and augment the post-order encoding B with rank/select data structures of Lemma 1 using extra $o(m)$ bits. We also conceptually define the so-called excess array $E[1..m]$ such that $E[i] := \mathsf{rank}_0(B,i) - \mathsf{rank}_1(B,i)$ for any $1 \leq i \leq m$. Since it holds that $E[i] > 0$ and $|E[i] - E[i-1]| = 1$, we can use the succinct data structure for a balanced parentheses sequence [11]. In particular, we augment B to compute in constant time bwdsearch(i,d) that returns the maximum position j with $j \leq i$ and $E[j] = E[i] + d$.

Now we show how to respond to the queries listed in our lemma. For isleaf(v), we return true if $B[v] = 0$ and false otherwise. Since the right child of v immediately precedes v in post order, we just return $v - 1$ for rchild(v). The rightmost leaf in the subtree rooted at v is at the maximum position j such that $j \le v$ and $B[j] = 0$, which can be computed by $\mathsf{select}_0(B, \mathsf{rank}_0(B, v))$. For lchild($v$), we look for the rightmost node to the left from v in B that is not in the subtree rooted at the right child of v. This can be computed by $\mathsf{bwdsearch}(v - 1, -1)$ as the subtree rooted at the right child of v is also a full binary tree in which the leaves exceed the internal nodes by exactly one in number. For a leaf v, we can compute leafrank(v) by $\mathsf{rank}_0(B, v)$. □

2.4 Symmetric Centroid Decomposition

Here we briefly review the symmetric centroid decomposition proposed in [5]. Let us work on the DAG of $\mathcal{G}$ for a string of length N. For every node v we consider the pair $(\lfloor \lg \dot{v} \rfloor, \lfloor \lg \ddot{v} \rfloor)$ of values, where $\dot{v}$ and $\ddot{v}$ are the numbers of paths from the root to v and from v to the leaves, respectively. Note that $\dot{v}$ and $\ddot{v}$ are both upper bounded by N. An edge from u to v is called an *SC-edge* if and only if $(\lfloor \lg \dot{u} \rfloor, \lfloor \lg \ddot{u} \rfloor) = (\lfloor \lg \dot{v} \rfloor, \lfloor \lg \ddot{v} \rfloor)$. By definition every node has at most one outgoing SC-edge and at most one incoming SC-edge (see Lemma 2.1 of [5]). Hence a maximal subgraph connected by SC-edges forms a path, which we call an *SC-path*. Note that SC-paths include an empty path, which consists of a single node. Lemma 2.1 of [5] also states that every path from the root to a leaf contains at most $2 \lg N$ non-SC-edges.

3 New Encoding of SLPs

In this section we prove Theorem 1. In what follows, $\mathcal{T}, \sigma, N, \mathcal{G}, n$ and n' are used as defined in the theorem.

3.1 Strategy to Achieve $O(\log N)$-Time Random Access

In this subsection we give a high-level strategy to achieve $O(\log N)$-time random access.

Given a position p on $\mathcal{T}$, we simulate on $\mathcal{D}_{\mathcal{G}}$ the traversal from the root to the p-th leaf of the derivation tree of $\mathcal{G}$. As described in Sub Sect. 2.4, the traversal contains at most $2 \lg N$ non-SC-edges, say $(x_1^{\mathsf{out}}, x_2^{\mathsf{in}}), (x_2^{\mathsf{out}}, x_3^{\mathsf{in}}), \ldots, (x_e^{\mathsf{out}}, x_{e+1}^{\mathsf{in}})$ with $e \le 2 \lg N$, where x_i^{in} and x_i^{out} are variables on the same SC-path for all $1 \le i \le e$ (for convenience let x_1^{in} be the root node of $\mathcal{D}_{\mathcal{G}}$, i.e., the starting variable) and $x_{e+1}^{\mathsf{in}} = \mathcal{T}[p]$. See Fig. 1 for an illustration. If we can move from x_i^{in} to x_{i+1}^{in} in a way of length-weighted biased search in $O(1 + \log |\langle x_i^{\mathsf{in}} \rangle| - \log |\langle x_{i+1}^{\mathsf{in}} \rangle|)$ time, $O(\log N)$-time random access can be achieved because $O(\sum_{i=1}^{e} (1 + \log |\langle x_i^{\mathsf{in}} \rangle| - \log |\langle x_{i+1}^{\mathsf{in}} \rangle|)) = O(\log N + \log |\langle x_1^{\mathsf{in}} \rangle| - \log |\langle x_{e+1}^{\mathsf{in}} \rangle|) = O(\log N)$.

For an SC-path $(u_1, u_2, \ldots, u_m)$ with m nodes, there are $m + 1$ non-SC-edges branching out from the SC-path, which splits $\langle u_1 \rangle$ into $m + 1$ subsrings. A

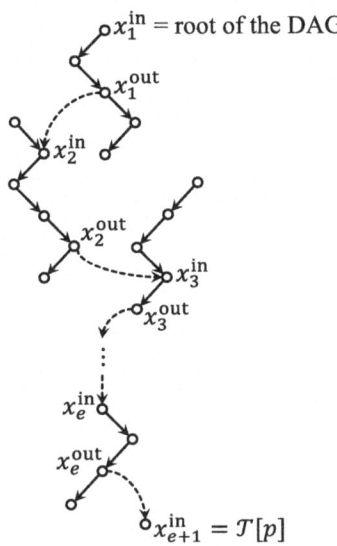

Fig. 1. Illustration for the high-level strategy to achieve $O(\log N)$-time random access. The path from the root to the target leaf $x_{e+1}^{\text{in}} = T[p]$ contains e ($\leq 2\lg N$) non-SC-edges $(x_1^{\text{out}}, x_2^{\text{in}})$, $(x_2^{\text{out}}, x_3^{\text{in}})$, ... and $(x_e^{\text{out}}, x_{e+1}^{\text{in}})$ depicted by dashed arrows. The components connected by plain arrows are SC-paths. Our sub-goal is to move from x_i^{in} to x_{i+1}^{in} efficiently in $O(1 + \log|\langle x_i^{\text{in}}\rangle| - \log|\langle x_{i+1}^{\text{in}}\rangle|)$ time.

subproblem in question is to find the non-SC-edge (u_j, v) such that v contains the target position in time $O(1 + \log|\langle u_1\rangle| - \log|\langle v\rangle|)$. In Sub sect. 3.2 we take this subproblem as a general interval search problem and show how to solve it.

3.2 Compacted Binary Tries for Interval-Biased Search

In this subsection we consider the following problem.

Problem 1 (Interval Search Problem). Preprocess a sequence $g_1, g_2, \ldots, g_m$ of integers such that $g_0 = 0 < g_1 < g_2 < \cdots < g_m$, to support interval search queries that ask, given an integer $0 < p \leq g_m$, to compute k with $p \in (g_k..g_{k+1}]$.

Data structures for this problem have been extensively studied in the context of the predecessor search problem (e.g. y-fast trie [16]).

Bille et al. [2] proposed the interval-biased search tree to answer the interval search query in $O(1 + \log g_m - \log(g_{i+1} - g_i))$ time. The interval-biased search tree for m intervals $(g_0..g_1], (g_1..g_2], \ldots, (g_{m-1}..g_m]$ is a binary tree with m nodes defined as follows: The root is set to be the interval $(g_{m'}..g_{m'+1}]$ that contains the position $(g_m - g_0)/2$ and its left (resp. right) child subtree is defined recursively for $(g_0..g_1], \ldots, (g_{m'-1}..g_{m'}]$ (resp. $(g_{m'+1}..g_{m'+2}], \ldots, (g_{m-1}..g_m]$) if $m' > 0$ (resp. $m' < m-1$).

Unfortunately it does not seem straightforward to work on the succinct tree representation of the interval-biased search tree: Although one can observe that the node with in-order rank $i+1$ (defined in a reasonable way for trees in which each edge is categorized into left or right) corresponds to interval $(g_i, g_{i+1}]$ in the interval-biased search tree, the computation is not supported by the succinct tree representation of [11].[1]

In this paper we instead show that the compacted binary trie (also known as compressed binary trie or Patricia trie [10]) for the binary representations of integers $g_1, g_2, \ldots, g_m$ can be used for interval-biased search.

Lemma 3. *For Problem 1, there is a data structure of $m\lceil \lg g_m \rceil + 2m + o(m)$ bits to support interval search queries in $O(1 + \log g_m - \log(g_{k+1} - g_k))$ time, where k is the answer of the query.*

Proof. We store $g_1, g_2, \ldots, g_m$ naively in an array $A[1..m] \in [1..g_m]^m$ using $m\lceil \lg g_m \rceil$ bits of space. For any $1 \leq i \leq m$, let b_i denote the ($\lceil \lg g_m \rceil$)-bit binary representation of g_i, and c_i be the bit string that is obtained from b_i by removing the most significant bits common to both b_1 and b_m. The compacted binary trie for $c_1, c_2, \ldots, c_m$ forms a full binary tree with m leaves and $m - 1$ internal nodes and its tree topology can be represented in $2m + o(m)$ bits by the data structure of Lemma 2. Note that the i-th leaf v_i corresponds to c_i (and also $g_i = A[i]$), and the lowest common ancestor u_i of v_i and v_{i+1} corresponds to the interval $(g_i..g_{i+1}]$. The crucial point for our interval-biased search is that the depth of u_i in the tree is $O(\log g_m - \log(g_{i+1} - g_i))$ because the height of u_i in the "uncompacted" binary trie is at least $\lg(g_{i+1} - g_i)$ and the distance from the root to u_i is upper bounded by $\lg g_m - \lg(g_{i+1} - g_i)$. Hence, we can answer interval search queries in $O(1 + \log g_m - \log(g_{i+1} - g_i))$ time by a simple traversal from the root: When we arrive at an internal node u (given by its post-order rank), we can compute in $O(1)$ time its corresponding interval $(A[i]..A[i+1]]$ using $i = \mathsf{leafrank}(\mathsf{rmleaf}(\mathsf{lchild}(u)))$. We return i if $p \in (A[i]..A[i+1]]$, and otherwise, we move to the left child $\mathsf{lchild}(u)$ (resp. right child $\mathsf{rchild}(u)$) if $p \leq A[i]$ (resp. $A[i+1] < p$). A final remark is that we return 0 if $p \leq g_1 = A[1]$, i.e., $p \in (g_0..g_1]$, which should be checked prior to the binary search. □

3.3 Details of New Encoding

First of all, we arrange the variables of $\mathcal{G}$ so that $v = u + 1$ holds for every SC-edge (u, v) on $\mathcal{D}_\mathcal{G}$. As a result we can now assume that the variables on the same SC-path are given consecutive integers while distinct SC-paths are ordered arbitrarily. Let n' be the number of SC-paths of $\mathcal{D}_\mathcal{G}$ excluding the ones that consist of the terminal symbols.

Our encoding stores a bit string $\mathsf{P}[1..n]$ to separate $[1..n]$ into n' segments of SC-paths, i.e., for any variable u ($1 \leq u \leq n$), $\mathsf{P}[u] = 1$ if and only if u is the last node of an SC-path. We remark that P contains n' ones and $n - n'$ zeros.

Next we show how to encode the information of R. For any u with $\mathsf{P}[u] = 0$, one of its children is $u + 1$, and hence, we can store R in $(n + n')\lceil \lg(n + \sigma) \rceil + n -$

---
[1] Note that $\mathsf{in\_rank}$ queries in [11] work only for the nodes that have at least two children.

n' bits in addition to n bits of P as follows. Let $\mathsf{D}[1..n-n']$ be a bit string of length $n - n'$ that indicates the direction of the non-SC-edge (u, v) branching out from every node u with $\mathsf{P}[u] = 0$, and $\mathsf{R}_1[1..n-n'] \in [1..n+\sigma]^{n-n'}$ be an integer array of length $n - n'$ that stores v's. More precisely, for any variable u $(1 \leq u \leq n)$ with $\mathsf{P}[u] = 0$, we set $\mathsf{D}[\mathsf{rank}_0(\mathsf{P}, u)]$ to be 0 if and only if the non-SC-edge (u, v) branching out from u is the left child, and store v in $\mathsf{R}_1[\mathsf{rank}_0(\mathsf{P}, u)]$. For the nodes u with $\mathsf{P}[u] = 1$, we create another integer array $\mathsf{R}_2[1..2n'] \in [1..n+\sigma]^{2n'}$ such that $\mathsf{R}_2[2 \cdot \mathsf{rank}_1(\mathsf{P}, u) - 1..2 \cdot \mathsf{rank}_1(\mathsf{P}, u)]$ store $R(u)$.

Using P, D, R_1 and R_2, we can compute in $O(1)$ time the children $R(u)$ of a given variable u if we augment P with the data structure of Lemma 1 for rank/select queries. Also, $r = \mathsf{rank}_1(\mathsf{P}, u - 1) + 1$ tells us that u is on the r-th SC-path on P and the interval for the SC-path u belongs can be computed in $O(1)$ time by $[\mathsf{select}_1(\mathsf{P}, r - 1) + 1..\mathsf{select}_1(\mathsf{P}, r)]$.

Finally we add $(n\lceil \lg N \rceil + 2n - n' + o(n))$-bits data structure to achieve $O(\log N)$-time random access. From now on we focus on the r-th SC-path. Let $m := \mathsf{select}_1(\mathsf{P}, r) - \mathsf{select}_1(\mathsf{P}, r - 1)$ and $u_i := \mathsf{select}_1(\mathsf{P}, r - 1) + i$ for any $0 \leq i \leq m$, then $(u_1, u_2, \ldots, u_m)$ is the sequence of nodes on the r-th SC-path. Let t be the number of non-SC-edges (u_i, v) such that v is the left child of u_i for some $1 \leq i < m$, which can be computed by $\mathsf{rank}_0(\mathsf{D}, d_0 + m - 1) - \mathsf{rank}_0(\mathsf{D}, d_0)$ with $d_0 := \mathsf{rank}_0(\mathsf{P}, u_0)$. Let $v_1, v_2, \ldots, v_{m+1}$ be the endpoints of the non-SC-edges branching out from the SC-path sorted by the preorder of left-to-right traversal from u_1 so that $\langle u_1 \rangle = \langle v_1 \rangle \langle v_2 \rangle \cdots \langle v_{m+1} \rangle$. For any $1 \leq i \leq m+1$ it holds that

$$v_i = \begin{cases} \mathsf{R}_1[\mathsf{select}_0(\mathsf{D}, \mathsf{rank}_0(\mathsf{D}, d_0) + i)] & \text{if } i \leq t, \\ \mathsf{R}_2[2 \cdot \mathsf{rank}_1(\mathsf{P}, u_0) + i - t] & \text{if } t < i \leq t+2, \\ \mathsf{R}_1[\mathsf{select}_1(\mathsf{D}, \mathsf{rank}_1(\mathsf{D}, d_0) + m + 2 - i)] & \text{if } t+2 < i. \end{cases} \quad (1)$$

Hence, if we augment D with the data structure of Lemma 1 for rank/select queries, we can compute v_i in $O(1)$ time.

In order to support interval-biased search on the SC-path, we can use the data structure of Lemma 3 for the sequence $(\sum_{i'=1}^{i} |\langle v_{i'} \rangle|)_{i=1}^{m+1}$ of prefix sums of $|\langle v_i \rangle|$'s. However, if we store $m + 1$ integers for the SC-path of m nodes, the number adds up to $n + n'$ in total for all SC-paths. An easy way to reduce this number to n is to exclude the largest prefix sum $\sum_{i'=1}^{m+1} |\langle v_{i'} \rangle| = |\langle u_1 \rangle|$ for every SC-path. $|\langle u_1 \rangle|$ is not needed for efficient random access queries as we can immediately proceed to v_{m+1} if a target position is greater than $\sum_{i'=1}^{m} |\langle v_{i'} \rangle|$. One drawback of not storing $|\langle u_1 \rangle|$ explicitly is that the data structure cannot answer $|\langle u_i \rangle|$ efficiently for a given variable u_i if the path from u_1 to u_i consists only of right edges. Since such queries might be useful in some scenarios, we show an alternative way in the following.

We decompose $\langle u_1 \rangle$ into m strings $\langle v_1 \rangle, \ldots, \langle v_k \rangle, \langle u_m \rangle, \langle v_{k+3} \rangle, \ldots, \langle v_{m+1} \rangle$ and construct the data structure of Lemma 3 for the sequence $g_1, g_2, \ldots, g_m$ of prefix sums of these m string lengths. Now we can compute $|\langle u_i \rangle|$ in $O(1)$ time by $g_{i'} - g_{i''}$, where $i' = m - (\mathsf{rank}_1(\mathsf{D}, d_0 + i - 1) - \mathsf{rank}_1(\mathsf{D}, d_0))$ and $i'' = \mathsf{rank}_0(\mathsf{D}, d_0 + i - 1) - \mathsf{rank}_0(\mathsf{D}, d_0)$. During a search on the SC-path, if the target position falls in $\langle u_m \rangle = \langle v_{t+1} \rangle \langle v_{t+2} \rangle$, we can decide which of $\langle v_{t+1} \rangle$ and $\langle v_{t+2} \rangle$ contains the target position using $|\langle v_{t+1} \rangle|$. We can also compute in $O(1)$ time the prefix sum of $|\langle v_{i'} \rangle|$'s up to i ($1 \leq i \leq m+1$) as

$$\sum_{i'=1}^{i} |\langle v_{i'} \rangle| = \begin{cases} g_i & \text{if } i \leq t, \\ g_t + |\langle v_{t+1} \rangle| & \text{if } i = t+1, \\ g_{i-1} & \text{if } t+1 < i. \end{cases} \qquad (2)$$

We use a global integer array $\mathsf{G}[1..n] \in [1..N]^n$ to store $g_1, g_2, \ldots, g_m$, in $\mathsf{G}[\mathsf{select}_1(\mathsf{P}, r-1)+1..\mathsf{select}_1(\mathsf{P}, r)]$. We also consider concatenating all post-order encodings for compacted binary tries to work on a single bit string: We store a bit string $\mathsf{B} := B_1 B_2 \cdots B_{n'}$ of length $2n - n'$, where B_r is the post-order encoding of the compacted binary trie for the r-th SC-path on P, which consists of m zeros (leaves) and $m-1$ ones (internal nodes). Since every prefix of B has larger zeros than ones, B can be augmented by the succinct data structure of [11] to support queries of Lemma 2 for each compacted binary trie. In short, interval-biased search on the r-th SC-path can be performed using the information stored in $B_r = \mathsf{B}[2 \cdot \mathsf{select}_1(\mathsf{P}, r-1) + 2 - r..2 \cdot \mathsf{select}_1(\mathsf{P}, r) + 1 - r] = \mathsf{B}[2u_1 - r..2u_m + 1 - r]$ and $\mathsf{G}[\mathsf{select}_1(\mathsf{P}, r-1) + 1..\mathsf{select}_1(\mathsf{P}, r)] = \mathsf{G}[u_1..u_m]$.

We are now ready to prove Theorem 1:

Proof (Theorem 1). Our encoding consists of P, D, R_1, R_2, G and B (see Fig. 2 for an illustration), and succinct data structures built on P, D and B, which clearly fits in $n\lceil \lg N \rceil + (n + n')\lceil \lg(n + \sigma) \rceil + 4n - 2n' + o(n)$ bits in total.

We follow the strategy described in Sub Sect. 3.1 and use the variables introduced there in the following explanation. Suppose that we now arrive at x_i^{in} for some $1 \leq i < e$ and we know that the relative target position in $\langle x_i^{\mathsf{in}} \rangle$ is p_i. Our task is to find the non-SC-edge $(x_i^{\mathsf{out}}, x_{i+1}^{\mathsf{in}})$ such that $\langle x_{i+1}^{\mathsf{in}} \rangle$ contains $\langle x_i^{\mathsf{in}} \rangle[p_i]$ by interval-biased search in $O(1 + \log|\langle x_i^{\mathsf{in}} \rangle| - \log|\langle x_{i+1}^{\mathsf{in}} \rangle|)$ time on the SC-path $(u_1, u_2, \ldots, u_m) = (\mathsf{select}_1(\mathsf{P}, r-1)+1, \mathsf{select}_1(\mathsf{P}, r-1)+2, \ldots, \mathsf{select}_1(\mathsf{P}, r))$ with $r = \mathsf{rank}_1(\mathsf{P}, x_{i+1}^{\mathsf{in}} - 1) + 1$. Recall that $\lfloor \lg |\langle u_m \rangle| \rfloor = \lfloor \lg |\langle u_1 \rangle| \rfloor$ by the definition of the symmetric centroid decomposition. Since $O(\log|\langle x_i^{\mathsf{in}} \rangle| - \log|\langle x_{i+1}^{\mathsf{in}} \rangle|) = O(\log|\langle u_1 \rangle| - \log|\langle x_{i+1}^{\mathsf{in}} \rangle|)$, it is fine to start interval-biased search from the root of the compacted binary trie for the SC-path to the relative target position p_i' in $\langle u_1 \rangle$. We set $p_i' = p_i + \mathsf{G}[u_1 + t' - 1]$ if $t' > 0$, and otherwise $p_i' = p_i$, where t' is the number of non-SC-edge branching out to the left from the path $(u_1, u_2, \ldots, u_{j-1})$, which can be computed by $t' = \mathsf{rank}_0(\mathsf{D}, d_0 + x_i^{\mathsf{in}} - u_1) - \mathsf{rank}_0(\mathsf{D}, d_0)$ with $d_0 = \mathsf{rank}_0(\mathsf{P}, u_1 - 1)$.

Let $v_1, v_2, \ldots, v_{m+1}$ be the endpoints of the non-SC-edges branching out from the SC-path sorted by the preorder of left-to-right traversal from u_1 so that $\langle u_1 \rangle = \langle v_1 \rangle \langle v_2 \rangle \cdots \langle v_{m+1} \rangle$. Interval-biased search (and possibly some additional work if the target position falls in the child of u_m) finds the index s with $x_{i+1}^{\mathsf{in}} = v_s$ from which we can compute the value of x_{i+1}^{in} using Eq. 1. We also compute the relative target position p_{i+1} in $\langle x_{i+1}^{\mathsf{in}} \rangle$ as $p_{i+1} = p_i' - \sum_{i'=1}^{s-1} |\langle v_{i'} \rangle|$ using Eq. 2. In order to retrieve $q - p + 1$ consecutive characters from p efficiently, we push the pair of index s and z in a stack before moving to v_s, where v_z is

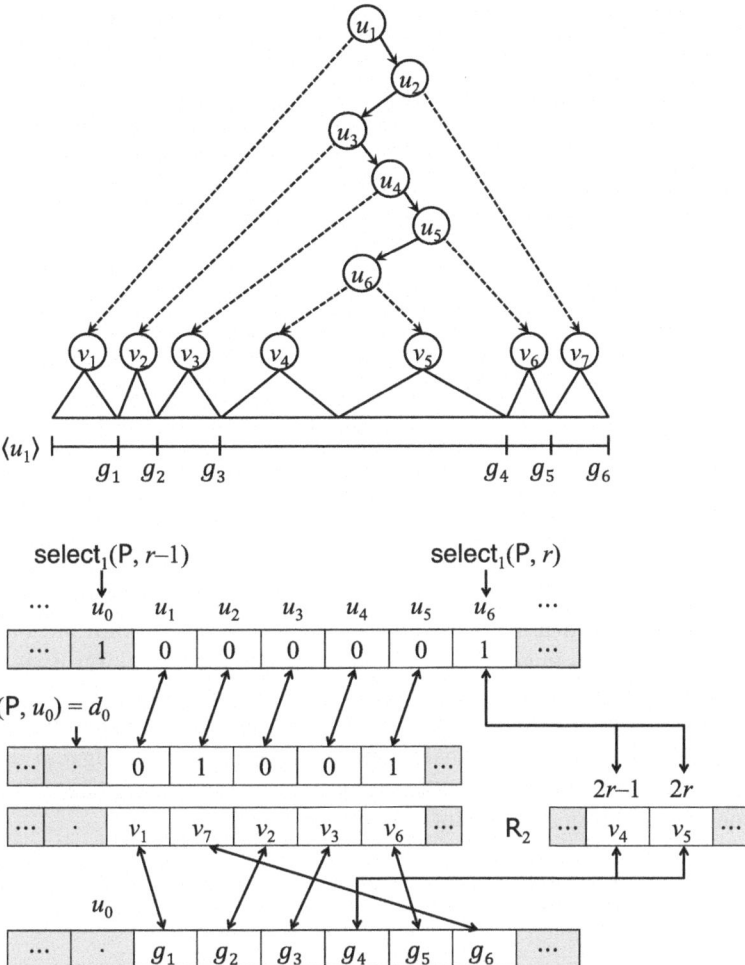

Fig. 2. Illustration for our encoding. Supposing that the r-th SC-path has 6 nodes $(u_1, u_2, u_3, u_4, u_5, u_6)$ in the form depicted above, the layout of the information for this SC-path in P, D, R_1, R_2, G and B is shown below.

the rightmost variable in the subtree rooted at x_i^{in}, which can be computed by $z = m + 1 - (\text{rank}_1(\mathsf{D}, d_0 + x_i^{\text{in}} - u_1) - \text{rank}_1(\mathsf{D}, d_0))$. If $|\langle v_s \rangle[p_{i+1}..]| < q - p + 1$, we will be back to this SC-path after computing $\langle v_s \rangle[p_{i+1}..] = \mathcal{T}[p..p + |\langle v_s \rangle| - p_{i+1}]$ below v_s: We pop s and z, and move to v_{s+1} to expand the prefix of $\langle v_{s+1} \rangle$ if $s + 1 \leq z$, and otherwise go back to the previous SC-path. This process goes on until we get $q - p + 1$ consecutive characters.

Note that, for any variable x, we can expand $\langle x \rangle$ in $O(|\langle x \rangle|)$ time because the derivation tree of x has $O(|\langle x \rangle|)$ nodes. During the computation of $\mathcal{T}[p..q]$, we meet $O(\log N)$ variables that partially contribute to $\mathcal{T}[p..q]$ (such as x_i^{in}'s) along the paths from the root to p and q based on SC-paths. Once we find these marginal paths in $O(\log N)$ time, the incompletely expanded variables can be decomposed into the sequence of fully expanded variables for which the cost of expansion can be charged to $O(q - p)$. Hence, we can compute $\mathcal{T}[p..q]$ in $O(\log N + q - p)$ time. □

Using the same random access procedure described in the proof of Theorem 1, we can extract $\langle x \rangle[p..q]$ in $O(\log N + q - p)$ time for a given variable x and positions p and q in $\langle x \rangle$. Unfortunately, the $\log N$ term cannot be reduced to $\log |\langle x \rangle|$ because there could still be $O(\log N)$ non-SC-edges on the path from x to the target leaf.

4 Conclusions and Future Work

In this paper, we presented a novel space-efficient SLP encoding that supports fast random access. P, D, R_1 and R_2 hold the information of R in $(n+n')\lceil \lg(n+\sigma) \rceil + 2n - n'$ bits, which asymptotically approaches to the information-theoretic lower bound for SLPs in general if $n' \in o(n)$. Even if n' is as large as n, the size is just n bits away from $2n \lceil \lg(n+\sigma) \rceil$ bits of the naive encoding of production rules. For the most space consuming term $n \lceil \lg N \rceil$ bits of G in our encoding, it might be effective to think about storing each value of $g_1, g_2, \ldots, g_m$ in $\lceil \lg g_m \rceil$ bits instead of $\lceil \lg N \rceil$ bits. Future work would include devising practical implementation based on the theoretical result of this paper.

Acknowledgements. This work was supported by JSPS KAKENHI (Grant Numbers 22K11907 and 24K02899) and JST AIP Acceleration Research JPMJCR24U4, Japan.

References

1. Belazzougui, D., Cording, P.H., Puglisi, S.J., Tabei, Y.: Access, rank, and select in grammar-compressed strings. In: Proceedings 23rd Annual European Symposium on Algorithms (ESA) 2015, pp. 142–154 (2015)
2. Bille, P., Landau, G.M., Raman, R., Sadakane, K., Satti, S.R., Weimann, O.: Random access to grammar-compressed strings and trees. SIAM J. Comput. **44**(3), 513–539 (2015)
3. Gagie, T., et al.: Practical random access to SLP-compressed texts. In: Proc. 27th International Symposium on String Processing and Information Retrieval (SPIRE) 2020. Lecture Notes in Computer Science, vol. 12303, pp. 221–231. Springer (2020). https://doi.org/10.1007/978-3-030-59212-7_16
4. Ganardi, M.: Compression by contracting straight-line programs. In: Mutzel, P., Pagh, R., Herman, G. (eds.) Proceedings 29th Annual European Symposium on Algorithms (ESA) 2021. LIPIcs, vol. 204, pp. 45:1–45:16. Schloss Dagstuhl - Leibniz-Zentrum für Informatik (2021). https://doi.org/10.4230/LIPIcs.ESA.2021.45, https://doi.org/10.4230/LIPIcs.ESA.2021.45

5. Ganardi, M., Jez, A., Lohrey, M.: Balancing straight-line programs. J. ACM **68**(4), 271–2740 (2021)
6. Larsson, N.J., Moffat, A.: Offline dictionary-based compression. In: Proceedings Data Compression Conference (DCC) 1999, pp. 296–305 (1999). https://doi.org/10.1109/DCC.1999.755679, https://doi.org/10.1109/DCC.1999.755679
7. Lempel, A., Ziv, J.: On the complexity of finite sequences. IEEE Trans. Inf. Theory **22**(1), 75–81 (1976). https://doi.org/10.1109/TIT.1976.1055501, https://doi.org/10.1109/TIT.1976.1055501
8. Lohrey, M.: Algorithmics on SLP-compressed strings: a survey. Groups Complex. Cryptology **4**(2), 241–299 (2012)
9. Maruyama, S., Tabei, Y., Sakamoto, H., Sadakane, K.: Fully-online grammar compression. In: Proceedings 20th International Symposium on String Processing and Information Retrieval (SPIRE) 2013, pp. 218–229 (2013). https://doi.org/10.1007/978-3-319-02432-5_25
10. Morrison, D.R.: PATRICIA - practical algorithm to retrieve information coded in alphanumeric. J. ACM **15**(4), 514–534 (1968). https://doi.org/10.1145/321479.321481
11. Navarro, G., Sadakane, K.: Fully functional static and dynamic succinct trees. ACM Trans. Algorithms **10**(3), 16 (2014). https://doi.org/10.1145/2601073
12. Raman, R., Raman, V., Satti, S.R.: Succinct indexable dictionaries with applications to encoding k-ARY trees, prefix sums and multisets. ACM Trans. Algorithms **3**(4) (2007). https://doi.org/10.1145/1290672.1290680
13. Tabei, Y., Takabatake, Y., Sakamoto, H.: A succinct grammar compression. In: Proceedings 24th Annual Symposium on Combinatorial Pattern Matching (CPM) 2013. Lecture Notes in Computer Science, vol. 7922, pp. 235–246. Springer (2013). https://doi.org/10.1007/978-3-642-38905-4_23
14. Takabatake, Y., I, T., Sakamoto, H.: A space-optimal grammar compression. In: Proceedings 25th Annual European Symposium on Algorithms (ESA) 2017. LIPIcs, vol. 87, pp. 671–6715. Schloss Dagstuhl - Leibniz-Zentrum für Informatik (2017). https://doi.org/10.4230/LIPICS.ESA.2017.67
15. Verbin, E., Yu, W.: Data structure lower bounds on random access to grammar-compressed strings. In: Proceedings 24th Annual Symposium on Combinatorial Pattern Matching (CPM) 2013, pp. 247–258 (2013)
16. Willard, D.E.: Log-logarithmic worst-case range queries are possible in space theta(n). Inf. Process. Lett. **17**(2), 81–84 (1983). https://doi.org/10.1016/0020-0190(83)90075-3
17. Ziv, J., Lempel, A.: A universal algorithm for sequential data compression. IEEE Trans. Inf. Theory **23**(3), 337–343 (1977). https://doi.org/10.1109/TIT.1977.1055714

Simple Linear-Time Repetition Factorization

Yuki Yonemoto[1] and Shunsuke Inenaga[2]

[1] Department of Information Science and Technology, Kyushu University, Fukuoka, Japan
yonemoto.yuuki.240@s.kyushu-u.ac.jp
[2] Department of Informatics, Kyushu University, Fukuoka, Japan
inenaga.shunsuke.380@m.kyushu-u.ac.jp

Abstract. A factorization $f_1, \ldots, f_m$ of a string w of length n is called a *repetition factorization* of w if f_i is a repetition, i.e., f_i is a form of $x^k x'$ where x is a non-empty string, x' is a (possibly-empty) proper prefix of x, and $k \geq 2$. Dumitran et al. [SPIRE 2015] presented an $O(n)$-time and space algorithm for computing an arbitrary repetition factorization of a given string of length n. Their algorithm heavily relies on the *Union-Find* data structure on trees proposed by Gabow and Tarjan [JCSS 1985] that works in linear time on the word RAM model, and an interval stabbing data structure of Schmidt [ISAAC 2009]. In this paper, we explore more combinatorial insights into the problem, and present a simple algorithm to compute an arbitrary repetition factorization of a given string of length n in $O(n)$ time, without relying on data structures for Union-Find and interval stabbing. Our algorithm follows the approach by Inoue et al. [ToCS 2022] that computes the smallest/largest repetition factorization in $O(n \log n)$ time.

Keywords: repetitions · runs · string factorization

1 Introduction

A factorization of a string w is a sequence $f_1, \ldots, f_k$ of non-empty strings of w such that $w = f_1 \cdots f_k$. The length k of such a factorization is called the size of the factorization. String factorizations have widely been used for data compression [18,19]. Also, factorizing a given string into combinatorial objects, including palindromes [3,5,10,12], squares [7,16], repetitions [7,13,14], Lyndon words [8], and closed words [1,2]. Among others, we focus on *repetition factorizations* of strings first proposed by Dumitran et al. [7], which has a close relation to the *runs* (a.k.a. maximal repetitions) [4,9,15], which are one of the most significant combinatorial structures in stringology.

A period of a string w of length n is a positive integer $0 < p < n$ satisfying $w[i] = w[i+p]$ for any $1 \leq i \leq |w| - p$, and we represent the smallest period of a string w as $\mathsf{p}(w)$. Also, a substring s of a string w is called a repetition in w if

$2p(s) \leq |s|$. Then, a factorization $f_1, \ldots, f_m$ of a string w is called a repetition factorization of w if each factor f_i is a repetition. Dumitran et al. [7] presented an algorithm for computing an arbitrary repetition factorization of a string w of length n in $O(n)$ time and space. Their algorithm heavily relies on the *Union-Find* data structure on trees proposed by Gabow and Tarjan [11] that works in $O(N + m)$ time in the word RAM model, where N is the size of the tree and m is the number of queries and operations. In Dumitran et al.'s algorithm [7] the tree is a list of length n and $m = O(n)$, and thus their algorithm runs in $O(n)$ time in the word RAM model. Dumitran et al. [7] also use the *interval stabbing* data structure of Schmidt [17] that is applied to the precomputed sets of runs.

In this paper, we explore more combinatorial insights into the problem of finding an arbitrary repetition factorization, and show that neither Union-Find structures nor interval stabbing structures are necessary. Our resulting algorithm is quite simple and works in $O(n)$ time and space. Our algorithm is based on the idea from Inoue et al. [13] that computes the repetition factorizations of the smallest/largest size in $O(n \log n)$ time. Their algorithm builds a *repetition graph* over an input string w, that is a weighted directed graph representing all repetitions in w. A shortest (resp. longest) source-to-sink path of the repetition graph corresponds to a smallest (resp. largest) repetition factorization of w. The bottleneck is that the number of nodes and edges of the repetition graph is $\Omega(n \log n)$ [13] in the worst case. In this paper, we propose a sparse version of the graph, called the *arbitrary repetition factorization graph* (*ARF* graph), of size $O(n)$ for any string of length n, which is a key data structure for our $O(n)$-time solution.

The rest of this paper is organized as follows: In Sect. 2, we present preliminaries of this paper. Especially, in Sect. 2.2, we present preliminaries of repetitive structures in strings, and define the problem considered in this paper. In Sect. 3, we describe our method. Especially, in Sect. 3.1, we recall previous work and present the main idea of our algorithm. Also, in Sect. 3.5, we present our algorithm and prove the following:

Theorem 1. *Given a string w of length n, we can compute an arbitrary repetition factorization of w in $O(n)$ time and $O(n)$ space without bit operations in the word RAM model.*

2 Preliminary

2.1 Strings

Let Σ be an alphabet. An element of Σ^* is called a *string*. The *length* of a string w is denoted by $|w|$. The *empty string*, denoted by ε, is a string of length 0. Let $\Sigma^+ = \Sigma^* \setminus \{\varepsilon\}$. For a string $w = xyz$ with $x, y, z \in \Sigma^*$, strings x, y, and z are called a *prefix*, *substring*, and *suffix* of string w, respectively. For a string w, let Prefix(w), Substr(w) and Suffix(w) denote the sets of prefixes, substrings and suffixes of w, respectively. The strings in Prefix(w) \ $\{w\}$ (resp. Suffix(w) \ $\{w\}$) are called *proper prefixes* (resp. *proper suffixes*) of w. The i-th character of a string w is denoted by $w[i]$ for $1 \leq i \leq |w|$, and the substring of w that begins at position i and ends at position j is denoted by $w[i..j]$ for $1 \leq i \leq j \leq |w|$. For convenience, let $w[i..j] = \varepsilon$ for $i > j$.

2.2 Repetitive Structures

A *period* of a string w is a positive integer $0 < p < n$ satisfying $w[i] = w[i+p]$ for any $1 \leq i \leq |w| - p$. Let $\mathsf{p}(w)$ denote the *smallest* period of a string w. A string w is called *primitive* if w cannot be expressed as $w = x^k$ with any string x and integer $k \geq 2$. A non-empty string w is called a *square* if $w = x^2$ for some string x. A square x^2 is called a *primitively rooted square* if x is primitive. A substring s of another string w is called a *repetition* in w if $2\mathsf{p}(s) \leq |s|$. Alternatively, a repetition s has a form $s = x^k x'$ where $k \geq 2$ is an integer and x' is a proper prefix of x. In what follows, we consider an arbitrarily fixed string w, and we represent a repetition $s = w[\mathsf{beg}..\mathsf{end}]$ in w by a tuple $(\mathsf{beg}, \mathsf{end}, \mathsf{p}(s))$.

A *run* in a string w is a repetition $s = w[\mathsf{beg}..\mathsf{end}]$ satisfying the following two conditions:

- $\mathsf{beg} = 1$ or $w[\mathsf{beg} - 1] \neq w[\mathsf{beg} + \mathsf{p}(s) - 1]$ and,
- $\mathsf{end} = |w|$ or $w[\mathsf{end} + 1] \neq w[\mathsf{end} - \mathsf{p}(s) + 1]$.

Let $\mathsf{Runs}(w)$ denote the set of the runs in a string w, and let $|\mathsf{Runs}(w)|$ denote the number of runs in $\mathsf{Runs}(w)$. Kolpakov and Kucherov [15] showed that $|\mathsf{Runs}(w)| = O(n)$ for any string w of length n, and later Bannai et al. [4] proved that $|\mathsf{Runs}(w)| < n$. Ellert and Fischer [9] showed how to compute $\mathsf{Runs}(w)$ for a given string w of length n over an general ordered alphabet in $O(n)$ time. We can then sort $\mathsf{Runs}(w)$ based on their beginning positions of runs in non-decreasing order in linear time by bucket sort. Let $r_i = (\mathsf{beg}_i, \mathsf{end}_i, \mathsf{p}_i)$ denote the i-th run in $\mathsf{Runs}(w)$ sorted in increasing order of beg_i, and of $\mathsf{end}_j - 2\mathsf{p}_j$ if draw.

2.3 Repetition Factorizations

For a given string w of length n, a sequence $f_1, \ldots, f_m$ of non-empty strings is said to be a *repetition factorization* of w if the sequence satisfies the following two conditions:

- $w = f_1 \cdots f_m$, and
- each f_i ($1 \leq i \leq n$) is a repetition in w.

Each f_i is called a *factor*, and m is called the size of the factorization of w. For example, string $w = $ abaabababbabaabaab has a repetition factorization f_1, f_2, f_3 with $f_1 = $ abaaba, $f_2 = $ baba, and $f_3 = $ baabaab. Note also that some strings do not have a repetition factorization.

In this paper, we tackle the following problem:

Problem 1 (Computing an arbitrary repetition factorization).

   Input: A string w of length n.

Output: An arbitrary repetition factorization of w if w has a repetition factorization, and "no" otherwise.

For a reference, we introduce the problem considered by Inoue et al. [13]:

Problem 2 (Computing smallest/largest repetition factorization).

Input: A string w of length n.
Output: A repetition factorization of smallest/largest size if w has a repetition factorization, and "no" otherwise.

For example, for string $w = $ aabaabaacbbcbbcbb, a smallest repetition factorization of w is f_1, f_2 of size 2 with $f_1 = $ aabaabaa and $f_2 = $ cbbcbbcbb, and a largest repetition factorization of w is g_1, g_2, g_3, g_4 of size 4 with $g_1 = $ aabaab, $g_2 = $ aa, $g_3 = $ cbbcbbc, and $g_4 = $ bb.

3 Simple Linear-Time Repetition Factorization Algorithm

In this section, we present a simple linear-time algorithm for computing an arbitrary repetition factorization of a given string, which does not use bit operations in the word RAM model.

3.1 Overview of Our Approach

In this subsection, we describe the basic ideas of our algorithm, following the notations from Inoue et al. [13]: For an input string w of length n, Inoue et al. introduced the repetition graph G_w of w of size $O(n \log n)$, such that the smallest/largest repetition factorization problem is reducible to the shortest/longest path problem on G_w. This solves Problem 2 in $O(n \log n)$ time.

To solve Problem 1 in linear time, we introduce the *arbitrary repetition factorization graph* (*ARF-graph*) which is a compact version of the repetition graph for finding an arbitrary repetition factorization. While the ARF-graph is still of size $O(n \log n)$, it permits us to compute a solution to Problem 1 in $O(n)$ time without explicitly constructing the whole graph.

3.2 Repetition Graphs

Let us recall the method of Inoue et al. [13]. The repetition graph $G_w = (V, E)$ of a string w is the following directed acyclic edge-weighted graph with weight function $w : V \to \mathcal{N}$. The set $V = V' \cup V''$ of nodes consists of two disjoint sets of nodes V' and V'' such that

$$V' = \{(0, j) \mid 0 \le j \le n\},$$

$$V'' = \bigcup_{i=1}^{|\mathsf{Runs}(w)|} V''_i,$$

where $V_i'' = \{(i,j) \mid \mathsf{beg}_i + 2\mathsf{p}_i - 1 \le j \le \mathsf{end}_i\}$ for each $r_i = (\mathsf{beg}_i, \mathsf{end}_i, \mathsf{p}_i) \in \mathsf{Runs}(w)$. Intuitively, the subset V' is the nodes representing all $n+1$ positions in w, including 0 that is the source of G_w. We call these nodes as *black nodes* (see also Fig. 1). The other subset V'' of nodes represents the positions included in the runs from $\mathsf{Runs}(w)$, but excluding the first $2p - 1$ positions for each run $r = x^k x'$ with $|x| = p$ (the white nodes in Fig. 1). We call these nodes as *white nodes*. For each node $v = (i,j) \in V$, j is said to be the *position* of the node v and we denote it by $p(v) = j$.

The set $E = E' \cup E'' \cup E'''$ of edges consists of three disjoint sets of edges E', E'', and E''' such that

$$E' = \{((i_1, j_1), (i_2, j_2)) \mid j_2 - j_1 = 2\mathsf{p}_{i_2}, (i_1, j_1) \in V', (i_2, j_2) \in V''\},$$
$$E'' = \{((i_1, j_1), (i_2, j_2)) \mid i_1 = i_2, j_1 + 1 = j_2, (i_1, j_1), (i_2, j_2) \in V''\},$$
$$E''' = \{((i_1, j_1), (i_2, j_2)) \mid j_1 = j_2, (i_1, j_1) \in V'', (i_2, j_2) \in V'\}.$$

For each edge $e \in E$, $\mathsf{w}(e)$ is defined as follows:

$$\mathsf{w}(e) = \begin{cases} 1 & \text{if } e \in E', \\ 0 & \text{otherwise.} \end{cases}$$

Intuitively, the nodes and edges of G_w have the following structure:

- Each black node $(0, j) \in V'$ represents the position j of each character in w.
- Each white node $(i, k) \in V''$ represents the end position k of a repetition contained in the ith run r_i.
- Each edge $(u, v) \in E'$ with black node $u = (0, j-1) \in V'$ and white node $v = (i, k) \in V''$ represents a square of period $(k - j + 1)/2$ that begins at position j and ends at position k in the ith run r_i. By setting the origin node of each edge in E' to the immediately preceeding position $j - 1$, any two consecutive repetitions from different runs are connected and form a path in G_w.
- Each edge in E'' connecting white nodes represents an operation of extending a repetition to the right by one character within the corresponding run.
- Each edge in $(u, v) \in E'''$ with white node $u = (i, j) \in V''$ and black node $v = (0, j) \in V'$ represents a (possible) boundary position i of a repetition factorization in w.

Figure 1 shows an example of a repetition graph. For example, the repetition $w[14..20] = \mathsf{bcbbcbb}$ in w is represented by the path from the black node $(0, 13)$ to the black node $(0, 20)$ which goes through the diagonal edge from $(0, 13)$ to the white node $(6, 19)$, the horizontal edge from $(6, 19)$ to $(6, 20)$, and the vertical edge from $(6, 20)$ to $(0, 20)$.

Clearly $|V'| = n$. Inoue et al. [13] showed that $|V''| = O(n \log n)$ using Lemma 1 below. Since $|E| = O(|V|)$, the size of the repetition graph is $O(n \log n)$.

Lemma 1 *([6]). The number of primitively rooted squares which appear in suffixes of $w[1..i]$ is $O(\log i)$ for a string w.*

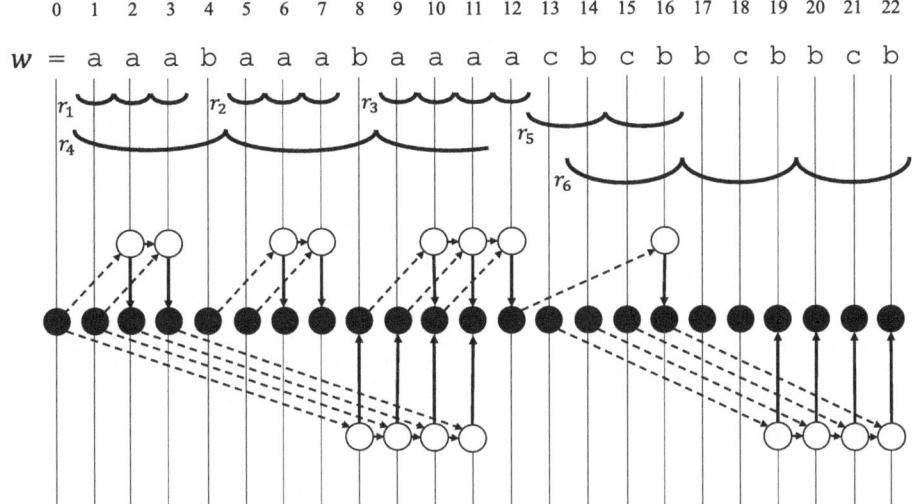

Fig. 1. This figure shows the repetition graph of a string $w =$ aaabaaabaaaacbcbbcbbcb. The concatenations of curves at the top represent each run of w. The nodes in V' are illustrated as black circles, whose positions correspond to each character of w, the nodes in V'' are illustrated as white circles for each run, the edges in E' are illustrated as dashed arrows, and the edges in E'' and E''' are illustrated as horizontal and vertical solid arrows, respectively. (Color figure online)

For convenience, we will denote $(0,0) \in V'$ by v_s. Inoue et al. [13] showed a correspondence between a path on G_w and a repetition factorization of w by proving Lemma 2 shown below.

Lemma 2 *([13]). For any string w of length n and integer $1 \leq t \leq n$, repetition factorizations of $w[1..t]$ and $s-t$ paths have a one-to-one correspondence, where a $s-t$ path is a path from v_s to $(0,t) \in V'$ in G_w.*

From Lemma 2, we can compute the smallest/largest repetition factorizations in $O(n \log n)$ time and space based on dynamic programming and backtracking on the repetition graph, such that each node stores the maximum/minimum value of the weight of a path starting at v_s.

Also, we can compute an arbitrary repetition factorization in $O(n \log n)$ time and space based on dynamic programming and backtracking on the repetition graph, such that each node stores the 0/1 value depending on whether it is unreachable/reachable from v_s.

3.3 ARF-Graphs

When it comes to computing an *arbitrary* repetition factorization, there is some redundant information in the repetition graph. See Fig. 2a, where we consider a part of the repetition graph of a string w that has substring cbaabaabaac, in

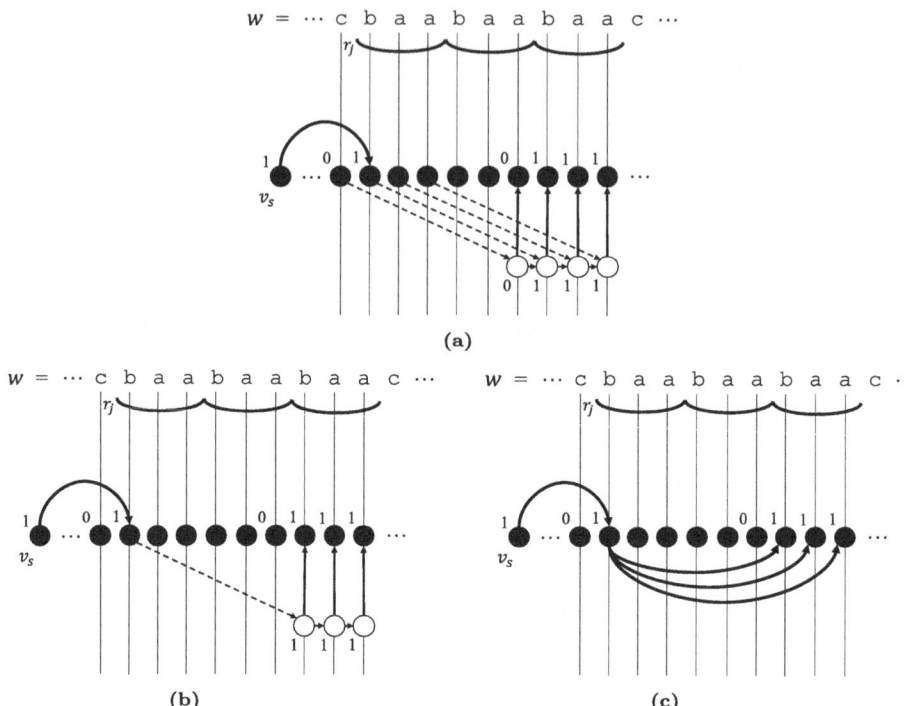

Fig. 2. Parts of repetition graph G_w (Fig. 2a), intermediate graph H_w (Fig. 2b), and ARF-graph $\hat{G}_w$ (Fig. 2c).

which there is a run r_j baabaabaa. Let us assign a one bit information b to each node v of the repetition graph G_w, such that $\mathsf{b}(v) = 1$ iff v is reachable from the source v_s. For each subset V_i'' of white nodes corresponding to a run r_i, we focus on the leftmost white node w_i with $\mathsf{b}(w_i) = 1$. It is clear that any nodes y in V_i'' to the left of w_i are redundant (unreachable from v_s), and any other nodes z to the right of w_i in V_i'' are reachable from v_s. Thus we can remove all such nodes y. Also, since there is an alternative path from v_s to each z via u, we can also remove the diagonal edge to z. This gives us an intermediate graph H_w shown in Fig. 2b. Further, we represent the path that begins with the diagonal edge (u_i, w_i) and arrives at a black node x with a single edge from the origin black node u_i to the target black node x. Note that node x remains reachable from the source v_s. The resulting graph is the ARF-graph $\hat{G}_w$, illustarted in Fig. 2c.

Simple Linear-Time Repetition Factorization 355

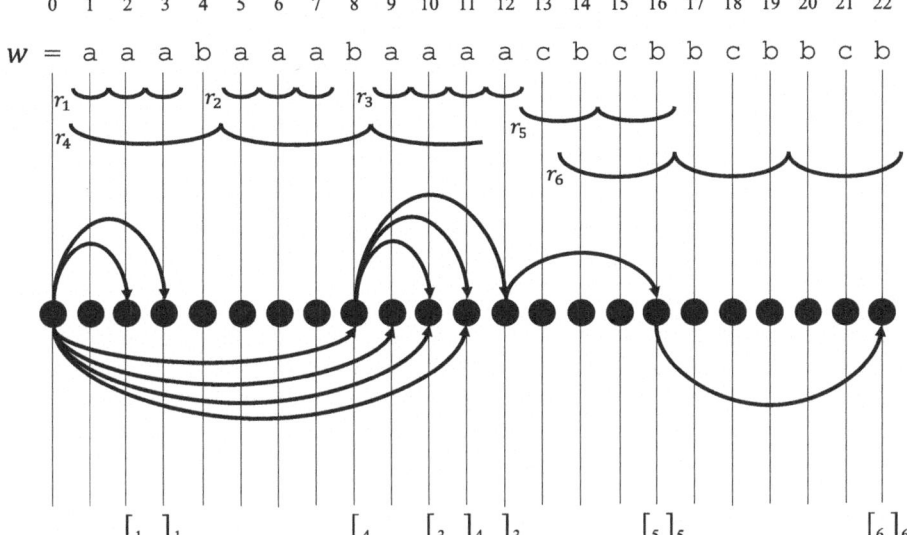

Fig. 3. The figure shows the ARF-graph factorization of a string aaabaaabaaaa cbcbbcbbcb. The square brackets at the bottom represent the intervals for the consecutive incoming edges.

Formally, the ARF-graph $\hat{G}_w = (\hat{V}, \hat{E})$ for a string w is given by

$$\hat{V} = \{v_j \mid 0 \leq j \leq n\},$$

$$\hat{E} = \bigcup_{i=1}^{|\mathsf{Runs}(w)|} \hat{E}_i$$

$$= \bigcup_{i=1}^{|\mathsf{Runs}(w)|} \{(u_i, x) \mid p(u_i) + 2\mathsf{p}_i \leq p(x) \leq \mathsf{end}_i\},$$

where, for each run $r_i = (\mathsf{beg}_i, \mathsf{end}_i, \mathsf{p}_i)$, $\hat{E}_i$ is the set of edges from the leftmost black node u_i in the position range $[\mathsf{beg}_i - 1, \mathsf{end}_i - 2\mathsf{p}_i]$ reachable from the source (i.e. $\mathsf{b}(u_i) = 1$), to the target black nodes x within the run r_i that are at least $2\mathsf{p}_i$ aray from u_i. Figure 3 illustrates a concrete example of the ARF-graph.

3.4 How to Exclude Redundant Edges from the ARF-Graph

The size of the ARF-graph remains $O(n \log n)$ since the number of edges remains $O(n \log n)$. Therefore, we cannot afford to explicitly constructing all the edges of the ARF-graph $\hat{G}_w$ for a given string w of length n. Still, our new method allows one to compute a solution in linear time.

Before describing our new algorithm, we revisit the previous method by Dumitran et al. [7] and recall their algorithm in terms of the ARF-graph. We emphasize that the ARF-graph was not present in their paper [7].

Algorithm of Dumitran et al. The algorithm of Dumitran et al. [7] computes a solution (i.e. an arbitrary repetition factorization of w) in linear time, with a help of the interval Union-find data structure [11] and without explicitly constructing all the edges of the ARF-graph. For the sake of clarity, we represent each element in the Union-Find structure is an interval $[a, b]$ ($1 \leq a \leq b \leq n$). Initially, there are exactly n intervals of form $[i, i]$ for all $1 \leq i \leq n$. One executes the following two basic operations and an additional operation in this structure. One is the union operation for two intervals, and the other is the find operation which returns the start and end positions of the interval containing a given position on the ARF-graph. In Gabow and Tarjan's algorithm [11], the time complexity of executing these operations is $O(n + m)$ for n intervals and m operations. The additional one is marking operations for each interval of the above structure, where all the intervals are initially unmarked. These operations are done so that later one can determine whether the value of each node has already been updated or not (corresponding to being marked or unmarked, respectively).

Initially, set $\mathsf{b}(v_s) \leftarrow 1$ and $\mathsf{b}(v) \leftarrow 0$ for all nodes v of $\hat{G}_w$ except for v_s. The one bit information $\mathsf{b}(v)$ is then updated to 1 iff v is reachable from the source, in $O(n)$ total time, with the Union-Find structure as follows:

Initially set $R = \mathsf{Runs}(w)$. The algorithm executes the following queries and operations for node v_j at each position j with $\mathsf{b}(v_j) = 1$ for $1 \leq j \leq n$ in increasing order:

- Retrieve all runs $r_i = (\mathsf{beg}_i, \mathsf{end}_i, \mathsf{p}_i)$ satisfying $\mathsf{beg}_i - 1 \leq j \leq \mathsf{end}_i - 2\mathsf{p}_i + 1$ from R, and execute the following to each computed run r_i:

  • For the interval $[s, t] = [j + 2\mathsf{p}_i, \mathsf{end}_i]$ (this interval is not an interval in the Union-find structure) of the positions of the nodes that can be reached from v_j by the edges of r_i, execute a find query for the position s. Let $[c, c']$ be the obtained interval.
  • If $[c, c']$ is unmarked, mark the interval and update the value of the node of position t to 1. Execute this operation for all the intervals obtained by the find queries.
  • If $[c, c']$ does not contain t, execute a find query for given position $c' + 1$, and let $[c' + 1, c'']$ be the obtained interval.
  • Execute a union operation to $[c, c']$ and $[c' + 1, c'']$.

- If the new interval $[c, c'']$ does not contain t, execute the union operation for the new interval $[c, c'']$ and the interval $[c'', c''']$ containing c''. Continue executing this series of operations until one obtains the interval containing t.
- Remove the run r_i from R.

After this computation, all values of the nodes of positions between s and t are 1. Since the number of the union operations for each position is at most one, the total time complexity for the union operations is $O(n)$. Also, for all computations for all runs and positions, the number of times to update the values of nodes is $O(n)$.

In what follows, we present another idea for computing an arbitrary repetition factorization of a given string based on the ARF-graph.

3.5 New Algorithm

Excluding Union-Find. The ending nodes of the edges that correspond to the same run are continuous in the ARF-graph, and we decide whether each node can be reached from v_s or not using these edges. Therefore, the edges for a run which can be reached from v_s can be expressed as an interval. The start point of the interval is the left-maximal node which has in-coming edges in V_r, and the end point of the interval is the right-maximal node which has in-coming edges in V_r, where V_r is the set of nodes of the ARF-graph for run r. Figure 3 shows examples of interval expressions for edges as square brackets. For example, the interval $[8..11]$ expresses the edges of the run r_4. Thus, the node can be reached from v_s, iff the position of that node is in any intervals. Therefore, we can compute the value of each node by counting the number of the start points and end points of intervals by the following method. For each position i, we first compute whether i is in any intervals or not, and if it is, we update the value of the node of position i to 1. We can compute the conditions dynamically by counting the number of the start/end points of intervals from the position 0 to i. Then, we execute the following operations for each position i of the ARF-graph such that the value of the node of position i is 1: we compute the list of the runs $r_j = (\text{beg}_j, \text{end}_j, \text{p}_j)$ satisfying $\text{beg}_j - 1 \leq i \leq \text{end}_j - 2\text{p}_j + 1$, and execute the following operations for each run $r_j = (\text{beg}_j, \text{end}_j, \text{p}_j)$ in the list: we mark the start and end point of the interval $[i + 2\text{p}_j, \text{end}_j]$ to nodes of positions $i + 2\text{p}_j$ and end_j, and remove r_j.

Algorithm. Our algorithm has two steps which are the step of deciding solution existence and the step of reconstructing a solution. First, we present the step for deciding whether there is a solution.

From the above idea, our algorithm for the step of deciding solution existence is shown in Algorithm 1. In this algorithm, the value of each node is stored in an array M whose index begins with 0, and the number of the start/end points of intervals for each node is stored in an array L/R, and an array P stores the start node of edges in the ARF-graph for each run, which is used to obtain the solution in the latter step.

If there are no edges for run r_j, $P[j] = -1$ (the elements of P are initialized -1). C is a counter for deciding that each node is in any interval. L and R are updated dynamically from M, namely, each interval is computed dynamically. In the lines 5 and 14 in Algorithm 1, we compute all runs $r_j = (\mathsf{beg}_j, \mathsf{end}_j, \mathsf{p}_j)$ satisfying $\mathsf{beg}_j - 1 \leq i \leq \mathsf{end}_j - 2\mathsf{p}_j$ for any $1 \leq i \leq n$. Dumitran et al. showed how to compute these runs based on the interval stabbing algorithm [17] in total $O(n+k)$ time ($k = O(n)$ that is the number of runs). On the other hand, we can compute such runs in our algorithm by only the operations on arrays without the complicated structures in the interval stabbing algorithm, since each run is checked only once in our algorithm.

We now describe this method. Assume that all runs $r_j = (\mathsf{beg}_j, \mathsf{end}_j, \mathsf{p}_j)$ are sorted for beg_j, and for $\mathsf{end}_j - 2\mathsf{p}_j$ in increasing order. We maintain the array SubRuns which stores runs r_j satisfying $\mathsf{beg}_j - 1 \leq i \leq \mathsf{end}_j - 2\mathsf{p}_j$ for $1 \leq i \leq n$. We describe the state of SubRuns which stores runs satisfying $\mathsf{beg}_j - 1 \leq i \leq \mathsf{end}_j - 2\mathsf{p}_j$ as $\mathsf{SubRuns}_i$. In addition, we maintain the array P whose j-th element stores the index of run r_j in SubRuns. For each step of the computation of the i-th node in our algorithm ($1 \leq i \leq n$), we execute the following operations:

- All runs r_j satisfying $\mathsf{beg}_j - 1 \leq i \leq \mathsf{end}_j - 2\mathsf{p}_j$ are appended to $\mathsf{SubRuns}_i$.
- We update $P[j]$ to the index in $\mathsf{SubRuns}_i$ of added run r_j.
- We update $\mathsf{SubRuns}_i[P[j]]$ to -1 for all r_j satisfying $\mathsf{end}_j - 2\mathsf{p}_j = i - 1$.
- If the value of node i is 1, we compute all runs in $\mathsf{SubRuns}_i$ and delete the runs. After that, we delete all elements of $\mathsf{SubRuns}_i$.

This method is executed in $O(n)$ time, since each run is added, updated, and deleted at most once, and the number of runs is $O(n)$.

After all computations, if $M[n] = 1$ then any repetition factorization exists, and does not otherwise. Thus, the following lemma clearly holds.

Lemma 3 *Algorithm 1 can decide whether w has any repetition factorizations in $O(n)$ time and space.*

Algorithm 1. Decide whether string w has a repetition factorization
Input: string w ($|w| = n$)
Output: Yes or No

1: compute all runs $r_j (1 \leq j \leq |\mathsf{Runs}(w)|)$
2: an integer array P of size $|\mathsf{Runs}(w)|$ is initialized with -1
3: integer array M of size $n+1$ whose element $M[0]$ is initialized with 1 and element $M[i]$ is initialized with 0 for $1 \leq i \leq n+1$ (The index of M begins with 0.)
4: integer arrays L, R of size n are initialized with 0
5: **foreach** $r_j = (\mathsf{beg}_j, \mathsf{end}_j, \mathsf{p}_j)$ **satisfying** $\mathsf{beg}_j = 1$ **do**
6:    $L[2\mathsf{p}_j] \leftarrow L[2\mathsf{p}_j] + 1$, $R[\mathsf{end}_j] \leftarrow R[\mathsf{end}_j] + 1$, $P[j] \leftarrow 0$
7:    remove r_j
8: **end for**
9: $C \leftarrow 0$
10: **for** $i \leftarrow 1$ **to** n **do**
11:    $C \leftarrow C + L[i]$
12:    **if** $C > 0$ **then**
13:      $M[i] \leftarrow 1$
14:      **foreach** $r_j = (\mathsf{beg}_j, \mathsf{end}_j, \mathsf{p}_j)$ **satisfying** $\mathsf{beg}_j - 1 \leq i \leq \mathsf{end}_j - 2\mathsf{p}_j$ ($\mathsf{SubRuns}_i$) **do**
15:        $L[i + 2\mathsf{p}_j] \leftarrow L[i + 2\mathsf{p}_j] + 1$, $R[\mathsf{end}_j] \leftarrow R[\mathsf{end}_j] + 1$, $P[j] \leftarrow i$
16:        remove r_j
17:      **end for**
18:    **end if**
19:    $C \leftarrow C - R[i]$
20: **end for**
21: If $M[n] = 1$ then a solution exists, and does not otherwise.

Second, we present the step of reconstructing a solution in Algorithm 2.

Algorithm 2 simply joins the sorted runs whose P elements are not -1 from backward. The sorting can be done in $O(n)$ time and space by a bucket sort. Thus, the following lemma clearly holds.

Lemma 4 *Algorithm 2 can reconstruct a repetition factorization of w in $O(n)$ time and space from computation of Algorithm 1.*

Theorem 1 follows from Lemma 3 and Lemma 4.

Algorithm 2. Compute a repetition factorization of w

Input: string w ($|w| = n$), and an integer array P computed in Algorithm 1
Output: a set I of intervals for a repetition factorization of w

1: $I \leftarrow \emptyset$
2: the array F stores sorted runs $r_j = (\mathsf{beg}_j, \mathsf{end}_j, \mathsf{p}_j)$ satisfying $P[j] \neq -1$ in w in descending order of end_j, and of be_j if draw
3: $k \leftarrow n, i \leftarrow 0$
4: **while** $k > 0$ **do**
5: $\quad r_j = F[i]$
6: $\quad$ **if** $P[j] + 2\mathsf{p}_j \leq k \leq \mathsf{end}_j$ **then**
7: $\quad\quad$ add an interval $[P[j]+1, k]$ to I
8: $\quad\quad k = P[j]$
9: $\quad$ **else**
10: $\quad\quad i \leftarrow i + 1$
11: $\quad$ **end if**
12: **end while**

4 Conclusions and Future Work

We proposed a simple algorithm which computes an arbitrary repetition factorization in $O(n)$ time and space without relying on data structures for Union-Find and interval stabbing. The idea of our algorithm is based on combinatorial insights of repetition factorizations, which allows us to omit redundant edges and nodes of repetition graphs. Also, both our algorithm and Dumitran et al.'s algorithm for an arbitrary repetition factorization can compute a factorization into repetitions for all prefixes $T[1..i]$ of the given string T, by solving the 0-to-n reachability on the ARF-graph.

Matsuoka et al. [16] proposed an $O(n)$-time algorithm for computing a square factorization based on the reduction to the reachability on a similar but different run-oriented DAG. The method of Matsuoka et al. [16] makes heavy use of bit operations in the word RAM model. It is an interesting question whether we can compute a square factorization without bit operations in the word RAM model in linear time, in a similar way to our algorithm proposed in this paper.

References

1. Alzamel, M., Iliopoulos, C.S., Smyth, W.F., Sung, W.: Off-line and on-line algorithms for closed string factorization. Theoritical Comput. Sci. **792**, 12–19 (2019)
2. Badkobeh, G., et al.: Closed factorization. Discrete Appl. Math. **212**, 23–29 (2016)
3. Bannai, H., et al.: Diverse palindromic factorization is NP-complete. Int. J. Found. Comput. Sci. **29**(2), 143–164 (2018)
4. Bannai, H., I, T., Inenaga, S., Nakashima, Y., Takeda, M., Tsuruta, K.: The runs theorem. SIAM J. Comput. **46**(5), 1501–1514 (2017)
5. Borozdin, K., Kosolobov, D., Rubinchik, M., Shur, A.M.: Palindromic length in linear time. In: 28th Annual Symposium on Combinatorial Pattern Matching (CPM 2017), pp. 231–2312 (2017)

6. Crochemore, M., Rytter, W.: Squares, cubes, and time-space efficient string searching. Algorithmica **13**(5), 405–425 (1995)
7. Dumitran, M., Manea, F., Nowotka, D.: On prefix/suffix-square free words. In: 22th International Symposium on String Processing and Information Retrieval (SPIRE 2015), pp. 54–66. Springer (2015). https://doi.org/10.1007/978-3-319-23826-5_6
8. Duval, J.: Factorizing words over an ordered alphabet. J. Algorithms **4**(4), 363–381 (1983)
9. Ellert, J., Fischer, J.: Linear time runs over general ordered alphabets. In: 48th International Colloquium on Automata, Languages and Programming (ICALP 2021). LIPIcs, vol. 198, pp. 631–6316 (2021)
10. Fici, G., Gagie, T., Kärkkäinen, J., Kempa, D.: A subquadratic algorithm for minimum palindromic factorization. J. Discrete Algorithms **28**, 41–48 (2014)
11. Gabow, H.N., Tarjan, R.E.: A linear-time algorithm for a special case of disjoint set union. J. Comput. Syst. Sci. **30**(2), 209–221 (1985)
12. I, T., Sugimoto, S., Inenaga, S., Bannai, H., Takeda, M.: Computing palindromic factorizations and palindromic covers on-line. In: Proceedings 25th Annual Symposium on Combinatorial Pattern Matching (CPM 2014), pp. 150–161 (2014)
13. Inoue, H., Matsuoka, Y., Nakashima, Y., Inenaga, S., Bannai, H., Takeda, M.: Factorizing strings into repetitions. Theory Comput. Syst. **66**(2), 484–501 (2022)
14. Kishi, K., Nakashima, Y., Inenaga, S.: Largest repetition factorization of Fibonacci words. In: 30th String Processing and Information Retrieval (SPIRE 2023). Lecture Notes in Computer Science, vol. 14240, pp. 284–296 (2023)
15. Kolpakov, R.M., Kucherov, G.: Finding maximal repetitions in a word in linear time. In: Proceedings 40th Annual Symposium on Foundations of Computer Science (FOCS 1999), pp. 596–604 (1999)
16. Matsuoka, Y., Inenaga, S., Bannai, H., Takeda, M., Manea, F.: Factorizing a string into squares in linear time. In: Proceedings 27th Annual Symposium on Combinatorial Pattern Matching (CPM 2016), pp. 271–2712 (2016)
17. Schmidt, J.M.: Interval stabbing problems in small integer ranges. In: 30th International Symposium on Algorithms and Computation (ISAAC 2009), pp. 163–172 (2009)
18. Ziv, J., Lempel, A.: A universal algorithm for sequential data compression. IEEE Trans. Inf. Theory **IT-23**(3), 337–349 (1977)
19. Ziv, J., Lempel, A.: Compression of individual sequences via variable-length coding. IEEE Trans. Inf. Theory **24**(5), 530–536 (1978)

Author Index

A
Amir, Amihood 1

B
Badkobeh, Golnaz 16
Bannai, Hideo 16, 174
Becker, Ruben 26
Boneh, Itai 41
Boucher, Christina 184
Büchler, Thomas 249

C
Carfagna, Lorenzo 57
Cenzato, Davide 73
Cleary, Alan M. 88

D
Delabre, Mattéo 127
Dood, Jordan 88
Duyster, Anouk 102

E
El-Mabrouk, Nadia 127

G
Gagie, Travis 184
Ganguly, Arnab 118
Gascon, Mathieu 127
Gibney, Daniel 118, 143
Golan, Shay 41
Guo, Peaker 159

H
Hamai, Rikuya 174
Hong, Aaron 184
Hossen, Md Helal 143

I
I, Tomohiro 336
Inenaga, Shunsuke 88, 174, 192, 321, 348

K
Kikuchi, Masaru 192
Kim, Sung-Hwan 26
Kociumaka, Tomasz 102
Kondratovsky, Eitan 1
Köppl, Dominik 16, 289

L
Levy, Avivit 41
Li, Yansong 184

M
MacNichol, Paul 118
Manzini, Giovanni 57
Marcus, Shoshana 1, 306
Mieno, Takuya 321

N
Nakashima, Yuto 174
Navarro, Gonzalo 204, 218
Nikolaev, Maksim S. 233

O
Ohlebusch, Enno 249
Olbrich, Jannik 249
Olivares, Francisco 73

P
Porat, Ely 41
Prezza, Nicola 26, 73

R
Radoszewski, Jakub 257, 272
Robert, Josefa 218
Romana, Giuseppe 57
Rytter, Wojciech 257

S
Sciortino, Marinella 57
Shalom, B. Riva 41

© The Editor(s) (if applicable) and The Author(s), under exclusive license to Springer Nature Switzerland AG 2025
Z. Lipták et al. (Eds.): SPIRE 2024, LNCS 14899, pp. 363–364, 2025.
https://doi.org/10.1007/978-3-031-72200-4

Shibata, Hiroki   289
Sokol, Dina   1, 306
Sumiyoshi, Wataru   321

T
Takasaka, Akito   336
Taketsugu, Kazushi   174
Thankachan, Sharma V.   118
Tosoni, Carlo   26

U
Umboh, Seeun William   159
Urbina, Cristian   57

W
Waleń, Tomasz   257
Winjum, Joseph   88
Wirth, Anthony   159

Y
Yonemoto, Yuki   348

Z
Zeh, Norbert   184
Zelikovitz, Sarah   306
Zobel, Justin   159
Zuba, Wiktor   272

SPRINGER NATURE

GPSR Compliance

The European Union's (EU) General Product Safety Regulation (GPSR) is a set of rules that requires consumer products to be safe and our obligations to ensure this.

If you have any concerns about our products, you can contact us on ProductSafety@springernature.com

In case Publisher is established outside the EU, the EU authorized representative is:

Springer Nature Customer Service Center GmbH
Europaplatz 3
69115 Heidelberg, Germany

The manufacturer's authorised representative in the EU is Springer Nature Customer Service Centre GmbH, Europaplatz 3, 69115 Heidelberg, Germany. If you have any concerns regarding our products, please contact ProductSafety@springernature.com

Printed and bound by CPI Group (UK) Ltd, Croydon, CR0 4YY
26/03/2026
02078962-0010